NEGATIVE IONS, BEAMS AND SOURCES

To learn more about AIP Conference Proceedings, including the Conference Proceedings Series, please visit the webpage **http://proceedings.aip.org/proceedings**

NEGATIVE IONS, BEAMS AND SOURCES

1st International Symposium on Negative Ions, Beams and Sources

Aix-en-Provence, France   9 – 12 September 2008

EDITORS
Elizabeth Surrey
EURATOM/UKAEA Fusion Association
Oxfordshire, United Kingdom

Alain Simonin
IRFM CEA Cadarache
Saint-Paul-Lez-Durance, France

All papers have been peer reviewed

SPONSORING ORGANIZATIONS
Association EURATOM-CEA
Aix Regional Community

AMERICAN INSTITUTE OF PHYSICS   Melville, New York, 2009
AIP CONFERENCE PROCEEDINGS ■ VOLUME 1097

Editors:

Elizabeth Surrey
EURATOM/UKAEA Fusion Association
Culham Science Center
Abingdon Oxfordshire OX14 3DB
United Kingdom

E-mail: Elizabeth.Surrey@jet.uk

Alain Simonin
IRFM
CEA Cadrache
F-13108 Saint-Paul-Lez-Durance
France

E-mail: alain.simonin@cea.fr

The articles on pp. 89 – 98 and 402 – 411 were prepared by UKAEA and retain Crown Copyright. Therefore these articles are not covered by the copyright mentioned below.

The copyright for the article on pp. 480 – 490 belongs to ITER.

Authorization to photocopy items for internal or personal use, beyond the free copying permitted under the 1978 U.S. Copyright Law (see statement below), is granted by the American Institute of Physics for users registered with the Copyright Clearance Center (CCC) Transactional Reporting Service, provided that the base fee of $25.00 per copy is paid directly to CCC, 222 Rosewood Drive, Danvers, MA 01923, USA. For those organizations that have been granted a photocopy license by CCC, a separate system of payment has been arranged. The fee code for users of the Transactional Reporting Services is: 978-0-7354-0630-8/09/$25.00.

© 2009 American Institute of Physics

Permission is granted to quote from the AIP Conference Proceedings with the customary acknowledgment of the source. Republication of an article or portions thereof (e.g., extensive excerpts, figures, tables, etc.) in original form or in translation, as well as other types of reuse (e.g., in course packs) require formal permission from AIP and may be subject to fees. As a courtesy, the author of the original proceedings article should be informed of any request for republication/reuse. Permission may be obtained online using Rightslink. Locate the article online at http://proceedings.aip.org, then simply click on the Rightslink icon/"Permission for Reuse" link found in the article abstract. You may also address requests to: AIP Office of Rights and Permissions, Suite 1NO1, 2 Huntington Quadrangle, Melville, NY 11747-4502, USA; Fax: 516-576-2450; Tel.: 516-576-2268; E-mail: rights@aip.org.

L.C. Catalog Card No. 2008944315
ISBN 978-0-7354-0630-8
ISSN 0094-243X
Printed in the United States of America

CONTENTS

Preface ... ix
Symposium Organization ... xi

FUNDAMENTAL PROCESSES

Monte Carlo Simulation of H⁻ Ion Transport ... 3
   P. Diomede, S. Longo, and M. Capitelli
2D Model of a Tandem Plasma Source: The Role of the Transport Processes ... 12
   T. V. Paunska, A. P. Shivarova, K. Ts. Tarnev, and T. V. Tsankova
The Cascaded Arc: High Flows of Rovibrationally Excited H$_2$ and Its Impact on H⁻ Ion Formation ... 22
   O. Gabriel, W. E. N. van Harskamp, D. C. Schram, M. C. M. van de Sanden, and R. Engeln
The Electronegative Plasma Pre-Sheath in Magnetic Field and Extraction of Negative Ions ... 31
   B. M. Annaratone and J. E. Allen
Modeling of the Plasma Electrode Bias in the Negative Ion Sources with 1D PIC Method ... 38
   D. Matsushita, S. Kuppel, A. Hatayama, A. Fukano, and M. Bacal
Study of the Plasma near the Plasma Electrode by Probes and Photodetachment in ECR-Driven Negative Ion Source ... 47
   M. Bacal, P. Svarnas, S. Béchu, and J. Pelletier
Numerical Analysis of Electronegative Plasma near the Extraction Grid in Negative Ion Sources ... 55
   S. Kuppel, D. Matsushita, A. Hatayama, and M. Bacal
Plasma Structure in the Extraction Region of a Hybrid Negative Ion Source ... 65
   F. Taccogna, S. Longo, M. Capitelli, and R. Schneider
Production of H⁻ Ions by Surface Mechanisms in Cs-Free Multi-Dipolar Microwave Plasma ... 74
   S. Béchu, D. Lemoine, M. Bacal, A. Bès, and J. Pelletier
Caesium Free Negative Ion Sources for Neutral Beam Injectors: A Study of Negative Ion Production on Graphite Surface in Hydrogen and Deuterium Plasma ... 84
   L. Schiesko, M. Carrere, G. Cartry, and J.-M. Layet
Surface Production of Negative Ions by Positive Ions and Atoms in the Electron Suppressor Region ... 89
   R. McAdams and E. Surrey
Spatial Distribution of the Plasma Characteristics of a Tandem Plasma Source ... 99
   T. V. Paunska, A. P. Shivarova, K. Ts. Tarnev, and T. V. Tsankov
Characteristics of the RF Negative Ion Source Using a Mesh Grid Bias Method ... 109
   O. Fukumasa, J. Okada, Y. Tauchi, W. Oohara, K. Tsumori, and Y. Takeiri
Enhancement of D⁻ Negative Ion Volume Production in Pure Deuterium Plasmas ... 118
   O. Fukumasa, T. Nakano, S. Mori, W. Oohara, K. Tsumori, and Y. Takeiri
Experiments on the Detection of Negative Hydrogen Ions in a Small-Size Tandem Plasma Source ... 127
   S. St. Lishev, A. P. Shivarova, and T. V. Tsankov
Study of Multi-Cusp Magnetic Field in Cylindrical Geometry for H⁻ Ion Source ... 137
   A. Kumar, V. K. Senecha, and R. M. Vadjikar
Development of Small Multiaperture Negative Ion Beam Sources and Related Simulation Tools ... 149
   M. Cavenago, V. Antoni, T. Kulevoy, S. Petrenko, G. Serianni, and P. Veltri

ION SOURCES FOR ACCELERATORS

H⁻ Ion Source Development for the LANSCE Accelerator Systems ... 161
   R. Keller, O. Tarvainen, E. Chacon-Golcher, E. G. Geros, K. F. Johnson, G. Rouleau, J. E. Stelzer, and T. J. Zaugg
The NEW DESY RF-Driven Multicusp H⁻ Ion Source ... 171
   J. Peters

Next Generation H⁻ Ion Sources for the SNS 181
    R. F. Welton, M. P. Stockli, S. N. Murray, D. Crisp, J. Carmichael, R. H. Goulding, B. Han, O. Tarvainen,
    T. Pennisi, and M. Santana

A Proposal for a Novel H⁻ Ion Source Based on Electron Cyclotron Resonance Plasma
Heating and Surface Ionization 191
    O. Tarvainen and S. Kurennoy

First Results with a Surface Conversion H⁻ Ion Source Based on Helicon Wave Mode-Driven
Plasma Discharge 199
    O. Tarvainen, E. G. Geros, R. Keller, G. Rouleau, and T. Zaugg

Negative Hydrogen Ion Source with Inverse Gas Magnetron Geometry 208
    V. A. Baturin, P. A. Litvinov, and S. A. Pustovoitov

A 15 mA CW H- Source for Accelerators 214
    Yu. Belchenko, A. Sanin, and A. Ivanov

Ramping Up the SNS Beam Power with the LBNL Baseline H⁻ Source 223
    M. P. Stockli, B. X. Han, S. N. Murray, D. Newland, T. R. Pennisi, M. Santana, and R. F. Welton

The HERA Magnetron: 24 Years of Experience, a World Record Run and a New Design ... 236
    J. Peters

Commissioning the Front End Test Stand High Performance H⁻ Ion Source at RAL 243
    D. C. Faircloth, S. Lawrie, A. P. Letchford, C. Gabor, P. Wise, M. Whitehead, T. Wood, M. Perkins,
    M. Bates, P. J. Savage, D. A. Lee, and J. K. Pozimski

Redesign of the Analysing Magnet in the ISIS H⁻ Penning Ion Source 253
    S. R. Lawrie, D. C. Faircloth, A. P. Letchford, M. Westall, M. O. Whitehead, T. Wood, and J. Pozimski

ION SOURCES FOR FUSION

Plasma and Beam Homogeneity of the RF-Driven Negative Hydrogen Ion Source for ITER
NBI 265
    U. Fantz, P. Franzen, W. Kraus, D. Wünderlich, R. Gutser, M. Berger and the NNBI Team

Long Pulse H⁻ Beam Extraction with a RF Driven Ion Source with Low Fraction of
Co-Extracted Electrons 275
    W. Kraus, M. Berger, U. Fantz, P. Franzen, M. Fröschle, B. Heinemann, R. Riedl, E. Speth, A. Stäbler,
    and D. Wünderlich

Multi-Antenna RF Ion Source at a High RF Power Level 282
    Y. Oka, T. Shoji, O. Kaneko, Y. Takeiri, K. Tsumori, M. Osakabe, K. Ikeda, K. Nagaoka, E. Asano,
    M. Sato, T. Kondo, M. Shibuya, and S. Komada

Characteristics of rf H⁻ Ion Source by Using FET Power Source 291
    A. Ando, C. H. Moon, J. Komuro, K. Tsumori, and Y. Takeiri

Simulations for the Generation and Extraction of Negative Hydrogen Ions in RF-Driven Ion
Sources 297
    R. Gutser, D. Wünderlich, U. Fantz, P. Franzen, B. Heinemann, R. Nocentini and the NNBI Team

BEAM FORMATION, ACCELERATION, NEUTRALIZATION AND TRANSPORT

Aperture Size Effect on Extracted Negative Ion Current Density 309
    H. P. L. de Esch, L. Svensson, and D. Riz

How to Find Valid Parameters for the Modelling of H⁻ and D⁻ Ion Extraction with nIGUN[(C)] 319
    R. Becker

Design of a Low Voltage, High Current Extraction System for the ITER Ion Source 325
    P. Agostinetti, V. Antoni, M. Cavenago, H. P. L. de Esch, G. Fubiani, D. Marcuzzi, S. Petrenko, N. Pilan,
    W. Rigato, G. Serianni, M. Singh, P. Sonato, P. Veltri, and P. Zaccaria

Development of 1 MeV H⁻ Accelerator at JAEA for ITER NB 335
    M. Taniguchi, H. P. L. de Esch, L. Svensson, N. Umeda, M. Kashiwagi, K. Watanabe, H. Tobari,
    M. Dairaku, K. Sakamoto, and T. Inoue

Electrons in the Negative-Ion-Based NBI on JT-60U 344
    M. Kisaki, M. Hanada, M. Kamada, Y. Tanaka, K. Kobayashi, and M. Sasao

Results of the SINGAP Neutral Beam Accelerator Experiment at JAEA 353
    H. P. L. de Esch, L. Svensson, T. Inoue, M. Taniguchi, N. Umeda, M. Kashiwagi, and G. Fubiani

Lithium Jet Neutralizer to Improve Negative Ion Neutral Beam Performance 364
    L. R. Grisham

Kinetic Study of the Secondary Plasma Created in the ITER Neutraliser 374
    F. Dure, A. Lifschitz, J. Bretagne, G. Maynard, K. Katsonis, A. Simonin, and T. Minea

Photo-Neutralization of Negative Ion Beam for Future Fusion Reactor 385
    W. Chaibi, C. Blondel, L. Cabaret, C. Delsart, C. Drag, and A. Simonin

Model of a SNS Electrostatic LEBT with a Near-Ground Beam Chopper 395
    B. X. Han and M. P. Stockli

Beam Induced Effects in the ITER Electrostatic Residual Ion Dump 402
    E. Surrey, A. J. T. Holmes, and T. T. C. Jones

Steering of Multiple Beamlets in the JT-60U Negative Ion Source 412
    M. Kamada, M. Hanada, Y. Ikeda, and L. R. Grisham

Compensation of Beamlet Repulsion in a Large Negative Ion Source with a Multi Aperture
Accelerator 421
    M. Kashiwagi, T. Inoue, L. R. Grisham, M. Hanada, M. Kamada, M. Taniguchi, N. Umeda, and
    K. Watanabe

Purification of Radioactive Ion Beams by Photodetachment in a RF Quadrupole Ion Beam
Cooler 431
    Y. Liu, C. C. Havener, T. L. Lewis, A. Galindo-Uribarri, and J. R. Beene

Characteristics of a He⁻ Beam Produced in Lithium Vapor 443
    N. Tanaka, T. Nagamura, M. Kikuchi, A. Okamoto, T. Kobuchi, S. Kitajima, M. Sasao, H. Yamaoka, and
    M. Wada

BEAMLINES AND FACILITIES

Physical and Experimental Background of the Design of the ELISE Test Facility 451
    P. Franzen, U. Fantz, W. Kraus, H. Falter, B. Heinemann, R. Nocentini and the NNBI Team

A Test Stand for Ion Sources of Ultimate Reliability 461
    R. Enparantza, L. Uriarte, F. J. Bermejo, V. Etxebarria, J. Lucas, J. M. Del Rio, A. Letchford,
    D. Faircloth, M. Stockli, P. Romano, J. Alonso, I. Ariz, and M. Egiraun

Recent Progress in the Negative-Ion-Based Neutral Beam Injectors in Large Helical Device 470
    Y. Takeiri, K. Tsumori, K. Ikeda, M. Osakabe, K. Nagaoka, Y. Oka, E. Asano, T. Kondo, M. Sato,
    M. Shibuya, S. Komada, and O. Kaneko

Status of the Negative Ion Based Heating and Diagnostic Neutral Beams for ITER 480
    B. Schunke, D. Bora, R. Hemsworth, and A. Tanga

List of Participants 491
Author Index 497

Preface

In 2006, at Santa Fe, we bade farewell to the symposium *Production and Neutralization of Negative Ions and Beams*, a series which started in 1977 exclusively for hydrogen negative ions but evolved over the years to include other elements. Ion sources have always played a large role in this symposium and in 2006 the decision was made to change the name of the meeting to the *International Symposium on Negative Ions, Beams and Sources*, the final version of the name being determined by ballot of the International Programme Committee in 2008. The new name better reflects the content of the meeting, which is the only international forum dedicated to all aspects of negative ion physics from formation to application.

The new Symposium had an auspicious start at its first meeting at Aix-en-Provence in September 2008, with over sixty submissions on subjects ranging from negative ion formation to purification of radioactive ion beams. The most striking feature of these proceedings was the explosion in modelling work in all aspects of plasma and beam generation. One area that received particular attention was sheath formation in the presence of heavy, negative charges. This is of particular importance in fusion applications where a bias grid (positive with respect to the anode) is commonly employed to reduce the co-extracted electron component in the beam. It is clear that advanced modelling techniques and computational resources are enabling 2D- and 3D- representations of negative ion systems to be created but the community must not forget the importance of validation against real data. It is therefore essential that test facilities equipped with adequate diagnostics for both plasma and beam measurements are maintained. So it was gratifying to hear of a new test facility being created in Spain which will enable different sources to be compared by a standard set of diagnostics. The facility is open to international cooperation, so important in the world of today, and I believe the community will support this endeavour wholeheartedly.

It was encouraging that the level of support for the Symposium was such that it proved necessary to introduce a poster session, the oral presentation time being significantly over-subscribed. Whilst this is welcome as indicative of the popularity of the Symposium, a poster session always lacks the communion of the spoken presentation, so perhaps the International Programme Committee might consider extending the length of the next meeting.

On behalf of all the delegates I would like to thank our hosts, Association EURATOM-CEA and IRFM-CEA-Cadarache, the sponsors, Communauté Pays d'Aix and of course the local organizing committee for such a successful Symposium and without whom these proceedings would not exist.

I am pleased to announce that the 2010 Symposium will be hosted by the National Institute for Fusion Science, Toki, Japan and send best wishes to Takeiri-san and his Local Organizing Committee for a successful *2nd International Symposium on Negative Ions, Beams and Sources*.

Elizabeth Surrey,
Proceedings Editor

EURATOM/UKAEA Fusion Association
Culham Science Centre, U.K.

1st International Symposium on Negative Ions, Beams and Sources

International Programme Committee

M. Bacal, Ecole Polytechnique Palaiseau (France)

Y. Belchenko, INP Novosibirsk (Russia)

D. Faircloth, STFC (UK)

R. Gobin, CEA Saclay (France)

R. Hemsworth, ITER Cadarache

Y. Hwang, SNU (Korea)

T. Inoue, JAEA Naka (Japan)

R. Keller, LANL (USA)

W. Kraus, IPP Garching (Germany)

K.N. Leung, LBNL Berkeley (USA)

J. Peters DESY Hamburg (Germany)

M. Stockli, ORNL (USA)

E. Surrey, UKAEA Culham (UK)

Y. Takeiri, NIFS Toki (Japan)

1st International Symposium on Negative Ions, Beams and Sources

Local Organizing Committee

A. Simonin (CEA), Chairman

R. Hemsworth (ITER), Co-chairman

D. Boilson (DCU)

T. Hutter (CEA)

P. Lotte (CEA)

P. Vatblé, Secretary

V. Bernadac (Promo Sciences)

FUNDAMENTAL PROCESSES

Monte Carlo Simulation Of H⁻ Ion Transport

P. Diomede[a], S. Longo[a,b] and M. Capitelli[a,b]

[a] *Dipartimento di Chimica dell'Universita' di Bari, Via Orabona 4, 70126 Bari, Italy*
[b] *IMIP/CNR, Bari Section, Via Orabona 4, 70126 Bari, Italy*

Abstract. In this work we study in detail the kinetics of H⁻ ion swarms in velocity space: this provides a useful contrast to the usual literature in the field, where device features in configuration space are often included in detail but kinetic distributions are only marginally considered. To this aim a Monte Carlo model is applied, which includes several collision processes of H⁻ ions with neutral particles as well as Coulomb collisions with positive ions. We characterize the full velocity distribution i.e. including its anisotropy, for different values of E/N, the atomic fraction and the H⁺ mole fraction, which makes our results of interest for both source modeling and beam formation. A simple analytical theory, for highly dissociated hydrogen is formulated and checked by Monte Carlo calculations.

Keywords: Monte Carlo methods, Ion transport.
PACS: 52.20.Hv, 52.65.Pp

INTRODUCTION

In spite of the importance of transport models for H⁻ ions in ion sources and other kinds of discharges in hydrogen containing mixtures, very little attention has been given in the past to the theoretical descriptions of the ion kinetics itself, i.e. in simple conditions where the details of specific sources, like the presence of boundaries with more or less complex geometries or source-specific space distributions of electric and magnetic fields are eliminated. As a result, the choice of material available in the literature for the benchmarking of new collision modules to be incorporated in numerical codes is rather poor. This situation strikingly contrasts with the state and history of discharge studies, where typically many detailed studies of the electron and ion energy distribution function, the anisotropy of the velocity distributions, the rates for electron-neutral and ion-neutral processes were presented and mutually compared before coming to full geometry studies and self-consistent Particle in Cell – Monte Carlo methods [1, 2]. We have therefore decided to model swarm experiments, where the ions propagate in a uniform gas under the action of a uniform and constant electric field, using a Monte Carlo method suitable for all non relativistic ion energies including the thermal range and rigorously derived from the appropriate linear transport equation [3]. In addition to the very important role they can play in the comparison of results and the establishment of a common understanding of the role of different collision processes on ion transport, swarm results provide ground for the discussion of crucial experiments and are of direct interest for ion extraction. Here we report a summary of methods and results of our first campaign of theoretical swarm studies devoted to H⁻ ions. A more detailed survey with related discussion can be found on ref. [4].

MODEL DESCRIPTION

By using a modeling procedure described in ref. [3] in this work we calculate the velocity distribution for H⁻ ions in H/H$_2$ mixtures of different composition and characterize its anisotropy. While the method can be applied to any degree of spatial non uniformity, in order to reduce the number of parameters whose effect is to be discussed, here we assume a uniform plasma model, where the velocity distribution is supposed to result from the collisional kinetics of negative ions with neutral particles, here H atoms and H$_2$ molecules, and H$^+$ ions under a uniform electric field.

We present here the solution of the transport equation for negative ions using realistic cross sections to include the following four collision processes

H⁻ + H$_2$ → H⁻ + H$_2$ (elastic collision)
H⁻ + H → H + H⁻   (charge exchange)
H⁻ + H$_2$ → e + H + H$_2$ (detachment)
H⁻ + H$^+$ → H⁻ + H$^+$ (elastic collision)

In the specific case the Boltzmann equation is written in the following form:

$$\frac{\partial f(\mathbf{v},t)}{\partial t} + q\mathbf{E}/m \cdot \frac{\partial f(\mathbf{v},t)}{\partial \mathbf{v}} = \sum_p C_p \qquad (1)$$

where C$_p$ is the contribution of the p-th process to the Boltzmann collision term.

As shown in ref. [3] the expression of the solution of a transport equation like (1) above in terms of Poisson distribution, which arises from application of the formalism of time dependent perturbation theory, can be numerically evaluated by a null collision test particle Monte Carlo method which takes into account the thermal distribution of the target particles. To this aim the dynamics of an ensemble of mathematical point particles is simulated: particles are subjected to the electric force q**E** and undergo collisions with neutral particles with a frequency given by

$$\nu_{max} = \max_g (\sum_p n_{s(p)} \sigma_p(g) g) \qquad (2)$$

where $n_{s(p)}$ and σ_p are, respectively, the number density of the target species s and the cross section for the p process, and g is the relative speed.

Since the E/N is assumed uniform, only the trajectory in velocity space is relevant and the ion velocity variation is given by

$$\mathbf{v}_i(t) = \mathbf{v}_i(t - t_c) - e\mathbf{E}t_c/m \qquad (3)$$

4

where v_i is the velocity vector of the i-th ion, and t_c is the time of the last collision before t (different for all particles) which is calculated by the usual relation

$$t_c = -(1/v_{max})\ln \eta \qquad (4)$$

where η is a random number from a uniform distribution, 0<η≤1.

The real collision frequency, taking into account the thermal velocity distribution of target particles, is matched by removing excess collisions (null collisions) using appropriate collision probabilities. To this aim candidate target particles (hydrogen molecules, ions and positive ions) for any collision process are generated: the velocity of the target particle is selected from a Maxwell distribution.

The collision probabilities for any of the collision processes are:

$$P_p = \frac{n_{s(p)} \sigma_p(g) g}{v_{max}} \qquad (5)$$

As regards the simulation of collision processes, elastic collisions with the hydrogen molecule are treated assuming isotropic scattering in the center-of-mass frame.

Charge-exchange collisions are treated according to their definition.

Elastic collisions with H^+ ions are treated as elastic, using the momentum transfer cross section for screened Coulomb scattering and an appropriate value of the Coulomb logarithm.

The H^-/H_2 detachment process is treated as an inelastic one: this approximation is justified by the high ratio between elastic and reactive collision frequencies in typical conditions for ion sources. Also direct check by our code shows that the effect of this approximation on the overall distribution is rather small in the conditions analyzed.

The H^-/H detachment process is not included in calculations since it has a cross section which is about one order of magnitude lower than the charge exchange one with the same target species.

Details about the cross sections used are found in ref.[4].

The simulations were performed using 10^5 particles and sampling the results 10 times once the steady state was reached.

RESULTS AND DISCUSSION

In the present set of simulations we consider a uniform electric field applied along the x direction.

The simulation has been performed for a gas and H atom temperature of 300K, while T_{H+}=0.25eV.

The knowledge of the ion energy distribution function (iedf, see definition in the next section) allows a simple determination of inelastic collision frequencies. However, in order to calculate transport coefficients, the unreduced velocity distribution is needed. In the general case, the graphical representation of this function is a very complex task. In the present test case, where transversal magnetic fields or thermal gradients are absent, the velocity distribution is axially symmetric with respect to the velocity vector associated with translation in the direction of the electric field, in this case the v_x axis.

Under such conditions, a full account of this two-dimensional velocity distribution can be given by contour plots of $f(v_x, v\perp)$, where the x axis is parallel to the applied electric field and $v\perp = \sqrt{v_y^2 + v_z^2}$.

Samples of such plots are shown is fig.1 for different values of the reduced electric field, the H atom number density and the H$^+$ mole fraction. The values of the reduced electric field E/N (evaluated accounting for both the atomic and the molecular number densities in the denominator) in the different test cases have been chosen in order to have the same average kinetic energy of the distribution, that is 0.4eV.

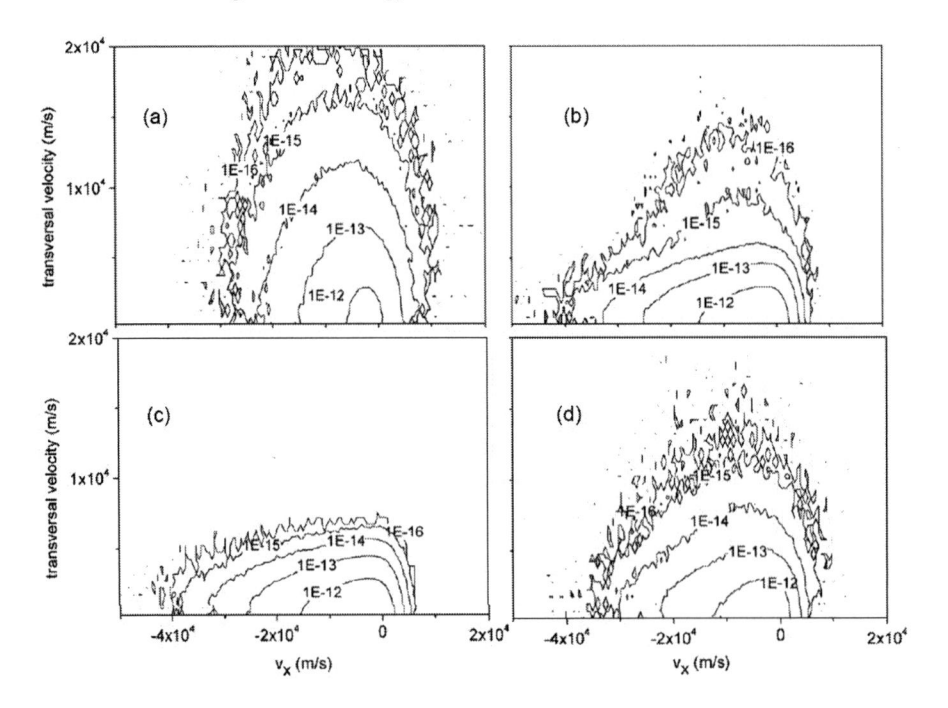

FIGURE 1. H velocity distribution $f(v_x, v\perp)$ for: (a) a non dissociated gas, $E_x/N = 49$Td, (b) $n_H = n_{H2}$, $E_x/N = 1250$Td, (c) a fully dissociated gas, $E_x/N = 2450$Td, (d) $n_H = n_{H2}$, $E_x/N = 1250$Td, $\chi_{H^+} = 10^{-3}$.

The distribution in fig.1(a) appears as a weakly deformed Maxwellian, while in fig.1(b) two components appear: an almost isotropic bulk and a strong anisotropic tail extending to a higher relative speed.

In fig. 1(c) a different feature seems to emerge: a low temperature source distribution, due to charge-exchange collisions, is stretched towards the (-x) direction by the effect of the electric force producing a highly anisotropic distribution. In the next section we will show by analytical considerations that a better description of the distribution implies a one-dimensional Maxwellian distribution with a low energy peak due to the kinematics of ion acceleration.

While in figs. 1(a)-(c) a low ionization degree is assumed, so that Coulomb collisions can be neglected, in fig.1(d) a typical value of the H^+ mole fraction is considered, and the Coulomb logarithm is estimated for a gas pressure of 3 mTorr. This last plot is otherwise in the same conditions of 1(b) and can be compared to this last: it can be noted that the Coulomb collisions have a noticeable but not substantial effect on the velocity distribution which retains the same overall shape.

An alternative description of the full velocity distribution is obtained as follows: preliminarily the variables are changed to ε and $\cos\theta$, with θ the angle between the velocity vector and the direction (+x) of the electric field.

The angular dependence of the resulting function is expanded in Legendre polynomials

$$f(\varepsilon, \cos\vartheta) = \sum_{l=0}^{\infty} f_l(\varepsilon) P_l(\cos\theta) \qquad (6)$$

The functions $f_l(\varepsilon)$ are determined using the orthonormality relations.

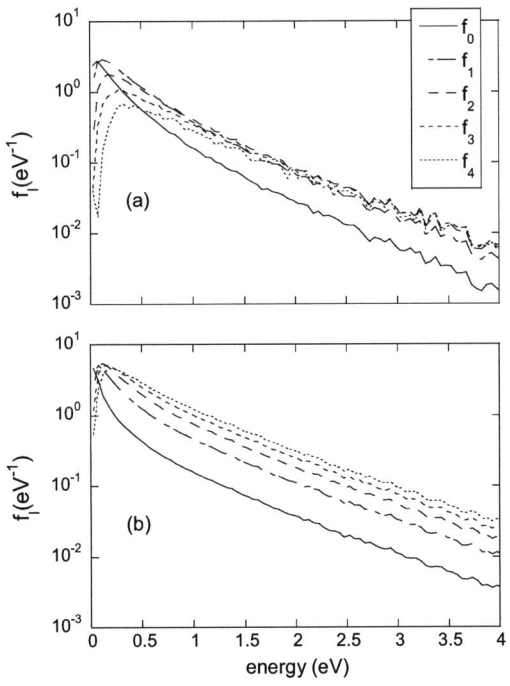

FIGURE 2. Absolute values of Legendre polynomials expansion coefficients for: (a) a non dissociated gas, E_x/N = 49Td, (b) a fully dissociated gas, E_x/N = 2450Td. Signs are positive for even harmonics and negative for odd ones, except, in case (a) for the f_3-f_4 at very low energy (<0.07 eV).

In fig. 2 the Legendre polynomial expansion coefficients for some test cases in fig. 1 are reported.

From fig. 2 it can be noted that the anisotropy of the ion distribution is much higher than in the case of electrons, where usually only the first two or three Legendre expansion coefficients are not negligible and besides as a rule $f_0 \gg |f_{(l>0)}|$ with the exception of energies very close to the threshold of inelastic collisions.

Contrarily to the case of electron transport, it cannot be said that the first anisotropic coefficient alone gives a satisfactory estimate of the ion mobility, therefore the use of transport theories assuming weak anisotropy is ruled out for this case.

In the presence of significantly dissociated gas, the velocity distribution is actually strongly anisotropic and it is better described as an ion beam with a kinetic energy distribution given by the iedf. This is confirmed by the ratios of the Legendre coefficients which, for any ion energy but very low ones, approach the relations $f_l / f_0 = (-1)^l (2l+1)$ valid for an ion beam moving in the (-x) direction.

The results of Monte Carlo modeling in the case of strongly dissociated hydrogen suggest the possibility of an analytical theory, based on a mono-directional velocity distribution and the inclusion of a single collision process, namely the charge exchange. This is shown in the next section.

ANALYTICAL THEORY FOR THE IEDF

We apply here analytical considerations to the determination of the ion energy distribution function (iedf): the iedf is obtained from the f(v) by averaging over the direction of the velocity and expressing the speed by means of the kinetic energy; the result is then divided by n_{H^-}.

In the case of highly dissociated H_2 the low energy component of the iedf is essentially due to charge exchange collisions of negative ions with the atomic fraction. Under such conditions the only important collision process is charge exchange H/H⁻, i.e. any H⁻ ion basically starts from a Maxwellian distribution at temperature T and constantly accelerates under the force –eE until a collision leads it back to thermal energy. Since kT is much smaller than the average ion energy we can assume that the post collision velocity is negligible and the iedf becomes the product of two factors: the relative residence time of the ion in the energy interval dε and an absorption factor, which represents the collision process, i.e.

$$n(\varepsilon) \propto \frac{dt}{d\varepsilon} \exp(-\int \frac{1}{\lambda(\varepsilon)} dx), \qquad (7)$$

where $\lambda(\varepsilon)$ is the mean free path for a monochromatic beam of ions moving at speed $\sqrt{2\varepsilon/m}$ which is given by $1/(\sigma_{ce} n_H)$ since the target velocity is neglected.

By simple calculations the following expression (in the integral $dx = d\varepsilon/eE$) is found:

$$n(\varepsilon) \propto \frac{1}{\sqrt{\varepsilon}} \cdot \exp\left(-\frac{1}{eE/n_H}\int_0^\varepsilon \sigma_{ce}(\xi)d\xi\right). \tag{8}$$

In Fig. 3 one can find a comparison between the H⁻ iedf calculated by means of the Monte Carlo technique, for a fully dissociated gas, and the iedf obtained with the approximate theory for H plasma for different values of the reduced electric field: a very good agreement is shown. Following a traditional usage all results are divided by $\sqrt{\varepsilon}$ (in order to obtain a straight line in a semi-log plot for the case of a Maxwell distribution) therefore n(ε) is measured in eV$^{-3/2}$.

The overall coefficient of the analytical form is selected here to match the high energy tails of the MC results: very slight deviations are barely noticeable at low energy. An alternative is to fix the overall coefficient by imposing normalization (which is always possible for the energy dependence given by eq.(8)).

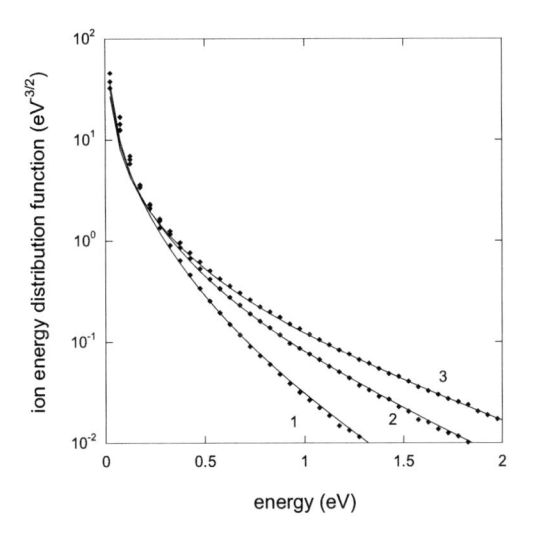

FIGURE 3. Comparison between the H⁻ iedf calculated by means of the Monte Carlo technique (dots), for a fully dissociated gas, and the iedf obtained with the approximate theory for H plasma for different values of the reduced electric field (full lines). The values of the reduced electric field, for the curves from 1 to 3, are, respectively, 1000, 1500 and 2000Td.

Furthermore, since the charge exchange cross section is approximately constant in the energy range considered eq.(8) can be written as

$$\frac{1}{\varepsilon} \cdot \sqrt{\varepsilon}\exp\left(-\frac{\varepsilon}{k\Theta}\right), \quad \Theta = \frac{eE}{k\sigma_{ce}n_H} \tag{9}$$

i.e. as a Maxwell distribution modulated by a factor $1/\varepsilon$ which produces a low energy component: this last is of kinematic origin and has no connection to the gas temperature.

This distribution is close to the two temperature distribution which has been postulated [5] to fit the results of photodetachment determination of ion velocity distribution in sources [6] and has the advantage of containing only a single parameter instead of the three (the two T's and the ratio of the two ion populations) of the two-T model.

At the same time our results show that the ion velocity distribution produced by an electric field is strongly anisotropic and this should therefore be taken into account in the modeling of experiments, while the Ballistic theory [6] assumes isotropy of the ion velocity distribution.

CONCLUSIONS

The most relevant aspects of our contribution can be summarized as follows:

(1) A Monte Carlo solution of the kinetic equation for H⁻ transport is used to analyze H⁻ ion swarms in velocity space in H/H_2 mixtures, including the thermal motion of target particles and Coulomb collisions.

(2) Swarm results for the energy distribution, the full velocity distribution and its expansion in spherical harmonics provide a ground for comparison to future calculations, since they are not affected by source design details

(3) At a constant ion energy, the degree of anisotropy of the velocity distribution is strongly enhanced when the dissociation degree of the gas is increased.

(4) A simple analytical model is obtained for energy distribution in the case of a fully dissociated H plasma and uniform reduced electric field which agrees with the MC solution.

(5) Results can be relevant to the discussion of photodetachment experiments.

The main limit of the present calculation is that all results are space-averaged. New calculations should take into account the space dependencies while starting from an ion source or initial condition for the ion distribution: in these improved calculations the loss terms due to ion detachment or recombination will have to be included.

ACKNOWLEDGEMENTS

This work was partially supported by Progetto Strategico Regione Puglia: "Sviluppo di un rivelatore a film di diamante per radiazione ultravioletta" and by MIUR-PRIN07 (2007H9S8SW_003).

REFERENCES

1. L. C. Pitchford, B. V. McKoy, A. Chutjian, and S. Trajmar ed. *Swarm Studies and Inelastic Electron-Molecule Collisions: Proceedings of the Meeting of the Fourth International Swarm Seminar and the Inelastic Electron-Molecule Collisions Symposium*, Springer, 1987
2. M. Capitelli and J. Norman Bardsley, *Nonequilibrium Processes in Partially Ionized Gases (NATO Science Series: B)*, Springer, 1990
3. S. Longo, P. Diomede, *Eur. Phys. J. Appl. Phys.* **26**, 177-185 (2004)
4. P. Diomede, S. Longo, *IEEE Trans. Plasma Sci*, in press
5. A. A. Ivanov, Jr., *Rev. Sci. Instrum.* **75**, 1754-1756 (2004)
6. M. Bacal, P. Berlemont, A. M. Bruneteau, R. Leroy, and R. A. Stern, *J. Appl. Phys* **70**, 1212-1219 (1991)

2D Model of a Tandem Plasma Source: The Role of the Transport Processes

Tsvetelina V. Paunska[a], Antonia P. Shivarova[a], Khristo Ts. Tarnev[b] and Tsanko V. Tsankov[a]

[a]*Faculty of Physics, Sofia University, BG-1164 Sofia, Bulgaria*
[b]*Department of Applied Physics, Technical University–Sofia, BG-1000 Sofia, Bulgaria*

Abstract. A 2D fluid-plasma model of a tandem two-chamber plasma source operating in hydrogen gas is presented in the study. The configuration of the source is of the type of the inductively-driven sources developed for neutral beam injection heating of fusion plasmas: the rf power deposition is in the first smaller-radius chamber of the source and the second bigger-radius chamber is a volume for plasma expansion from the driver. Plasma maintenance in the source governed by transport processes – charged-particle and electron-energy fluxes – is the main conclusion stemming from the results. Moreover, the metal walls of the source and its different dimensions in the longitudinal and transverse directions determine a regime of a rf discharge characterized by a dc current in the source: electrons and ions flowing in different directions determine a solenoidal current in the discharge. This is accompanied by a strong drop of electron temperature, charged-particle and hydrogen-atom densities and plasma potential at the transition between the two chambers of the source. In general, the results from the model lead to the conclusion that analysis of the transport processes should be in the basis of the optimization of the tandem-plasma-source operation.

Keywords: 2D discharge modeling, tandem two-chamber plasma sources, hydrogen discharges
PACS: 52.50.-b, 52.25.Fi, 52.30.-q

INTRODUCTION

The current active research [1, 2, 3, 4] on 2D modeling of plasma sources with complicated design is forced by their use in the plasma processing technologies and as particle sources. In the discussions on the operation of the sources, the spatial (and temporal) distributions of the plasma parameters are the results usually presented. The charged-particle and electron-energy fluxes have been stressed on as results from recent modeling [5] of a tandem two-chamber plasma source operating in an argon gas, showing that the transport processes determine the mechanisms of the source operation. Moreover, a gas-discharge regime with a net dc current in rf discharges has been established in the gas-pressure range considered to be governed by ambipolar diffusion.

This study presents an extension of the 2D model [5] of a rf-driven tandem two-chamber plasma source towards operation of the source in a hydrogen gas. Stressing on the gas-discharge regime – a regime with a net dc current due to appearance of a solenoidal net flow $\vec{\Gamma}$ (div$\vec{\Gamma} = 0$ with $\vec{\Gamma} = \vec{\Gamma}_e - \vec{\Gamma}_i$ being the difference of the electron and ion fluxes) – the discussion on the results displays the mechanisms governing discharge maintenance of expanding hydrogen plasmas.

FORMULATION OF THE PROBLEM

The design of the source (Fig. 1) is the same, however, in a smaller size, as of the inductively-driven two-chamber plasma sources constructed [6] regarding ITER applications. In the model, the rf power deposition for the discharge maintenance is shaped by a super-Gaussian profile in the axial (z) direction $P_{\text{ext}} = P_0 \exp\left[-\frac{1}{2}\left(\frac{z-z_0}{\sigma}\right)^{2m}\right]$ centered at $z = z_0$; $P_0 = P_{\text{ext}}(z = z_0)$ and σ scales the width of the profile.

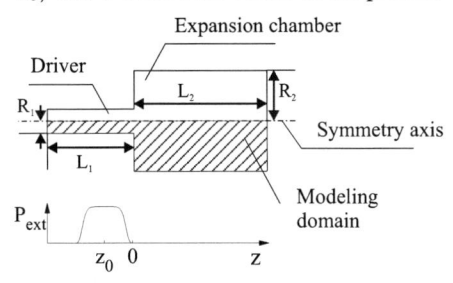

FIGURE 1. Configuration of the source. The modeling domain and the location of the input power are denoted; the ($z=0$)-position is at the transition between the two chambers. Dimensions: $R_1 = 2.25$ cm, $L_1 = 20$ cm, $R_2 = 11$ cm and $L_2 = 30$ cm. Parameters of the power input: $z_0 = -7.5$ cm, $\sigma = 4.8$ cm and $m = 4$.

The description is within the fluid plasma model involving the continuity equations

$$\frac{\partial n_e}{\partial t} + \nabla \cdot \vec{\Gamma}_e = \frac{\delta n_e}{\delta t}, \quad \frac{\partial n_{ij}}{\partial t} + \nabla \cdot \vec{\Gamma}_{ij} = \frac{\delta n_{ij}}{\delta t} \quad (1)$$

of electrons (e) and positive ions (i), with $j = 1,2,3$ denoting the H^+-, H_2^+- and H_3^+-ions, respectively, the balance of the hydrogen atoms H

$$\frac{\partial N_1}{\partial t} + \nabla \cdot \vec{\Gamma}_H = \frac{\delta N_1}{\delta t}, \quad (2)$$

the electron energy balance

$$\frac{3}{2}\frac{\partial (n_e T_e)}{\partial t} + \nabla \cdot \vec{J}_e = P_{\text{ext}} - P_a^{(\text{coll})} - P_m^{(\text{coll})} + P_{\text{dc}}, \quad (3)$$

the Poisson equation

$$\Delta \Phi = -\frac{e}{\varepsilon_0}\left(\sum_{j=1}^{3} n_{ij} - n_e\right) \quad (4)$$

and the expression $p = \kappa T_g(N_1 + N_2)$ for the gas pressure.

In (1)-(4), the charged particle densities and fluxes are denoted by n and $\vec{\Gamma}$, respectively, N_1 and N_2 are the concentrations of atoms and molecules, T_e is the electron temperature (in energy units) and Φ is the potential of the dc electric field ($\vec{E}_{\text{dc}} = -\nabla\Phi$) formed in the discharge; the other notation is standard. The electron and ion fluxes in (1) are

$$\vec{\Gamma}_e = -D_e \nabla n_e - b_e n_e \vec{E}_{\text{dc}} - D_e^T n_e (\nabla T_e/T_e), \quad \vec{\Gamma}_{ij} = -D_{ij}\nabla n_{ij} + b_{ij} n_{ij} \vec{E}_{\text{dc}}, \quad (5)$$

TABLE 1. Processes of volume production and losses of charged and neutral particles

Number	Process	Rate coefficient	Ref.
1	$e + H_2(X^1\Sigma_g^+) \to e + H_2^+(v) + e$	a_1	[7]
2	$e + H_2(X^1\Sigma_g^+) \to e + H(1s) + H(1s)$	a_2	[7]
3	$e + H_2^+(v) \to e + H^+ + H(1s)$	a_3	[7]
4	$H_2^+ + H_2 \to H_3^+ + H(1s)$	a_4	[8]
5	$e + H(1s) \to e + H^+ + e$	a_5	[7]
6	$e + H_3^+ \to H + H + H$	a_6	[7]
7	$e + H_3^+ \to H^+ + 2H(1s) + e$	a_7	[7]

the former consisting of a diffusion ($\vec{\Gamma}_e^{(d)} = -D_e \nabla n_e$), drift ($\vec{\Gamma}_e^{(E)} = -b_e n_e \vec{E}_{dc}$) and thermal-diffusion ($\vec{\Gamma}_e^{(td)} = -D_e^T n_e (\nabla T_e / T_e)$) fluxes and the latter being drift-diffusion fluxes. The electron mobility b_e and diffusion coefficient D_e are the ordinary ones, defined through the frequencies of elastic collisions whereas the ion mobilities and diffusion coefficients are effective ones, involving both elastic and inelastic collisions, the latter through $\delta n_{ij}/\delta t$. The production and volume losses ($\delta n_e/\delta t$, $\delta n_{ij}/\delta t$ and $\delta N_1/\delta t$) of the charged particles and hydrogen atoms are according to the reactions in Table 1. In (3), the electron energy flux

$$\vec{J}_e = -\chi_e \nabla T_e + \frac{5}{2} T_e \vec{\Gamma}_e \qquad (6)$$

involves both the conductive flux ($\vec{J}_e^{(cond)} = -\chi_e \nabla T_e$, with $\chi_e = (5/2) n_e D_e$ being the thermal conductivity coefficient) and the convective flux ($\vec{J}_e^{(conv)} = (5/2) T_e \vec{\Gamma}_e$). $P_{dc} = -e\vec{\Gamma}_e \cdot \vec{E}_{dc}$ is the term usually called electron energy losses for maintenance of the dc electric field $\vec{E}_{dc}$ in the discharge. (Here it will be shown that P_{dc} could appear also as a power input.) $P_a^{(coll)}$ and $P_m^{(coll)}$ in (3) are the electron energy losses in collisions with atoms (a) and molecules (m). The former ($P_a^{(coll)}$) includes losses for atom ionization and excitation (to the $n = 2, 3$ and 4 excited states) as well as in elastic electron-atom collisions. The latter ($P_m^{(coll)}$) accounts for excitation of vibrational ($v = 1, 2$) and singlet ($H_2^*(B^1\Sigma_u^+ 2p\sigma)$, $H_2^*(C^1\Pi_u 2p\pi)$ and $H_2^*(E, F^1\Sigma_g^+)$) states of the molecule, its dissociation and ionization as well as for elastic electron-molecule collisions.

The two chambers of the source (Fig. 1) are with metal walls (like in [6]) which imposes $\Phi = 0$ as a boundary condition. The boundary conditions for the fluxes at the walls are those [5, 9] usually used: $\vec{n} \cdot \vec{\Gamma}_e = (1/4) n_e v_e^{th}$ and $\vec{n} \cdot \vec{\Gamma}_{ij} = (1/2) n_{ij} v_{ij}^{th} + b_{ij} n_{ij} (\vec{n} \cdot \vec{E}_{dc})$ for electrons and ions for which, respectively, thermal and directed motions are the crucial ones, and $\vec{n} \cdot \vec{\Gamma}_H = (1/4) \gamma N_1 v_H^{th}$ for the hydrogen atoms. (Here the thermal velocities are denoted by v^{th}, $\gamma = 4 \times 10^{-3}$ is the coefficient of the wall recombination losses of the hydrogen atoms and $\vec{n}$ is the outward unit vector.) The boundary condition for the electron energy flux is $\vec{n} \cdot \vec{J}_e = (5/2) T_e (\vec{n} \cdot \vec{\Gamma}_e)$. The symmetry on the axis is also taken into account.

The results presented in the next section are for a cw regime of discharge maintenance.

RESULTS AND DISCUSSIONS

A configuration of a source like that in Fig. 1 supposes remote plasma maintenance in the second chamber of the source and, respectively, importance of the charged-particle and electron-energy fluxes in the entire source. Thus, *the basic question* which arises is about *the regime of the discharge maintenance in the source*. The results presented here for hydrogen discharges confirm previous results from 2D modeling of argon discharges in sources with the same configuration [5] and show a discharge regime with a net dc current flowing through plasmas sustained by rf power deposition in the driver region. This dc current results from electron and ion fluxes, different in their magnitude and directions, i.e. it is caused by the appearance of a net flux $\vec{\Gamma} = \vec{\Gamma}_e - \vec{\Gamma}_i$ (where, in hydrogen discharges, $\vec{\Gamma}_i = \Sigma_{j=1-3}\vec{\Gamma}_{ij}$ is the total ion flux, i.e. the sum of the fluxes of the H^+-, H_2^+- and H_3^+-ions). According to (1), $\vec{\Gamma}$ is a solenoidal flux (div$\vec{\Gamma} = 0$). The two cases discussed here display discharge behavior under conditions of different modifications of this flux: streamlines of $\vec{\Gamma}$ going directly to the discharge walls and streamlines of $\vec{\Gamma}$ circulating in the plasma volume.

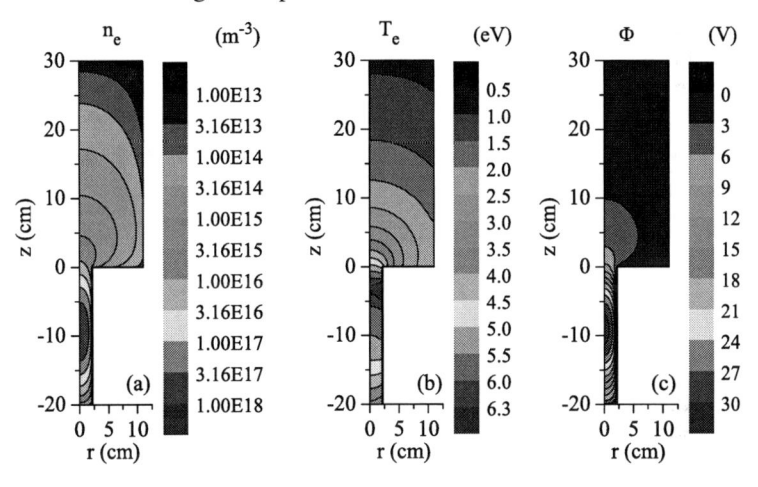

FIGURE 2. Contours of constant electron density (a), electron temperature (b) and potential of the dc field (c); $p = 20$ mTorr, $Q = 500$ W.

The first case – a case when all the streamlines of $\vec{\Gamma} = \vec{\Gamma}_e - \vec{\Gamma}_i$ are closed through the discharge walls – is displayed for gas discharge conditions specified by $p = 20$ mTorr and total applied power $Q = 500$ W. The spatial distribution of the plasma parameters (Figs. 2 and 3(a)-(c)) shows strong variation – over the two chambers of the discharge vessel – of the electron (n_e) and ion (n_{ij}) densities (which change in orders of magnitude) as well as of the electron temperature T_e and of the potential Φ of the dc electric field. Thus, the very low plasma density and the low electron temperature in the second chamber of the source is *the first conclusion* which should be stressed on. The concentration N_1 of the hydrogen atoms also drops at the transition between the two chambers and this determines an increase of the concentration N_2 of the molecular hydrogen in the second chamber. The asymmetry of the profiles of n_e, n_{ij}, T_e and Φ in the driver

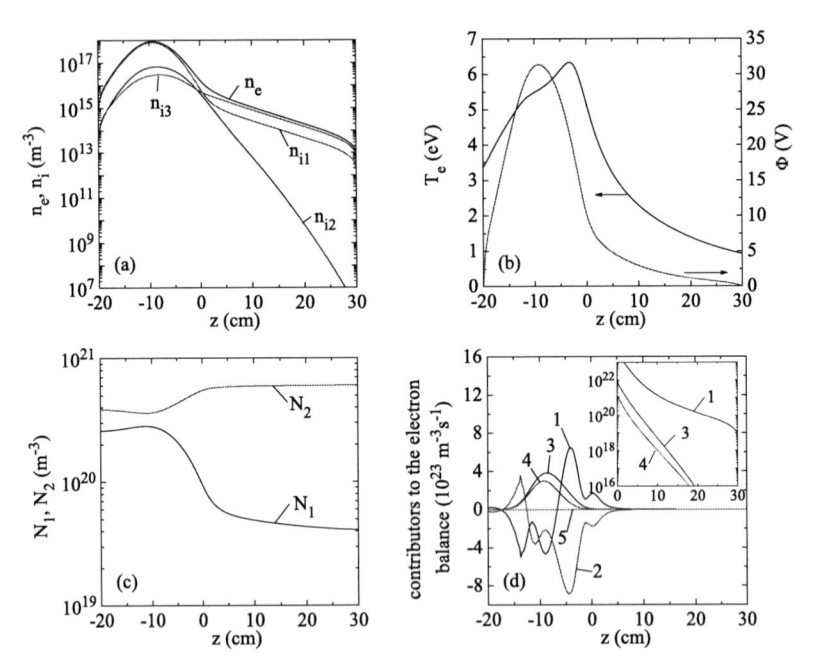

FIGURE 3. Axial profiles – on the discharge axis ($r = 0$) – of the concentration of electrons, H^+-, H_2^+- and H_3^+- ions in (a), of the electron temperature and the potential of the dc field in (b), of the concentrations N_1 and N_2 of hydrogen atoms and molecules in (c) and of the contributors ($-\partial\Gamma_{ez}/\partial z$) (1), $-(1/r)(\partial(r\Gamma_{er})/\partial r)$ (2), ionization of molecules (3) and atoms (4) and dissociative recombination of H_3^+-ions (5)) to the electron balance in (d); $p = 20$ mTorr, $Q = 500$ W.

(which is the most pronounced for T_e (Figs. 2(b) and 3(b))) is due to the asymmetry in the N_1-distribution (Fig. 3(c)). This is commented on in more details in [10] based on the mechanism of self-consistency (the generalized Schottky condition [11]) of hydrogen discharges in regions with external power deposition. Due to the high n_e in the driver region, the H^+-ions are with the highest concentration there (Fig. 3(a)).

What maintains the discharge in the second chamber of the source? The answer is that *the charged-particle fluxes from the driver* – in fact, their axial components (the radial fluxes are charged-particle losses) – *are responsible for the plasma existence in the second chamber* (Fig. 3(d)). This concerns both electrons and ions. The flux of the H_3^+-ions is the largest one among the ion fluxes and these are the ions with the highest concentration in the second chamber (Fig. 3(a)).

How do the charged particle fluxes in the source behave? The electron flux $\vec{\Gamma}_e$ (Fig. 4(a)) is formed, as always, by a fine balance between diffusion ($\vec{\Gamma}_e^{(d)}$, Fig. 4(b)) and drift ($\vec{\Gamma}_e^{(E)}$, Fig. 4(c)) fluxes in which, here, due to the strong variation of T_e (Figs. 2(b) and 3(b)), the thermal-diffusion flux ($\vec{\Gamma}_e^{(td)}$, Fig. 4(d)) is involved. The behavior of the latter is complicated in the first chamber, because of the maximum of T_e (Fig. 2(b)) formed due to an electron density decrease. However, with respect to the total electron

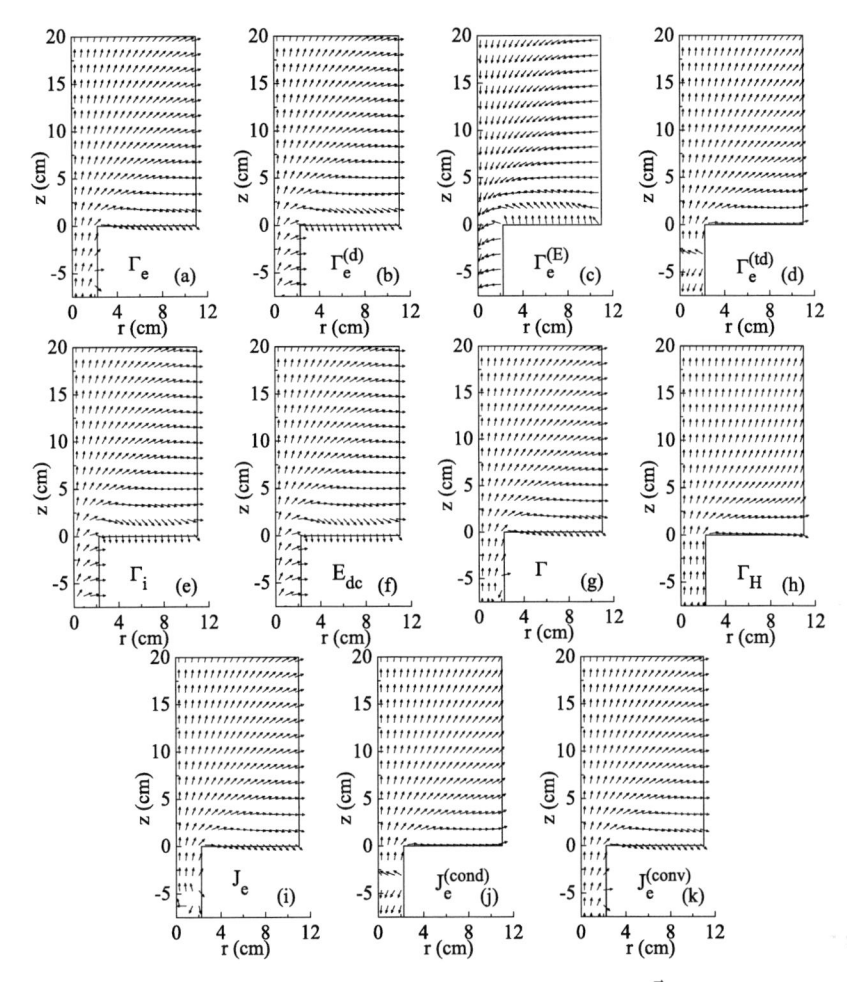

FIGURE 4. Arrow plot presentation of the direction of the electron flux $\vec{\Gamma}_e$ (a) and its components (diffusion (b), drift (c) and thermal-diffusion (d) fluxes), the total ion flux $\vec{\Gamma}_i$ (e), the electric field $\vec{E}_{dc}$ (f), the flux $\vec{\Gamma} = \vec{\Gamma}_e - \vec{\Gamma}_i$ (g) and the flux $\vec{\Gamma}_H$ (h) of the hydrogen atoms as well as of the electron energy flux $\vec{J}_e$ (i) and its components (conductive $\vec{J}_e^{(cond)}$ and convective $\vec{J}_e^{(conv)}$ fluxes in (j) and (k), respectively); $p = 20$ mTorr, $Q = 500$ W.

flux from the first chamber towards the second one, $\vec{\Gamma}_e^{(td)}$ is a forwards flux like the drift-diffusion flux $\vec{\Gamma}_e^{(dd)} = \vec{\Gamma}_e^{(d)} + \vec{\Gamma}_e^{(E)}$ (Fig. 5(a)). (Neglecting the thermal diffusion flux in the model leads to redistribution of the diffusion and drift fluxes resulting in almost the same total electron flux.) Since $\vec{E}_{dc}$ is an accelerating field for the ions, their flux (Fig. 4(e)), following $\vec{E}_{dc}$ (Fig. 4(f)), is also a forward flux, from the driver to the second chamber of the source. However, the ion flux is much smaller than the electron flux and $\Gamma_z \simeq \Gamma_{ez}$ (Fig. 5(b)). Hydrogen atoms also flow (Figs. 4(h) and 5(a)) from the first chamber, where they are mainly produced, towards the second one.

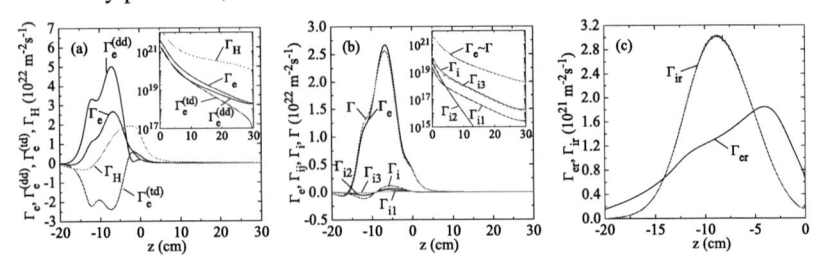

FIGURE 5. (a) Axial variations on the discharge axis ($r = 0$) of the total electron flux Γ_e and its components (the drift-diffusion $\Gamma_e^{(dd)}$ and thermal-diffusion $\Gamma_e^{(td)}$ fluxes) as well as of the flux Γ_H of the hydrogen atoms. (b) Comparison of the axial variations at $r = 0$ of the electron (Γ_e) and ion (Γ_i) fluxes and of the magnitude of the flux $\vec{\Gamma} = \vec{\Gamma}_e - \vec{\Gamma}_i$. (c) Axial variations on the side wall of the first chamber of the radial components of the electron (Γ_{er}) and ion (Γ_{ir}) fluxes. $p = 20$ mTorr, $Q = 500$ W.

The plasma potential (Figs. 2(c) and 3(b)) is above the wall potential, as usual. However, at that point the specificity of the source (Fig. 1) – its metal walls and larger axial size than the radial one – brings the behavior of the fluxes into play determining *the discharge regime*. The dc electric field built in the discharge has lower axial component than the radial one, except for the on-axis region. Since $\vec{E}_{dc}$ (Fig. 4(f)) is a retarding field for the electrons and an accelerating one for the ions, such a configuration of $\vec{E}_{dc}$ favors an electron motion predominantly in the axial direction (Fig. 4(a)) and an ion motion predominantly in the radial direction (Figs. 4(e) and 5(c)). This results in the appearance of the flux $\vec{\Gamma} = \vec{\Gamma}_e - \vec{\Gamma}_i$ (Fig. 4(g)) and, respectively, of a dc current in the discharge, which is short-circuited through the metal walls of the source (a dc current flowing, due to electron motion, from the walls of the second chamber towards the driver where it continues towards the side wall of the driver, owing to a radial ion flux). Thus, except for the radial component Γ_r in the driver, where $\Gamma_r \simeq \Gamma_{ir}$, all over the source $\vec{\Gamma} \simeq \vec{\Gamma}_e$. This determines *a regime of the discharge* – a rf-driven discharge – *with a net dc current flowing in the source*.

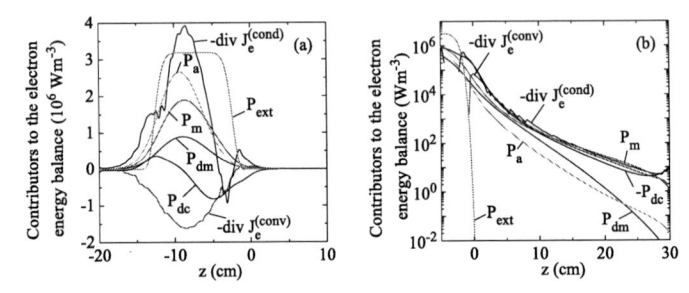

FIGURE 6. Variation of the contributors to the electron energy balance: power input externally applied (P_{ext}) and via $\mathrm{div}\vec{J}_e^{(cond)}$ both in the driver (a) and in the second chamber (b), power losses in the driver and power input to the second chamber via $\mathrm{div}\vec{J}_e^{(conv)}$ and power losses in the two chambers via electron collisions with atoms (P_a) and molecules (P_m), excluding the dissociation of the molecules given separately (P_{dm}) and for maintenance of the dc electric field (P_{dc}); $p = 20$ mTorr, $Q = 500$ W.

Plasma maintenance in the second chamber of the source needs also *electron energy input* there, *for compensating the electron energy losses*. It is ensured by both the convective flux $\vec{J}_e^{(conv)}$ (Fig. 4(k)) and the conductive flux $\vec{J}_e^{(cond)}$ (Fig. 4(j)), the latter due to the decrease of T_e (Figs. 2(b) and 3(b)) towards the second chamber of the source. The complicated behavior of $\vec{J}_e^{(cond)}$ in the first chamber close to the transition to the second one, influencing also the behavior of the total flux $\vec{J}_e$ there (Fig. 4(i)), is due to the maximum of T_e (Fig. 2(b)) already commented on. The power input to the second chamber of the source ensured by $\vec{J}_e^{(cond)}$ and $\vec{J}_e^{(conv)}$ goes for compensating electron energy losses mainly for vibrational excitation of the molecules and for maintenance of the dc field (Fig. 6). The calculations show that 97% of the input power goes for compensating the losses in the first chamber and only 3% in the second one.

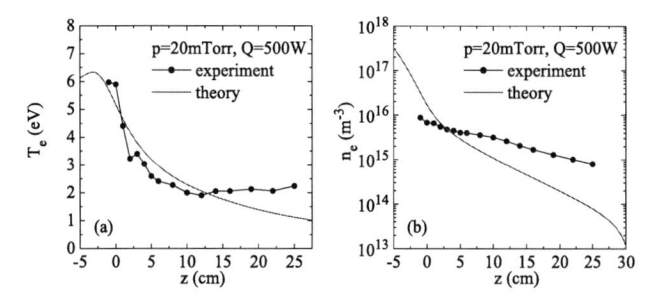

FIGURE 7. Comparison of theoretical and experimental results for the axial profiles at $r = 0$ of the electron temperature (a) and density (b); $p = 20$ mTorr, $Q = 500$ W.

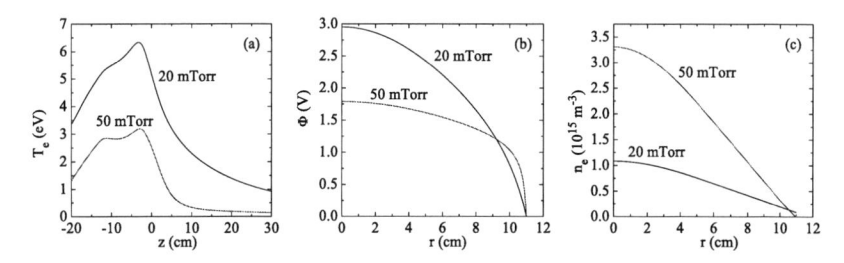

FIGURE 8. Axial profiles of the electron temperature at $r = 0$ over the total length of the source (a) and radial profiles of the plasma potential (b) and the electron density (c) in the second chamber at $z = 10$ cm for $p = 50$ mTorr compared with the corresponding quantities for $p = 20$ mTorr; $Q = 500$ W.

Experiments [12] carried out under the same conditions ($p = 20$ mTorr, $Q = 500$ W) in a plasma source similar to that in Fig. 1 provide a possibility for comparison of results from the model with experimental ones, for the axial profiles of T_e and n_e in the expanding plasma volume of the source. The agreement is very good for T_e (Fig. 7(a)). The deviation in the n_e-profiles (Fig. 7(b)) is probably due to the dielectric walls of the first chamber and the complicated design of the second chamber in the experiment.

The second case – a case with a vortex-type flux $\vec{\Gamma}$ in the plasma volume – demonstrated for conditions of $p = 50$ mTorr and $Q = 500$ W comes to show *the uniqueness in the self-organization of the gas discharges regarding self-optimization which ensures*

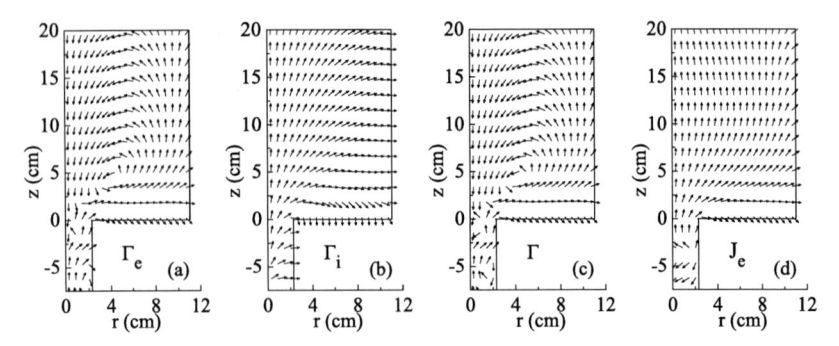

FIGURE 9. Arrow plot presentation of the direction of the electron flux $\vec{\Gamma}_e$ (a), of the total ion flux $\vec{\Gamma}_i$ (b) and of the flux $\vec{\Gamma} = \vec{\Gamma}_e - \vec{\Gamma}_i$ (c) as well as of the total electron energy flux $\vec{J}_e$ (d); $p = 50$ mTorr, $Q = 500$ W.

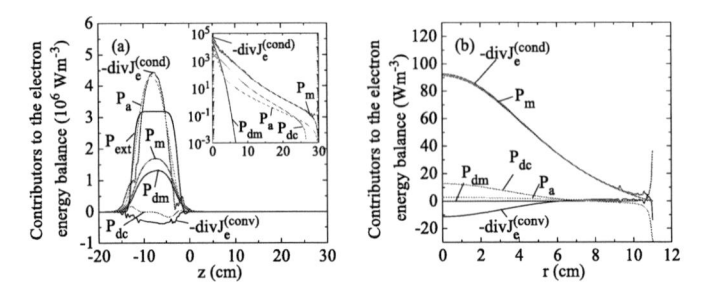

FIGURE 10. The same as in Fig. 6 except for P_{dc} and $\vec{J}_e^{(conv)}$ in the second chamber, the former appearing as a power input and the latter being power losses; $p = 50$ mTorr, $Q = 500$ W. The radial profiles in (b) are at $z = 10$ cm.

their existence. With the gas pressure increase, the electron temperature in the driver strongly decreases (Fig. 8(a)) and the electron energy flux to the second chamber needed for compensating the electron energy losses there also decreases. However, the dc electric field formed in the discharge in a self-consistent manner with the plasma density and the electron temperature manages to deal, via redistribution of its potential, with the problem. The orientation of the field is as before (Fig. 4(f)). However, a well-pronounced wall sheath with a strong drop of the dc potential (Fig. 8(b)) appears which takes care for reflecting the electrons back into the plasma volume (Fig. 9(a)) and to ensure their stay there via circulation. The lower electron density in the sheath (Fig. 8(c)) comes to confirm this. The total ion flux is, as before, a forward flux (Fig. 9(b)). Since the flux $\vec{\Gamma}$ is, like in the first case, almost equal to $\vec{\Gamma}_e$, its streamlines also circulate in the plasma volume (Fig. 9(c)). Although, the convective flux $\vec{J}_e^{(conv)}$ follows the $\vec{\Gamma}_e$-flux (Fig. 9(a)) the total electron-energy flux $\vec{J}_e$ (Fig. 9(d)) determined mainly by the conductive flux $\vec{J}_e^{(cond)}$ is a forward one. The changes in the spatial distribution of the potential Φ of the dc field and the formation of a circular electron flux in the second chamber of the source lead to redistribution of the contributors to the electron energy balance there. The conductive flux manages to compensate the electron energy losses in collisions with molecules and for maintenance of the dc field in the wall sheath (Fig. 10). Now, P_{dc} and $\vec{J}_e^{(conv)}$ are in an interplay. Due to the circulation of the electron flux $\vec{\Gamma}_e$, the convective flux $\vec{J}_e^{(conv)}$ appears as losses. However – owing to the turn back of $\vec{\Gamma}_e$ – the P_{dc}-term in (3) transforms from electron energy losses for maintenance of the dc field into electron heating by this field in the plasma volume and it compensates the losses due to $\vec{J}_e^{(conv)}$.

CONCLUSIONS

The study presents a 2D model – within the fluid plasma theory – of hydrogen discharges in a tandem two-chamber plasma source. The results show a discharge regime in the source – an rf-driven source – with a net dc current. The discussion stresses on the discharge maintenance in the second chamber of the source since it outlines the behavior of the expanding plasmas. The latter is characterized by: (i) a strong drop of the electron temperature, the plasma density and the potential of the dc electric field formed in the discharge, (ii) discharge behavior governed by charged-particle and electron-energy fluxes from the driver and (iii) spatial redistribution of the potential of the dc field aiming to ensure plasma existence in the expanding plasma volume. The main conclusion is that the optimization of such type of sources should be based on analysis of the transport processes.

ACKNOWLEDGMENTS

The work is within the programme of the Bulgarian Association EURATOM/INRNE (task P2).

REFERENCES

1. B. Ramamurthi and D. J. Economou, *Plasma Sources Sci. Technol.* **11**, 324–332 (2002).
2. G. L. M. Hagelaar, J. Bareilles, L. Garrigues and J.-P. Boeuf, *J. Appl. Phys.* **91**, 5592–5598 (2002).
3. P. Subramonium and M. J. Kushner, *J. Vac. Sci. Technol. A* **20**, 313–324 (2002).
4. G. L. M. Hagelaar, *Plasma Sources Sci. Technol.* **16**, S57–S66 (2007).
5. St. Kolev, A. Shivarova, Kh. Tarnev and Ts. Tsankov, *Plasma Sources Sci. Technol.* **17**, 035017 (1–13) (2008).
6. E. Speth *et al.*, *Nucl. Fusion* **46**, S220–S238 (2006).
7. R. K. Janev, W. D. Louger, K. Evants, Jr. and D. E. Post, Jr., *Elementary Processes in Hydrogen-Helium Plasmas* (Springer, Berlin, 1987).
8. R. H. Neynaber and S. M. Trujillo, *Phys. Rev.* **167**, 63–66 (1968).
9. G. I. M. Hagelaar, F. J. de Hoog and G. M. W. Kroesen, *Phys. Rev. E* **62**, 1452–1454 (2000).
10. Ts. Paunska, A. Shivarova, Kh. Tarnev and Ts. Tsankov, *"Spatial distribution of the plasma characteristics of a tandem plasma source"*, AIP Conf. Proc. (1st Int. Conf. on Negative Ions, Beams and Sources, Aix en Provence, 2008), submitted.
11. H. Schlüter and A. Shivarova, *Phys. Rep.* **443**, 121–255 (2007).
12. N. Djermanova, Zh. Kiss'ovski, St. Lishev and Ts. Tsankov, *J. Phys.: Conf. Series* **63**, 012013 (1–6) (2007).

The Cascaded Arc: High Flows of Rovibrationally Excited H$_2$ and its Impact on H$^-$ Ion Formation

O. Gabriel, W.E.N. van Harskamp, D. C. Schram,
M. C. M. van de Sanden and R. Engeln

*Department of Applied Physics, Plasma and Materials Processing,
Eindhoven University of Technology, The Netherlands*

Abstract. The cascaded arc is a plasma source providing high fluxes of excited and reactive species such as ions, radicals and rovibrationally excited molecules. The plasma is produced under pressures of some kPa in a direct current arc with electrical powers up to 10 kW. The plasma leaves the arc channel through a nozzle and expands with supersonic velocity into a vacuum-chamber kept by pumps at low pressures. We investigated the case of a pure hydrogen plasma jet with and without an applied axial magnetic field that confines ions and electrons in the jet. Highly excited molecules and atoms were detected by means of laser-induced fluorescence and optical emission spectroscopy. In case of an applied magnetic field the atomic state distribution of hydrogen atoms shows an overpopulation between the electronic states $p = 5$, 4 and 3. The influence of the highly excited hydrogen molecules on H$^-$ ion formation and a possible mechanism involving this negative ion and producing atomic hydrogen in state p=3 will be discussed.

Keywords: hydrogen, arc plasma, gas expansion, magnetic field, population inversion, recombining plasma.
PACS: 34.50.Fa; 39.30.+w; 52.20.-j; 52.70.Kz; 52.70.-m

INTRODUCTION

Gas phase plasma chemistry is determined strongly by radicals and excited atoms and molecules, because the rate coefficients of many reactions increase with an increase of the internal energy of the reaction partners. The aim of plasma source design is therefore an optimization of excited species production. These species have been produced in discharges driven by alternating electric fields, e. g., radio-frequency, microwaves or electron cyclotron resonance. While these kinds of discharges require in general low pressures, direct current discharges like arc discharges can be operated at higher pressures. The output of excited species is correspondingly high.

Hydrogen is of interest as well as in fundamental plasma research as for technological applications. H atoms often serve as primary reactive particles and are used for surface modification or thin film deposition, e. g., the deposition of nanocrystalline silicon layers and diamond [1]. In these processes atomic hydrogen plays a dominant role in surface crystallization processes or the formation of radicals from precursor molecules. Moreover, hydrogen isotopes deuterium and tritium will be used as fuel in future nuclear fusion reactors. Fusion experiments at JET in Culham showed the possibility of this technology, while these experiments will be extended in the coming ITER reactor in Cadarache.

One of the challenging technological problems in fusion technology is the design of a plasma heating source to sustain the fusion plasma. A neutral beam heater is one of the key concepts, which requires the formation of negative hydrogen ions (H^-, D^-) in large amounts, i. e., ion currents of 40 A [2]. The two main ways to achieve the corresponding high formation rates are gas phase formation via dissociative attachment (DA) of rovibrationally excited hydrogen molecules ($H_2 + e \rightarrow H + H^-$) and enhanced surface formation using caesium. The latter way is currently proposed for ITER [2], in spite of the problems caused by the usage of caesium.

Our research is focused on the role of excited hydrogen molecules and atoms in gas phase processes in an expanding thermal hydrogen plasma jet. With the PLEXIS setup hydrogen gas is ionized and dissociated under high gas flow rates in a cascaded arc and expands, after leaving the source, supersonically into a vacuum vessel. Thereby, a high flow of rovibrationally excited molecules is achieved.

An increase of the internal energy of H_2 leads to an increase of the H^- formation rate via DA by several orders of magnitude [3]. On the other hand, negative ions may contribute significantly to the formation of excited atoms via mutual ion-ion recombination. We detected rovibrationally excited H_2 and electronically excited H atoms by means of laser-induced fluorescence and optical emission spectroscopy, while negative ions are indirectly detected by photo-detachment. In this paper we want to reveal some of the relations between excited molecules and atoms in the plasma jet.

EXPERIMENTAL SETUP

The plasma source is a wall-stabilized cascaded arc, through which hydrogen gas flows under a pressure of about 9 kPa with flow rates in the order of 10^{21} s^{-1}. The arc channel is 4 cm long with a diameter of 4 mm (see Fig. 1). The plasma is partially ionized and dissociated by applying a direct current of 60 A with a power input of 9 kW between three cathode tips and the anode plate. The kinetic energy of species within the arc channel is in the order of 1 eV. Cascade and anode plates are made from copper and are water cooled.

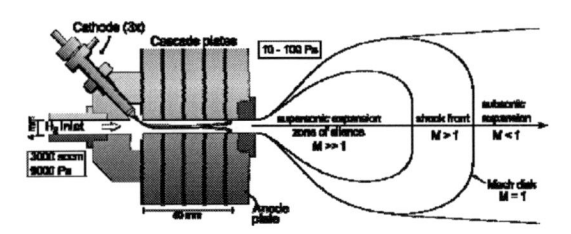

FIGURE 1. Scheme of the cascaded arc and the plasma expansion.

The plasma expands at the exit of the channel through a copper nozzle supersonically into a vacuum chamber (2 m long and 0.3 m in diameter), where pumps keep the back ground pressure between 5 and several 100 Pa. These conditions lead to a sonic exit velocity of about 5000 m/s. A shock front is formed some centimeters downstream of the nozzle, where the plasma expansion undergoes a transition from supersonic to subsonic flow.

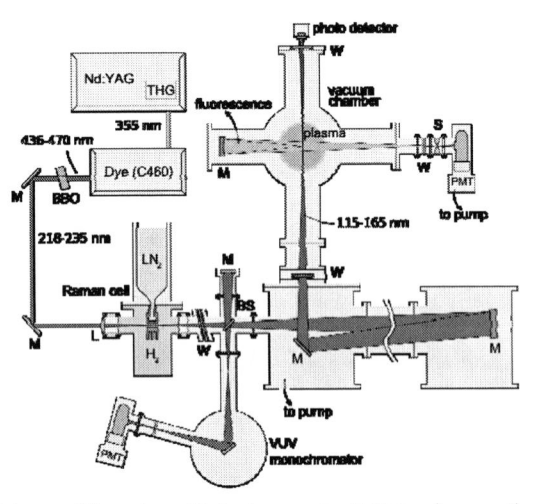

FIGURE 2. Scheme of the setup with the laser and anti-Stokes beam paths within the vuv spectrometer and the vacuum chamber (M - mirror, L - lens, BS - beam splitter, W - MgF$_2$ window, S - slit, PMT - photo multiplier tube).

Excited hydrogen molecules by means of laser induced fluorescence. Since transitions within the electronic ground state $X^1\Sigma_g^+$ of molecular hydrogen are dipole-forbidden, a direct detection of these states requires an excitation to the next higher electronic state ($X^1\Sigma_g^+ \rightarrow B^1\Sigma_u^+$), i. e., radiation in the vacuum-uv spectral range. We produce vuv radiation by stimulated anti-Stokes Raman scattering (SARS) [4,5]. A scheme of the beam path and the spectroscopic setup is shown in Fig. 2. Tunable laser radiation in a wavelength range from 436 nm to 470 nm is provided by a dye laser that is pumped by a frequency tripled Nd:YAG laser. The dye laser radiation is frequency doubled in a BBO crystal resulting in laser pulses of 5 ns duration in the spectral range from 218 nm to 235 nm. This beam is focused into a Raman cell, filled with 250 kPa of hydrogen gas, cooled by liquid nitrogen to enhance the SARS process. By the SARS process AS beams are subsequently shifted by the vibrational Raman shift of H$_2$ (4155.22 cm^{-1}) towards the vuv. All Raman orders together are focused into the center of the vacuum chamber. Emitted light by spontaneous emission is focused onto a photo multiplier tube. Spectra analysis is performed using spectroscopic data for the H$_2$ Lyman band provided by Abgrall et al. [6,7]. Depending on the fluorescence yield, the minimal detectable H$_2$ state density is about 10^{13} m^{-3}.

Densities of excited H atoms are measured by means of optical emission spectroscopy on the Balmer line series. The emitted light from the plasma jet is focused by a lens on an optical fiber of the optical emission spectrometer. To obtain absolute population densities a calibration of the optical system is carried out by comparison measurements using a tungsten ribbon lamp.

Negative hydrogen ions H⁻ are detected indirectly by photo-detachment [8]. The extra electron of the ion is detached by a laser photon (1064 nm), forming a ground state atom and a free electron (H⁻+ hv → H(1) + e). The response of the emitted light of the plasma jet to a short laser pulse (5-10 ns) is detected by optical emission spectroscopy on hydrogen Balmer lines.

EXCITED HYDROGEN MOLECULES

An example distribution of state densities of rovibrationally excited H_2 molecules measured in the supersonic part of the jet is shown in Fig. 3. The dashed lines represent a model for the $H_2(v,J)$ state distribution. The points represent densities obtained from the integrated area under pronounced peaks in the measured spectra. The density of H_2 with internal energies higher than 1 eV is in the order of a few percent of the total density. Therefore, the source is producing these high excited molecules with rates in the order of 10^{19} molecules × s⁻¹.

The reason of this high production is not revealed yet, since kinetic energies in the order of 1 eV within the arc channel are not high enough for a direct excitation via e-V or E-V processes [9]. Surface association of H atoms at the channel walls may play a dominant role, since such reactions lead directly to high excited H_2 [10,11].

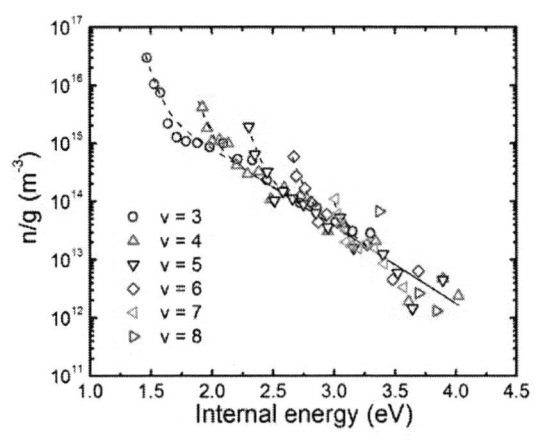

FIGURE 3. Weighted state densities of H_2 over the internal energy (8 mm from the exit of the plasma source, 100 Pa, 3000 sccm H_2/D_2, 9 kW).

The H_2 density distribution is a superposition of two different Boltzmann distributions for rovibrational states. Lower rotational states ($J < 7$) of each vibrational state follow a Boltzmann distribution described by a temperature of 700 K, i. e., close to the temperature of the background gas. However, higher rotational states follow a Boltzmann-like distribution with a much hotter temperature in the order of 3000 K. The same temperature can be used to describe the distribution of the vibrational states.

Therefore, we will call this temperature T_{vib}, although one should be aware of the fact that the term "temperature" may be inappropriate in case of non-Boltzmann distributions. Similar distributions were measured also by others and in other types of hydrogen discharges (see [9,12] and references therein). The exact reason for this shape of the distribution still remains unknown. However, relaxation by means of gas phase collision of H_2 molecules [9] and/or collisions with a cooled surface [5] may play an important role.

We also want to emphasize the high rovibrational excitation. The highest detected excited states are only a few 0.1 eV below the dissociation limit at 4.5 eV. Molecules in states with internal energies higher than the detected ones are most likely dissociated by gas phase collisions with molecules and atoms, since the necessary energy of a few 0.1 eV is always available at kinetic temperatures of some 1000 K.

The result in Fig. 3 was obtained in a 50:50 mixture of H_2 and D_2 gas. We have first indications that the distributions of HD and D_2 are very similar to H_2 [13], although their distributions follow slightly lower vibrational temperatures than the one of H_2.

The contribution of each measured rovibrational H_2 state density to the formation of H^- due to dissociative attachment by collisions with electrons is shown in Fig. 4. The necessary state resolved DA rate coefficients are calculated using published cross section data by Horáček et al. [3] and assuming a Maxwellian EEDF. Plotted are production terms of H^- for three different electron temperatures, which are typical for our plasma expansion. Multiplied with the electron density the plotted values would result in the production rate of H^- resolved for each state (v,J) of H_2.

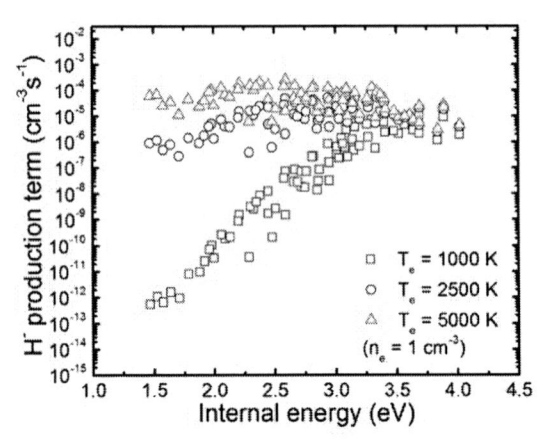

FIGURE 4. Formation of H^- ions by DA in dependence of the internal energy of H_2 for three different electron temperatures (the H_2 state densities are the same as in Fig. 3).

Although the state density decreases by several orders of magnitude with increasing internal energy of H_2 (see Fig. 3), the formation of negative ions by higher excited molecules is larger in total compared to lower excited molecules. The reason for this are the DA rate coefficients, which increase much stronger with the internal energy compared to the decrease of densities.

EXCITED HYDROGEN ATOMS

The pressures used in this paper are low enough that the transport is not longer limited by collisions with neutrals. Therefore, ions and electrons are confined in an applied magnetic field. A photo of the magnetized hydrogen jet is shown in Fig 5.

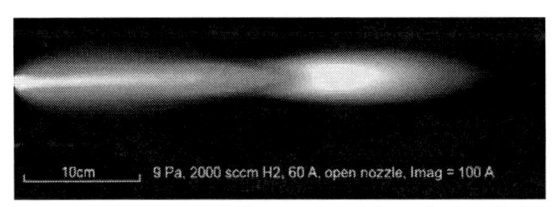

FIGURE 5. The magnetized hydrogen jet. The nozzle exit is on the left (just outside of the picture). The strength of the magnetic field on the expansion axis is 0.04 T

To visualize the plasma jet more clearly, the experimental conditions in Fig. 5 were chosen differently from the conditions of the experiments discussed below. However, the difference in conditions merely influences the exact position of the red-to-blue transition region (flow, magnetic field strength and background pressure) and the absolute light intensity emitted by the plasma jet.

The emission of the magnetized hydrogen plasma jet is dominated by red H_α light in the beginning of the jet. Since electron energies in the jet are too low (1 eV and less) to excite atomic hydrogen to the state $p = 3$, a possible formation route is via mutual recombination of atomic ions: $H^+ + H^- \rightarrow H + H(p = 2, 3)$. After 20 to 30 cm a transition occurs from a red to a blue emission [14]. This is due to a population inversion between higher excited atomic states of $p = 4, 5$ and 6 in respect to $p = 3$ (Fig. 6). The reaction proposed to be responsible for this overpopulation is the formation of highly excited hydrogen atoms via mutual ion-ion recombination: $H_2^+ + H^- \rightarrow H_2 + H(p \geq 2)$ [15]. However, this reaction is still under some debate, e. g., it can also produce electronically excited H_2 molecules instead of excited hydrogen atoms [16,17].

Figure 6 shows the densities per statistical weight of the atomic hydrogen levels $p = 3$ to 6 as a function of the axial position z, i. e., the distance from the exit of the source. The first 22 centimeters of the jet is dominated by H_α light. After that, H_β to H_δ light becomes dominant and the correspondent weighted densities n_p/g_p become higher than the one for $p = 3$. Note that also the emission of H_α increases by some amount in the blue region again. There is also a small region between the red and blue where the densities of all Balmer lines are low. This dark space at $z = 23$ cm is also visible in Fig. 5.

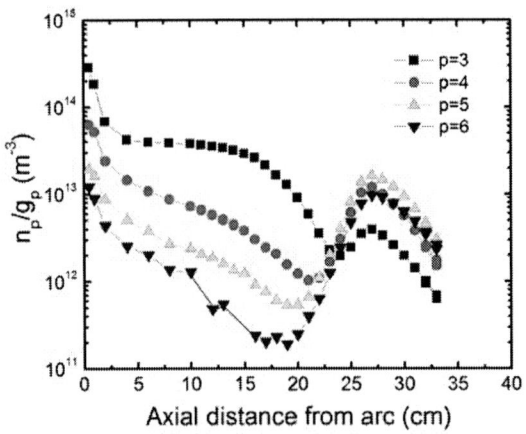

FIGURE 6. Development of absolute densities per statistical weight of the excited atomic states $p = 3$ to 6. The experimental conditions are as follows: Φ_{H2}=3000 sccm, p=10 Pa, B=0.012 T, P_{arc}=6.8 kW.

Densities of excited states $p = 3$ to 12 of atomic hydrogen are plotted in Fig. 7 as function of the difference between the ionization energy at 13.6 eV and the energy to the states $p = 3$ to 12. For distances up to 23 cm no population inversion is observed. For larger distances, states $p = 4$ to 6 are overpopulated in respect to $p = 3$. Visible is also the drastic change in the atomic state distribution around 23 cm. A reason for this change could be the decay of the electron temperature [18] in the expanding jet. A lower electron temperature increases both the population rates of high excited H atoms due to three-body recombination ($H^+ + 2e \rightarrow H + e$) and the formation of H^- due to DA and, thus, the mutual ion-ion recombination leading to excited H atoms.

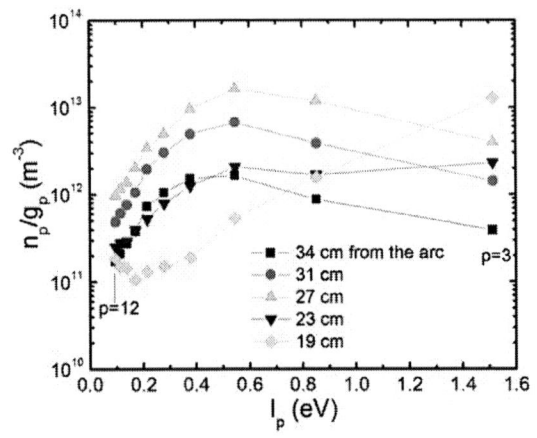

FIGURE 7. The atomic hydrogen population density per statistical weight. The conditions are as follows: Φ_{H2}=3000 sccm, p=10 Pa, B=0.012 T, P_{arc}=6.8 kW.

NEGATIVE HYDROGEN IONS

Photo-detachment is used to determine the role of H⁻ in the population of excited states of atomic hydrogen. We here focus on the effect of the p=3 state. Laser radiation with a wavelength of 1064 nm, a pulse duration of 5 ns and a beam diameter of 10 mm is directed perpendicular to the direction of the jet. The corresponding photon energy is sufficient to remove the extra electron of the H⁻ ion that is attached to the atom with a binding energy of 0.75 eV [8]. If we assume that the population into the excited H-atom states is partly due to mutual ion-ion recombination, the destruction of H⁻ ions, through the laser-induced photo-detachment process, will disturb the plasma chemistry and thus the emission originating from the excited states. The change in emission of the Balmer-α line is recorded at the center of the crossing of the plasma jet and the laser beam.

The emission of the Balmer-α line is constant during steady state operation. When laser radiation is directed into the plasma jet, the emission intensity of the Balmer-α line temporally decreases. This decrease is up to 50 % of the original intensity, meaning that the H-atom density in state $p = 3$ decreases by 50 % due to a laser-induced disturbance of a mechanism populating p = 3. After a few hundred nanoseconds of the laser pulse the emission reaches its steady-state level. To exclude photo-ionization as a possible mechanism depopulating p=3, the depopulation of atomic levels was also tested on excited Ar in a pure argon plasma with a $4p$ line at 696.54 nm. However, no depopulation of this Ar level was observed. This indicates that photo-ionization can be neglected as a disturbing mechanism. The results tentatively show that the temporal change in the emission on the Balmer-α wavelength after irradiating the plasma jet with 1064-nm photons, could be used to detect negative hydrogen ions. Experiments involving the recording of other Balmer transitions are performed to test these results. A collisional-radiative model is constructed to underpin the proposed mutual-recombination mechanism involving H⁻ ions and leading to the population of excited H-atoms.

CONCLUSIONS

The cascaded arc is an effective source of high flows of rovibrationally excited hydrogen molecules. These excited molecules are the dominant precursors for gas phase formation of H⁻ via dissociative attachment. During the expansion of the plasma jet, kinetic temperatures of molecules, atoms and electrons decrease, leading to a further increase of DA rates. From the temporal behavior of the Balmer-α emission after irradiating the plasma jet with 1064-nm photons, it is inferred that negative ions can play an important role in the formation of excited hydrogen atoms in the electronic state $p = 3$ via mutual ion-ion recombination.

ACKNOWLEDGMENTS

This work is part of the research program of the Dutch Foundation for Fundamental Research on Matter FOM. The work is also supported by the Euratom Foundation.

The authors appreciate the skillful technical assistance of M. J. F. van de Sande, J. J. A. Zeebregts and H. M. M. de Jong. The authors thank W. M. Soliman and B. Delplanque for their assistance in the measurements.

REFERENCES

[1] M. N. van den Donker, B. Rech, F. Finger, W. M. M. Kessels, and M. C. M. van de Sanden, *Appl. Phys. Lett.* **87**, 263503 (2005)
[2] U. Fantz, P. Franzen, W. Kraus, M. Berger, S. Christ-Koch, M. Fröschle, R. Gutser, B. Heinemann, C. Martens, P. McNeely, R. Riedl, E. Speth, and D. Wünderlich, Plasma Phys. Control. Fusion 49, B563 (2007)
[3] J. Horáček, M. Čízek, K. Houfek, and P. Kolorenč, *Phys. Rev. A* **70**, 052712 (2004)
[4] T. Mosbach, *Plasma Sources Sci. Technol.* **4**, 610 (2005)
[5] O. Gabriel, D. C. Schram and R. Engeln, *Phys. Rev. E* **78**, 016407 (2008)
[6] H. Abgrall, E. Roueff, F. Launay, J. Roncin, and J. Subtil, *Astron. Astrophys., Suppl. Ser.* **101**, 273 (1993)
[7] H. Abgrall, E. Roueff, X. Liu, D. E. Shemansky, and G. James, *J. Phys. B: At. Mol. Opt. Phys.* **32**, 3813 (1999)
[8] M. Bacal, G. Hamilton, A. Bruneteau, H. Doucet and J. Taillet, *Rev. Sci. Instr.* **50**, 719 (1979)
[9] P. Vankan, D. C. Schram and R. Engeln, *J. Chem. Phys.* **121**, 9876 (2004)
[10] C. T. Rettner and D. J. Auerbach, *J. Chem. Phys.* **104**, 2732 (1996)
[11] R. F. G. Meulenbroeks, D. C. Schram, M. C. M. van de Sanden, and J. A. M. van der Mullen, *Phys. Rev. Lett.* **76**, 1840 (1996)
[12] P. Vankan, D. C. Schram and R. Engeln, *Chem. Phys. Lett.* **400**, 196 (2004)
[13] O. Gabriel, J. J. A. van den Dungen, W. M. Soliman, D. C. Schram, and R. Engeln, *Chem. Phys. Lett.* **451**, 204 (2008)
[14] J. J. A. van den Dungen, O. Gabriel, H. S. M. M. Elhamali, D. C. Schram, and R. Engeln, *IEEE T. Plasma Sci.* **36**, 1028 (2008)
[15] M. J. J. Eerden, M. C. M. van de Sanden, D. K. Otorbaev, and D. C. Schram, *Phys. Rev. A* **51**, 3362 (1995)
[16] R. K. Janev, D. Reiter, and U. Samm, Collision Processes in Low-Temperature Hydrogen Plasmas, *Forschungszentrum Jülich, Report No. JUEL-4105* (2003), p. 92
[17] C. L. Liu, J. G. Wang and R. K. Janev, *J. Phys. B: At. Mol. Opt. Phys.* **39** 1223 (2006)
[18] Z. Qing, D. K. Otorbaev, G. J. H. Brussaard, M. C. M. van de Sanden, and D. C. Schram, *J. Appl. Phys.* **80**, 1312 (1996)

The Electronegative Plasma Pre-Sheath in Magnetic Field and Extraction of Negative Ions

B. M. Annaratone[a], J.E.Allen[b]

*[a] Laboratoire de Physique des Interactions Ioniques et Moléculaires,
CNRS/Université de Provence, case 321, 13397, Marseille, France. .
[b] University College, Oxford OX1 4BH, UK*

Abstract. We derive the modified Bohm criterion for the typical conditions of negative ion extraction from producing cells.

Keywords: RF-ICP negative ion sources, extraction region, negative ions.
PACS: 52.20.j, - 52.27.Cm

INTRODUCTION

Understanding the physics of negative deuterium ions sources is crucial for the optimisation of neutral beam, NB, injectors for fusion reactors. This research and development is particularly challenging because of the demanding constraints. These involve uniformity, tight mechanical specifications on the grid, stringent ion optics, low gas pressure (to avoid collisions), high negative ion density and a high ratio of negative ions to electrons extracted. In prototype sources, to avoid the spread of plasma outside the chamber, a magnetic field is applied parallel to the orifices. Accordingly the study of the electronegative plasma sheath in magnetic field is fundamental to control the extraction of negative ions from the producing sources; notwithstanding very little theoretical work can be found in literature. This is so because of the mixing of two already difficult topics, dealt from different communities: the electronegative sheath, domain of the low temperature plasma community and the magnetic sheath, typically fusion research.

The development of highly efficient NB systems still depends essentially on the quality and intensity of the ion source in a determinant way. At present the experimental approach, with several noticeable contributions, among them [1-5] precedes the theoretical analysis that, when developed, will remove empiricisms and lead to optimisation.

It is well known that the Bohm criterion is modified by the presence of negative ions [6, 7]. For a certain range of the negative ion density and of the temperature of the negative species the modified Bohm criterion breaks down in multiple solutions. Physically a discontinuity in the electrostatic potential may arise because the colder, usually heavier, negative species is confined in the plasma whilst the electrons extend further in the sheath. When ionisation happens secondary electropositive plasmas may form down in the already established plasma sheath, see for example the experimental proof in Oxygen plasma [8]. The effect is also called stratification of the discharge; it will give discontinuity in the negative ion current behaviour as function of the extraction grid potential.

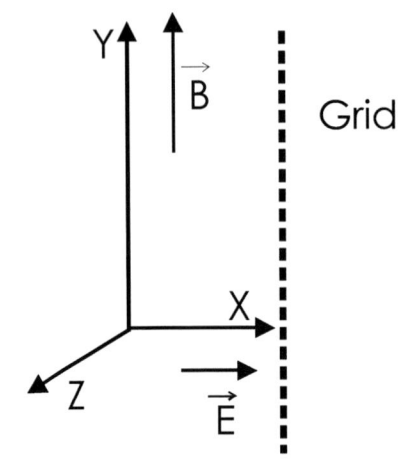

From another point of view the magnetic plasma pre-sheath has been dealt extensively but, intriguingly, with the magnetic force acting on the ions and not on the electrons [9, 10]! Here we mean the net force per unit volume; individual electrons do experience a magnetic force. In the regime $r_{ce} \sim \lambda_{De} \ll r_{ci}$, with r_c the gyration radius, positive and negative ions traverse the sheath while the electrons are confined. This regime is relevant also in a different context, i.e. Hall thrusters; see for example ref. [11].

FIGURE 1. Geometry of the system

One reference dealing with the electronegative plasma sheath in magnetic field is [12]. Also in this reference the electrons are not influenced by the magnetic field, while the ions are.

In this contribution we demonstrate the validity of the Bohm criterion, pre-sheath calculation, in moderately magnetised electronegative plasma and the parameters for stratification for the typical sources. The above criterion is modified to take in account different ion species and their temperatures. The ratio of electron to the negative ions temperature should in general be kept below ~10; this may be achieved letting the discharge diffuse through grids as in the tandem source. Actually the control of the electron temperature favours the electron attachment so increasing the negative ion density.

ANALYSIS

We consider the one dimensional case of electronegative plasma with a uniform magnetic field parallel to the surface (grid from which the negative ions are extracted). The role of the grid is to confine the electrons, while letting as many negative ions as possible to exit to form the beam. The geometry is chosen as in ref. [10] keeping zero angle, see fig. 1. We assume a magnetic field of intermediate strength so that the electrons are subject to the Lorentz force while the magnetic deflection of the ions (positive and negative) can be neglected. The electrons drifting to the wall experience a force in the z direction, the ExB direction, and it is the flow of electrons in this direction that produces a back force in the x direction.

Whether or not electrons can flow in the direction z depends on the problem being considered. Much of the published literature does not consider a real situation. In our case there are two limiting configurations, either the electron current can flow, limited by collisions, or it may build up an opposing electric field as in the Hall effect.

THE ELECTRON BEHAVIOUR

For the electrons in the pre-sheath plasma we consider the momentum equations in the x and z directions:

$$mn_e \frac{dv_{ex}}{dx} + kT_e \frac{dn_e}{dx} - en_e \frac{d\phi}{dx} - en_e v_{ez} B = 0 \qquad (1)$$

$$mn_e v_{ex} \frac{dv_{ez}}{dx} + en_e v_{ex} B + n_e m v_{ez}(Z + \nu_e) = 0 \qquad (2)$$

where m, v and ν are respectively mass, velocity and collision frequency for the electrons, Z is the ionisation frequency and ϕ is the electrical potential. The loss of momentum due to ionisation and friction can be neglected in the x direction (but not in the z direction). Also the inertia is negligible unless for very fast transients.
When ionization and collisions are important, $x >> v_x/(Z+\nu)$ the balance of forces is local and the driving force in the z direction, ev_xB is balanced by the friction. In this case

$$n_e = n_0 \exp \frac{1}{kT_e}\left(eV - \int_0^x \frac{m\omega_{ce}^2 v_{ex}}{\nu_{e-total}} dx \right) \qquad (3)$$

The values of the other terms that contribute to the electron density are given in appendix I. We note here that although the magnetic field has a very important contribution in the electron density it it will not affect its gradient at the plasma boundary, where dx/dV tends to zero, as demonstrated by Allen in ref. [13].

THE BOHM CRITERION FOR ION BEAM SOURCES

The equations defining the behaviour of ions in the pre-sheath region of electronegative plasmas in presence of a moderate magnetic field follow.
Continuity equation for positive ions:

$$\frac{d}{dx}(n_i v_x) = Zn_e \qquad (4)$$

Momentum equation for cold ions with negligible deflection due to the magnetic field, in the x direction:

$$Mn_i v_{ix} \frac{dv_{ix}}{dx} + en_i \frac{d\phi}{dx} + MZn_e v_{ix} + n_i F_{ix} = 0 \qquad (5)$$

where F_{ix} is the frictional force on the ion due to collisions.

Momentum equation for negative ions with negligible deflection due to the magnetic field, in the x direction:

$$-en_-\frac{d\phi}{dx} + kT_-\frac{dn_-}{dx} = 0 \qquad (6)$$

The momentum equation for electrons in the x direction is eq. 1.
The (quasi)neutrality eq. is:

$$n_i = n_e + n_- \qquad (7)$$

and definitions, as in ref. [6]:

$$\alpha = \frac{n_-}{n_e} \qquad \gamma = \frac{T_e}{T_-} \qquad (8)$$

Here alpha depends on position but gamma is assumed to be constant
Adding eq.s 1, 5 and 6 in plasma, using eq. 7 and substituting dv_{ix}/dx derived from eq. 4:

$$\frac{dn_i}{dx} = -\frac{n_i}{\left(v_{ix}^2 - \dfrac{1 + \dfrac{1}{\gamma}\dfrac{dn_-}{dn_e}}{1 + \dfrac{dn_-}{dn_e}}\dfrac{kT_e}{M}\right)}\left[\frac{2Zv_{ix}}{(1+\alpha)} - \frac{\omega_{ci}v_{ez}}{(1+\alpha)} + \frac{F_{ix}}{M}\right] \qquad (9)$$

$$\frac{d\phi}{dx} = -\frac{\dfrac{1 + \dfrac{1}{\gamma}\dfrac{dn_-}{dn_e}}{1 + \dfrac{dn_-}{dn_e}}\dfrac{kT_e}{e}}{\left(\dfrac{1 + \dfrac{1}{\gamma}\dfrac{dn_-}{dn_e}}{1 + \dfrac{dn_-}{dn_e}}\dfrac{kT_e}{M} - v_{ix}^2\right)}\left[\frac{2Zv_{ix}}{(1+\alpha)} - \frac{\omega_{ci}v_{ez}}{(1+\alpha)} + \frac{F_{ix}}{M}\right] - \frac{en_iv_{ez}B}{1+\alpha} \qquad (10)$$

These equations are true at the sheath edge as well as in all the pre-sheath and may be used for future plasma calculation. Dividing eq.9 by eq.10 we find that, as the denominator of the first term tends to zero, the following relation is found to hold at the plasma boundary:

$$\frac{dn_i}{d\phi} = n_i \frac{e}{kT_e} \frac{1 + \frac{dn_-}{dn_e}}{1 + \frac{1}{\gamma}\frac{dn_-}{dn_e}} \quad (11)$$

In eq. 9 and 10 there is a singularity of the gradient of density and of the electric field when the ion velocity reaches a modified ion acoustic velocity. We have calculated the energy which the positive ions have when they arrive at the plasma boundary. This may be considered another formulation of the Bohm criterion. At the plasma boundary the electric field "takes over" (in the limit dV/dx tending to infinity the magnetic forces are ignorable), with the result that dn_-/dn_e becomes equal to $\alpha\gamma$, see eq.s 3 and 6. The expression for v_i^2 derived from the divergences of eq.s 10 and 11 then becomes equal to the result obtained by Braithwaite and Allen, ref. [6], see the unnumbered equation just before eqn. (10) in that paper. The normalized ion energy at the plasma edge, $\eta_s = eV/kT_e$ is then:

$$\eta_s = \frac{1}{2}\frac{(1+\alpha_s)}{(1+\gamma\alpha_s)} \quad (12)$$

The magnetic field does not modify the velocity of the positive ions at the Bohm point even if the electrostatic potential at this place has to be calculated taking in account magnetic field and collisions. The relative density of electrons and negative ions at the boundary is much lower than in the un-magnetised case, see eq. 3. Note that the criterion deals with the local values of α and γ; to work out the value for these quantities in the far away plasma we need to solve the full set of equations.

The negative ions in the extraction region have a spread in energy due to their production mechanism and to the potential difference w.r.t. the place where they have formed. Here we consider the two negative ions species [8], the volume produced ions (with density α_v and temperature, normalised to Te, γ_v) and surface produced ions, accelerated by the sheath voltage, which in plasma are normally fast (represented by α_f and γ_f). In this condition the modified Bohm criterion, eq. 12, can be re-written [8] as:

$$\eta_s = \frac{1}{2}\frac{(1+\alpha^v_s + \alpha^f_s)}{(1+\gamma^v\alpha^v_s + \gamma^f\alpha^f_s)} \quad (13)$$

DISCUSSION

Following [2] the volume ions have energy 0.1-0.2eV while the surface produced negative ions have in plasma the energy of the wall where they are produced. Plasma potential in typical sources is of the order of several volts, see, for example [3]. The electron temperature varies typically between 1 and 10 eV. With these data, to give an example, for $\alpha_v = 2$, $\gamma_v = 50$ and $\alpha_f = 0.12$, $\gamma_f = 1$, in plasma and considering a reduction in the electron density due to the magnetic field of ~10^{-3}, see appendix I, multiple solutions are unlikely to appear. Vice versa at the Bohm point the quasineutrality is given mainly by positive and negative ions, the transition between a three specie plasma and ion-ion plasma being continuous. This effect should be considered in the extraction of negative ions by biasing the grid together with the back influence of the grid bias on the plasma potential and the dependence of the anode-like potential profile in front of the positive, w.r.t. plasma, surfaces.

CONCLUSION

The Bohm criterion is not modified by the superposition of a moderate magnetic field to an electronegative discharge. As the magnetic field introduces a sharp reduction in the electron density in certain cases the Bohm point may be at the edge of negative – positive ion plasma, the electrons density being almost negligible. This later statement depends on our approximation of uniform magnetic field. Experimentally extracted negative currents have been reported to be in ratio 1 (same current of negative ions and electrons); clearly n_e is much lower than n_- because the ratio of the thermal velocities of electrons over hydrogen ions is 30. The extraction of negative ions is strongly dependent on all the plasma parameter and not related only to the extraction region. The plasma extension plays a role as seen experimentally in [1].

This contribution is only a step in the comprehension of the physical factors governing the extraction of negative ions from sources, to be considered together with the chemistry, see for example [14, 15]. Further work will include a space dependent magnetic field, the positive sheath and the plasma behaviour in the vicinity of the orifices.

APPENDIX I

In order to find the correct law for the density of the electrons we compare the terms in the equation for the electron density, eq. 3. We assume B=100G so that $\omega_{ce}=1.7\ 10^9 \text{s}^{-1}$. $v_{x\ \text{drift electrons}}$ must be comparable with the $v_{x\ \text{drift ions}}$ if the fluxes are the same on the wall. $v_{x\ \text{drift electrons}}$ is small, as the Bohm speed ~10^3ms^{-1} (vice versa $v_{y\ \text{drift}}$ may be much higher, say 10^6 ms^{-1}, comparable to the thermal speed). For the dimension of the discharge we assume half of the chamber, x=0.2m. For the total cross section (including ionization) we assume 10^{-19}m^2. Hence the second term in the exponential, in the approximation v_x constant, is ~10^{-10}J.

The electrical potential V must be between 1 and 10V so the electrostatic energy is ~10^{-19}-10^{-18}J. Despite the small relative value we can not ignore the electrostatic potential because it is the only force driving positive ions out of the plasma.

REFERENCES

1. J. R. Hiskes and A. M. Karo, J. Appl. Phys. 56, 1927 (1984)
2. R. Leroy, M. Bacal, P. Berlemont, C. Courteille and R. A. Stern, Rev. Sci. Instr. **63**, 2686 (1992).
3. B. Crowley, D. Homfray, S. J. Cox, D. Boilson, H. P. L. de Esch and R. S. Hemsworth, Nucl. Fusion 46, S307 (2006)

4. O. Fukumasa and S. Mori, Nucl. Fusion 46, S287 (2006)
5. S. Miyamoto, F. Kanayama, T. Minami, T. Mune and H. Hiriike, Nucl. Fusion 46, S313 (2006)
6. N.St.J. Braithwaite and J.E. Allen , J. Phys.D: Appl.Phys., **21**, 1733 (1988).
7. J. E. Allen, Plasma Sources Sci. Technol, **13**, 48 (2004).
8. B. M. Annaratone, T. Antonova, H. M. Thomas and G. E. Morfill, Phys. Rev. L. **93**, 185001 (2004).
9. H. Schmitz, K.U. Riemann and Th. Daube, Phys. Plasmas, **3**, 2486 (1996).
10. T. M. G. Zimmermann, M. Coppins and J. E. Allen, Physics of plasmas **15**, 072301 (2008).
11. E. Ahedo and D. Escobar, Physics of plasmas **15**, 033504 (2008).
12 X. Zou, J. Y. Liu, Z. X. Wang, Y. Gong, Y. Liu, X.G. Wang, Chin. Phys. Lett. 21, 1572, (2004).
13. J. E. Allen, Contribution to plasma physics 48, 400 (2008).
14 M. Bacal, Nucl. Fusion 46, S250 (2006).
15. M. Capitelli, M. Cacciatore, R. Celiberto, O. de Pascale, P. Diomede, F. Esposito, A. Giquel, C. Gorse, K. Hassouni, A. Laricchiuta, S. Longo, D. Pagano and M. Rutigliano, Nucl. Fusion 46, S260 (2006)

Modeling of the Plasma Electrode Bias in the Negative Ion Sources with 1D PIC Method

D. Matsushita[a], S. Kuppel[a], A. Hatayama[a], A. Fukano[b] and M. Bacal[c]

[a] *Graduate School of Science and Technology, Keio University, 3-14-1 Hiyoshi, Kouhoku-ku, Yokohama 223-8522, Japan*
[b] *Mechanical Systems Engineering Cource, Monozukuri Department, Tokyo Metropolitan College of Industrial Technology, Higashioi, Shinagawa, Tokyo 140-0011, Japan*
[c] *Laboratoire de Physique et Technologie des Plasmas, UMR 7648 du CNRS, Ecole Polytechnique, 91128 Palaiseau, France*

Abstract. The effect of the plasma electrode bias voltage in the negative ion sources is modeled and investigated with one-dimensional plasma simulation. A particle-in-cell (PIC) method is applied to simulate the motion of charged particles in their self-consistent electric field. In the simulation, the electron current density is fixed to produce the bias voltage. The tendency of current-voltage characteristics obtained in the simulation show agreement with the one obtained from a simple probe theory. In addition, the H⁻ ion density peak appears at the bias voltage close to the plasma potential as observed in the experiment. The physical mechanism of this peak H⁻ ion density is discussed.

Keywords: Negative ion source, Particle-in-cell modeling, PE bias
PACS: 29.25.Ni, 52.65. Rr,

INTRODUCTION

Neutral beam injection based on a negative ion beam is very promising for heating and current drive in future fusion reactors. To generate intense beams of H⁻ ions and optimize the negative ion sources, analysis of plasma characteristics in the extraction region is indispensable.

In the experiment, a positive DC bias voltage with respect to the ion source chamber is applied to the plasma electrode (PE) for the optimization of negative ion extraction. This bias voltage is supposed to modify the plasma potential profile near the PE and consequently improve the negative ion extraction efficiency. Svarnas *et al.* have indicated that the H⁻ ion density in the vicinity of the PE has a peak value when the applied bias voltage is close to the plasma potential [1]. The following mechanism has been proposed to explain this peak H⁻ ion density in front of the PE. When the PE bias approaches the plasma potential, volume-produced H⁻ ion is attracted by the PE from the bulk plasma region. This results in the enhancement of H⁻ ion density. By contrast, further increased bias accelerates the H⁻ ion toward the PE and reduces the H⁻ ion residence time. As a result, the optimum H⁻ ion density appears at bias voltage close to the plasma potential.

The purpose of this study is to prove this hypothesis. For this purpose, it is essential to simulate the effect of the PE bias and analyze its effect upon plasma profiles near the PE. In this study, we focus on two main topics: (i) One-dimensional particle-in-cell (1D PIC) modeling of the PE bias and (ii) its effect on plasma characteristics in the vicinity of the PE. The modeling of the PE bias is validated by comparing with a simple probe theory [2]. In addition, the physical mechanism of peak negative ion density at certain bias voltage observed in the experiment is discussed.

THEORY AND SIMULATION MODEL

Particle-in-cell Modeling

The motion of charged particles (H$^-$ ions, H$^+$ ions, and electrons) in their self-consistent electric field is solved by 1D3V (one-dimensional in real space and three-dimensional in the velocity space) PIC method [3]. The trajectories of the particles are calculated by numerically solving the equation of motion,

$$m_s \frac{d\mathbf{v}}{dt} = q_s \mathbf{E}, \qquad (1)$$

where m_s, $\mathbf{v}$, q_s and $\mathbf{E}$ are the charged particle mass, velocity, charge and the electric field for particles of species s, respectively. Note that the magnetic field is not considered in this study for simplicity to focus on the effect of the PE bias. To obtain the charge density on a grid point, the particle charge is allocated to each grid point with linear interpolation. The particle densities (n_s) also are calculated from the charge densities on the grid points. Poisson's equation for the spatial electric potential ϕ,

$$\nabla^2 \phi = -\frac{q(n_{H^+} - n_{H^-} - n_e)}{\varepsilon_0}, \qquad (2)$$

is solved by the finite difference method, where ε_0 is the permittivity in vacuum.

In this study, the calculation time step dt is set to be 0.25 ω_p^{-1}, inverse of plasma frequency, to avoid numerical instability. The computational particles are weighted to be the $10^8 \sim 10^9$ number of particles (super particle). This super particle weight is determined by the trade-off between the calculation cost and statistical accuracy.

Model Geometry and Main Assumption

The model geometry used in the simulation is shown in Fig. 1. To analyze the plasma near the PE, x axis is set perpendicular to the PE toward inside of the source and the original point is at the PE, as shown in Fig. 1. The space grid dx is taken as 0.25 λ_D, Debye length. In this study, λ_D is calculated as ~ 1.05×10^{-4} m. The system length L is set to be 800 dx, i.e. 200 λ_D (~ 2.1×10^{-2} m). This L is sufficient to analyze the effect of the bias because the region influenced by the bias is supposed to be several tens of λ_D. The source region in which H^+ ions and electrons are produced is modeled in the region where $x \geq 0.8\ L$.

The initial plasma parameters (density, temperature) are given based on the experimental data in Camembert III and shown in table 1. Table 1 also shows the neutral parameters used in reaction processes. The density of H_2-molecule n_{H2} is calculated as $n_{H2} = P/kT_{H2}$, where k is Boltzmann constant. The density of H-atom n_H is calculated by using dissociation fraction (H/H$_2$). Reactions of H$^-$ ion production or destruction are taken into account in the simulation, as shown in table 2. The vibrational population of $H_2(v)$ is assumed by the use of ratio presented in the analytical work [4]. These background neutral parameters are fixed throughout this study.

TABLE 1. Parameters used in the simulation.

Electron temperature T_e	1.0 eV	Gas pressure P	0.3 Pa
H$^+$ ion temperature T_{H+}	0.5 eV	H$_2$ temperature T_{H2}	0.1 eV
Initial electron density n_e	5 x 10^{15} m^{-3}	H temperature T_H	0.2 eV
Initial H$^+$ ion density n_{H+}	5 x 10^{15} m^{-3}	Dissociation fraction (H/H$_2$)	0.1

TABLE 2. Reactions taken into account in the simulation

	Reactions	Reference
Dissociative attachment (DA)	$H_2(v) + e \rightarrow H^- + H$	[5]
Electron detachment (ED)	$H^- + e \rightarrow e + H(1s) + e$	[6]
Associative detachment (AD)	$H^- + H \rightarrow H + H + e$	[6]
	$H_2 + e$	
Mutual Neutralization (MN)	$H^- + H^+ \rightarrow H^*(n=2) + H(1s)$	[6]
	$H^*(n=3) + H(1s)$	

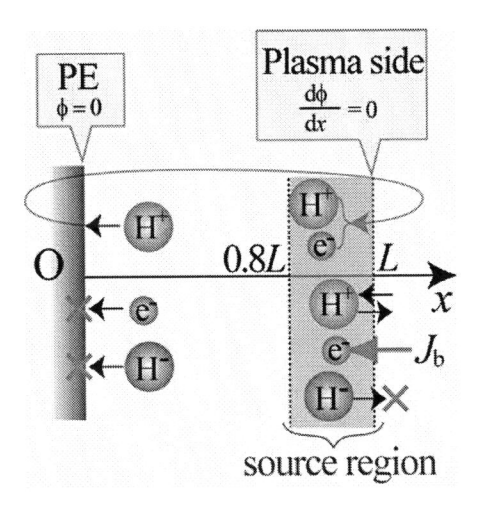

FIGURE 1. Model geometry used in the simulation

In Fig. 1, we call the left side boundary as the PE boundary and the right side boundary as the plasma boundary. At the PE boundary, the potential ϕ is fixed to be zero (Dirichlet condition), while the electric field $d\phi/dx$ is fixed to be zero at the plasma boundary (Neumann condition). Electrons and H$^+$ ions exiting through the plasma boundary are re-injected immediately with the velocity picked up from half-Maxwellian distribution with their temperature. This kind of "half open boundary" is often applied to model the limited region of plasma in contact with metallic wall and to save the calculation cost. In this boundary model, however, the plasma current is implicitly imposed to be zero at the plasma boundary. As a result, the potential difference between the PE boundary and plasma boundary is unchanged irrespective of the voltage applied to the PE. To simulate the bias voltage with using this half open boundary, the plasma current in the system should be specified. In this simulation, electron current density is fixed at the plasma boundary at J_b to produce bias voltage.

Collisions are treated with Monte Carlo method. Null collision method [7] is applied to treat electron collisions (ED, DA). Collisions between heavy particles (AD, MN) are treated with the maximum collision number method [8]. The energy loss in inelastic collisions is taken into account. To calculate the velocity after the collision, isotropic scattering is assumed.

The H$^+$ ions and electrons are initially loaded randomly in the system and their velocities are sampled from Maxwellian distribution with their temperatures. The H$^+$ ions that strike the PE are assumed to be recycled as neutral and then reionized in the source region. A pair of H$^+$ ion and electron is reloaded. The H$^+$ ion and electron velocities are chosen from Maxwellian, respectively, in the same manner as in the initial particle loading above. The electrons which reach the PE are simply removed from the calculation. The H$^-$ ions are produced after DA reaction (volume production). Surface-produced H$^-$ ion is not considered in this study. The H$^-$ ions which pass through both the boundaries are removed from the calculation. The calculation is repeated until the system reaches the steady state.

RESULTS AND DISCUSSION

The results are divided into two cases. First, the case without collision, i.e. only H$^+$ ion and electron case is presented. The results are compared with a simple probe theory to validate the modeling of the bias current. Second, the effect of the bias voltage is examined in the case with collision (production or destruction of H$^-$ ion is included). The bias voltage is applied in the same way in both the cases.

Validity of Modeling

Figure 2 shows the potential profiles with several bias current densities (J_b). The potential configuration is controlled by the fixed J_b. In addition, as J_b increases, the plasma potential (potential at the plasma boundary: V_p) becomes lower with respect to the PE. In floating case, i.e. $J_b = 0$ Am^{-2}, V_p is ~ 3.5 V and we define this value as the floating plasma potential V_{pf}. Figure 3 shows the current-voltage (I-V) characteristics; current corresponds to the total current density to the PE and voltage corresponds to V_b. Here, V_b is defined as

$$V_b = V_{pf} - V_p(J_b). \tag{3}$$

This definition is based on the assumption that the plasma potential is constant even if bias voltage changes.

The tendency of the characteristics is similar to the one obtained from the experiments and the theory. The total current density to the PE slightly increases when V_b is lower than the floating plasma potential V_{pf}, and abruptly increases toward electron saturation current density when the V_b reaches the V_{pf}. After the V_b exceeds the V_{pf}, the total current density to the PE is almost saturated. The electron saturation current density I_{esat} from the probe theory is calculated by using the parameters in this study as follows,

$$I_{esat} = \frac{1}{4} e n_e \langle v_e \rangle \sim 100 \text{ Am}^{-2}, \tag{4}$$

where e and $\langle v_e \rangle$ represent elementary charge and average velocity of electron, respectively. This value agrees well with the saturation current density observed in Fig. 2 (100 ~ 110 Am^{-2}).

Based on the study discussed above, the effect of the bias voltage is examined as described in the following chapter. In addition, numerical scheme developed here is also applied to the 2D problems [9].

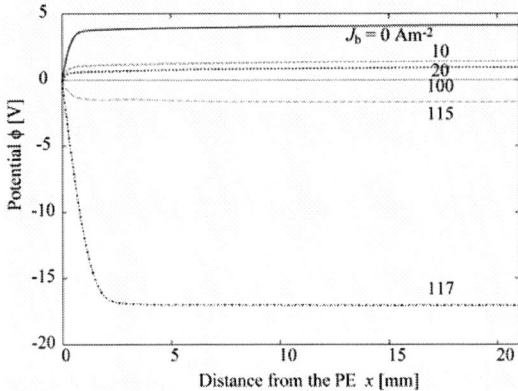

FIGURE 2. Potential profile with each bias current density J_b = 0, 10, 20, 100, 115, 117 Am^{-2} in the case without collision.

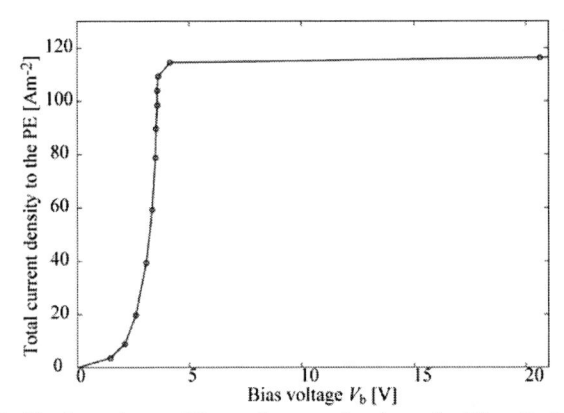

FIGURE 3. The dependence of the total current density to the PE on the bias voltage (I-V characteristics).

The Effect of the PE Bias Voltage

Figure 4 shows the potential profiles with the several bias current densities in the case with collision. The electron saturation current density I_{esat} is a little lower (~ 90 Am^{-2}) than previously because electron density is decreased (~ 10 %) by the presence of H$^-$ ions. In this case, floating plasma potential V_{pf} is ~ 8.6 V. Potential configuration is controlled in the same way as previous case. Figure 5 shows the density profiles of all the particle species with several bias voltages. In the case of any bias voltages, H$^-$ ion density is approximately 10 % of electron density. When the bias current density J_b is smaller than the I_{esat} (Fig. 5 (a), (b)), an ion sheath is formed, while an electron sheath is formed when the J_b is greater than the I_{esat} (Fig. 5 (d)). When the J_b is close to the I_{esat} (Fig. 5 (c)), no sheath appears and flat potential configuration is formed. The H$^-$ ion density varies with the bias voltage, especially in the region where x is less than 4 mm.

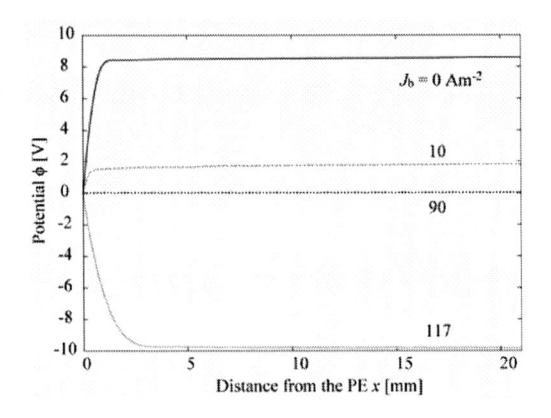

FIGURE 4. Potential profile with each bias current density J_b = 0, 10, 90, 117 Am^{-2} in the case with collision.

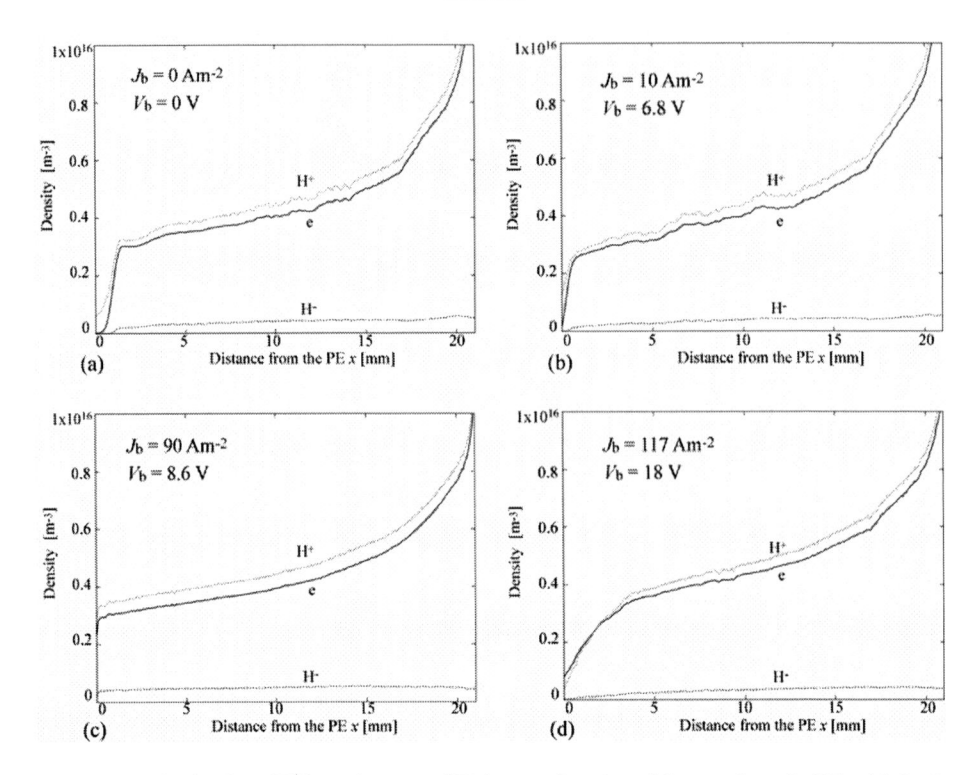

FIGURE 5. The density of H$^+$ ion, electron and H$^-$ ion as a function of distance from the PE at (a) J_b =0 Am^{-2}, V_b=0 V (b) J_b =10 Am^{-2}, V_b=6.8 V, (c) J_b =90 Am^{-2}, V_b=8.6 V, (d) J_b =117 Am^{-2}, V_b=18 V.

To see the variation of the H$^-$ ion density more clearly, H$^-$ ion density profiles near the PE with the several bias voltages is shown in Fig. 6. For V_b less than V_{pf}, H$^-$ ion cannot climb the potential wall in front of the PE and trapped in bulk plasma region. When the bias voltage V_b is close to the floating plasma potential V_{pf}, H$^-$ ion density keeps relatively high density in the region where x is less than 4 mm compared to other cases. This is explained by flat potential configuration formed when the V_b is roughly the V_{pf}. The H$^-$ ions are no longer trapped and thus distributed in the whole region. If the potential of the PE is higher than the plasma potential, H$^-$ ions near the PE are attracted immediately toward the PE. Consequently, the H$^-$ ion density becomes low when the V_b is larger than the V_p. As a result, the peak of H$^-$ ion density at the V_b close to the plasma potential can be seen as shown in Fig. 7, which shows the dependence of H$^-$ ion density on the bias voltage. This understanding is in agreement with the idea suggested by Svarnas et al. [1].

As far as plasma parameters used in this study, e.g. plasma density is the order of 10^{15} m^{-3}, H$^-$ ion destruction processes are not effective in this tiny region. We have done a series of test calculations where different vibrational distributions of H$_2$-molecules are applied. Such neutral parameters have influence on the total amount of H$^-$ ion. However, they do not affect the tendency of spatial distribution of H$^-$ ion. Therefore, we conclude that the physical mechanism of the peak H$^-$ ion density is mainly due to the potential configuration influenced by the PE potential.

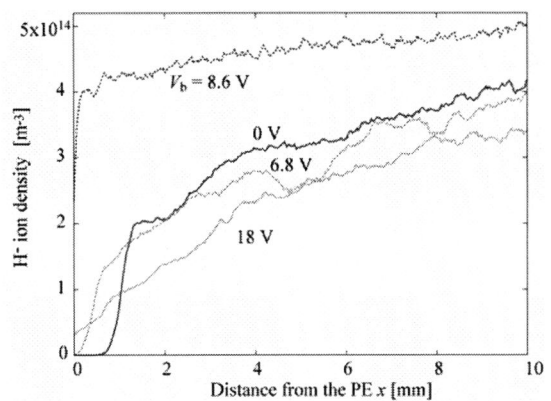

FIGURE 6. The H$^-$ ion density as a function of distance from the PE at bias voltage V_b = 0, 6.8, 8.6, 18 V.

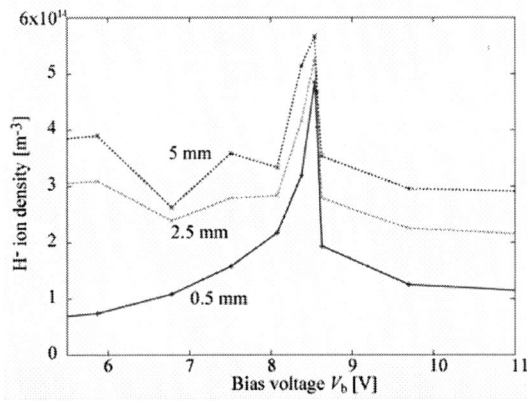

FIGURE 7. The H$^-$ ion density as a function of the bias voltage at x=0.5, 2.5, 5 mm.

CONCLUSIONS

The effect of the PE bias voltage on plasma profiles near the PE in the negative ion sources is modeled and studied with a 1D PIC method. The modeling of the bias voltage is validated by comparing the I-V characteristics with the probe theory.

The physical mechanism that H⁻ ion density peaks at the bias voltage close to the plasma potential is due to the variation of potential configuration. If the bias voltage is smaller than the plasma potential, H⁻ ions are trapped because of the potential drop in front of the PE. Then, as the bias voltage increases, the density of H⁻ ion near the PE increases until it reaches the plasma potential. When the bias voltage is close to the plasma potential, a flat potential configuration is formed and H⁻ ion can reach the region near the PE. This results in relatively high density of H⁻ ion near the PE at this bias voltage. If the applied bias voltage is larger than the plasma potential, a positive sheath is formed in front of the PE, and H⁻ ions are easily absorbed by the PE. Consequently, H⁻ ion density near the PE decreases for such a large bias voltage. This tendency does not depend on the neutral parameters.

For the purpose of focusing on only the effect of the PE bias voltage, the effect of the magnetic field is neglected in this study. To confirm the idea described above, the coupled effect of the PE bias voltage with the magnetic field should be considered. The analysis of 2D study which takes into account the magnetic field is under way [9]. In addition, for quantitative analysis, more precise treatments of neutrals which have great impact on the source rate of H⁻ ion and larger calculation domain are required.

REFERENCES

1. P. Svarnas, J. Breton, M. Bacal and R. Faulkner., IEEE Trans on Plasma Sci., vol. 35, no. 4, pp. 1156-1162, (2007).
2. P.C. Stangeby, *The Role and Properties of the Sheath in The Plasma Boundary of Magnetic Fusion Devices,* IOP Publishing, Bristol and Philadelphia, 2000, pp. 61-88.
3. C. K. Birdsall and A. B. Langdon, *Plasma physics via Computer Simulation* (McGraw-Hill, New York, 1985).
4. T. Mosbach, Plasma Sources Sci. Technol. 14 610 (2005).
5. R. K. Janev, D. Reiter, and U. Samm, *Collision Processes in Low-Temperature Hydrogen Plasmas* (Forschungszentrm-Juelich Report No. Juel-4105, 2003).
6. R. K. Janev, W. D. Langer, K. Evans, Jr., and D. E. Post, Jr., *Elementary Processes in Hydrogen-Helium Plasma, Cross Sections and Reaction Rate Coefficients,* Springer, Berlin (1987).
7. V. Vahedi, M. Surendra, Comp. Phys. Comm. 87 179 (1995).
8. K. Nanbu, IEEE Trans. Plasma Sci. 28, 971 (2000).
9. S. Kuppel, D. Matsushita, A. Hatayama and M. Bacal., 1st Int. Conf. on NIBS 2008 (to be presented).

Study of the Plasma near the Plasma Electrode by Probes and Photodetachment in ECR-driven Negative Ion Source

M. Bacal[a], P. Svarnas[a], S. Béchu[b], J. Pelletier[b]

[a] *LPTP, UMR CNRS 7648, Ecole Polytechnique, 91128 Palaiseau, France*
[b] *Laboratoire de Physique Subatomique et de Cosmologie, UMR CNRS 5821, 53 rue des Martyrs, 38026 Grenoble, France*

Abstract. The effect of the plasma electrode bias on the plasma characteristics near the extraction aperture in a large volume hybrid multicusp negative ion source, driven by 2.45 GHz microwaves, is reported. Spatially resolved negative ion and electron density measurements were performed under various pressures (1-4 mTorr) by means of electrostatic probe and photodetachment technique.

Keywords: Negative ions, plasma electrode bias, magnetic filter, extracted beams
PACS : 32.80.Gc, 52.27.Cm, 29.25.Ni, 52.80.Pi

INTRODUCTION

Contemporary negative ion sources operate with a magnetic filter field extending up to the plasma electrode (PE) which contains the extraction aperture [1]. It is generally accepted that a positive PE bias (V_{PE}) reduces the extracted electron current. In some experiments, an optimum V_{PE} for negative ion extraction was found [2-4]. A possible cause of these effects was suggested in [3]: electrons are trapped in the weak magnetic field and lost along the field lines, whereas positive and negative ions are not trapped because of their large Larmor radius. More negative ions arrive from the bulk plasma to ensure plasma neutrality. However, a definite explanation of the H⁻ ion current peak, when the PE bias is varied, is still not available.

The electrostatic particle simulation [1,5,6] without applied PE bias showed that the presence of the weak magnetic field produces important modifications in the positive ion flow and as a result in the structure of the plasma potential. A characteristic peak in the plasma potential (V_p) dependence on the distance from the PE (D) is predicted, where many negative ions are localized.

The purpose of this work is the experimental study of the spatial distribution of the plasma parameters in the vicinity of the PE, specifically on the axis of the extraction aperture, using electrostatic probes and laser photodetachment.

EXPERIMENTAL SETUP AND DIAGNOSTICS

The H⁻ ion source Camembert III has been described in [4] and a draft is shown in Fig. 1. The magnetic multipole chamber Camembert III was previously equipped with 16 filaments, installed on the top flange and operated in dc mode, similar to a filamented ion source. Since 2003, a 2-D network of seven elementary independent ECR plasma sources [7] has been installed on the top flange. They are powered by microwaves (2.45 GHz) at 0.9 kW, equally distributed at the seven elementary sources. Each source contains a permanent magnet which provides the magnetic field required for ECR (0.0875 T). This magnetic field also confines the fast electrons. It represents a first magnetic filter and the region around the magnet represents the "driver region".

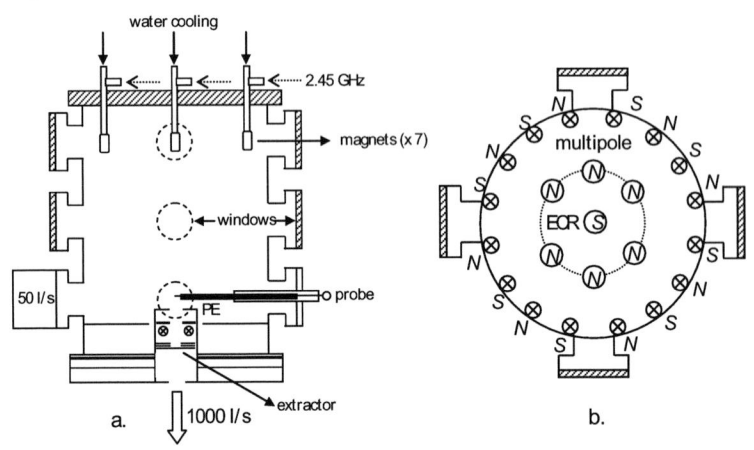

FIGURE 1. Schematic representation of the ECR-driven hybrid multicusp source Camembert III (a) and the magnets forming the multicusp configuration (b). Svarnas et al, IEEETPS [4] © 2008 IEEE

Opposite to the top flange there is a PE with the extraction hole, followed by the extraction electrode equipped with a pair of magnets for electron separation, and the acceleration electrodes. The stray magnetic field from the magnets located in the extraction electrode extends into the plasma and provides a second magnetic filter. This magnetic field attains a maximum of 0.002 T at a distance of 1.6 cm from the PE. This region can be denoted as the "extraction region". A large volume of cold plasma separates the above two extreme regions.

The various plasma parameters were measured with an electrostatic probe made of tungsten wire of 0.5 mm diameter, mounted on a three-axis manipulator installed on a flange at the PE level. A tip of 15 mm was exposed to the plasma, while the rest was shielded by a ceramic cylinder. This probe was L-shaped so that its longer part was coaxial with the laser beam, used for measuring the negative ion density by photodetachment [8]. The measurements were effected on the axis passing through the center of the extraction opening; they were not affected by the proximity to the PE surface, since the distance was long enough compared to the collection radius of the probe (a few Debye lengths). The probe was perpendicular to the weak magnetic filter field. A standard deviation of about 5% was roughly estimated for all the measurements.

A Nd-YAG laser (repetition rate 10 Hz, photon energy 1.2 eV) was used to illuminate the plasma, through a 6 mm diaphragm, and the photodetached electrons from the H⁻ ions were collected by the L-shaped, positively biased, probe, situated on the axis of the laser beam. When varying the distance D between the studied point and the PE, both the probe and the laser beam are displaced, so that the probe remains collinear with the laser.

The probe transient current, generated by the collected photodetached electrons, was recorded on a digital oscilloscope (500 MHz, 2 GSample/s) operating in averaging mode, via capacitive decoupling. Since previous studies showed that in the region near the PE the negative-ion relative density n_H^-/n_e is much higher than in the center of the source, the analysis of the probe signals was carried out according to the theory and recommendations of [9].

SPATIAL VARIATION OF THE PLASMA PARAMETERS

The results obtained at a pressure p of 3 mTorr are presented in Fig. 2 to 6. The electron density spatial variation is shown in Figure 2. The electron density goes down linearly with the distance to the PE. The higher the positive bias V_{PE}, the lower the electron density. This shows that the electrons are efficiently swept away from the PE neighborhood due to the applied positive PE bias.

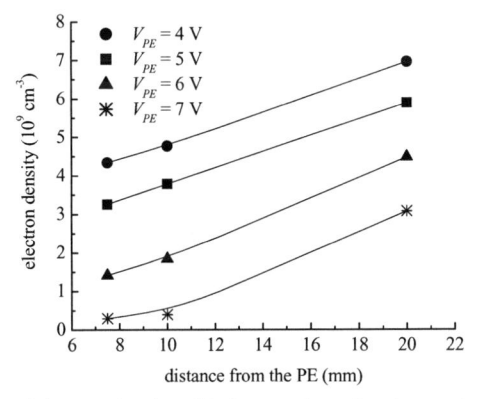

FIGURE 2. Variation of electron density with distance from the plasma electrode, for four values of the plasma electrode bias V_{PE}.

The spatial variation of the negative ion density (Fig. 3) clearly indicates its enhancement at distances close to the plasma electrode where the electron density tends to be reduced. The optimum PE bias is $V_{PE} = 5.5$ V.

Fig. 4 shows the variation of the relative negative ion density n_H^-/n_e versus D. Note that this relative density goes up with the PE bias and attains the value of 1.9 for the highest V_{PE} applied (+7V).

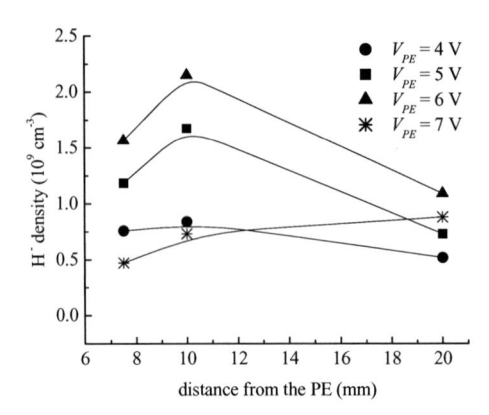

FIGURE 3. Variation of the negative ion density with distance from the plasma electrode, for four values of the plasma electrode bias V_{PE}.

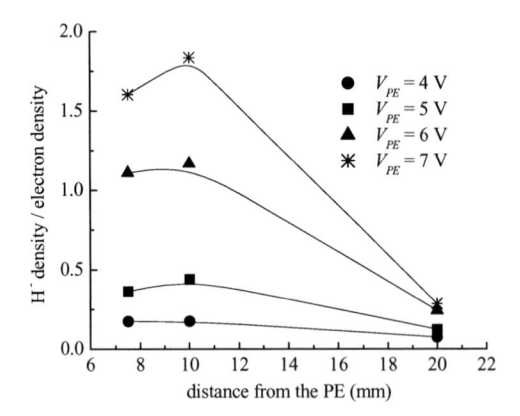

FIGURE 4. Variation of the relative negative ion density with distance from the plasma electrode, for four values of the plasma electrode bias V_{PE}.

An indication of a weak plasma potential peak versus the distance to the PE exists for $V_{PE} > 4$ V. This "maximum" is located at a distance of 1 cm (Fig. 5). The plasma potential varies abruptly near the PE for extreme bias values, when ion or electron sheaths are formed, but stays approximately constant for the medium value $V_{PE} = 5.5$ V, which is close to the plasma potential far from the PE. Let us denote this value V_{PE}^*. This value of the PE bias - V_{PE}^* - is important since the maximum negative ion density occurs at this bias (not shown in Fig. 3). The electron temperature values increase abruptly when $V_{PE} > V_{PE}^*$ (see Fig. 6).

The spatial variation of the electron temperature (Fig. 6) indicates that for $V_{PE} < 6$ V the electron temperature is low (0.3 - 0.5 eV) at all the distances studied, but goes up abruptly near the PE when $V_{PE} > 6$ V.

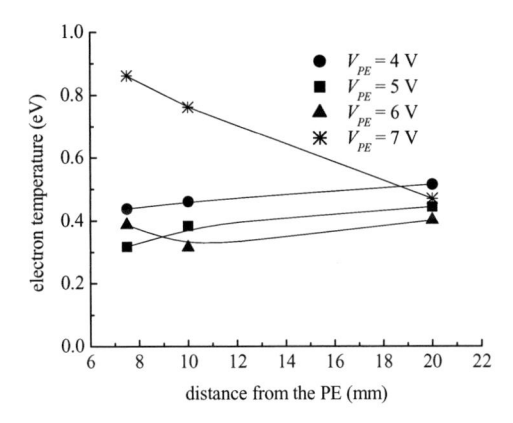

FIGURE 5. Variation of the plasma potential with distance from the plasma electrode, for four values of the plasma electrode bias V_{PE}.

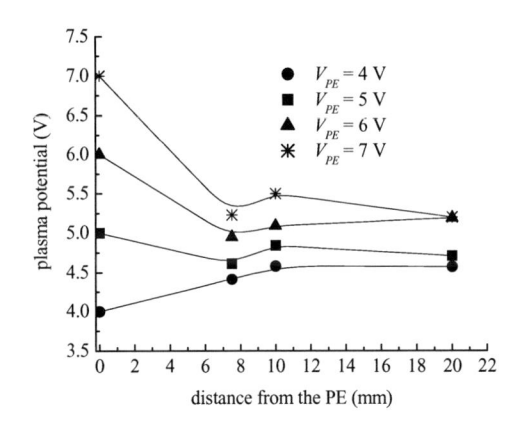

FIGURE 6. Variation of the electron temperature with distance from the plasma electrode, for four values of the plasma electrode bias V_{PE}.

A comparison of the pressure dependence of the negative ion densities at 150 mm and 7.5 mm from the PE shows that the latter is enhanced particularly at the lowest pressure (Fig. 7). Even at a distance of 20 mm the effect of the magnetic field and the PE bias starts to go down, as can be seen in Fig. 3.

The extracted negative ion current I_{H^-} is correlated better with the negative ion density near the extraction opening (Fig. 7) than with that in the center of the source (150 mm from the PE). The optimum pressure at the center of the source lies between 4 and 5 mTorr, as opposed to 1.5-3 mTorr at 7.5 mm from the PE. The latter approaches the optimum pressure for the extracted H⁻ ion current.

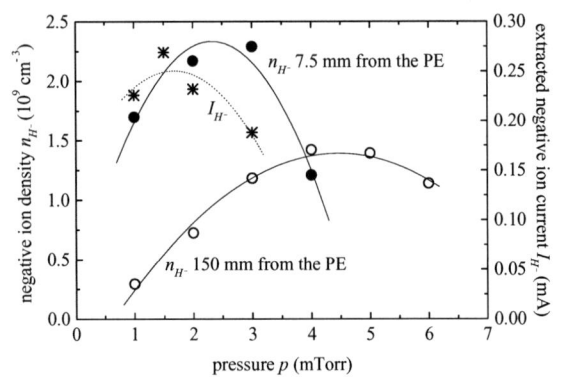

FIGURE 7. Comparison between the negative ion density close to the extractor aperture (solid symbols) and at the center of the source (open symbols) vs. pressure. The extracted negative ion current is also compared (dotted line). Svarnas et al, IEEETPS [4] © 2008 IEEE

Fig. 8 shows the ratio n_{H^-}/n_e between the negative-ion density and the electron density found at a distance of 7.5 mm from the PE when V_{PE} is varied. In the same figure the ratio I_{H^-}/I_e of the extracted currents is as well presented.

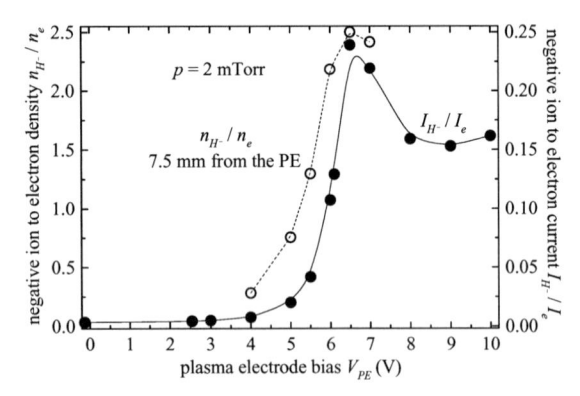

FIGURE 8. Ratios of the negative ion to electron density (dashed line) and the extracted negative ion to electron current (solid line) vs. plasma electrode bias. Svarnas et al, IEEETPS [4] © 2008 IEEE

CONCLUSION

Compared to the electron density at the center of the source (3×10^{10} cm^{-3} at 3 mTorr), the electron density at 1 cm from the plasma electrode is reduced by a factor 30, when a PE bias $V_{PE} = 6$ V is applied. The highest negative ion densities are found in the neighborhood of the plasma electrode, at distances where the electron density is reduced.

Note that the ratio n_{H^-}/n_e attains the value 1.9 (Fig. 4), which means that the negative ions constitute the majority of the negative species. We should remind that in the filamented version of Camembert III values of n_{H^-}/n_e as high as 12 were measured in a discharge of 2.5 kW [9]. It was observed that the maximum n_{H^-}/n_e ratio went up when the plasma density or discharge power was enhanced. This may explain the lower n_{H^-}/n_e value (*i.e.* ~2) found in this, ECR driven, version of CAMEMBERT III, where the plasma density is limited by the available microwave power.

In conclusion the explanation of the H⁻ ion density peak can be summarized as follows: the depletion of the electron population, when the positive bias of the PE is enhanced, disturbs the local plasma neutrality and new negatively charged particles should replace the electrons collected by the PE. In a collisionless situation, the electrons present in the bulk plasma are unable to cross the transverse magnetic field in front of the PE. Therefore, negative ions from the central plasma replace the depleted electrons in the magnetized region to ensure plasma neutrality, as suggested in [3]. Indeed, the measured H⁻ ion density increases close to the PE and peaks when V_{PE} is slightly greater than V_{PE}^*. This results in higher H⁻ ion extracted current. When the PE bias is further increased the H⁻ ions are accelerated toward the PE. The negative-ion residence time decreases due to the larger velocity and the H⁻ density near the PE as well as the extracted negative-ion current go down.

Two simulation efforts [10,11] confirm the above interpretations. The PIC 2D3V simulation by Kuppel et al [10] performed for the conditions of Camembert III provides information concerning the electron and negative ion density in the system. Figure 5a in Ref. 10 shows the 2D distribution of electron density for $V_{PE} = 0$. Electrons are trapped along the magnetic field lines, which act as a barrier keeping most of them far from the extraction aperture. Electrons approach the PE at some distance from the axis of the extraction opening, but they are notably collected by the latter only when V_{PE} exceeds the plasma potential. As a result, electron density is reduced in front of the extraction opening and the central region of the plasma electrode.

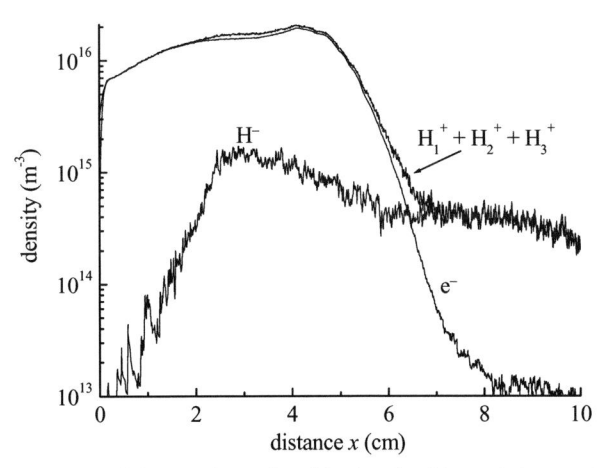

FIGURE 9. Variation of the negative and positive ion densities and electron density in the region between the PE and the magnetic filter. The PE is located at x = 10 cm. [11]

The results of the 1D3V PIC-MCC model reported by Kolev et al [11] (shown on Fig. 9) indicate the strong electron density depletion in the plasma region close to the PE at sufficiently high positive V_{PE}. They also show that the neutrality in this region is ensured by the negative and positive ions, in agreement with our observations and interpretation [3,4].

ACKNOWLEDGEMENTS

The authors acknowledge the collaboration of Professor A. Hatayama, S. Kuppel, and D. Matsushita in the simulation of this plasma device and useful discussions.

REFERENCES

1. M. Bacal, A. Hatayama, J. Peters, *IEEE Trans. Plasma Sci.,* **33**, N° 6, 2005, 1845
2. K.N. Leung, K.W. Ehlers, M. Bacal, *Rev. Sci. Instrum.*, **54**, 1983, 56
3. M. Bacal, J. Bruneteau, P. Devynck, *Rev. Sci. Instrum.*, **59**, 1988, 2152
4. P. Svarnas, J. Breton, M. Bacal, R. Faulkner, *IEEE Trans. Plasma Sci.*, **35**, N°4, 2007, 1156
5. T. Sakurabayashi, A. Hatayama, M. Bacal, *A.I.P. Conference Proceedings 763,* Editors J.D. Sherman, Y.I. Belchenko, 2005, 96
6. A. Hatayama, *Rev. Sci. Instrum.* **79**, 2008, 02B901
7. A. Lacoste, T. Lagarde, S. Béchu, Y. Arnal, J. Pelletier, *Plasma Sources Sci. & Technol.*, **11**, 2002, 407
8. M. Bacal, *Rev. Sci. Instrum.*, **71**, 2000, 3981
9. F.El Balghiti-Sube, F.G. Baksht, M. Bacal, *Rev. Sci. Instrum.*, **67**, 1996, 2221
10. S. Kuppel, D. Matsushita, A. Hatayama, M. Bacal, These Proceedings
11. St. Kolev, G.J.M. Hagelaar, J.P. Boeuf, Private communication.

Numerical Analysis of Electronegative Plasma near the Extraction Grid in Negative Ion Sources

S. Kuppel[a], D. Matsushita[a], A. Hatayama[a], and M. Bacal[b]

[a]Graduate School of Science and Technology, Keio University, 3-14-1 Hiyoshi, Kohoku-ku, 223-8522 Yokohama, Japan
[b]Laboratoire de Physique et Technologie des Plasmas, UMR 7648 du CNRS, Ecole Polytechnique, 91128 Palaiseau, France

Abstract. The effects of plasma electrode (PE) bias on the extraction process in a negative ion source are studied with a series of two-dimensional (2D) electrostatic particle simulations. Motion of charged particles in their self-consistent electric field is modeled by the Particle-In-Cell (PIC) method. The effect of a weak transverse magnetic field is also taken into account. Extracted electron current density gradually decreases as the PE bias increases, while the absolute value of PE net current density initially increases and is saturated for higher values of PE bias. The extracted negative ion current density reaches a weak peak when the PE bias approaches the bulk plasma potential (*Vpf*) in the case without the PE bias, but the decrease afterwards is mild. Physical mechanisms leading to these results are discussed.

Keywords: Negative ions, plasma electrode (PE) bias, particle-in-cell (PIC), two-dimensional, weak magnetic field, extraction.
PACS: 52.50.Gj, 52.65.Rr

INTRODUCTION

Using negative hydrogen ion sources is a very promising way to obtain the high energy neutral beams required for heating plasmas in the future nuclear fusion reactors. Among other R&D issues (such as, i) the improvement of H⁻ production efficiency, ii) spatial uniformity for volume/surface H⁻ production in large H⁻ sources, etc.), optimization of H⁻ extraction from the source is one of the most important issues for developing negative ion sources.

In order to understand the effect of week transverse magnetic field on the potential structure near the extraction aperture and the resultant H⁻ extraction, Sakurabayashi et al have done a systematic numerical study [1-2] based on the PIC (Particle-in-Cell) method [3]. The 2D3V (2D in the real space and 3D in the velocity space) PIC modeling shows that the difference in dynamics between electrons and ions under the presence of the "weak" magnetic field leads to a significant change of potential structure near the PE and this leads to the enhancement of H⁻ extraction. Being based on these basic studies, a different version of the 2D3V PIC code [4], also for the analysis of extraction of surface produced H⁻ ions, has been recently developed and the previous results above in Ref.1 and 2 have been reexamined and verified.

In the present study, we focus on the combined effect on H⁻ extraction of a bias voltage on the plasma electrode (PE) along with a weak transverse magnetic field described above. It has been shown in the Camembert III experiments [5] that the extracted H⁻ current I_{H^-} reaches a maximum for a certain value of the PE-bias voltage when using the weak magnetic field close to the extraction area. In order to understand the physical mechanism and the underlying physics which lead to such an optimum value for the I_{H^-} in the experiments, a numerical study based on the 2D3V PIC method has been done. For this purpose, the 2D3V PIC code already used to show the effect of the weak magnetic field in Ref.3 has been further improved, in particular to include PE bias effect. Also, volume production of negative ions, independently from extraction conditions, is taken into account. The effect of the PE bias on the potential structure, particles densities and electron and negative ion extraction is discussed below.

SIMULATION MODEL

Model Geometry and Basic Equations

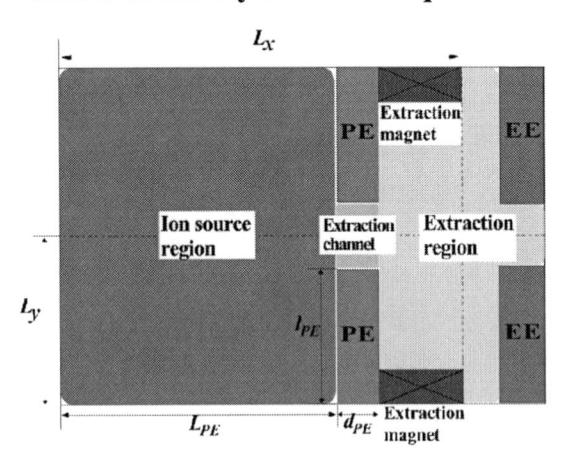

FIGURE 1. Geometry of the modeled system.

Figure 1 shows the model geometry used in the present PIC modeling. For simplicity, only the region close to the extraction opening has been modeled with a 2D slab geometry. The x-axis is taken to be parallel to the direction of the H⁻ beam extracted from the aperture, while the y-axis is parallel to the PE. A pair of magnets with an infinite length in the z direction is installed in the extraction region. The model geometry in Fig.1 is assumed to be periodically extended over the y direction with respect to $y = 0$ and $y = 2L_y$ boundaries. The equation of motion and Poisson's equation are given, respectively, by Eqs.(1) and (2):

$$m_j \frac{d\mathbf{v}_j}{dt} = q_j(\mathbf{E} + \mathbf{v}_j \times \mathbf{B}), \quad \mathbf{E} = -\nabla\phi \qquad (1)$$

$$\varphi = -\rho(x,y)/\varepsilon_0 \qquad (2)$$

In Eq.(1), the subscript j denotes the particle species (j= H$^+$ ion, electron, and H$^-$ ion) taken into account in the simulation, and m_j and q_j are respectively the mass and charge. The charge density in Eq.(2) is given by

$$\rho = e(n_{H^+} - n_e - n_{H^-}) \qquad (3)$$

where the symbol e is the unit charge and n_j is the number density of jth particle. These equations are solved self-consistently in the model geometry shown in Fig.1.

Normalization and Main Numerical Parameters

The space coordinates (x,y) and the time t are normalized in the following numerical simulations as follows: $\tilde{x} = x/\lambda_{De}$, $\tilde{y} = y/\lambda_{De}$, $\tilde{t} = \omega_{pe}t$, where λ_{De} and ω_{pe} are, respectively, the electron Debye length and the plasma frequency with the initial density n_e and temperature T_e. The grid-size and the time-step used are respectively $d\tilde{x} = d\tilde{y} = 0.5$ and $d\tilde{t} = 0.1$. The following normalized dimensions are used in the simulation:
1) System length: $\tilde{L}_x = 50$, $\tilde{L}_y = 25$, $\tilde{L}_{PE} = 35$
2) Dimensions of the PE: $\tilde{l}_{PE} = 20$, $\tilde{d}_{PE} = 5$

The electrostatic potential Φ, magnetic flux density B, and current density j are also normalized as follows: $\tilde{\Phi} = e\Phi/kT_e$, $\tilde{B} = B/(m_e\omega_e/e)$, $\tilde{J} = j/n_e e v_{th}$, where e, k, and v_{th} are respectively the unit charge, the Boltzmann constant and electron thermal speed.

Initial and Boundary Conditions

To reduce the calculation time, only the half region ($0 \leq \tilde{y} \leq \tilde{L}_y$) in the y direction is used as a calculation domain. To do so, we have imposed the symmetric boundary condition ($\partial\tilde{\Phi}/\partial y = 0$) on the left-hand-side (l.h.s: $\tilde{x} = 0$), the upper-half ($\tilde{y} = \tilde{L}_y$), and the bottom ($\tilde{x} = 0$) boundaries in Fig.1. Along the PE surface, $\tilde{\Phi}$ is fixed to be zero ($\tilde{\Phi} = 0$). At the right-hand-side (r.h.s) boundary ($\tilde{x} = \tilde{L}_x$), $\tilde{\Phi}$ is also fixed to be a given voltage $\tilde{V}_{applied}$, i.e., $\tilde{\Phi}(\tilde{L}_x, \tilde{y}) = \tilde{V}_{applied}$.

At $\tilde{t} = 0$, a large number of positive and negative charged particles are uniformly loaded in the source region ($0 < \tilde{x} < 0.2\,\tilde{L}_x$, $0 < \tilde{y} < \tilde{L}_y$). The initial population ratio is chosen as follows: $n_{H^+}:n_e:n_{H^-} = 10:9:1$. The initial velocity of each particle is chosen from the Maxwellian distribution with a given temperature of each species: $kT_e = 1.0\,eV$, $kT_{H^+} = kT_{H^-} = 0.25\,eV$.

H$^+$ ions and electrons across the l.h.s. boundary ($\tilde{x} = 0$) are reloaded, with a velocity chosen from half Maxwellian with a thermal speed determined by the initial temperature. H$^-$ ions crossing this boundary are simply removed from the system.

The H$^+$ ions that strike PE surface are assumed to be recycled as neutrals and then re-ionized in the source region. A pair of H$^+$ ion and electron is reloaded. The H$^+$ ion and the electron velocities are chosen, respectively, from Maxwellian in the same manner as in the initial particle loading above. On the other hand, electrons and H⁻ ions which reach the PE surface are simply removed from the system.

Particles across the boundary, $\tilde{y} = 0$ or $\tilde{y} = \tilde{L}_y$, are reloaded so as to satisfy the symmetric boundary condition with respect to these boundaries.

Finally, if particles reach to the r.h.s. boundary ($\tilde{x} = \tilde{L}_x$), they are assumed to be reloaded in source region with velocities chosen from Maxwellian profile.

The 2D spatial profile of the magnitude of the normalized magnetic flux density $|\tilde{B}| = \sqrt{\tilde{B}_x + \tilde{B}_y}$ is shown in Fig.2. The magnetic field is calculated at each point by using the analytical solution based on the surface magnetic charge model. The magnitude of the surface magnetic charge $\pm \Delta m$ is specified so as to satisfy the following condition: $\tilde{r}_{Le} < \tilde{L}_x, \tilde{L}_y \ll \tilde{r}_{LH^+}, \tilde{r}_{LH^+}$ where $\tilde{r}_{Lj}$ is the normalized Larmor radius of jth-species ($\tilde{r}_{Lj} = r_{Lj}/\lambda_{De}$), and $\tilde{L}_x, \tilde{L}_y$ are the normalized system lengths. Thus, electrons are magnetized, while positive and negative ions are not magnetized near the PE.

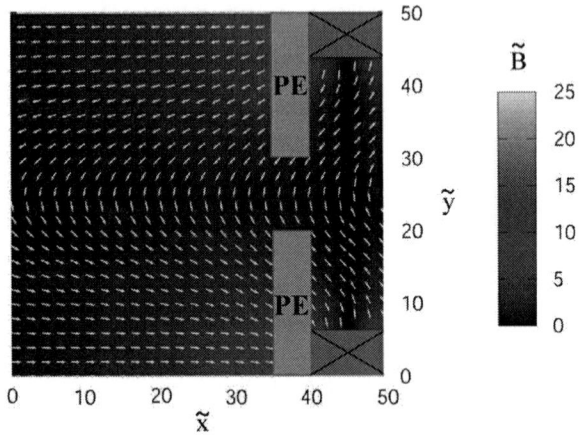

FIGURE 2. Spatial profile of normalized magnetic field amplitude used in the simulation. The field lines are also shown by arrows.

Negative ion volume production and PE bias method

In addition to the initial loading of particles, a constant number of negative ions are injected at each time step in the source region defined previously. Their velocity is chosen using Maxwellian distribution. The production rate is fixed and does not change in time nor depends on the other simulation parameters. This common value is chosen so that negative ion population always approximately reaches a steady state.

Concerning the PE bias $\tilde{\Phi}_{bias}$, it is not technically possible to model it by directly changing the voltage on the PE. As mentioned above, we have modeled mainly the extraction region to speed up the calculation, and we have employed the following boundary condition $\partial \tilde{\Phi}/\partial x = 0$ at the bulk plasma-side, i.e., the l.h.s boundary ($\tilde{x} = 0$). In the above scheme, i.e., if we directly change the voltage on the PE, the potential at the bulk plasma-side also starts changing. Finally, in the steady state, the positive current and negative current are almost balanced and no net current is drawn through the PE. As a result, the potential difference $\Delta\widetilde{\Phi}$ between the bulk plasma-side and the plasma electrode is almost unchanged irrespective of the voltage applied to the PE.

To overcome the problem, here, we adopt the alternative scheme in which a chosen number of electrons are injected at each time step from the bulk plasma-side (l.h.s.) boundary. Their velocity is chosen from the half-Maxwellian distribution with T_e, and the injection rate is kept constant in time. In this way, a net current can be forced to pass through the plasma. As a result, the potential difference $\Delta\widetilde{\Phi}$ between the bulk plasma and the PE decreases as if the PE was positively biased. The potential difference $\Delta\widetilde{\Phi}$ can be controlled by the injection rate of electrons. The detailed check of this scheme has been done with 1D PIC modeling [6]. The PIC results have been compared with the Langmuir probe theory and reasonable agreement has been obtained. Being based on these basic checks with 1D PIC modeling, here we adopt the above scheme.

As mentioned above, the potential difference between the PE and the r.h.s boundary ($\widetilde{x} = \widetilde{L}_x$) is kept constant as $\Delta\widetilde{\Phi} = 50$ throughout the present study.

RESULTS AND DISCUSSION

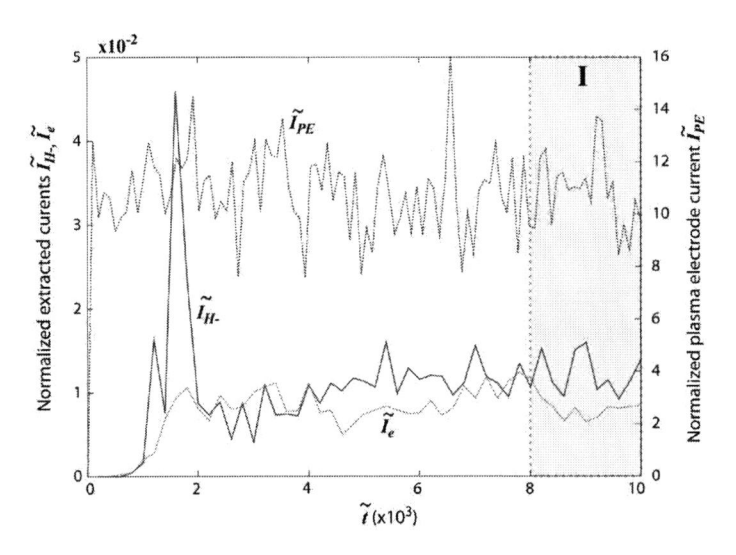

FIGURE 3. Typical time evolution of extracted negative-ion, extracted electron, and PE currents for a given value of PE bias. The time-average range used is shown by area I.

Figure 3 shows a typical time evolution of the following characteristic currents: the extracted H⁻ current $\widetilde{I}_{H^-}$, extracted electron current $\widetilde{I}_e$, and the absolute value of the net current through the PE $\widetilde{I}_{PE}$, respectively. After the transient phase, these currents reach almost the quasi-steady state as seen from Fig.3. The time-averaged values of

$\tilde{I}_{H^-}$, $\tilde{I}_e$, and $\tilde{I}_{PE}$ are shown in Fig.4 as a function of $\tilde{\Phi}_{bias}$. The time average has been taken over the period (I) in Fig.3 (the last 20000 time iterations in Fig.3).

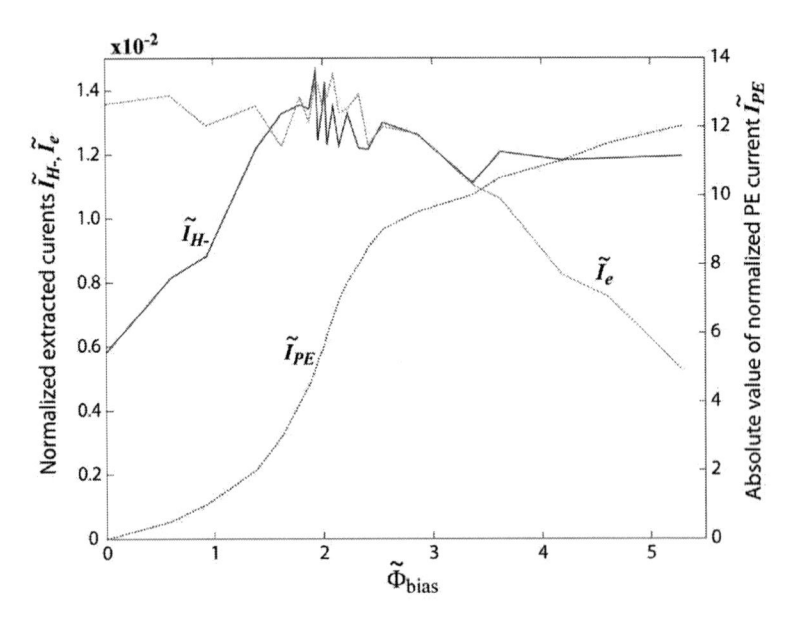

FIGURE 4. Dependence of extracted negative-ion, extracted electron, and PE current on PE bias $\tilde{\Phi}_{bias}$.

Effect of PE Bias Voltage on Electron Behavior

As seen from Fig.4, the PE current $\tilde{I}_{PE}$ increases rapidly when $\tilde{\Phi}_{bias}$ approaches the floating plasma potential $\tilde{V}_{pf}$ ($\tilde{V}_{pf} \sim 2.3$), and is almost saturated for the larger values of $\tilde{\Phi}_{bias}$. One can deduce that if $\tilde{\Phi}_{bias}$ becomes larger than $\tilde{V}_{pf}$, there is no potential barrier for electrons in front of the PE. Thus they are collected on the PE at a maximum rate: the electron saturation current density [7]. This feature is similar to the I-V characteristic in the Langmuir probe (LP) theory. A similar behavior has been also observed in the experiment in Ref.5.

Concerning $\tilde{I}_e$, one can see in Fig.4 that there is a clear decrease when $\tilde{\Phi}_{bias}$ increases.

Figure 5 provides information concerning the electron density in the system. Figure 5(a) shows the 2D distribution of electron density for $\tilde{\Phi}_{bias}=0$. As a complement, Fig. 5(c) and 5(d) allow seeing the influence of $\tilde{\Phi}_{bias}$ on the electron density profile along two lines (line A and line B) shown in Fig. 5(b).

First, Fig.5(a) shows the influence of the magnetic field upon electron density. Electrons are trapped along the magnetic field lines, which acts as a barrier keeping them far from the extraction aperture. As a result, an electron-depletion region is created in front of the extraction aperture. Besides, Fig. 5(c) shows that this depletion in the extraction channel axis is further advanced when $\tilde{\Phi}_{bias}$ increases. It can be deduced that more and more electrons present in the region close to extraction aperture are collected by the PE as $\widetilde{\Phi}_{bias}$ increases. This decay in electron population finally leads to the decreases of $\widetilde{I}_e$ observed in the results.

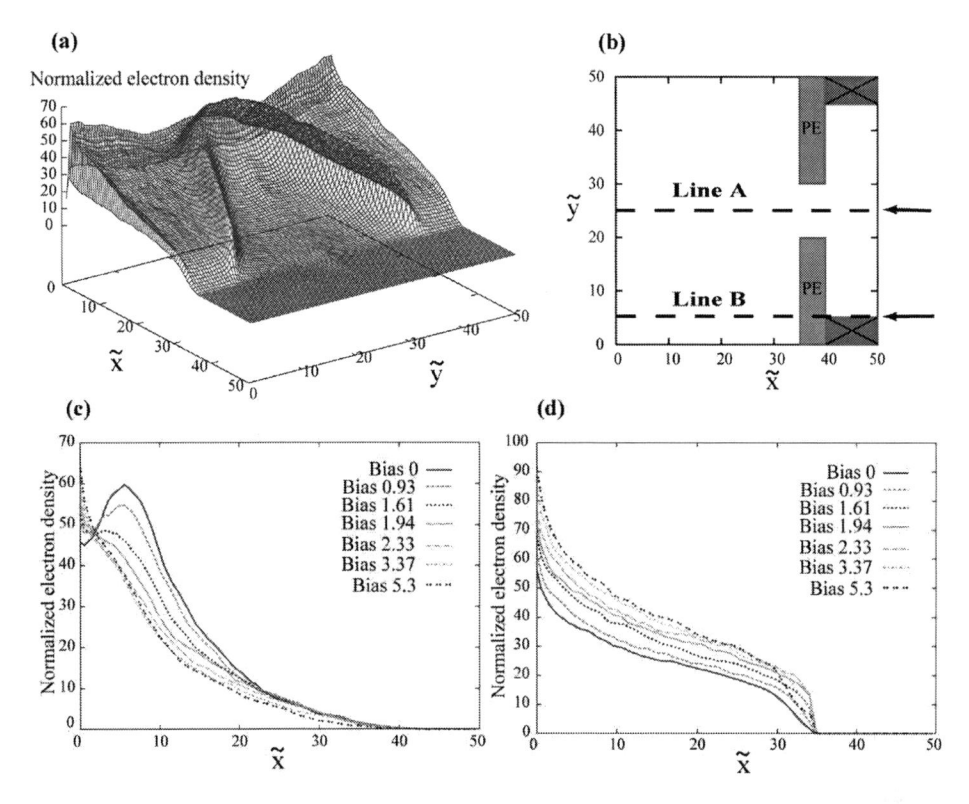

FIGURE 5. Electron densities : (a) Spatial distribution at PE bias=0, (b) Description of the two lines visualized in (c) the axis of extraction channel, and (d) far from the extraction axis.

Effect of PE Bias Voltage on Negative Ion Extraction

Coming back to Fig.4, it can be seen that $\widetilde{I}_{H^-}$ does not behave monotonically. It increases, and then reaches a weak peak for $\widetilde{\Phi}_{bias}$ around 2, to be finally almost constant at higher $\widetilde{\Phi}_{bias}$. This is why four ranges of $\widetilde{\Phi}_{bias}$ are distinguished in the following analysis. "A" stands for $\widetilde{\Phi}_{bias}=0$, "B" for $\widetilde{\Phi}_{bias}$ between 0 and 1.94, "C" for $\widetilde{\Phi}_{bias}$ between 1.94 and 2.87, and "D" for $\widetilde{\Phi}_{bias}$ larger than 2.87.

In the case A ($\widetilde{\Phi}_{bias} = 0$), the trapping of electrons by the weak magnetic field results in a local perturbation of potential. Due to the leak of negative charges, a positive peak is created in front the extraction aperture. Thus, negative ions are attracted in this region, since they are not trapped by the magnetic field due to their larger Larmor radius. This allows them to be extracted, thus $\widetilde{I}_{H^-}$ is not zero. Figure 6 shows the potential profiles for three different values of $\widetilde{\Phi}_{bias}$, and Fig 6(a) corresponds to the case $\widetilde{\Phi}_{bias} = 0$. One can see that the positive peak of potential still exists at steady state in front of extraction region.

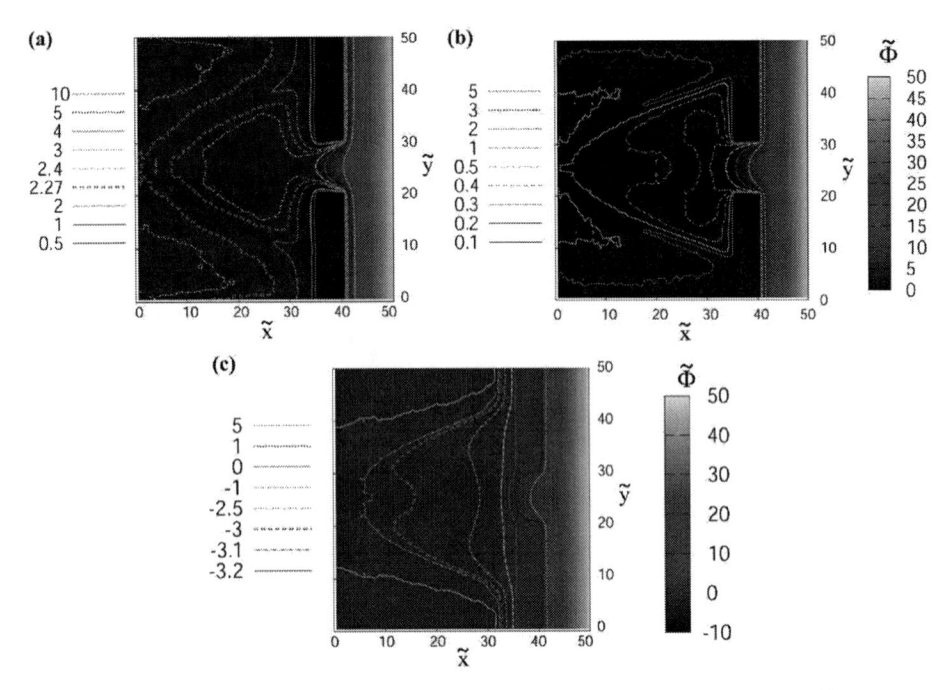

FIGURE 6. Potential profile at steady state: (a) PE bias = 0, (b) PE bias = 1.94, (c) PE bias = 5.3.

In the range B ($0 \leq \widetilde{\Phi}_{bias} \leq 1.94$), electrons depletion in front of extraction aperture gradually increases, as shown in Fig.5(c). Thus, an increasing number of negative ions are attracted to maintain the plasma neutrality. Some of the H⁻ ions are then accelerated towards the extraction aperture, and this finally results in an increase of H⁻ ion extracted current.

In the range C ($1.94 \leq \widetilde{\Phi}_{bias} \leq 2.87$), as shown on Fig. 4, a peak has been reached and $\tilde{I}_{H^-}$ starts to decrease. Besides, Fig. 7(c) shows the evolution with PE bias of different H⁻ ion densities, calculated in the extraction axis at different distances from the PE. All these densities present a peak for a certain value of $\widetilde{\Phi}_{bias}$, close to 2. This means that H⁻ ions are leaving the region in front of extraction aperture faster than they are attracted in it by electrons depletion. Such a mechanism could be explained by the fact that if $\widetilde{\Phi}_{bias}$ becomes close to $\widetilde{V}_{pf}$, the potential peak in front of the extraction aperture no longer appears. Consequently the residence time of H⁻ ions in the region in front of the PE decreases, as does the resulting density. However, these H⁻ ions are not all extracted since $\tilde{I}_{H^-}$ is decreasing. We can deduce that H⁻ ions are consequently necessarily collected by the PE where they are absorbed. Figure 6(b) shows the potential profile corresponding to $\widetilde{\Phi}_{bias} = 1.94$ and Fig.8 shows the evolution of the absolute value of H⁻ current on the PE. One can see on Fig. 6(b) that the equipotentials becomes narrower close to the PE, especially, close to the extraction aperture. The resulting strong electric field would tend to accelerate the H⁻ ions towards the PE. This is confirmed by Fig.8 where it can be seen that $\tilde{I}_{HPE}$ rapidly increases when $\tilde{\Phi}_{bias}$ is close to 2.

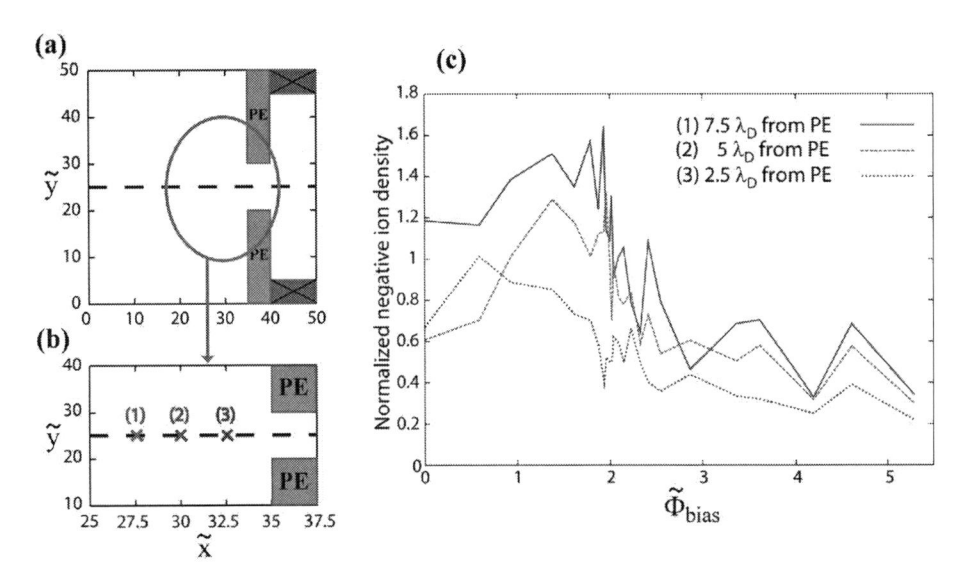

FIGURE 7. (a) Geometry of the system. (b) Zoom in the circled region in Fig.7(a). (c) Negative ion density in front of extraction channel (y=25) as a function of PE bias, at different distances from PE.

It should be also noted that in Fig. 7(c) the density peak is reached for different value of PE bias, depending on the distance from the PE. This result can be understood considering that the negative ions located closer to the extraction region are more influenced by the potential coming from the extraction electrode. Consequently, less PE bias is needed for them before being extracted.

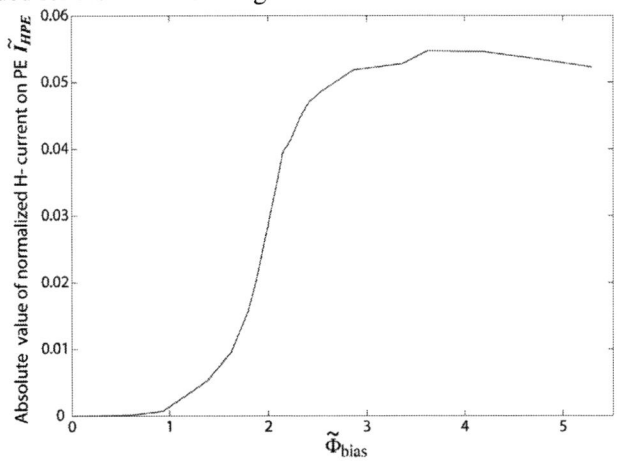

FIGURE 8. Normalized H- current on the PE as a function of the PE bias $\tilde{\Phi}_{bias}$.

In the range D (2.87 ≤ $\widetilde{\Phi}_{bias}$), the decrease stops and a stable value seems to be reached. This tendency is also found in Fig 7(c), since at high PE bias the densities decline becomes rather mild. Coming back to Fig.8, one can see that $\tilde{I}_{HPE}$ is saturated when $\widetilde{\Phi}_{bias}$ is slightly lower than 3. But meanwhile, the attraction of H⁻ ions in front of extraction aperture due to electrons depletion, already described above, is still active. Since collection of H⁻ ions on PE has reached a maximum rate, other H⁻ ions leaving the region in front of extraction aperture can only be extracted. This would be the reason why $\tilde{I}_{H^-}$ eventually stops decreasing. The corresponding potential, shown on Fig. 6(c), seems to confirm such a global mechanism. Far from the extraction axis, the narrow equipotentials close to the PE are parallel to the PE. But closer to the extraction axis, other narrow equipotentials induce electric field lines directed towards the extraction aperture (for negatively charged particles). This potential structure tends to support the aforementioned mechanism.

CONCLUSION

A two dimensional geometry of negative ion source extraction region has been modeled in order to study the influence of the plasma electrode bias voltage on the extraction efficiency, under a weak transverse magnetic field.

First, the effect of $\widetilde{\Phi}_{bias}$ on electron behavior is studied. A decrease in extracted electron current density is seen when PE bias increases, due to electron gradual depletion in front of the extraction aperture. Secondly, the study focuses on negative H⁻ ions. The extraction of these particles reaches a weak peak when $\widetilde{\Phi}_{bias}$ approaches $\tilde{V}_{pf}$. At first, electrons depletion attracts the H⁻ ions in front of extraction aperture, thus $\tilde{I}_{H^-}$ increases. When $\widetilde{\Phi}_{bias}$ approaches $\tilde{V}_{pf}$, H⁻ ions start being collected by the PE and $\tilde{I}_{H^-}$ decreases. Eventually, H⁻ current on the PE reaches a maximum, thus $\tilde{I}_{H^-}$ stops decreasing and reaches a stable value.

These results, although preliminary, seem to confirm the view proposed by Bacal *et al* [8] that the positive bias of the PE depletes the electron population near the PE. The plasma neutrality is maintained by H⁻ ions arriving from the plasma bulk and thus increases the negative ion density and extracted current. Besides, the difference observed with Ref.5 might be related to the 2D restriction. Particles here can be extracted at any z coordinate, leading to an overestimation of extraction currents.

REFERENCES

1. T. Sakurabayashi *et al*, *J. Appl. Phys.*, **95**, 3937 (2004).
2. A. Hatayama *et al*, *Rev. Sci. Instrum*, **77**, 03A530 (2006).
3. A. Hatayama, *Rev. Sci. Instrum.*, **79**, 02B901 (2008).
4. C.K. Birdsall and A.B. Langdon, *Plasma Physics via Computer Simulation,* McGraw-Hill, New York, 1985.
5. P. Svarnas *et al*, *IEEE TPS*, **35**, No. 4, 1156 (2007)
6. D. Matsushita *et al*, Proceedings of NIBS 2008 conference.
7. P.C. Stangeby, "The Role and Properties of the Sheath" in *The Plasma Boundary of Magnetic Fusion Devices*, IOP Publishing, Bristol and Philadelphia, 2000, pp. 61-88.
8. M. Bacal, J. Bruneteau and P. Devynck, *Rev. Sci. Instrum.*, **59**, 2152 (1988)

Plasma Structure in the Extraction Region of a Hybrid Negative Ion Source

F. Taccogna[a], S. Longo[a], M. Capitelli[a] and R. Schneider[b]

[a]*Istituto di Metodologie Inorganiche e dei Plasmi – CNR, via Amendola 122/D, 70126 - Bari, Italy.*
[b]*Max-Planck Institute für Plasmaphysik, Wendelsteinstrasse 1, D-17491- Greifswald, Germany.*

Abstract. Production, destruction and transport of negative ions in the extraction region of a hybrid negative ion source are investigated with a one-dimensional Particle-in-Cell electrostatic code. The influence of the plasma grid bias and of the magnetic filter on the plasma parameter profiles is taken into account. In particular, a transition from classical to complete reverse sheath is observed using a positively biased plasma grid, while the effect of the magnetic filter is relatively small proofing that H⁻ production is dominated by surface neutral conversion.

Keywords: RF-ICP negative ion source, extraction region, PIC-MCC model.
PACS: 52.20.-j, 52.27.Cm, 52.65.

INTRODUCTION

This work represents the simulation of the extraction region of the radio frequency inductively coupled plasma (RF-ICP) hybrid negative ion source developed at IPP Garching [1]. This source is considered hybrid: surface produced negative ions on the plasma grid PG surface (by positive ions and neutrals conversion) are added to the volume produced negative ions in the bulk (by electron attachment to vibrational excited molecular states).

Plasma parameters in the region next to the plasma grid PG (< 3 cm) are considered influential to the amount of effective extractable H⁻ current from an ion source. In this region, gradients due to the plasma sheath (modified by a biased PG and by the presence of a magnetic filter) create forces strongly affecting the free flight and collisional events involving H⁻ ions produced in the bulk and/or on the PG surface [2-6].

Many zero-dimensional codes [7,8] have been developed in the last two decades. These models take into account the most important production/destruction H⁻ volume processes. However, in these models, plasma profiles are averaged and self-consistent transport is completely missing. For this reason, different fluid models [9,10] and pure Monte Carlo codes [11-14] have been developed and applied to the analysis of the magnetic filter effect. In particular, the latter ones have been used to the analysis of H⁻ transport process in "real space" (calculation of the extraction probability) [11,12] and in "velocity space" (calculation of H⁻ velocity distribution and temperature) [13,14].

Only recently, the first Particle-in-Cell models have been developed [15-20]. Nevertheless, they suffer from some incompleteness. Previous PIC models used non-realistic plasma parameters and physical quantities (low plasma density and low magnetic field) or they do not take into account the H⁻ surface production self-consistently.

NUMERICAL MODEL

The region simulated (sketched in Figure 1) is one-dimensional representing the behavior normal to the PG surface and starting 3 cm upstream from the PG surface (indicated as plasma source line PSL). Uniformity in transverse x and y directions is considered.

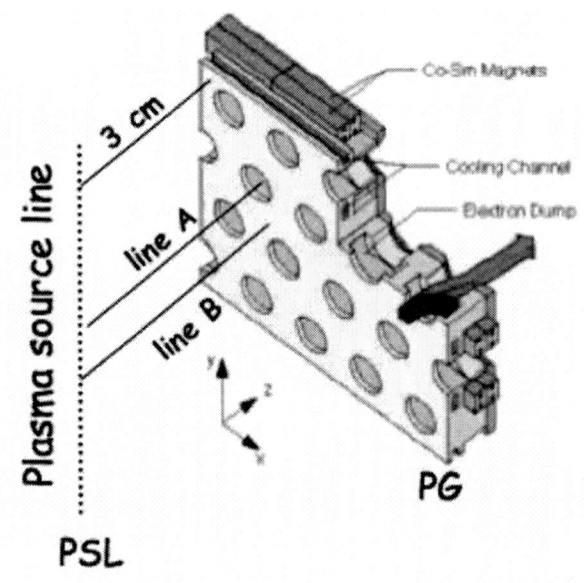

FIGURE 1. Schematic diagram of the simulation model showing the two different axial lines considered.

The choice of the simulation domain size D_z is due to different reasons:
a) quasi-neutrality is present at the entrance (D_z corresponds to 1000 Debye lenght);
b) the magnetic filter is included in this region;
c) due to an electron temperature larger than 3 eV in the bulk of the expansion region, the extraction probability of negative ions goes to zero if they are produced at a location greater than 3 cm from the PG, as demonstrated by previous models [11].

Only charged particles are moved and the code does not follow the dynamic of the neutrals but rather treats them as a fixed background with density, velocities and H_2 vibrational distribution profiles calculated from previous works [21]. The complete motion of electrons and different ion species (H^+, H_2^+ and H^-) is calculated in the self-consistent electric E_z and applied magnetic fields B_x by solving the equations of motion for the coordinate z and all the three velocity components:

$$\frac{dz}{dt} = v_z \tag{1.a}$$

$$m\frac{d\mathbf{v}}{dt} = q(\mathbf{E} + \mathbf{v} \times \mathbf{B}). \tag{1.b}$$

The time step is set equal to $0.3/\omega_p$ ($\omega_p=3.2\times10^{-11}$ s is the plasma frequency) and a uniform spatial grid (cell size smaller than Debye length: $\Delta z<\lambda_D=2.7\times10^{-5}$ m) is used for the electric potential calculation:

$$\frac{\partial^2 \phi(z)}{\partial z^2} = -\frac{\rho(z)}{\varepsilon_0}. \tag{2}$$

The direction of the magnetic filter field **B** is parallel to the PG. Its spatial profile is given by the Gaussian shape:

$$B_x(z) = B_{peak} \exp\left[-\left(\frac{z-z_{peak}}{l_z}\right)^2\right] \tag{3}$$

with $z_{peak}=0.017$ m, $l_z=3\times10^{-2}$ m (the axial position z is also in meter) and $B_{peak}=4$ mT ($\omega_c=1.4\times10^{9}$ s is the corresponding giro-frequency).

The initial condition for the calculation is an empty computational region. An ambipolar particle source is located at the plasma source line PSL and a Neumann boundary condition for the Poisson equation (2) is used there: $\partial\phi/\partial z|_{PSL}=0$. The velocities of injected particles (j=e,H$^+$,H$_2^+$) are chosen from an half-Maxwellian distribution function with bulk density $n_{0,p}=3\times10^{17}$ m^{-3}, source ratio $S_{0,j}$ (1/0.6/0.4) and source temperature $T_{0,j}$ (3/0.7/0.7 eV) taken from experimental measurements [22]. No negative ions are injected from the expansion region. Due to the high electron temperature all the possible negative ions created are supposed to be destroyed by electron detachment. In order to reproduce the plasma flow detected by measurements using a Mach probe [22], an axial drift component is added to the thermal one: $v_{z,drift}$=0.3 Mach. If particles cross the PSL from right to left, a refluxing method is used (particles are re-injected with bulk parameters).

As shown in Figure 1, two different axial lines are simulated: line A ending on one hole (open boundary condition: $\partial\phi/\partial z|_{PG}=0$ in order to reproduce the potential minimum) and line B ending on the PG surface (fixed potential ϕ_{PG}).

Volumetric collisional effects are included via different Monte Carlo techniques (see Ref. [21] for the complete list of collisions). Among the different processes, particular emphasis is pointed out to those involving production and destruction of negative ions:
- electron dissociative attachment of low-energy electrons (1 eV) to the vibrationally excited molecules in their ground electronic state;
- electron detachment in collisions with electrons;
- mutual neutralization in collision with positive ions;

- associative detachment in collisions with atoms.

Finally, important are also non-destructive collision reactions as elastic collisions with H$^+$ and charge exchange collisions with H by changing the transport and reducing the kinetic energy of the extracted H$^-$.

When a charged particle hits the PG wall (only in line B case) different processes are possible. Among them, the most important concerning caesiated surface is the formation of negative ions by positive ions. The numerical implementation of this process is accomplished expressing the H$^-$ yield as a function of positive ion incident energy E_{in} by the following formula [23]:

$$Y(E_{in}) = R_N \eta_0 \left(1 - \frac{E_{th}/R_E}{E_{in}}\right), \quad E_{in} \geq E_{th}/R_E. \quad (5)$$

The yield is completely characterized by two parameters: $R_N\eta_0$=0.3 and E_{th}/R_E=2 eV, where R_N and R_E are the particle and energy reflection coefficients [24], respectively, while η_0 is the height and E_{th} is the threshold energy for the electron transfer probability (approximated as an Heaviside function). The H$^-$ ions are launched from the surface with energy $R_E E_{in}$ and with a cosine distribution angle emission.

In addiction, a fixed flux of negative ions produced by neutral conversion is lunched from PG surface on the basis of previous calculations [21].

RESULTS AND DISCUSSIONS

In all the different cases studied, 10^{11} real charged particles are represented by one pseudo-particle. The maximum number of pseudo-particles used varies between 10^5 and 10^6. For the quasi-steady state to be reached, it takes about some ion transit times (~10 μs) (longer for the case of strong magnetic filter).

Axial Profiles towards PG Surface and Extraction Hole

Figure 2(a) shows the plasma potential for the two different lines considered, extraction hole (broken line A) and PG surface biased at ϕ_{PG}=0 V (full line B) using a magnetic filter strength B_{peak}=4 mT. Typically, the plasma potential reaches a flat value, the bulk value (~15 V) within 1 cm from PG in agreement with experimental results [22,25]. In the case of the extraction hole, the potential remains almost constant. In fact, due to the absence of plasma-wall interaction, the quasi-neutrality is kept all along the entire axial domain.

Figures 2(b) and 2(c) show the axial profiles of the corresponding charged particles density for the two different lines considered. The known phenomenology of plasma-wall transition region is retrieved: the plasma is neutral at the bulk, while a classical positive space charged region (the electron-repelling sheath) develops in contact to the grid. The most important visible effect comparing the two cases is the contribution of surface processes to the negative ion production. The negative ion density near the wall increases by a factor 5 in the case of PG line B, in agreement with recent experimental measurements [26]. For this reason the plasma density near the grid increases also, reducing the insulating properties of the sheath and the sheath size as well. However, it has to be pointed out that a complete model including transverse transport between lines A and B requires a complete 2D model.

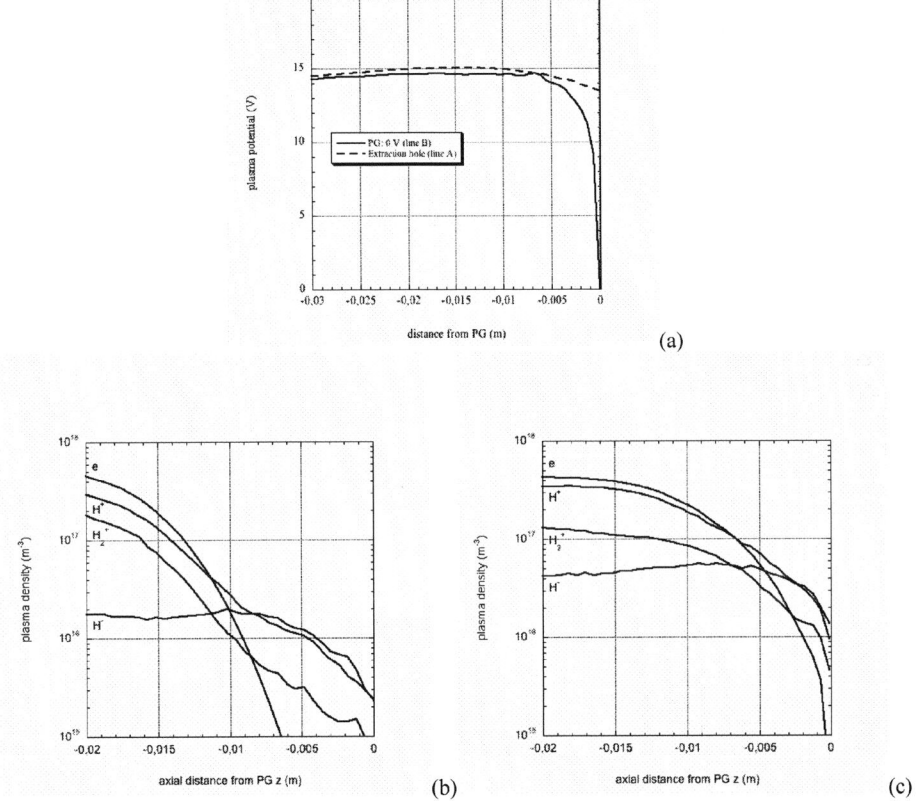

FIGURE 2. (a) Axial profiles of the plasma potential for the two different line considered: extraction hole (dashed line) and PG surface (full line). Axial profiles of the plasma density for the two different line considered: (b) extraction hole (line A) and (c) PG surface (line B). All the figures refer to the case B_{peak}=4 mT and Φ_{PG}=0 V.

Magnetic Filter Effect

In this section, we discuss the role of the magnetic filter, comparing three different simulations performed for maximum filter strengths B_{peak} of 0 mT, 4 mT and 8 mT.

It is well known experimentally [5,6] that the magnetic filter reduces the temperature and density of electrons in the extraction region compared to the value reached in the expansion chamber. In fact, hot electrons have a low diffusion coefficient compared with the cold electrons and therefore the electron density and temperature decrease through the filter region. This is confirmed in Figure 3(a), where the axial behaviour of electron temperature with and without the magnetic filter is plotted for the line B case. Most of the temperature drop takes place in region located beyond the magnetic field peak position ($z=-0.013$ m), while in the absence of magnetic filter, the electron temperature remains almost constant all along the entire extraction region dropping just very close to the PG surface.

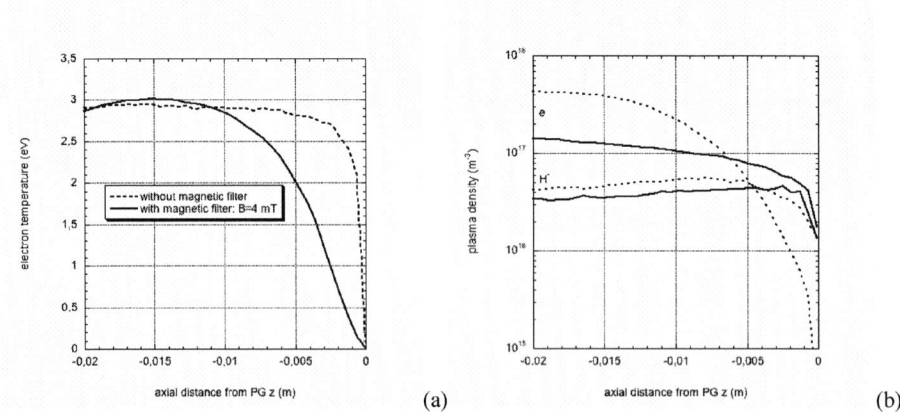

FIGURE 3. Axial profile of (a) electron temperature and (b) electron and H⁻ densities for line B case without magnetic filter (dashed line) and using a magnetic filter with $B_{peak}=4$ mT (full line). In both the cases $\Phi_{PG}=0$ V.

In Figure 3(b) we have compared the electron and negative ion H⁻ density profiles with and without magnetic filter. The electron density decreases with magnetic filter, due to the fact that the fraction of electrons able to transverse the filter falls off with increasing field strength as discussed before. On the other hand, the negative ion density profile remains almost unchanged. In fact, although the reaction rate for the dissociative attachment which produces bulk H⁻ increases with the decrease of the electron temperature, the production rate decreases also with the decrease of electron density. Concerning the surface production by positive ions, it is beneficial to have a larger plasma density near the plasma grid. Therefore, the unchanged H⁻ density profile of Figure 3(b) confirms that H⁻ production is dominated by surface neutral conversion, which is not influenced by the magnetic filter. In general, the magnetic filter does not enhance the negative ion density itself, but increases the negative ion to electron density ratio in the extraction region, helping the suppression of the electron current extracted.

Plasma Grid Bias Effect

The extracted negative ion current and electron current appeared to be sensitive to the bias potential applied to the PG surface [1,22,25,27]. Generally, it is observed that there is an optimal bias voltage for a particular source and this optimal bias potential is close to the plasma potential of that source. When the bias voltage of the PG is increased up to the plasma potential, negative ion current increases, while the electron current and the electron density decreases. But above that optimal bias voltage, both the currents and density decrease with different rate [27]. In fact, with no bias, the plasma potential in the bulk is positive relative to PG and the negative ions as well as the electrons are electrostatically trapped, *i.e.* the plasma is a potential well for them. In order to "climb out" this barrier (~$k_B T_e$ if there were only electrons and positive ions) they should have enough energy, unless PG is biased at a potential close to the plasma potential.

Furthermore, in a hybrid source, the reverse sheath helps not only the bulk-produced negative ions extraction, but also it push back in the direction of the extraction hole, the negative ions produced on the surface and moving otherwise into the bulk, that is in the direction opposite to the extraction grid.

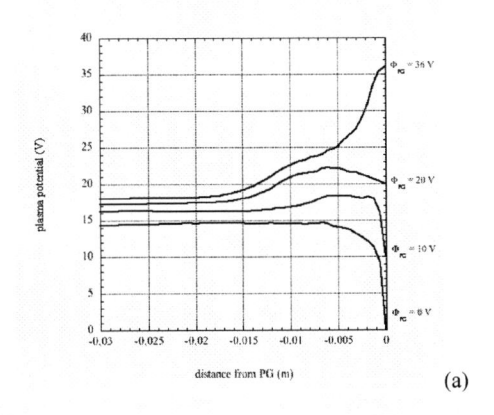

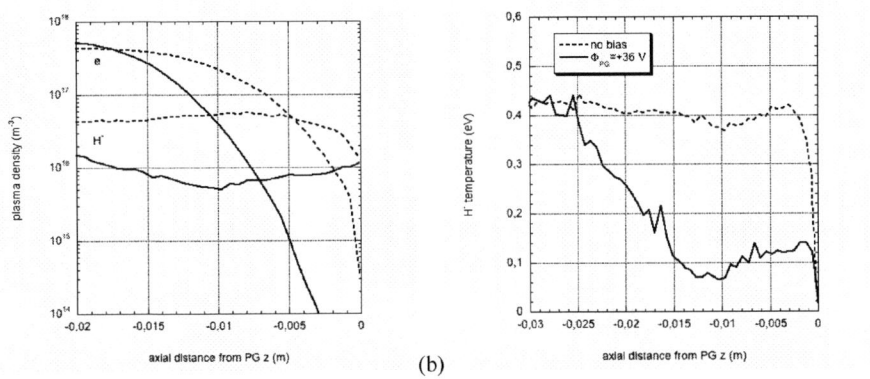

FIGURE 4. (a) Axial profiles of plasma potential for line B using four different PG bias Φ_{PG}: 0 V, 10 V, 20 V and 36 V. (b) Axial profiles of electron and H⁻ ion density for line B using two different PG bias: 0 V (dashed line) and 36 V (full line). (c) Axial profiles of H⁻ temperature for line B using two different PG bias: 0 V (dashed line) and 36 V (full line). In all the cases B_{peak}=4 mT.

In fact, as shown in Figure 4(a), where the electric potential profiles using different bias voltage Φ_{PG} (0V, +10 V, +20 V and +36 V) are shown, a transition from a classical sheath drop to a complete reversed sheath is visible using a positive bias. The potential profile acquires the characteristic curve already observed experimentally

[22,25] and in agreement with the hypothesis of the formation of a double layer in the region close to the PG, where the H⁻ ions are formed due to surface production process. The positive bias potential on the plasma grid can be considered as an extraction field for H⁻ ions. Furthermore, the plasma potential in the bulk is slightly affected by the PG bias due to the effect of the limited size of the simulation domain (3 cm from PG). In fact, in the presence of a magnetic filter, the plasma potential in the bulk remains almost unchanged by the PG bias as shown experimentally [22].

Figure 4(b) shows the electron and H⁻ ion density profiles in two cases: 0 V and +36 V PG biased. For positively biased grid, electron density is reduced drastically and the depletion is extended 1.5 cm inside the plasma from the PG surface. Even the H⁻ ion density is reduced. However, the ratio H⁻/e increases in the case of positively biased PG. Therefore, the experimentally result of negative ion current enhancement seems simply due to the transition of the negative ion type flow, from thermal flow in the case of no bias to directed flow in the case of positively PG bias. This is confirmed comparing the H⁻ axial temperature profile with and without bias reported in Figure 4(c). The effect of the bias is to reduce the H⁻ temperature, creating an H⁻ beam toward the PG surface. Furthermore, the reduction of the temperature of the negative ions is an important parameter of the negative ion source, since it affects the beam optics and determines the lower limit of the negative ion beam emittance.

However, the experimental fact that the negative ion density is not a monotonously increasing with Φ_{PG}, but it reaches a maximum in a limited range of variation of Φ_{PG}, near the plasma potential is an effect that could be visible only with a 2D simulation. In fact, it could be an effect of the transverse (parallel to the PG surface) dynamics. If the PG bias is greater than the transversal (along x and y) potential drop, the negative ions produced on the surface are reflected backward to the surface without having the possibility to go towards the extraction hole along x and y directions. This fact suggests that a negative potential bias between two holes could enhance the negative ion extraction current.

CONCLUSIONS

We have studied the negative ion formation and transport in the extraction region of a radio-frequency inductively coupled discharge using a fully kinetic self-consistent 1D(z)-3V electrostatic Particle-in-Cell simulation.

The influence of PG bias voltage and magnetic filter has been analyzed simulating two different cases: the biased PG surface and the open boundary extraction hole. A complete reversed sheath is recovered for a positively biased which helps the extraction of negative ions, while an irrelevant role is found for the magnetic filter proofing that H⁻ production is dominated by surface neutral conversion.

Future studies should include a second spatial dimension. In fact, in order to have a realistic model of the extracted electron and H⁻ ion current, a complete transport picture is needed in order to consider self-consistently the electron losses along the field lines of the filter magnetic field and the dynamics of negative ions. Furthermore, this will allow understanding the role of the **E**×**B** drift which causes a transverse non-uniformity through out the extraction area impacting on the beam optics and divergence.

ACKNOWLEDGMENTS

This work is supported by MIUR-PRIN 2005 under the contract n. 2005039049_005: "Dinamica dei processi elementari per la chimica e la fisica dei plasmi". R. Schneider acknowledges funding of the work by the Initiative and Networking Fund of the Helmholtz Association.

REFERENCES

1. U. Fantz, P. Franzen, W. Kraus, H. D. Falter, M. Berger, S. Christ-Koch, M. Fröschle, R. Gutser, B. Heinemann, C. Martens, P. McNeely, R. Riedl, E. Speth, D. Wünderlich, *Rev. Sci. Instrum.* **79**, 02A511 (2008).
2. K. N. Leung, K. W. Ehlers, M. Bacal, *Rev. Sci. Instrum.* **54(1)**, 56-61 (1983).
3. M. Bacal, F. Hillion, M. Nachman, *Rev. Sci. Instrum.* **56**, 649-654 (1985).
4. M. Bacal, J. Brunetau, P. Devynck, *Rev. Sci. Instrum.* **59**, 2152-2157 (1988).
5. T. Sakurabayashi, A. Hatayama, M. Bacal, *J. Appl. Phys.* **95(8)**, 3937-3942 (2004).
6. Y. Matsumoto, M. Nishiura, K. Matsuoka, M. Sasao, M. Wada, H. Yamaoka, *Thin Solid Films* **506-507**, 522-526 (2006).
7. W. G. Graham, *Plasma Sources Sci. Technol.* **4**, 281-292 (1995).
8. D. Pagano, C. Gorse, M. Capitelli, *IEEE Trans. Plasma Sci.* **35(5)**, 1247-1259 (2007).
9. A. J. T. Holmes, *Plasma Sources Sci. Technol.* **5**, 453-473 (1996).
10. Ts. Paunska, H. Schlüter, A. Shivarova, Kh. Tarnev, *Phys. Plasmas* **13**, 023504 (2006).
11. D. Riz, J. Pamela, *Rev. Sci. Instrum.* **69(2)**, 914-919 (1998).
12. M. Bandyopadhyay, R. Wilhelm, *Rev. Sci. Instrum.* **75(5)**, 1720-1722 (2004).
13. K. Makino, T. Sakurabayashi, A. Hatayama, K. Miyamoto, M. Ogasawara, *Rev. Sci. Instrum.* **73(2)**, 1051-1053 (2002).
14. P. Diomede, S. Longo, M. Capitelli, to appear on *IEEE Trans. Plasma Sci.*, (2008).
15. H. Naitou, O. Fukumasa, K. Sakachou, *Rev. Sci. Instrum.* **67(3)**, 1149-1151 (1996).
16. A. Hatayama, T. Matsumiya, T. Sakurabayashi, M. Bacal, *Rev. Sci. Instrum.* **77**, 03A530 (2006).
17. P. Diomede, S. Longo, M. Capitelli, *Rev. Sci. Instrum.* **77**, 03A503 (2006).
18. M. Turek, J. Sielanko, P. Franzen, E. Speth, in *Proc. International conference Plasma 2005*; AIP Conf. Proc. no. 812, 153-156 (2006).
19. D. Wunderlich, R. Gutser, U. Fantz, NNBI team, in *Proc. 11th International Symposium on Production and Neutralization of Negative Ions and Beams*, edited by M. P. Stockli, AIP Conf. Proc. no. 925, New York: American Institute of Physics, 46-57 (2007).
20. M. Capitelli, M. Cacciatore, R. Celiberto, O. De Pascale, P. Diomede, F. Esposito, A. Gicquel, C. Gorse, K. Hassouni, A. Laricchiuta, S. Longo, D. Pagano, M. Rutigliano, *Nucl. Fus.* **46(6)**, S260-S274, (2006).
21. F. Taccogna, R. Schneider, S. Longo, M. Capitelli, to appear on *Phys. Plasmas* (2008).
22. M. Bandyopadhyay, A. Tanga, H. D. Falter, P. Franzen, B. Heinemann, D. Holtum, W. Kraus, K. Lackner, P. McNeely, R. Riedl, E. Speth, R. Wilhelm, *J. Appl. Phys.* **96(8)**, 4107-4113 (2004).
23. M. Seidl, H. L. Cui, J. D. Isenberg, H. J. Know, B. S. Lee, S. T. Melnychuk, *J. Appl. Phys.* **79(6)**, 2896-2901 (1996).
24. W. Eckstein, J. P. Biersack, *Appl. Phys.* **A38**, 123-129 (1985).
25. U. Fantz, P. Franzen, W. Kraus, M. Berger, S. Christ-Koch, M. Fröschle, R. Gutser, B. Heinemann, C. Martens, P. McNeely, R. Riedl, E. Speth, D. Wünderlich, *Plasma Phys. Control. Fus.* **49(12)**, B563-580 (2007).
26. U. Fantz, H. Falter, P. Franzen, D. Wünderlich, M. Berger, A. Lorenz, W. Kraus, P. McNeely, R. Riedl, E. Speth, *Nucl. Fus.* **46(6)**, S297-S306 (2006).
27. P. Franzen, H. D. Falter, U. Fantz, W. Kraus, M. Berger, S. Christ-Koch, M. Fröschle, R. Gutser, B. Heinemann, S. Hilbert, S. Leyer, C. Martens, P. McNeely, R. Riedl, E. Speth, D. Wünderlich, *Nucl. Fus.* **47(4)**, 264-270 (2007).

Production of H⁻ Ions by Surface Mechanisms in Cs-free Multi-dipolar Microwave Plasma

Stéphane Béchu[1], Didier Lemoine[2], Marthe Bacal[3], Alexandre Bès[1] and Jacques Pelletier[1]

[1]*Laboratoire de Physique Subatomique et de Cosmologie (UJF, CNRS IN2P3 et ST2I, INPG), 53 rue des Martyrs, 38026 Grenoble Cedex, France*
[2]*Laboratoire Collisions Agrégats Réactivité, Unité Mixte de Recherche 5589 CNRS-Université Paul Sabatier Toulouse 3, 31062 Toulouse Cedex 9, France*
[3]*Laboratoire de Physique et Technologie des Plasmas, Ecole Polytechnique, Route de Saclay, 91128 Palaiseau Cedex, France*

Abstract. This study demonstrates the feasibility of the recently proposed idea of enhancing recombinations of the hydrogen atoms from the plasma on a surface in order to produce highly vibrationally excited molecules that can be attached and dissociated by the cold electrons of the plasma, hence creating negative ions that could be used as a Cs-free negative ion source. The negative ion density was obtained for a) two distinct materials, i.e. tantalum and stainless steel, and for b) two different degrees of molecular hydrogen dissociation, the higher degree of dissociation resulting from the cooling of the walls of the source. The relative negative ion density n-/n_e was measured by laser photodetachment and the electron density was obtained from Langmuir probe measurements. The pre-sheath was studied by emissive and conventional Langmuir probes to evaluate the potential drop near the surface.
Laser photodetachment measurements performed in the vicinity of the investigated material consisting of a disk inserted in the source, evidence the production of negative ions by surface mechanisms. With cooled walls a tantalum disk at floating potential increases the negative ion density by 60-100 % (depending on the probe distance to the disk) compared to a stainless steel disk under the same conditions. The observations strongly suggest surface-assisted recombination processes involving H atoms such as the Langmuir-Hinshelwood (diffusion), Eley-Rideal (direct impact) and hot atom (indirect collisional) mechanisms, followed by dissociative attachment.

Keywords: Multidipolar plasma, Hydrogen plasma, Negative ion source, Surface mechanisms, Laser photodetachment.
PACS: 29.25.Ni, 32.80.Gc, 52.27.Cm, 52.70.Kz, , 52.50.Sw, ,52.40.Kh

INTRODUCTION

When considering the state-of-the-art on H⁻/D⁻ ion source operation, ions are conventionally produced either inside the plasma volume by dissociative electron attachment (DA) [1,2], and/or by plasma-surface interactions.

The surface production of negative ions has, up to now, been induced by surface ionization on a low work function material. For this purpose, caesium has been used in many cases (LBNL, JAEA, and KEK facilities). Despite a real enhancement in the negative ion production [3] (by a factor 2.5 in the JAEA source), the use of Cs remains problematic and could lead to many drawbacks in the operation of ITER since Cs vapour escapes from the ion source between pulses and diffuses into the accelerator. Not only does this mean that the Cs has to be replenished continuously, but this can also cause breakdowns and lead to the production of negative ions inside the accelerator giving rise to beam halos (fractions of the beam with very high divergence). The use of Cs will require regular cleaning and maintenance of the ion source on the ITER injector, which will be difficult as the source and accelerator will quickly be highly activated and all such operations will have to be done via remote handling.

An innovative strategy [3-6] to enhance the production of H⁻/D⁻ ions, based on recent progress in the theory [7,8], lies in the possibility of recycling hydrogen atoms into highly vibrationally-excited molecular hydrogen by directing the atoms to a surface, let them stick and react in a recombinative desorption process that can be of three generic types: the Langmuir-Hinshelwood (thermal diffusion of two adsorbates), and more importantly in terms of energy deposition into the formed molecules, the Eley-Rideal (direct impact of an atom from the plasma with an adsorbate) and the hot atom (indirect collision consisting of several encounters with the surface and adsorbates) mechanisms (see e.g. [7] and references therein). The second step is the same as for volume production, that is the produced H_2 (X, v") molecules can be converted into H⁻ ions by dissociative attachment (DA) with slow electrons ($T_e < 1$ eV). Recent experiments suggest that this third production mechanism (recombinative desorption followed by DA) could be effective with specific surfaces (fresh tantalum on wall surface increases the negative ion density [9] by more than 60 %). The purpose of the present study is to find how to implement this third production mechanism by thoroughly investigating the interaction between hydrogen plasma and chosen materials.

In the present article, after briefly describing the experimental setup, we present the experimental results obtained using spectroscopy, emissive and conventional Langmuir probes as well as laser photodetachment. These measurements respectively yield: the degree of dissociation of molecular hydrogen, the plasma potentials, the electron density and temperature in the sample vicinity and the spatial profiles of n⁻. A conclusion is presented in the last section.

EXPERIMENTAL SET-UP AND DIAGNOSTICS

This study was performed in the "Camembert III" ion source extensively described in several papers (see [10] and references therein). Briefly, this experimental set-up uses 7 dipolar microwave (2.45 GHz) plasma sources [11] placed at the top of a stainless-steel cylinder of 0.44 m in diameter and 0.45 m in height. There are sixteen stainless steel tubes containing columns of samarium-cobalt magnets with the north and south poles alternatively facing the plasma. The tubes are welded to the source walls and are therefore at the same temperature as the walls. The wall temperature of the source could be modified by the use of external insulating foams and of a thermostat that regulates the coolant temperature of the sixteen columns of magnets. The cooling fluid temperature can thus be varied between -5 °C and 35 °C. The columns temperature range extends from a few degrees above 0 °C up to 60-70 °C.

Optical emission spectroscopy measurements are performed using a SPECTRAMAX 270 spectrometer from Jobin-Yvon with a 600 grooves per millimeter gratings blazed at 500 nm and mounted with a photodiode array. The spectral resolution is 0.25 nm for 10 μm entrance slit. The emission light was collected and transmitted through a fused optical fiber to the spectrometer.

Emissive Probe consists of a double bored alumina tube of 4 mm in diameter. At the tip of the alumina tube the probe filament made of tungsten wire (0.002 mm in diameter) is approximately 1 mm long and is spot welded to copper support wires placed in the ceramic bores. Emissive probes are unperturbed by magnetic field or plasma drift and permit accurate plasma potential measurement by the inflection-point method [12]. However, when used in the emitting mode, the heated tungsten filament emits not only electrons but also tungsten atoms. The surface of the studied sample is rapidly covered by tungsten atoms even if low current is applied to the probe. Hence, after several comparisons of potential profiles obtained with emissive and conventional Langmuir probes, such as displayed in Fig. 1, only the Langmuir probe (not emitting) was used. When the magnetic field was weak and the plasma drift low, the two types of probes yielded similar results (within 0.1 V).

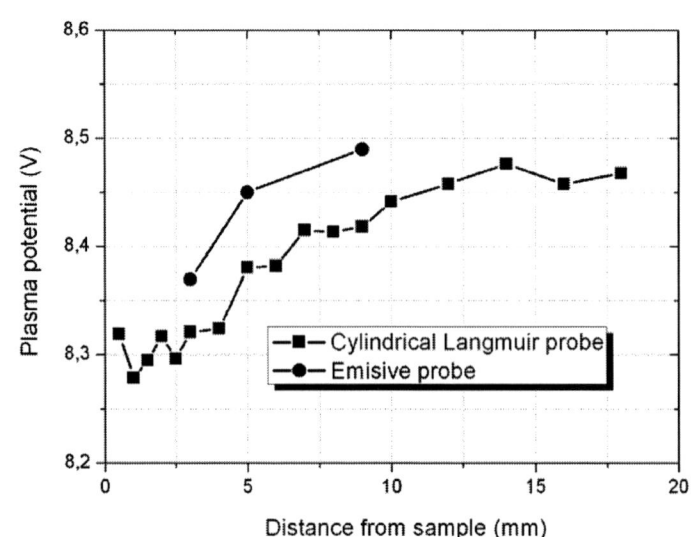

FIGURE 1. Plasma potential variations in the vicinity of a 7 cm-diameter disk sample at floating potential, in a H_2 discharge. Plasma conditions were 1.5 mTorr and 1200 W of microwave power.

The Langmuir probe is a cylindrical wire (0.5 mm in diameter) with an L bend. This particular geometry allows measurements near the sample without any modification of the surface itself (collecting mode only) and induces only a weak perturbation of the measured potential and density. The same Langmuir probe is used to collect photodetached electrons from negative ions during laser photodetachment measurements.

Photodetachment zone was located in the middle of the source, 85 mm below the dipolar sources. Nd-Yag laser of 170 mJ was used with adapting polarizer to restrict the power delivered in the plasma to 0.093 J/cm². According to the theory the saturation of the fraction of photodetached negative ions is reached for laser pulse energy above 0.05 J/cm² [13]. Hence, the selected power is large enough to ensure photodetachment of all the negative ions in the illuminated area. The whole laser beam, 9 mm in diameter, was used to induce the photodetachment of H⁻ ions around the Langmuir probe. The use of a large beam diameter enhances the plateau duration in the photodetachment signal. Fig. 2 presents photodetachment signals obtained at 10 mm above a stainless steel floating sample. The location of the overshot is shifted in time when large laser beam diameter is employed but the intensity of the collected current in the plateau region remains unchanged [13].

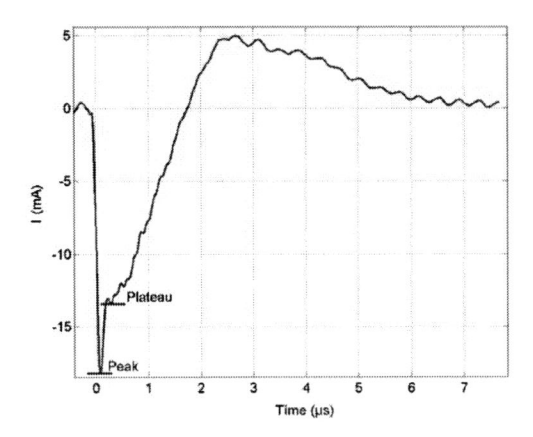

FIGURE 2. Probe current variation due to photodetachment with a voltage of + 40 V applied to the probe with respect to the ground. Experimental conditions were: 1 mTorr of pure H2, 1000 W of microwave power (2.45 GHz), 10 mm above the sample surface. The peak and plateau intensities are emphasized by horizontal lines.

The associated probe (L-bend shape) was biased at + 40 V to collect the detached electrons [13,14]. At this potential (with respect to the ground) the difference decreases between the relative densities obtained when the peak of the photodetachment signal was considered instead of the plateau.

RESULTS

Gas Temperature Determination

The method [15-18] used for the determination of the gas temperature is based on the H_2 emission bands. The accuracy of this diagnostic in low-pressure plasmas of pure hydrogen and its mixtures with rare gases (such as Ar) is established by comparison with values of translational temperature derived from Doppler broadening [16,19-20]. At low-pressure conditions, the α–Fulcher Q branch emission band (in the 623 - 630 nm range) is the most suitable system for the gas temperature estimation [21]. In a simple excitation-deactivation model the intensities of (2-2) Q lines of Fulcher-system of the H_2 molecule are connected with the ground state rotational temperature. The main requested assumptions are [21]:

i) the population distribution of the ground state obeys Boltzmann's law and the rotational temperature of the ground state is equivalent to the gas translational temperature.

ii) the excited state is mainly populated by electron impact excitation from the ground state. The rate coefficient is obtained from the adiabatic approximation and the transitions with a change of rotational angular momentum $|\Delta N| \geq 2$ can be neglected.

The method [21] for the calculation of the rotational temperature provides a quick estimation of the ground state rotational temperature $T^{rot}(H_2)$ obtained from the slope of the rotational population distribution as a function of the energy term. The gas temperature varies from 410 to

560 K ± 50 K when the coolant fluid temperature is varied from -5 °C to 20 °C. These results are presented in Fig. 3.

Degree of Dissociation of Hydrogen

The degree of dissociation for hydrogen plasma is [22]:

$$D = \frac{[H]}{2[H_2]+[H]} = \frac{[H]/[H_2]}{([H]/[H_2])+2} \quad (1)$$

In our experimental conditions, the electron temperature and the [H]/[H$_2$] ratio are outside the range of applicability of the method described by Lavrov et al [22,23]. This method uses the intensity ratio of H$_\alpha$ and H$_\beta$, and (2-2) Q1 lines for the determination of the degree of dissociation of H$_2$. The classical actinometry method described in [24,25] was used instead. For that purpose a small amount of argon (3 % in partial pressure) gas was introduced in the reactor. The results presented in Fig. 3 indicate that the lower the cooling temperature, the higher the degree of dissociation. The increase in the degree of dissociation of H$_2$ is roughly 15 % for the temperature range.

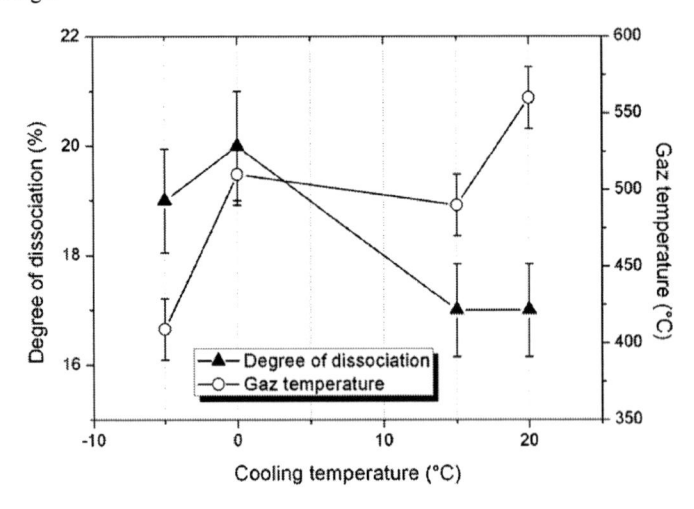

FIGURE 3. Measured gas temperature and degree of dissociation in percentage as a function of the cooling fluid temperature

Plasma Potential

We carefully analyzed the choice of the disk diameter to satisfy two requirements: a) the disk should be large enough so that the laser photodetachment diagnostic of negative ion density could detect any change in n⁻ due to the surface neighborhood, and b) the disk surface should be small enough to leave the plasma unperturbed. The first point will be discussed in section "Negative ion density". Perturbation of the plasma could be of two types: geometric and electrostatic. A disk of 7 cm in diameter was found suitable in these respects. In order to make sure that the plasma potential is not perturbed plasma and floating potentials were measured in front of the floating sample. Figure 4 displays the resulting spatial profiles. Near the floating sample at $V_f = + 6.06$ V the plasma potential abruptly increases over roughly 15 mm and then smoothly reaches + 9.6 V at 22 cm away from the surface, i.e. in the plasma center. The floating potential exhibits a similar profile. Thus, the difference between the plasma and floating potentials remains constant whatever the distance from the sample, which indicates that the electron losses are not significantly increased in the presence of the disk that is very small (0.008 m²) in comparison with the surface of the walls (0.7 m²).

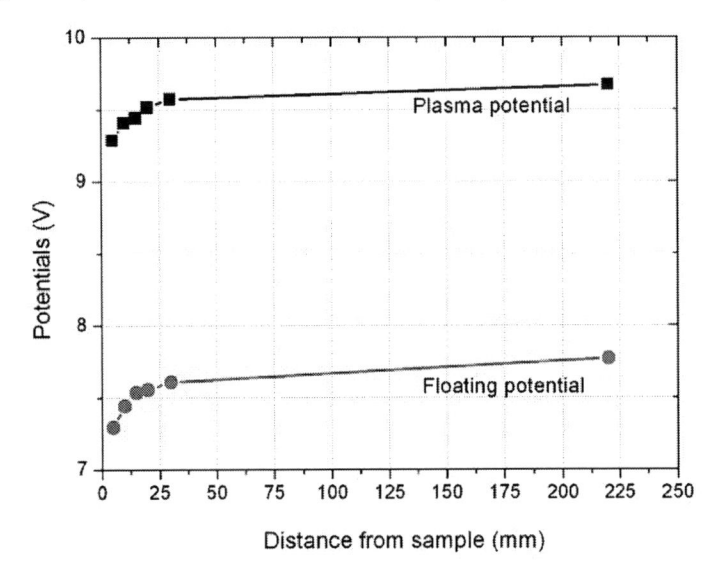

FIGURE 4. Spatial profiles of the plasma potential determined by Langmuir probe in the vicinity of a 7 cm-diameter disk sample at floating potential. Plasma conditions were 1.0 mTorr and 1000 W of microwave power.

Electron Temperature and Density

Since in our experimental conditions, the value of the n^-/n_e ratio remains below 0.15 (see photodetachment results in the next section) and the value of the T_e/T_i ratio is close to 10 [10], the sheath edge potential is not altered [26]. While this edge potential is not significantly modified and the Bohm criterion remains valid I(V) measurements were performed as in a conventional electropositive plasma in order to derive the electron density and temperature. The I(V) curves analysis is displayed in Fig. 5.

The pre-sheath built up in the surface neighbourhood is still a plasma (if we consider the definition of quasineutrality: $n_e + n^- = n_+$) but shows a lower density than the undisturbed bulk plasma. The decrease of the density is a direct consequence of diffusive mechanisms and electron losses on the sample surface. The sharp electron density decrease near the sample is visible in Fig. 5. At the centre of the plasma discharge the electron density is about 4×10^{16} m^{-3}

but only 2.5×10^{16} m^{-3} at 5 mm from the floating sample. The electron temperature presents a minimum at 15 mm from the sample and remains constant beyond 20 mm.

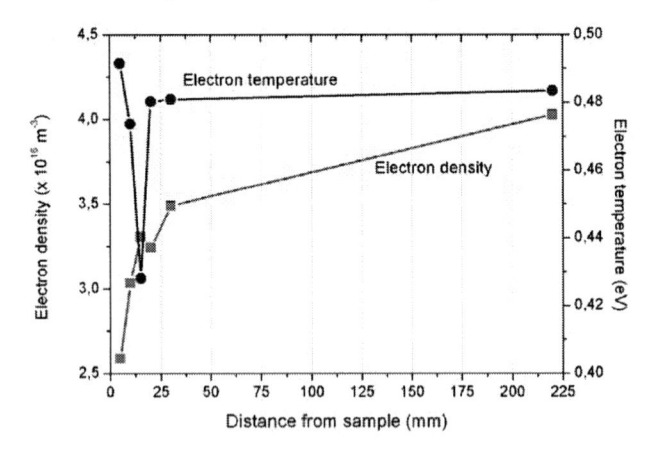

FIGURE 5. Spatial profiles of the a) electron temperature (circle marks) and b) electron density (square marks) for a 7 cm-diameter disk sample at floating potential. Plasma conditions were 1.0 mTorr and 1000 W of microwave power.

Negative Ion Density

In our experimental set-up tantalum surfaces smaller than 7 cm in diameter did not produce enough negative ions to observe a discrepancy of the measured photodetachment signal in cold or hot walls conditions. Hence in the following, 7 cm-diameter disks (300 μm thick) are considered for all experiments and the plateau intensity is considered to estimate the negative ion density. Figure 6 presents the n$^-$/n$_e$ spatial profiles in two different working conditions (cold and hot walls) and for a disk of either tantalum or stainless steel. Depending on the probe distance to the sample the use of tantalum versus stainless steel induces a significant increase of the n$^-$/n$_e$ ratio by 59-98 % with cold walls and by 92-125 % with hot walls.

The same electron density profile was obtained whatever the walls temperature (hot or cold) and the disk material (stainless steel or tantalum) were. It was found no significant difference. Fig. 7 presents the negative ion density profiles obtained by multiplying the n-/n$_e$ ratio measured by laser photodetachment and the electron density profile n$_e$ measured by Langmuir probe. The electron density profile is represented by a continuous bold line.

Table 1 reports the largest negative ion densities (n$^-$) obtained when modifying the walls temperature and the floating-disk material. With cold versus hot walls the n$^-$ densities at 30 mm from the sample are increased by roughly 50 % and 25 % for stainless steel and tantalum, respectively. In contrast, in the absence of both the sample and the sample holder, no temperature effect is observed on the n$^-$ densities. A notable enhancement of roughly 130 % is obtained with tantalum and cold walls versus stainless steel and hot walls (4.9×10^{14} m^{-3} instead of 2.1×10^{14} m^{-3}).

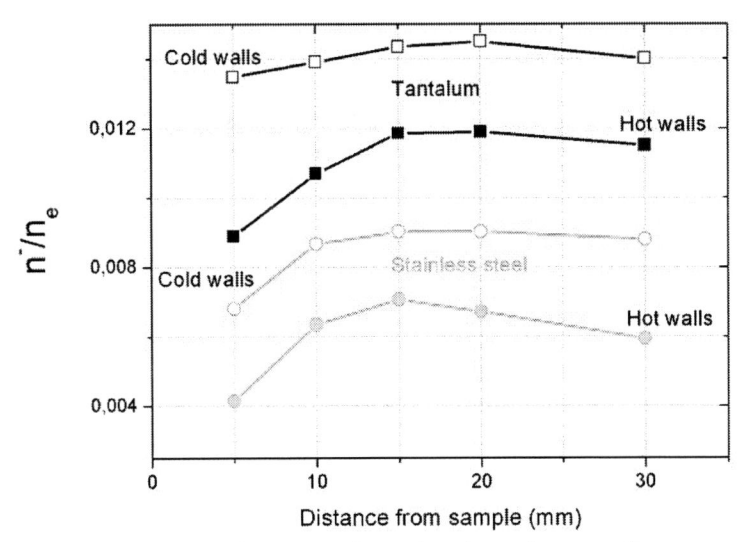

FIGURE 6. Variation of the n-/ne ratio versus the distance from the sample at two walls temperatures, cold (open marks) and hot (filled marks), for tantalum (square marks) and stainless steel (circle marks) floating disks. Plasma conditions were 1 mTorr of pure hydrogen and 1000 W of microwave power.

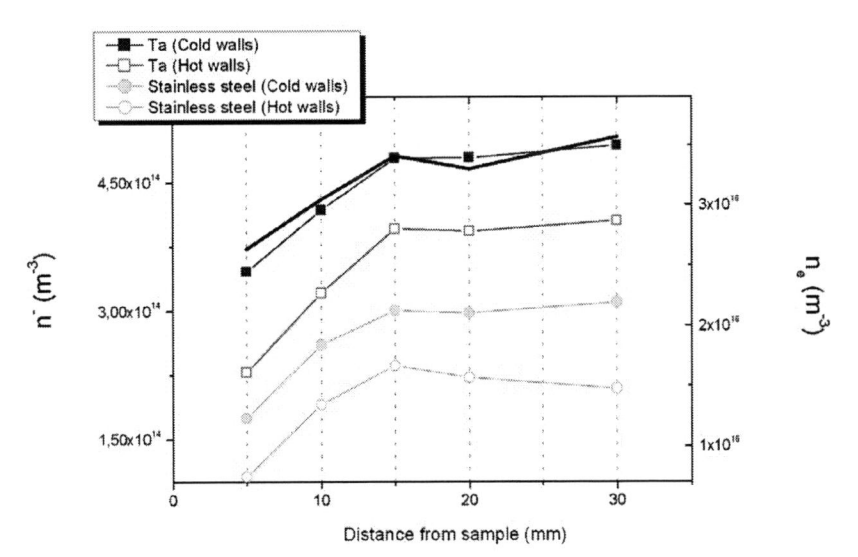

FIGURE 7. Variation versus the distance from the sample of the electron density ne (continuous bold line referring to the right scale) and of the negative ion densities n- (referring to the left scale) at two walls temperatures for tantalum and stainless steel disks at floating condition (see legend insert). Plasma conditions were 1 mTorr of pure hydrogen and 1000 W of microwave power.

TABLE 1. Negative ion density (n⁻) measured at 30 mm from the sample for tantalum and stainless steel disks of 7 cm in diameter. Empty ion source results were obtained when both the sample and the sample holder were removed

Material \ Working conditions	Cold walls	Hot walls
Stainless Steel	3.1×10^{14}	2.1×10^{14}
Tantalum	4.9×10^{14}	4×10^{14}
Empty ion source	3.3×10^{14}	3.3×10^{14}

The density profiles all have the same shape in Fig. 7: a continuous increase between 5 and 15 mm and a plateau-like region beyond. Changing the walls temperature or exchanging tantalum and stainless steel solely induces a vertical shift of these profiles. The similarity between the n⁻ and n_e suggests that the density of vibrationally-excited H_2 molecules is homogeneous in space since DA process yielding n⁻ is found proportional to n_e. Thus, the observed n⁻ variation in the vicinity of the disc sample is determined by the n_e variation, only. Yet, it is noteworthy that the molecule density strongly depends on sample materials and on experimental conditions.

Cooling the vessel walls increases the atomic hydrogen density in the plasma by up to 15 % in terms of degree of dissociation (see section "Degree of dissociation of hydrogen") and enhanced the n⁻ density by roughly 25 % (tantalum surface at floating potential). Since operating conditions were equivalent for all measurements in terms of pressure, gas flux and power intake, the n⁻ increase for tantalum versus stainless steel evidences the role of surface mechanisms (Eley-Rideal, hot atoms and to a lesser extent Langmuir-Hinshelwood) for the production of vibrationally-excited H_2 molecules.

CONCLUSION

An innovative strategy for enhancing negative ion production in plasma sources focuses on the conversion of H atoms sticking and reacting at a surface, yielding desorbing vibrationally-excited H_2 molecules that are dissociatively attached with the slow electrons of the plasma, thereby producing H⁻. Two distinct methods were used to demonstrate the relevance of this two-step mechanism: a) we enhanced the degree of dissociation of molecular hydrogen in the plasma and observed the effect on the negative ion density near the surface of a disk sample; b) we compared tantalum and stainless steel as the disk material in the enhancement of negative ion density near the sample.

a) The gas temperature and the degree of dissociation of H_2 were measured as a function of the temperature of the coolant fluid of the source walls, using optical emission spectroscopy. It was found that the higher the wall temperature, the higher the gas temperature and the lower the degree of dissociation (Fig.3). It was thus possible to increase the degree of dissociation by 15% by cooling the chamber walls.

b) In cold walls operating conditions, for which the degree of dissociation of H_2 is higher, the negative ion density increases by a factor 1.6 for a Ta sample disk instead of a same-size stainless steel disk (Fig. 7). The atom flux increase affects the negative ion density observed near both the tantalum and the stainless steel surfaces.

Since the work function (Φ) of tantalum is 4.1 eV (to be compared to 1.8 eV for Cs) surface ionisation seems very unlikely whereas other surface mechanisms (Eley-Rideal, hot atoms and to a lesser extent Langmuir-Hinshelwood) recycling atomic hydrogen are relevant. These neutral species (H atoms) easily cross the pre-sheath and sheath to recombine on the surface, which then produces negative ions with cold electrons, the so-called dissociative attachment (DA) mechanism, in the surrounding plasma.

This study shows that the H⁻ production can be greatly enhanced near a well chosen sample surface via the two-step mechanism: 1) vibrationally-excited H_2 molecules forming on and desorbing from the surface 2) followed by DA in the surface vicinity. It is different from surface production (direct surface ionization) or from volume production (molecule excitation by the hot electrons near the dipolar microwave plasma sources followed by DA). Therefore, the new two-step mechanism may be referred to as surface-vicinity production.

ACKNOWLEDGMENTS

This work, supported by the European Communities under the contract of Association between EURATOM, CEA and the French Research Federation for fusion studies, was carried out within the framework of the European Fusion Development Agreement. The views and opinions expressed herein do not necessarily reflect those of the European Commission.

REFERENCES

[1] J. R. Hiskes, J. Phys. **40**, C7-179 (1979)
[2] M. Bacal, Nucl. Fusion. **46**, 250-259 (2006)
[3] M. Bacal, A. Hatayama, J. Peters, IEEE Trans. Plasma Sci. **33**, 1845-1871 (2005)
[4] M. Bacal, Rev. Sci. Instrum. **79**, 02A516 (2008)
[5] M. Bacal, Nucl. Fusion. **46**, 250-259 (2006)
[6] S. Béchu et al. Rev. Sci. Instrum. **79**, 02A505 (2008)
[7] B. Jackson, D. Lemoine J. Chem. Phys. **114**, 474-482 (2001)
[8] X. Sha, B. Jackson, D. Lemoine, J. Chem. Phys. **116**, 7158-7169 (2002)
[9] M. Bacal et al., Rev. Sci. Instrum. **75**, 1699-1703 (2004)
[10] A.A. Ivanov Jr. et al., Rev. Sci. Instrum. **75**, 1754 (2004)
[11] A. Lacoste, T. Lagarde, S. Béchu, Y. Arnal, and J. Pelletier, Plasma Sources Sci. Technol. **11**, 407 (2002)
[12] J. R. Smith, N. Hershkowitz, P. Coakley, Rev. Sci. Instrum. **50**, 210-218 (1979)
[13] M. Bacal, Rev. Sci. Instrum. **71**, 3981- 4006 (2000)
[14] P. J. Eenshuistra et al., J. Appl. Phys. **67**, No. 1 85-96 (1990)
[15] S.A. Astashkevich et al., J. Quantitative Spectroscopy and Radiative Transfer. **56**, 725-751 (1996)
[16] B. P. Lavrov, D. K. Otorbaev, Sov Tech Phys Lett. **4**, 574-575 (1978)
[17] B. P. Lavrov, Opt. Spectrosc. **48**, 375-380 (1980)
[18] B. P. Lavrov, M. V. Tyutchev, Acta Phys Hungarica. **55**, 411-426 (1984)
[19] B.P. Lavrov et al., FLTP Diag. II, Bad Honnef, Conference Proceedings. 169 (1997)
[20] N. N. Sobolev, NewYork: Nova Science Publishers. 148-55 (1989)
[21] M. Osiac, B. P. Lavrov, J.Röpcke, JQSRT. **74**, 471-491 (2002)
[22] B. P Lavrov, A. V. Pipa, J. Röpcke, Plasma Sources Sci. Technol. **15**, 135-146 (2006)
[23] B. P Lavrov, A. V. Pipa, J. Röpcke, Plasma Sources Sci. Technol. **15**, 147-155 (2006)
[24] A. Gicquel et al., J. Appl. Phys. **83**, 7504-7521 (1998)
[25] M. Abdel-Rahman et al. Plasma Sources Sci. Technol. **15**, 620-626 (2006)
[26] N. St. J. Braithwaite, J. E. Allen, J. Phys. D: Appl. Phys. **21**, 1733-1737 (1988)

Caesium Free Negative Ion Sources for Neutral Beam Injectors: a Study of Negative Ion Production on Graphite Surface in Hydrogen and Deuterium Plasma

L. Schiesko, M. Carrere, G. Cartry and J.-M. Layet

Plasma Surface, PIIM, UMR 6633, Université de Provence-CNRS, Centre Scientifique de Saint Jérôme, case 241, 13397 Marseille Cedex 20, France

Abstract.
Negative ion generation on HOPG graphite surface has been studied in hydrogen and deuterium plasma. We measure Ion Distribution Function (IDF) of negative ions coming from graphite surface bombarded by positive ions in H_2/D_2 plasmas. We showed that negative ions flux was proportional to positive ion flux and was strongly dependant on impinging energy. IDF study shows two generation mechanisms are involved: sputtering of adsorbed H/D as negative ions and, in a less important way, double electron capture. We compare H_2/D_2 plasmas, and point out isotopic effect between H^- and D^- production.

Keywords: Plasma, negative ions, hydrogen, deuterium, HOPG
PACS: 52.40.-w, 52.40.Hf

INTRODUCTION

Neutral Beam Injector (NBI) is the most promising device for plasma heating. NBI deliver high flux, high energy deuterium neutral atom beam from a 1 MeV, 17 A D^- negative ion beam [1]. Basic design of ITER NBI was made in 1995 [2]. Principle of NBI based on negative ions is the following [3], [4]: negative ions are generated in a low pressure plasma source (0.3 Pa) [5], extracted and accelerated, and finally neutralized. In plasma, negative ions are generated either on surface, either in the bulk [6]. To increase surface generation, caesium is usually deposited on surfaces. However an important effort is presently being undertaken to develop caesium free negative ions sources, since caesium not only contaminates discharges but can also be responsible of beam halos and breakdowns in NBI. We focused our study on HOPG (Highly Oriented Pyrolytic Graphite) surface production, since it is one of the most promising material to be used as surface material. Indeed H atom surface recombination on graphite is expected to efficiently produce vibrationally excited hydrogen molecules allowing for an increase in the production of negative ions in the plasma volume [7],[8],[9]. This paper summarizes our results concerning hydrogen and deuterium negative ions generation on HOPG surface.

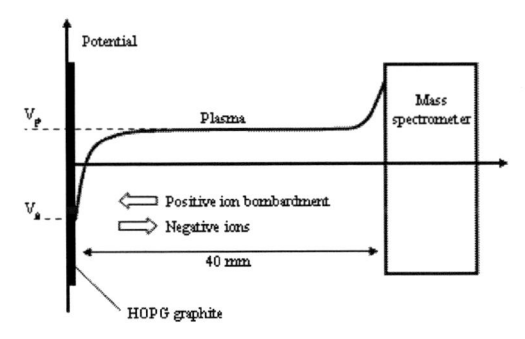

FIGURE 1. Sketch of sample and mass spectrometer arrangement

EXPERIMENTAL SETUP

Experimental setup was presented elsewhere [10]. It is a helicon reactor that consists of an upper source where plasma is generated by a Hüttinger PFG 1600 RF generator and a lower diffusion chamber. A base pressure lower than $10^{-5} Pa$ (Penning gauge limit) is achieved by means of a 150 l/s turbomolecular pump. Discharge pressure is set by a Baratron gauge during operation. Electron temperature and density is measured by Straatum (Smart Probe) Langmuir probe. Measurements were performed in low pressure (0.15 to 1 Pa) hydrogen and deuterium discharges and an injected power of 30 to 1 kW. A mass spectrometer (Hiden EQP 300), comprising a 45 electrostatic energy analyser is facing a 1 cm^2 HOPG sample at a distance of 40 mm. HOPG sample was biased down to -80 V with respect to the ground in this study.

RESULTS AND DISCUSSION

Figure 1 sketches sample and spectrometer arrangement. On figure 1, V_s and V_p are sample and plasma potential respectively. Sample is biased below plasma potential in order to attract positive ions and repel H^- and D^- formed on HOPG by cation bombardment. Negative ions are collected by mass spectrometer and analysed according to their energy. Figure 2 represents the negative ion flux measured by mass spectrometer versus the positive ion flux measured by Langmuir probe for 0.2 Pa H_2, 30, 100 and 300W discharges and different sample biases. It shows an almost linear dependance of negative ion flux with positive ion flux, demonstarting that cations bombardment causes negative ion surface generation. We also showed negative ions generation was strongly dependant of cations energy [11]. Indeed, we observed an increase of H^- relative intensity with positive ion energy, above a 20 eV threshold, followed by a maximum at around 60 eV and a strong decrease for higher energies. Figure 3 obtained for a deuterium plasma of 0.4 Pa and 100W of injected power shows the same behavior: a strong increase of D^- signal with positive ions energy, followed by a maximum at 80 eV and a decrease for higher energies. Figure 4 is an example of Ion Distribution Function (IDF) obtained for a 0.2 Pa H_2 plasma and an injected power of 100 W. By carefully

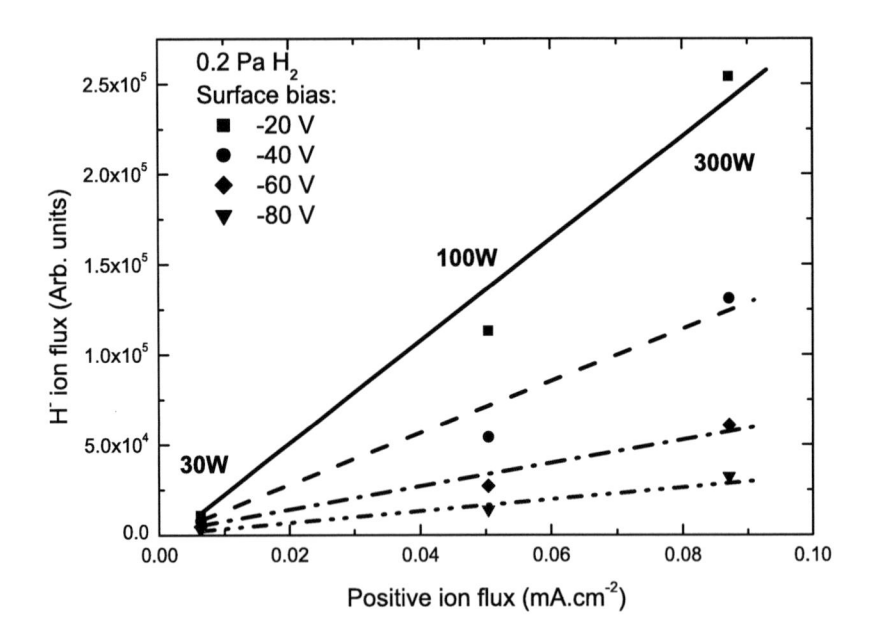

FIGURE 2. H^- flux as a function of the positive ion flux for different sample biases and 0.2 Pa H_2, 30, 100 and 300W discharges

studying IDF measurements, we aim at determining surface generation mechanisms. On this figure, a negative ion created at rest on surface is accelerated by the sheath and enters plasma with an energy equal to $E_0 = e(V_p - V_s)$ (105 eV on figure 4). Most of ions are not created at rest and we observe two features: a main peak followed by a high energy tail. We showed [12] that two mechanisms were responsible of negative ions generation:

- first, H^- of the IDF high energy tail are created via double capture (i.e. simultaneous capture of two electrons or two simple captures),
- second, H^- from main peak are created via sputtering as H^- of adsorbed hydrogen on HOPG, and in a less extent by double capture.

Let us first consider high energy tail of IDF. Positive ions striking sample in normal incidence are neutralized and dissociated, their energy being evenly split between nucleons. Nucleons are diffused through graphite, before possibly reemerge, as negative ions (see [11] for more details). This phenomenom is usually accompanied by a high energy loss on surface by nucleons. We observe that D^- maximum ion energy is lower than H^- maximum energy. We attribute this effect to better D_x^+ energy transfer to surface than H_x^+ because of mass difference [13], [14]. If now consider main peak, two mechanisms are superimposed: sputtering by positive ions of H/D adsorbed on HOPG as H^-/D^- and

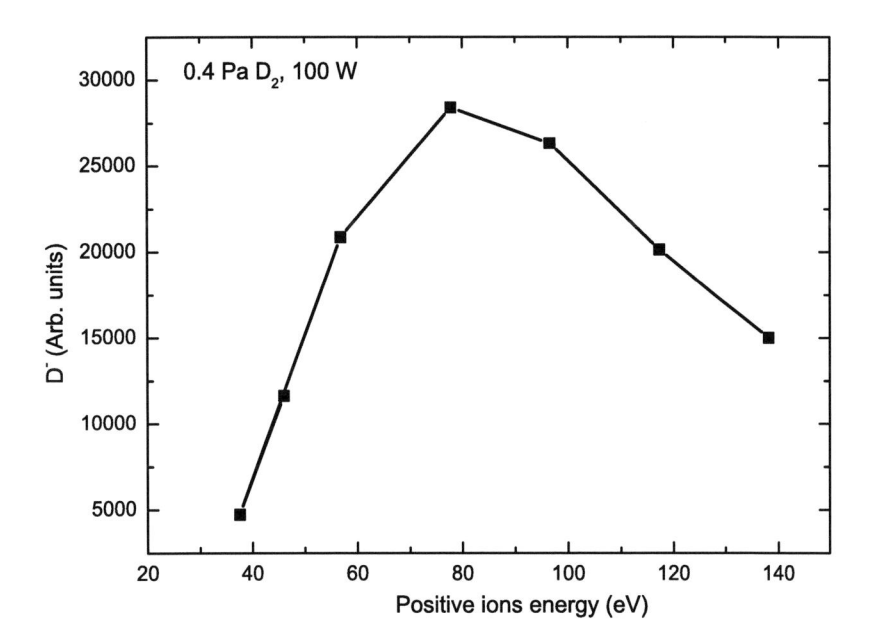

FIGURE 3. H^- intensity as a function of positive ion energy (sample biased from 0 to -80 V) for 0.4 Pa D_2 100W

double capture in a less important way. We were able to distinguish the relative intensity of negative ions generated via sputtering and double capture, showing that most of negative ions are created by sputtering [12]. Fit of IDF sputtering component by Thompson distribution [15], [16], which gives the energy distribution of sputtered species, is in good agreement with experimental data, for a 2.6 eV C-H or C-D binding energy. 2.6 eV is actually the value proposed by Atsumi [17] concerning most of C-H and C-D bonds on bombarded HOPG, presenting defects and vacancies.

CONCLUSION

We study Cs free surface negative ion generation in hydrogen and deuterium plasmas. We start our studies with HOPG material. We showed that negative ions flux was propotional to positive ions flux, and strongly depends on cation energy. IDF measurements exhibit two components: the first one which is responsible of the IDF high energy tail is due to cation double electron capture. Second mechanism is sputtering as H^- of adsorbed hydrogen and gives rise to the main peak. Fit of sputtering IDF part with Thompson distribution is in good agreement with experimental results.

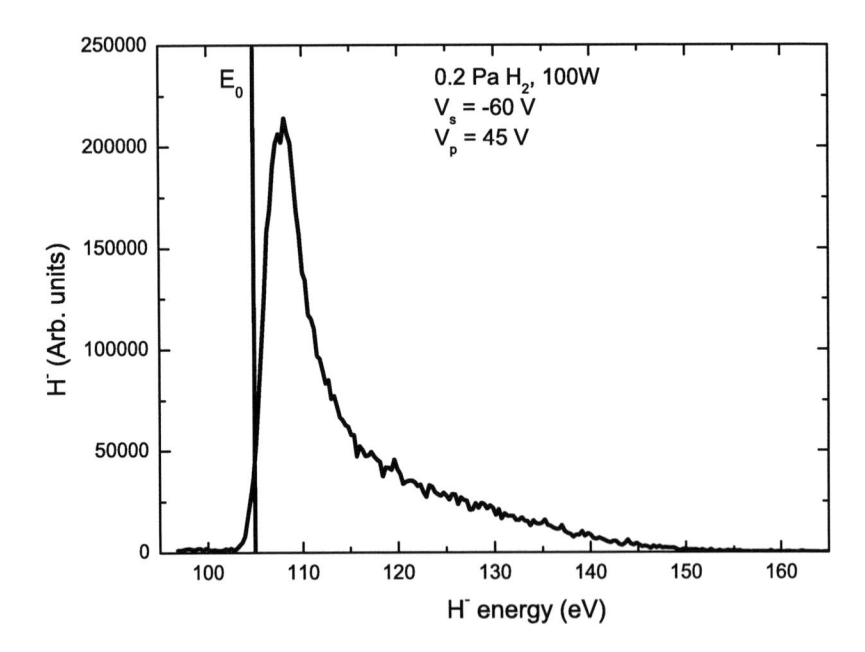

FIGURE 4. Ion distribution function obtained for a 0.2 Pa H_2 discharge, 100 W of injected power and a sample bias of -60V

REFERENCES

1. G. Fubiani, H. P. L. de Esch, and A. Simonin, *Physical Review Special Topics* **11**, 014202 (2008).
2. R. S. Hemsworth, and et al., *Review of Scientific Instruments* **67**, 1120 (2001).
3. R. S. Hemsworth, D. Boilson, B. Crowley, D. Homfray, H. P. L. de Esch, A. Krylov, and L. Svensson, *Review of Scientific Instruments* **77**, 03A511 (2006).
4. R. S. Hemsworth, and A. V. Tanga, A. and, *Review of Scientific Instruments* **79**, 02C109 (2007).
5. U. Fantz, and et al, *Review of Scientific Instruments* **79**, 02A511 (2008).
6. M. Bacal, *Nuclear Fusion* **46**, S250 (2006).
7. L. D. Jackson, B. and, *Journal of Chemical Physics* **114**, 474 (2001).
8. J. B. a. L. D. Sha, X. and, *Journal of Chemical Physics* **116**, 7158 (2002).
9. R. M. Cacciatore, M. and, *International Journal of Quantum Chemistry* **106**, 631 (2006).
10. V. Kaeppelin, M. Carrère, and J. B. Faure, *Review of Scientific Instruments* **72**, 7377 (2001).
11. L. Schiesko, M. Carrère, G. Cartry, and J.-M. Layet, *Plasma Sources Science and Technology* **17**, 035023 (2008).
12. L. Schiesko, M. Carrère, G. Cartry, and J.-M. Layet, *To be published* (2008).
13. Y. Yao, and et al., *Physical Review Letters* **81**, 550 (1998).
14. M. Mayer, M. Balden, and R. Behrisch, *Journal of Nuclear Materials* **252**, 55 (1997).
15. M. W. Thompson, *Physics Reports* **69**, 335 (1981).
16. A. Goehlich, D. Gillman, and H. F. Döbele, *Nuclear Instruments and Methods in Physics Research B* **164**, 834 (2000).
17. H.Atsumi and K.Tauchi, Journal of Alloys Compounds 77,233401 (2008).

Surface Production of Negative Ions by Positive Ions and Atoms in the Electron Suppressor Region

R McAdams and E Surrey

Tokamak Operations Department, EURATOM/UKAEA Fusion Association, Culham Science Centre, Abingdon, Oxfordshire, OX14 3DB, U.K.

Abstract. A simple model of the production of negative ions by positive ions and atoms from the source plasma striking a caesiated surface is presented. The model combines calculated flux rates for the ions and atoms at a surface with negative ion yields to determine the surface production rates. The model takes into account the relative potential between the plasma and the electron suppressor in calculating the surface production rates. By making the assumption that the surface production rates can be directly compared with the extracted beam current then the model is compared with data where the beam current has been measured as a function of electron suppressor bias and knowledge of the plasma parameters of the source also exists. At high bias voltages relative to the plasma potential only volume produced negative ions can be extracted as negative ions created on the surface cannot cross the sheath. The model reproduces features of the ion current dependence on electron suppressor bias. At high bias voltages where negative ions cannot leave the surface there is still an enhancement of the non-caesiated performance which may be attributed to volume production enhancement due to the presence of caesium in the plasma. This has implications for the replacement of caesium with other low work function materials.

Keywords: Fusion, negative ion beam, negative ion source, surface production.
PACS: 52.50.Dg, 52.20.Hv, 52.27.Cm

INTRODUCTION

Conventional volume negative ion sources represent a straightforward method of producing modest current densities of H⁻ and D⁻ of 5–20 mA/cm² (see [1,2] for example). However the heating and diagnostics beams for ITER require current densities of 30 mA/cm² of D⁻ and 35 mA/cm² of H⁻. In order to meet these requirements, the performance of the volume source is enhanced by the introduction of caesium into the source. Use of caesium can enhance the volume source performance by increasing the extracted current densities by a factor of ~2-10 [1,2]. The use of caesium also has the advantage of reducing the co-extracted electron flux to a level where it is approximately equal to the negative ion flux. The enhancement due to caesium is generally attributed to the surface production of negative ions on the low work function caesiated surfaces close to the extraction apertures.

Figure 1 shows the extracted ion current as a function of bias potential on the plasma grid for both non-caesiated and caesiated operation of the IPP BATMAN RF negative ion source [3,4]. Biasing the plasma grid is used to suppress the co-extracted electrons. The source operating conditions are approximately the same in each case with RF powers of 80-85 kW with source pressures of 0.4-0.45 Pa. The negative ion enhancement is clearly seen. Other examples [5,6,7] show similar behaviour of the ion current as the bias potential is changed.

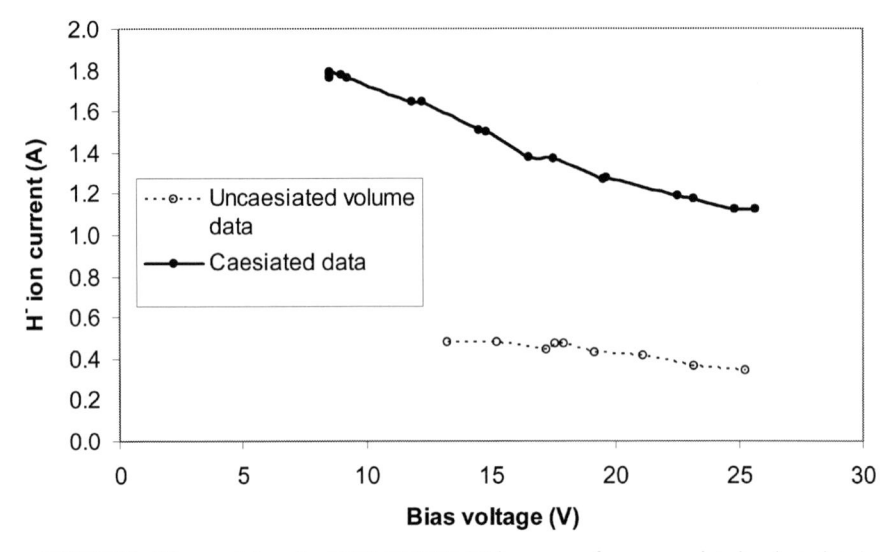

FIGURE 1. H⁻ current from the IPP BATMAN RF ion source for non-caesiated and caesiated operation as a function of the electron suppressor (plasma grid) bias. The RF powers are 80-85 kW and source pressures are 0.4-0.45 Pa.

SURFACE PRODUCTION OF NEGATIVE IONS

Figure 2 shows a simple schematic of the production of negative ions by atom and positive ions on a surface (for a detailed description, see reference [8] and references therein) with Fermi level ε_f and work function ϕ. Ions are rapidly neutralised by Auger interaction with the surface and then are treated as atoms. Molecules are dissociated into atoms. The atoms have an electron affinity level of energy, E_A, into which electrons from the surface will be captured to form the negative ion. The figure shows that as the atom approaches the surface, this affinity level moves to higher energies and broadens due to the proximity of the metal. Close to the surface the affinity level then overlaps with the metal electronic states. Electrons from the metal can tunnel into the affinity level to form the negative ion. As the ion leaves the surface there is also the possibility of the additional electron tunneling back into the metal. The lower the work function the greater the overlap of the atomic levels with the metal states and the greater the probability of negative ion formation. The process has a threshold energy of $E_{thr} = \phi - E_A$. For caesium this work function can be as low as 1.5 eV [8].

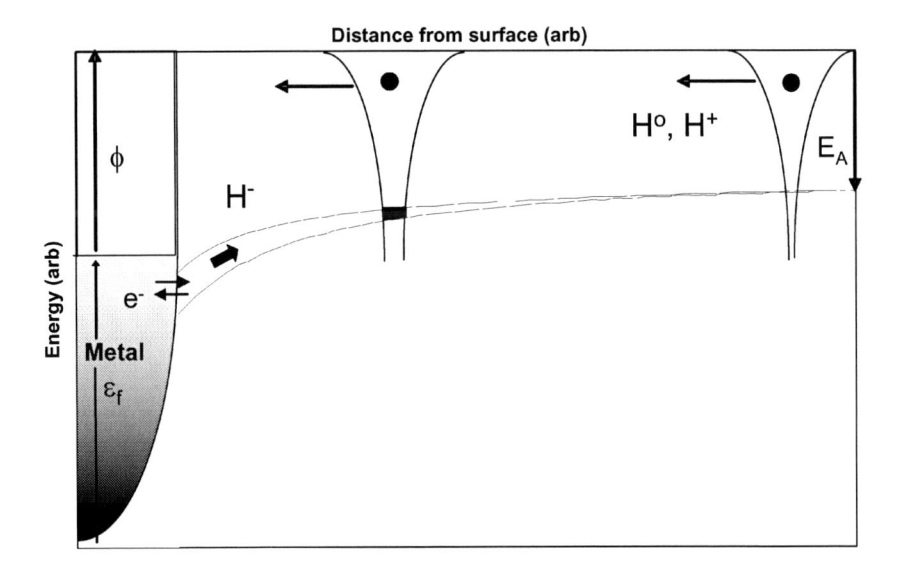

FIGURE 2. Schematic of surface production of negative ions by atoms and ions.

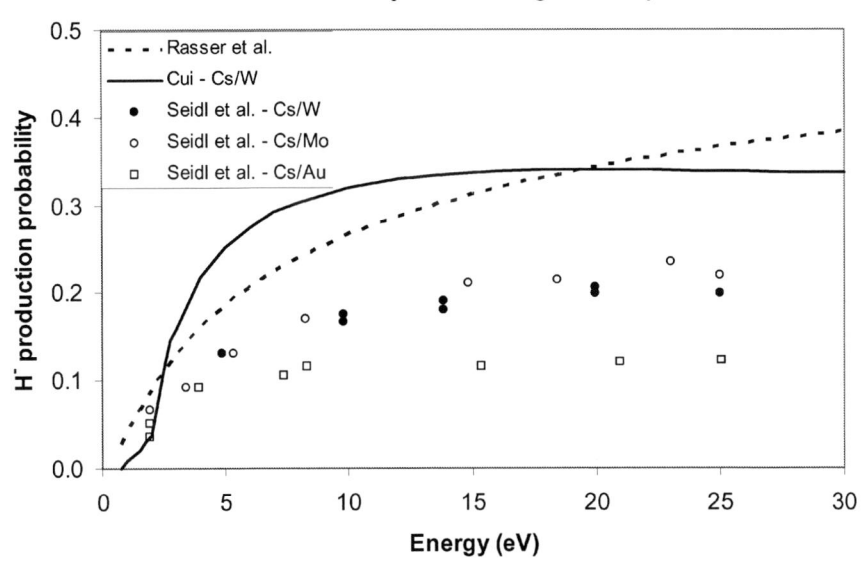

FIGURE 3. Production of negative ions by proton impact on caesiated surfaces

The probability, P, of negative ion production on a metal surface has been calculated by Rasser et al. [9,10] as

$$P = \frac{2}{\pi}\exp\left[-\frac{\pi(\phi - E_A)}{2av}\right] \quad (1)$$

91

where v is the atom velocity and a is a screening constant. In Figure 3 data for the yield of negative ions for hydrogen atom or proton impact from various caesiated surfaces are shown [11]. Also shown in the figure is the calculation by Rasser (eqn 1 above) and a quantum mechanical calculation by Cui [12] taken from Seidl et al. [11] where it has been adapted to include the energy reflection coefficient and the fraction of particles reflected together with a cosine angular scattering distribution.

In the case of production of negative ions by atoms from a plasma the yield must be averaged over a thermal (Maxwellian) distribution. The yield from [12], plotted as the reciprocal of the atomic thermal temperature, is shown in Figure 4 together with some experimental data [11].

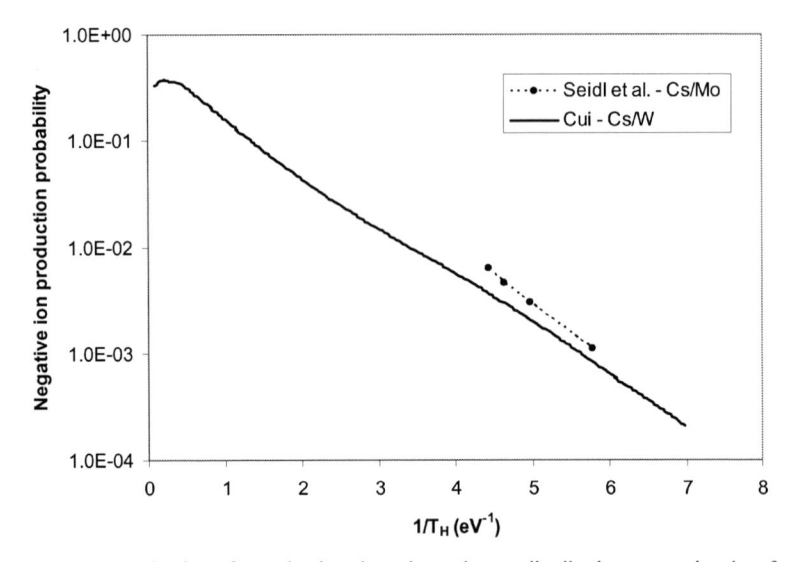

FIGURE 4. Production of negative ions by a thermal atom distribution on caesiated surfaces

SIMPLE PICTURE OF SURFACE PRODUCTION IN THE ELECTRON SUPPRESSOR REGION

An idealised schematic of the caesiated ion source is shown in Figure 5. The plasma has a potential V_p relative to the source body and the plasma grid is biased at a positive voltage V_{bias} with respect to the source. Caesium covers the plasma grid on which negative ions are formed by ions and atoms. The ions return to the plasma before being extracted. Caesium may be found on other surfaces but the destruction rates for negative ions by mutual neutralisation with positive ions and electron detachment by electrons indicate that only negative ions formed within a few centimeters of the extraction aperture can be extracted. Thus only negative ions formed on the plasma grid are considered. Magnetic field effects due to the filter field are ignored.

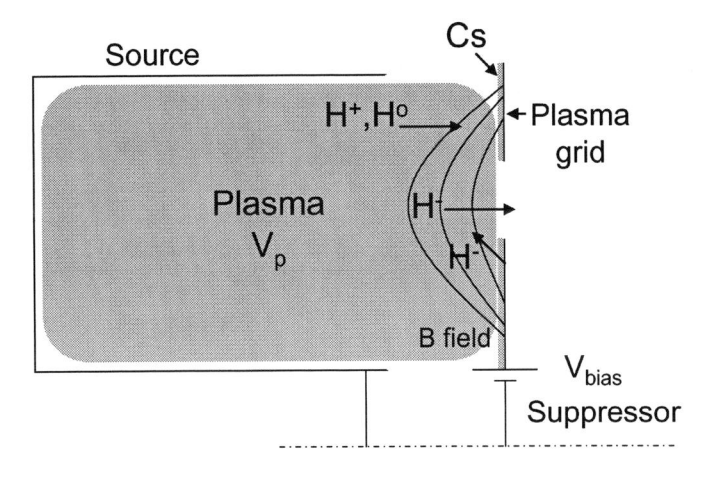

FIGURE 5. Idealised schematic of the ion source

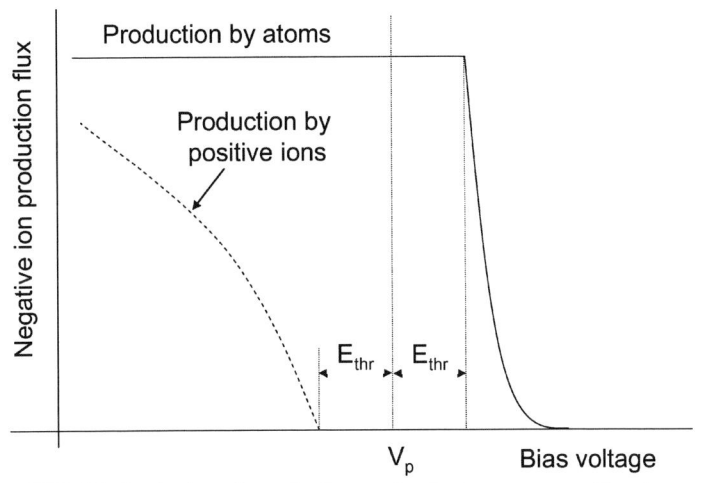

FIGURE 6. Production of negative ions as the electron suppressor bias changes

The behaviour of the production rates of negative ions by ions and atoms, as the bias voltage is changed, is shown in Figure 6. In the case of positive ions, if V_{bias} is less than the plasma potential then ions are accelerated across the sheath to the caesium surface with energy V_p-V_{bias}. As the bias voltage increases the ion energy decreases and the production of negative ions decreases in accordance with the production probability shown in Figure 3. When the ion energy is less than the threshold energy the production is zero. The surface production rate of negative ions due to positive ions, $\Gamma^+_{H^-}$, can be written

$$\Gamma^+_{H^-} = Y(E)\Gamma_+ \qquad (2)$$

where Y(E) is the yield of negative ions at an energy E for an incident flux Γ_+. Negative ions formed by ions can return to the plasma due to the potential difference.

Atoms cross from the plasma to the plasma grid irrespective of the potential difference. At $V_{bias} \leq V_p$ negative ions formed on the grid can return to the plasma. At $V_{bias} > V_p$ and $V_{bias} - V_p \leq E_{thr}$ all the negative ions can return to the plasma since they are created with a minimum energy of E_{thr} (neglecting energy loss on reflection) but for values of $V_{bias}-V_p > E_{thr}$ only negative ions with energy greater than this can return and this decreases as the potential difference increases.. Thus the yield of negative ions produced on the surface that can return to the plasma, $\Gamma^0_{H^-}$, can be written

$$\Gamma^o_{H^-} = Y(T_H)\Gamma_H K \qquad (3)$$

where $Y(T_H)$ is the yield for hydrogen atoms at a temperature T_H, as shown in Figure 4, Γ_H is the flux of atoms and K is the fraction returning to the plasma and is given by

$$K = \frac{2}{\pi^{\frac{1}{2}} T_H^{\frac{3}{2}}} \int_{E_{min}}^{\infty} E^{\frac{1}{2}} \exp(-\frac{E}{T_H}) dE \qquad (4)$$

where $E_{min} = E_{thr}$ for $V_{bias} \leq V_p + E_{thr}$ and $E_{min} = V_{bias} - V_p$ for $V_{bias} > V_p + E_{thr}$. The integral for K is over a standard Maxwellian energy distribution of the negative ions due to the distribution of the incident atoms. The lower integration limit defines only those ions with sufficient kinetic energy to cross back to the plasma as described above, as discussed in relation to the yield and as shown in Figure 6.

The positive ion and neutral fluxes to the wall are given by the standard equations

$$\Gamma_+ = 0.6 n_+ \sqrt{\frac{eT_e}{m_+}} \quad \text{and} \quad \Gamma_H = \frac{1}{4} n_H \sqrt{\frac{8eT_H}{\pi m_H}} \qquad (5)$$

where n_+ is the positive ion density, T_e the electron temperature, m_+ the proton mass, n_H the atomic hydrogen density, T_H the atomic hydrogen temperature and m_H the mass of atomic hydrogen. Thus the surface production rates of negative ions can be calculated using the yields and plasma parameters.

APPLICATION TO SOURCE AND BEAM DATA

This simple model can now be applied to ion source and beam data such as that shown in Figure 1. Unfortunately it was not possible to identify a consistent set of beam and source data i.e. measurements of both beam (caesiated and non-caesiated) and plasma data as a function of V_{bias}. Instead, representative source data [2,13] in the extraction region has been used and is shown in Table 1. The yield values from Cui [11,12] have also been used as they give values over a wide ion energy and atom temperature albeit for caesium on tungsten. Thus at this stage the model can only be expected to show qualitative agreement until consistent data is available.

TABLE 1. Plasma source data for the IPP BATMAN facility

Parameter	Value
Electron temperature (eV)	2.0
Positive ion density (m^{-3})	5.0 x 10^{17}
Atomic hydrogen density (m^{-3})	1.0 x 10^{19}
Floating potential (V)	8.55
Plasma potential (V)	15.25
Atomic hydrogen temperature (eV)	Left as free parameter

The floating potential, V$_f$, has been taken directly from the IPP data and the plasma potential, V$_p$, has been calculated from the equation

$$V_p = V_f - T_e \ln\left[0.6\sqrt{\frac{2\pi m_e}{m_+}}\right] \qquad (6)$$

where m$_e$ is the electron mass. The electron affinity of hydrogen is 0.75V and a work function of 1.5eV for caesium has been used giving a threshold energy of 0.75eV.

Figure 7 shows the ratio of atomic to ion flux at the wall calculated with these plasma parameters as a function of the atomic hydrogen temperature. The atomic flux is greater than the ionic flux for atomic temperatures greater than ~ 0.025eV.

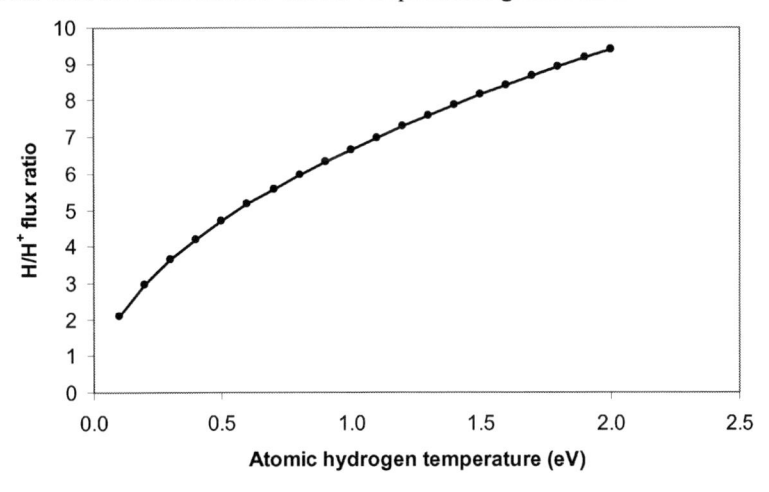

FIGURE 7. Ratio of atomic to ionic flux using the representative source parameters of Table 1

The total surface production rate is shown in Figure 8 as the bias is changed for some values of atomic hydrogen temperature. The regions of ion and atomic production are clearly identified. The ratio of production of negative ions by atoms and ions can be estimated from the value of the total production rate at zero bias and subtracting the value at the plasma potential which is the maximum surface production rate to give the ionic production rate as shown in Figure 9. For atomic hydrogen temperatures greater than ~ 0.75eV the production of negative ions becomes dominated by atomic bombardment of the caesiated surface.

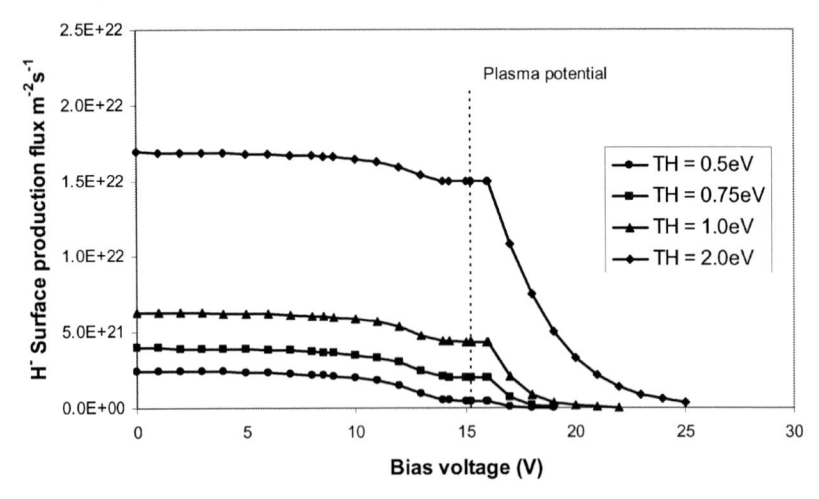

FIGURE 8. Total surface production flux of negative ions as a function of the bias voltage

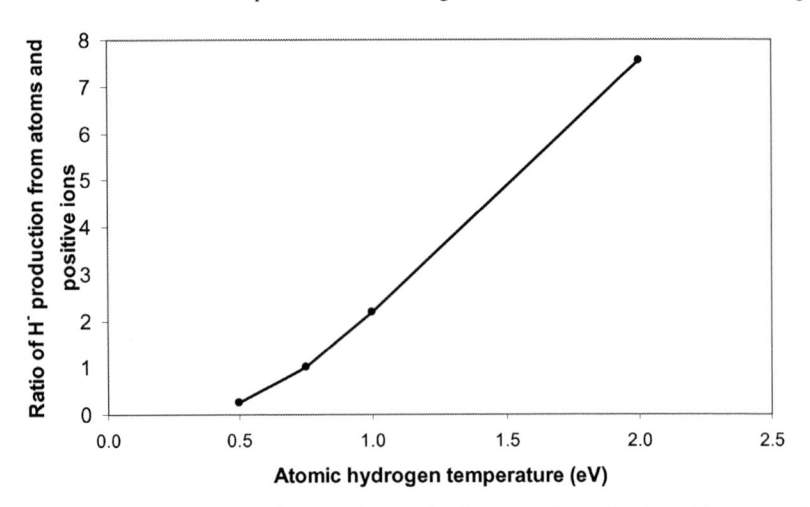

FIGURE 9. Ratio of atomic and ionic surface production rates of negative ions. The ratio is formed from surface production rates at zero bias and at the plasma potential.

The total caesiated surface production rates can be compared to the experimental extracted ion current data (from Figure 1) as shown in Figure 10 by the following method. It is assumed that a constant proportion of the surface produced ions are extracted as the bias voltage is varied. At $V_{bias} \gg V_p$ no negative ions can leave the surface in this simple picture and so the extracted current from the surface is zero at these values. The non-surface produced ions in the beam are assumed to have a parallel linear trend to the volume produced ions in the uncaesiated case. The difference between the measured current and the non surfaced produced ions represents the surface produced fraction of the beam. The data in Figure 8 is then normalized to the surface produced ion current at the floating potential (which is the lowest bias voltage at which measurements were made [4]). The total current can then be easily calculated. The model gives qualitative agreement with the data.

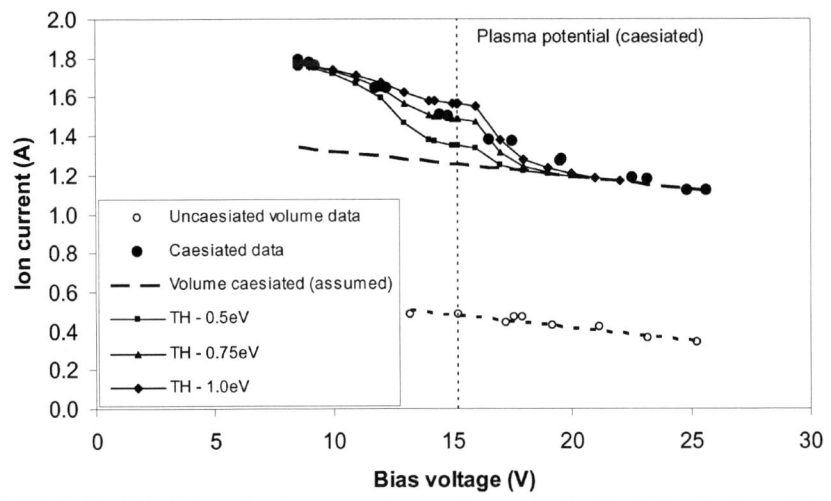

FIGURE 10. Calculation of the ion current from surface production for different values of atomic hydrogen temperature

DISCUSSION AND IMPLICATIONS

Although this simple model qualitatively reproduces the dependence of ion current on the bias voltage and shows that this may depend on atomic hydrogen temperature it can be improved. A consistent set of data relating extracted ion current with the plasma parameters is required together with negative ion yield data for the correct materials. It has been assumed that the plasma potential is constant but data from the BATMAN facility [2] shows that the plasma potential can increase as the bias voltage increases. This will maintain production by ions out to higher bias voltages and improve agreement with the data. In this model at $V_{bias} >> V_p$ no negative ions can leave the plasma grid. Thus, from Figure 10, at these values (and following the assumed trend to low V_{bias}) the difference in current with the uncaesiated data cannot originate at a surface and some other production mechanism or enhancement must be taking place. It has been shown [14] that increasing the amount of caesium added to a volume source at higher pressures results initially in a lowering of the electron temperature, electron density, and plasma potential accompanied by an increase in the negative ion density in the plasma until a minimum is reached and then the values of these parameters begin to increase while the negative ion density which continues to increase. There is an optimal coverage [15] of caesium when the yield of negative ions reaches a maximum corresponding to the lowest value of the surface work function. Thus the interpretation of this work was that as the negative ion density increases monotonically it is not totally due to a surface production mechanism but to enhanced volume production. This model could then support this interpretation. It is reported [2] that ~99% of the caesium is ionized and this raises the question as to what properties of caesium could produce this enhancement since seeding a source with xenon, which has a mass close to that of caesium, does not enhance the negative ion current.

One other possibility is that backstreaming ions are striking caesiated surfaces outside the source near the extraction aperture. This would depend on the beam current from the source but its magnitude is not known

The use of caesium is problematic. If the enhancement was due totally to surface production then it could be replaced by an alternative low work function material. In this simple picture this material would show ion current enhancement due to surface production but the enhanced volume performance would not be obtained. At high bias the ion current would approach the pure volume performance.

ACKNOWLEDGMENTS

This work was funded jointly by the United Kingdom Engineering and Physical Sciences Research Council and by EURATOM under the contract of association between EURATOM and UKAEA

REFERENCES

1. R McAdams et al., "Pure and cesiated volume source performance at the Culham Ion Source Test Stand" in *Production and Neutralisation of Negative Ions and Beams, Sixth International Symposium 1992*, edited by J Alessi and A Herschovitch, AIP Conference Proceedings 287, American Institute of Physics, Melville, NY, 1994, pp. 353-367
2. U Fantz et al., *Plasma Phys. Control. Fusion* **49**, B563–B580 ((2007)
3. U Fantz, CCNB Meeting, Cadarache, 2006 (unpublished)
4. P Franzen et al., *Nucl. Fusion* **47**, 264–270 (2007)
5. Y Okumura et al., "Cesium mixing in the Multi-Ampere Volume H⁻ Ion Source" in *Production and Neutralisation of Negative Ions and Beams, Fifth International Symposium 1990*, edited by A Herschovitch, AIP Conference Proceedings 210, American Institute of Physics, Melville, NY, 1994, pp. 169-183
6. K Watanabe et al.,, "Cs effects in a Large Multi-Ampere D⁻ Ion Source" in *Production and Neutralisation of Negative Ions and Beams, Sixth International Symposium 1992*, edited by J Alessi and A Herschovitch, AIP Conference Proceedings 287, American Institute of Physics, Melville, NY, 1994, pp. 327-336
7. T Takanashi et al., *Rev. Sci. Instr.* **67**(3), 1024-1026 (1996)
8. P W van Amersfoort et al., *J. Appl. Phys.* **58**(9), 3566-3572 (1985)
9. B Rasser et al., *Surf. Sci.* **118**, 697-710 (1982)
10. R F Welton et al., "Enhancing surface ionization and beam formation in volume type H⁻ sources" in *Proceedings of the European Particle Accelerator Conference*, Paris 2002.
11. M Seidl et al. *J. Appl. Phys.* **79**(6), 2896-2901 (1996)
12. H L Cui, *J. Vac. Sci. Technol.* **A9**(3) 1823-1827
13. U Fantz et al., *Nucl. Fusion* **46** S297-S306 (2006)
14. M Bacal et al., *Rev. Sci. Instr.* 71(2) 1082-1085 (2000)
15. J Los and J J C Geerlings, *Physics Reports* **190** (3) 133-190 (1990)

Spatial Distribution of the Plasma Characteristics of a Tandem Plasma Source

Tsvetelina V. Paunska[a], Antonia P. Shivarova[a], Khristo Ts. Tarnev[b] and Tsanko V. Tsankov[a]

[a]*Faculty of Physics, Sofia University, BG-1164 Sofia, Bulgaria*
[b]*Department of Applied Physics, Technical University–Sofia, BG-1000 Sofia, Bulgaria*

Abstract. The study presents results from a 2D model of hydrogen discharges in a tandem type of a plasma source. Analysis of the particle and electron-energy balance outlines the pattern of the source operation: (i) discharge production in the driver governed by the mechanisms ensuring self-consistency of hydrogen discharges in regions with rf power deposition, however, here influenced by the fluxes going to the second chamber of the source, and (ii) plasma existence in the expansion volume of the source completely determined by the particle fluxes from the first chamber. The influence of varying applied power and gas pressure is discussed.

Keywords: 2D discharge modeling, tandem two-chamber plasma sources, hydrogen discharges
PACS: 52.50.-b, 52.25.Fi, 52.30.-q

INTRODUCTION

The recent increasing interest in discharges in molecular gases and, in particular, in hydrogen is provoked by their applications in the technology and as ion sources for additional heating in fusion plasmas. The complications in the modeling [1-4] of hydrogen discharges come from the changing – with the gas-discharge conditions – concentrations of atoms and molecules, the presence of different types of ions and the variation of their concentrations with the concentrations of the neutral species. In addition, the rf-sources developed for plasma-processing and fusion applications are usually two-chamber plasma sources [5-7] (known in the literature related to fusion as tandem-type plasma sources [8]) consisting of a driver where the rf power producing the discharge is applied and a volume for plasma expansion from the driver. Such a design of the sources also creates complications since their description requires at least 2D models, accounting for the complex spatial distribution of the plasma parameters and for the particle and energy fluxes playing very important role [9, 10].

This study presents results from a 2D model [10] of hydrogen discharges in a tandem plasma source and analyses of the mechanisms ensuring plasma existence in the source. The spatial distribution of the plasma parameters is discussed based on the particle and electron-energy balance. The discharge production in the driver shows evidence of the basis [3, 4, 11] of self-consistency of hydrogen discharges in regions with rf power deposition (e.g., by external antenna). The plasma in the second chamber exists due to particle fluxes from the first chamber. The influence of the gas pressure and of the applied power is also discussed.

BRIEF DESCRIPTION OF THE MODEL

The source studied (Fig. 1) is a two chamber plasma source with metal walls consisting of a small radius tube (radius $R_1 = 2.25$ cm and length $L_1 = 20$ cm) where the driver – the rf energy deposition region – is located and a plasma expansion chamber ($R_2 = 11$ cm and $L_2 = 30$ cm are the corresponding dimensions).

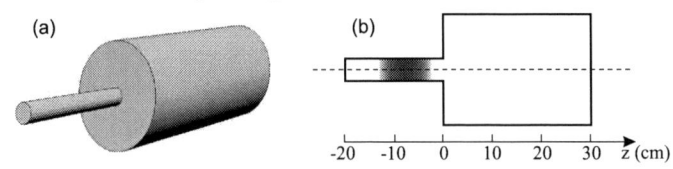

FIGURE 1. Configuration of the tandem type of a plasma source modeled (a) and its cross section (b). The region of the rf power deposition is marked on (b). Because of the azimuthal symmetry, the presented results are for half of the cross section of the source; $z = 0$ is at the transition between the two chambers.

The model [10] is based on the continuity equations for electrons, atomic (H^+) and molecular (H_2^+ and H_3^+) ions and atoms:

$$\mathrm{div}\vec{\Gamma}_{e,ij,a} = \frac{\delta n_{e,ij,a}}{\delta t}, \tag{1}$$

the electron energy balance

$$\mathrm{div}\vec{J}_e = P_{ext} + P_a^{(coll)} + P_m^{(coll)} + P_{dc} \tag{2}$$

and the Poisson equation for the potential Φ of the dc electric field $\vec{E}_{dc}$ in the discharge. In (1), $\vec{\Gamma}_{e,ij,a}$ are the fluxes of the electrons ($\vec{\Gamma}_e$), of the different types of ions ($\vec{\Gamma}_{ij}$ with $j = 1, 2$ and 3 for H^+-, H_2^+- and H_3^+-ions, respectively) and atoms ($\vec{\Gamma}_a$) and $\delta n_{e,ij,a}/\delta t$ represent the change of the concentrations (n_e, n_{ij} and n_a) of the corresponding particles due to volume processes. In (2), $\vec{J}_e$ is the electron energy flux, including both conductive $\vec{J}_e^{(cond)}$ and convective $\vec{J}_e^{(conv)}$ fluxes and $P_a^{(coll)}$ and $P_m^{(coll)}$ are the electron energy losses in collisions with atoms and molecules, respectively. Although $P_{dc} = -e\vec{\Gamma}_e \cdot \vec{E}_{dc}$ in (2) usually appears as losses for sustaining the dc electric field in the discharge it could be also an energy input to the expanding plasma volume of the source [10]. A radially homogeneous power deposition to the driver with a super-Gaussian profile in the axial (z) direction $P_{ext} = P_0 \exp\left[-\frac{1}{2}\left(\frac{z-z_0}{\sigma}\right)^{2m}\right]$ (where $P_0 = P_{ext}(z = z_0)$, $z_0 = -7.5$ cm, $\sigma = 4.8$ cm and $m = 4$) ensures almost constant power density between $z = -11$ cm and $z = -4$ cm and very fast power density decrease outside this region. The boundary conditions are for charged-particle and electron-energy fluxes absorbed on the walls and for a zero dc potential there. More detailed description of the model is given in [10].

SPATIAL DISTRIBUTION OF THE PLASMA PARAMETERS

Figure 2 shows the obtained spatial distribution of the plasma parameters at gas pressure $p = 50$ mTorr and total applied power $Q = 500$ W.

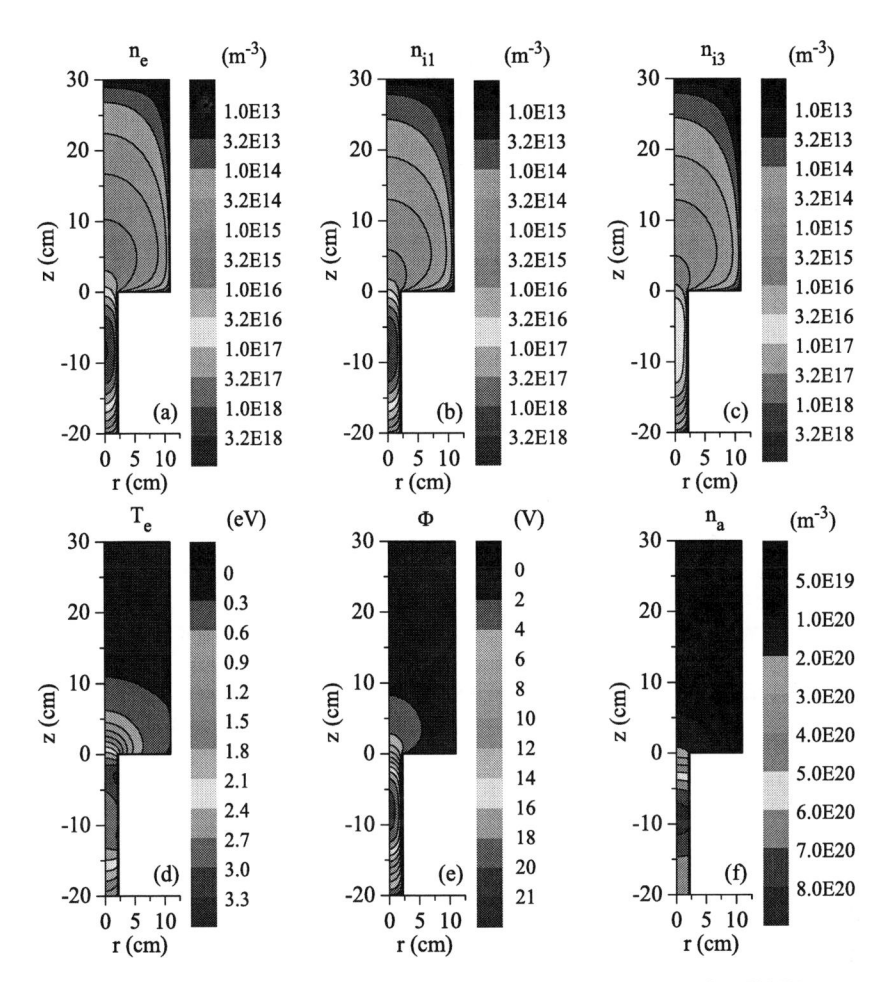

FIGURE 2. Spatial distribution of the plasma parameters for $p = 50$ mTorr and $Q = 500$ W: concentrations of electrons (a), atomic H^+ (b) and molecular H_3^+ ions (c), electron temperature (d), dc potential (e) and concentration of the atoms (f).

The concentrations of all the particles, the electron temperature T_e and the dc potential Φ have maxima in the driver and strongly decrease in the plasma expansion chamber. However, the positions of the maxima of the concentrations of the different species do not coincide to each other and to the position of the center ($z_0 = -7.5$ cm) of the rf power deposition P_{ext}. The hydrogen atoms are the key for understanding this behavior. The losses of hydrogen atoms on the back wall ($z = -20$ cm) are smaller than the losses due to their flux to the second chamber of the source. As a result, the maximum of the atom concentration n_a does not coincide with the center of P_{ext}. It is shifted towards the back wall (Fig. 2(f)). This leads to an increased production – via atom ionization – of atomic ions H^+ and electrons in the region towards the back wall of the source

(Figs. 2(a) and (b)). The lower diffusion losses of H^+-ions, due to the high elastic collision frequency of atomic ions with atoms (higher than that with molecules), also leads to higher concentration of H^+-ions and, thus, of electrons, and determine lower T_e (Fig. 2(d)) in the regions of higher n_a, n_e and n_{i1} formed towards the back wall. In general, the spatial distribution of the plasma parameters in the driver demonstrates the mechanism of discharge self-consistency expressed through the relation between T_e and n_e according to the generalized Schottky condition [11] which in hydrogen discharges – due to the efficient dissociation – passes through the concentration of the atomic hydrogen and is controlled by the atomic ions [3, 4, 11]. In opposite, the maximum of the concentration n_{i3} of the H_3^+-ions is slightly shifted to the second chamber (Fig. 2(c)), following the maximum of T_e and, respectively, the decrease of n_e. Such a behavior is expected [4, 11] as the importance of the H_3^+-ions is higher when n_e is lower.

In regions with a constant power deposition the electrons gain more energy where n_e is lower. As a result, T_e is higher near the wall of the driver (n_e is low there) and at the edges – in the z-direction – of the P_{ext}-profile. Because of the shift of the maximum of n_e to the back wall of the source, the local maximum of T_e at $z = -2.5$ cm is more pronounced than that at $z = -12.5$ cm (Fig. 2(d)).

Thus, the asymmetry in the axial distribution of the plasma parameters in the driver, appearing due to the flux of hydrogen atoms to the second chamber of the source, is in accordance with the generalized Schottky condition for self-consistency of hydrogen discharges.

The axial variation of Φ (Fig. 2(e)) follows that of n_e. Due to the different dimensions, the radial gradient of Φ in the driver is higher than the axial one which determines a higher radial dc field and a higher radial ion flux compared to the electron flux. The weaker axial dc field permits stronger axial flux of electrons determining a regime with a net dc current in the discharge due to the appearance of a solenoidal flux $\vec{\Gamma} = \vec{\Gamma}_e - \vec{\Gamma}_i$ [10].

INFLUENCE OF THE GAS PRESSURE AND OF THE POWER

Figure 3 shows the axial variations of the plasma parameters on the axis of the discharge vessel, for two values of the gas pressure ($p = 20$ and 50 mTorr) and two values of the total applied power ($Q = 200$ and 500 W).

In the driver, the changes of the plasma parameters with varying p and Q are – in broad terms – according to the general trends of the gas-discharge behavior and the mechanisms of discharge self-consistency [3, 4, 11]. Since the charged particle losses control the discharge, the increased diffusion losses for lower p (due to reduced elastic ion-neutral collision frequencies) lead to a T_e-increase (Fig. 3(e)) needed for ensuring enough ionization. This determines the reduction of the concentration of the charged particles (Figs. 3(a)-(c)), when Q is constant, and also the increase of the plasma potential (Fig. 3(f)). The dependence of T_e on Q is an evidence of the discharge self-consistency, i.e. of the ($n_e \Longleftrightarrow T_e$)-relation already discussed. The Q-increase causing an increase of n_e (in hydrogen discharges, related to an increase of the concentrations of the hydrogen atoms and of the atomic ions) leads to a decrease of T_e. Figure 3 also shows the asymmetry of the axial profiles of the plasma parameters in the driver, commented

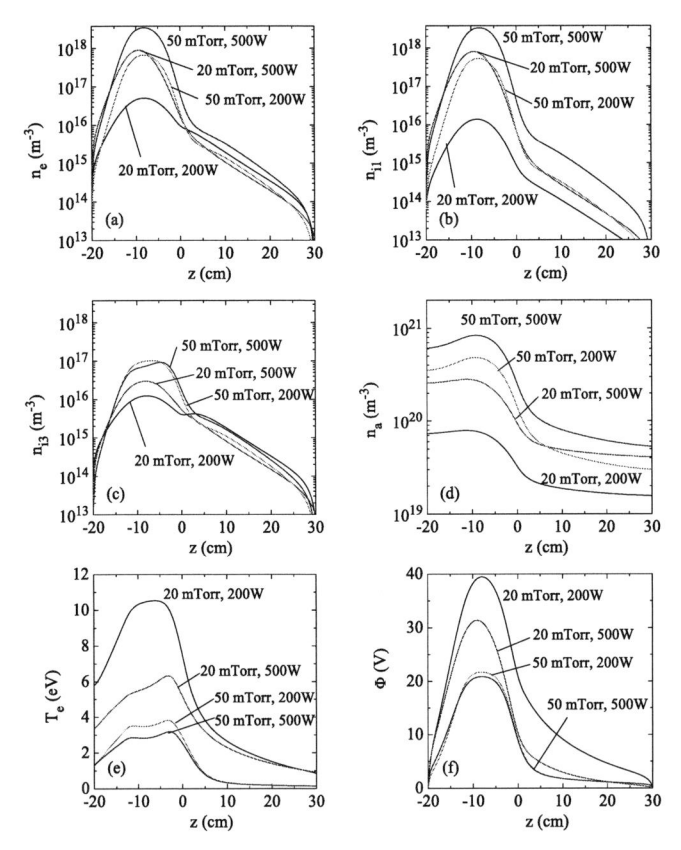

FIGURE 3. Axial distribution of the plasma parameters: concentrations of electrons (a), atomic ions H$^+$ (b), molecular ions H$_3^+$ (c) and atoms (d), electron temperature (e) and dc potential (f). The values of p and Q are given on the figures.

on in the previous section. A comparatively flat n_e-profile in the driver (the case of $p = $ 20 mTorr and $Q = 200$ W in (Fig. 3(a)) is accompanied by a monotonic decrease of T_e from the center of the driver (Fig. 3(e)). The increased diffusion for lower p determines a monotonic decrease of n_a towards the second chamber (Fig. 3(d)) with dissapearance of the maximum in the n_a-profile showing evidence for the higher pressure value studied. The maxima of Φ (Fig. 3(f)) in the driver coincide with the maxima of n_e.

In the expanding plasma volume (the second chamber of the source) the changes of n_e with varying Q and p are weak (Fig. 3(a)). The concentration of the atomic ions (Fig. 3(b)) keeps the dependence on p and Q it shows in the driver. The changes of the H$_3^+$-ion concentration n_{i3} with p and Q are weak (Fig. 3(c)). A significant volume production of H$_3^+$-ions at the beginning of the plasma expansion chamber results into a plateau region in the n_{i3}-profile, well pronounced at $z \simeq 5$ cm in the case of $p = 20$ mTorr and $Q = 200$ W. The concentration n_{i2} of the H$_2^+$-ions is lower than that of the other types of ions. Their losses in the plasma expansion chamber are mainly for formation of H_3^+-ions. In the second chamber Φ increases with the p-decrease (Fig. 3(f)).

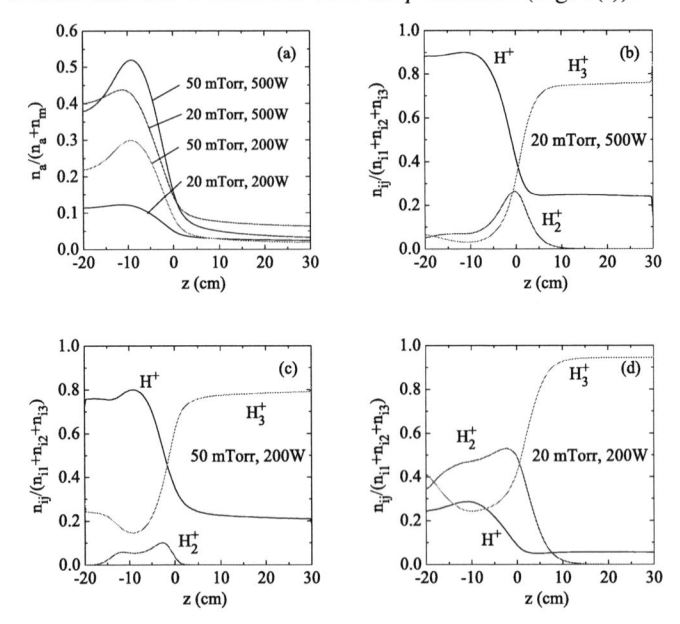

FIGURE 4. Relative particle concentrations for different values of p and Q: of the atomic hydrogen in (a) and of the ions n_{ij} in (b)-(d).

As it has been already discussed, the hydrogen atoms play a key role in the self-consistency of the discharge. The degree of dissociation in the driver increases with Q and also with p (Fig. 4(a)). High values of n_a determine high n_{i1} in the driver: Figures 4(b)-(d) show an increase of the relative concentration of the atomic ions with the increase of the degree of dissociation (Fig. 4(a)). The ratio of the concentrations of the H_2^+- and H_3^+-ions in the driver depends strongly on the pressure. With the p-increase the role of the process of transformation of H_2^+-ions into H_3^+-ions increases and n_{i3}/n_{i2} increases. Out of the region of the power deposition n_e and T_e decrease leading to a decrease of the degree of dissociation and of the concentration of the H^+-ions. In the second chamber the H_3^+-ions strongly dominate. They are more than 95% of all the ions at $p = 20$ mTorr and $Q = 200$ W, decreasing both with the p- and Q- increase.

PARTICLE AND ENERGY BALANCE

The regime of the rf discharge is with a net dc current flowing in the plasma [10], due to the flux $\vec{\Gamma}$ – a solenoidal flux (div$\vec{\Gamma} = 0$) – which appears as a difference ($\vec{\Gamma} = \vec{\Gamma}_e - \vec{\Gamma}_i$) of the electron ($\vec{\Gamma}_e$) and ion ($\vec{\Gamma}_i = \sum_{j=1}^{3} \vec{\Gamma}_{ij}$) fluxes. Moreover, two types of behavior of the flux $\vec{\Gamma}$ show evidence in the expansion chamber: (i) a case with all the streamlines of $\vec{\Gamma}$ closed through the discharge walls, and (ii) a case with a vortex-type flux $\vec{\Gamma}$ appearing in the plasma volume. Since $\vec{\Gamma} \simeq \vec{\Gamma}_e$, except for its radial component in the driver, $\vec{\Gamma}_e$ shows up with a similar behavior. In the first case $\vec{\Gamma}_e$ is a forward flux, from the driver to the walls of the plasma expansion chamber, while in the second case a backward $\vec{\Gamma}_e$-flux appears in the on-axis region of the second chamber.

In the presentation of radial profiles of the plasma parameters further on, an axial position $z = 15$ cm is chosen as this position is away from the vortex region in Fig. 5(a) and the $\vec{\Gamma}_e$-flux is directed to the walls. Thus, the plasma parameters at this position show a behavior corresponding to the first case. On Fig. 5(b) the same position is in the center of the vortex with the typical for the second case backward flux in the on-axis region.

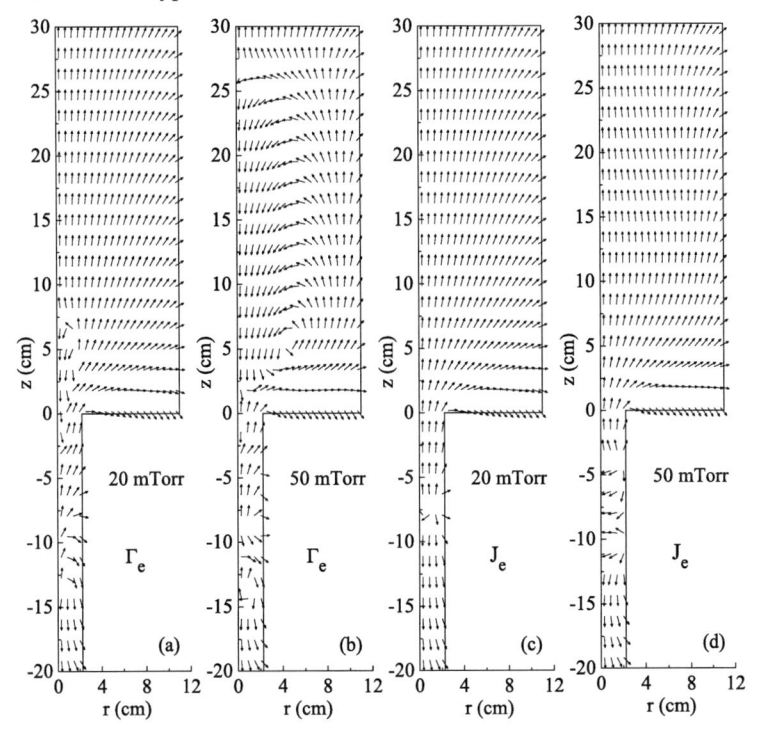

FIGURE 5. Electron fluxes in (a) and (b) and electron-energy fluxes in (c) and (d). The pressure values are given on the figures; $Q = 200$ W.

Figure 6 shows the radial changes of the contributors to the particle balance in the driver ($z = -8$ cm) and in the second chamber ($z = 15$ cm) for the two cases shown in Figs. 5(a) and (b). In the driver (Figs. 6(a) and (b)) the electrons are produced mainly by ionization of molecules. The relative contribution of the atom ionization increases with p, due to the increase of the degree of dissociation. Radial and axial fluxes are the main mechanisms of losses. Deep in the plasma expansion chamber (Figs. 6(c) and (d)) the plasma exists due to the fluxes from the driver. In the case of a forward $\vec{\Gamma}_e$-flux (Fig. 6(c)), the axial flux ensures electron income and the radial flux is the dominating mechanism of losses. In the case of a backward $\vec{\Gamma}_e$-flux (Fig. 6(d)) the radial flux carries electrons to the axis and the axial flux is a mechanism of losses. Near the walls of the

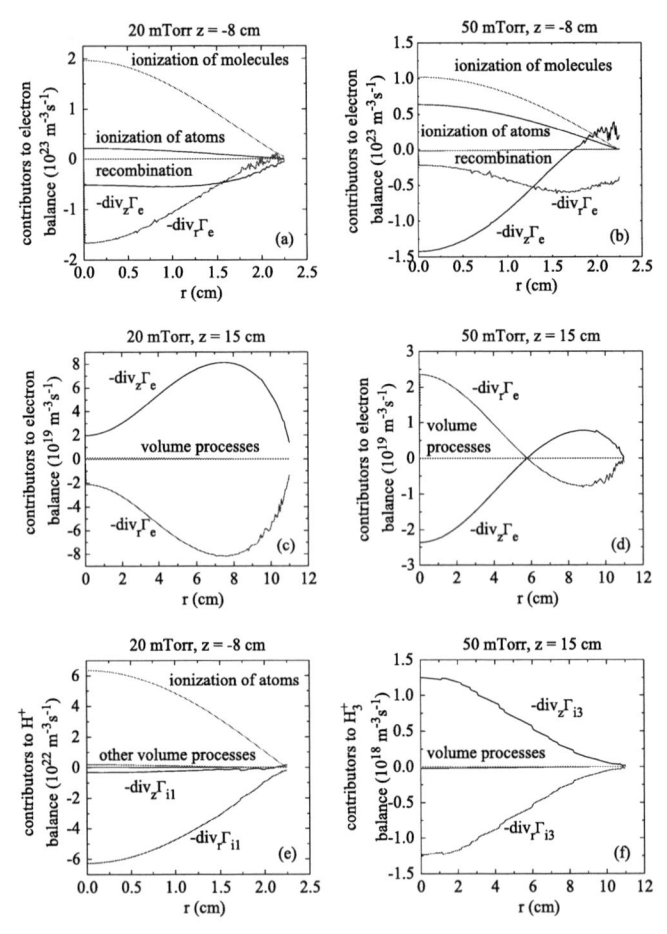

FIGURE 6. Radial variations of the contributors to the particle balance of the electrons at $z = -8$ cm for 20 mTorr (a) and 50 mTorr (b) and at $z = 15$ cm for 20 mTorr (c) and 50 mTorr (d) and of the dominating ions in the driver (H^+-ions, (e)) and in the second chamber (H_3^+-ions, (f)); $Q = 200$ W.

chamber the role of the fluxes is the opposite. This behavior is visible also in the arrow presentation of $\vec{\Gamma}_e$ in Fig. 5(b).

Figure 6(e) shows a typical balance of the H^+-ions, the dominating ions in the driver. The ions are produced mainly by atom ionization and lost by their radial flux. The role of the axial flux is minor, much smaller compared to its role in the electron balance (Figs. 6(a) and (b)). A typical behavior of the H_3^+-ions – the dominating type of ions in the plasma expansion chamber – is shown in Fig. 6(f). The axial flux provides income of the ions and the losses are through their radial flux. In the second chamber $\vec{\Gamma}_i$ is always a forward flux, regardless of the direction of $\vec{\Gamma}_e$.

The results from the particle balance confirm the conclusion in [9, 10]: the plasma in the expansion chamber exists due to the particle fluxes from the driver, not due to energy fluxes sustaining ionization there.

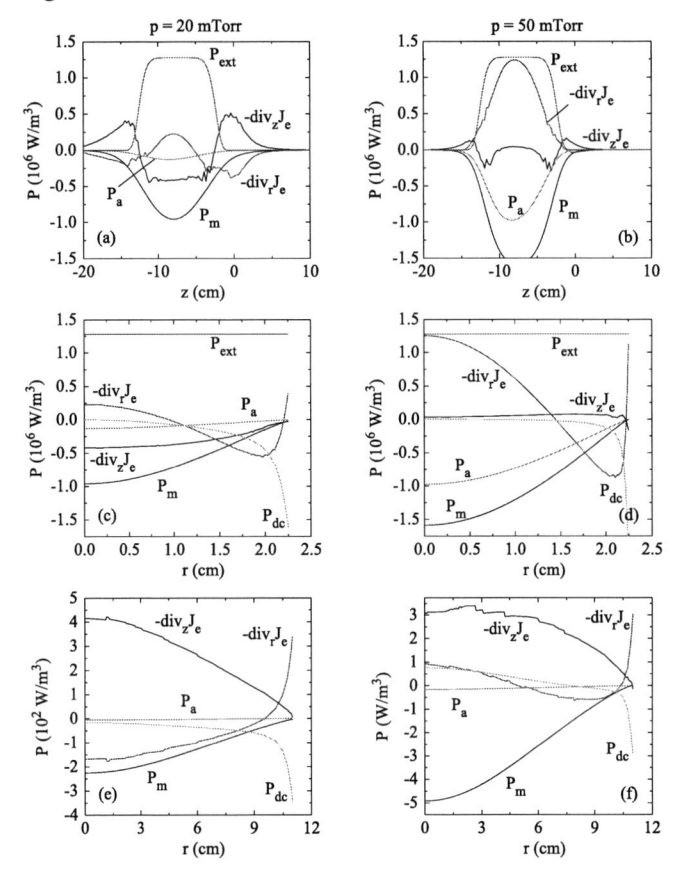

FIGURE 7. Variation of the contributors to the electron energy balance at $Q = 200$ W for $p = 20$ mTorr in (a), (c), (e) and $p = 50$ mTorr in (b), (d), (f). Axial profiles at $r = 0$ (a) and (b) and radial profiles at $z = -8$ cm (c) and (d) and at $z = 15$ cm (e) and (f).

The rf power applied to the first chamber is dissipated via energy fluxes (Figs. 5(c), (d)) and volume processes. For $p = 20$ mTorr (Fig. 7(a)), the axial flux is a mechanism of losses on the axis of the driver transporting energy to the plasma expansion region. Outside the driver, the role of this flux changes supplying the rest of the plasma volume with energy. The radial flux transfers energy from the region near the wall of the driver where T_e is higher to the axis. Outside the driver, the radial flux carries energy towards the wall and it is a loss mechanism on the axis. Electron collisions with molecules are more important as volume losses than collisions with atoms. With the pressure increase (Fig. 7(b)), the energy losses in collisions increase.

The radial electron energy balance in the driver (Figs. 7(c) and (d)) shows that near the walls the losses for sustaining the dc field dominate. These losses are compensated by the radial energy flux. In the plasma expansion chamber (Figs. 7(e) and (f)), at the position $z = 15$ cm chosen, the different behavior of the axial $\vec{\Gamma}_{ez}$-flux (a forward one and a backward one, respectively, in Figs. 5(a) and (b)) determines different role of the contributors to the electron energy balance. In both cases the total axial flux ($J_{ez} = J_{ez}^{(cond)} + J_{ez}^{(conv)}$) is an energy input. However, as it is commented on in [10], P_{dc} appears either as energy losses (Fig. 7(e)) or as an energy input (Fig. 7(f)) in the plasma interior, depending on the direction of $\vec{\Gamma}_e$ (Figs. 5(a) and (b)). Also the total radial electron energy flux changes its role. In the case of a forward $\vec{\Gamma}_e$-flux, the radial energy flux (Fig. 7(e)) is energy loss in the plasma interior. In the case of a backward $\vec{\Gamma}_e$-flux, although the total $\vec{J}_e^{(conv)}$ appears as losses [10], its radial component, due to the electron flux directed to the axis, carries energy there (Fig. 7(f)).

CONCLUSIONS

The spatial distribution of the plasma parameters and of the contributors to the particle and electron-energy balance in a tandem plasma source described in the study shows that: (i) the discharge behavior in the driver is in conformity with the mechanisms of self-consistency of hydrogen discharges, and (ii) plasma existence in the second chamber of the source is due to particle fluxes from the driver. The model will be further extended by including the negative ions. However, even at the current stage the results suggest that the optimization of the source regarding negative ion production should be based on analysis of the transport processes.

ACKNOWLEDGMENTS

The work is within the programme of the Bulgarian Association EURATOM/INRNE (task P2).

REFERENCES

1. A. Garscadden and R. Nagpal, *Plasma Sources Sci. Technol.* **4**, 268–280 (1995).
2. P. Diomede, M. Capitelli and S. Longo, *Plasma Sources Sci. Technol.* **14**, 459–466 (2005).
3. I. Koleva, Ts. Paunska, H. Schlüter, A. Shivarova and Kh. Tarnev, *Plasma Sources Sci. Technol.* **12**, 597–607 (2003).
4. Ts. Paunska, H. Schlüter, A. Shivarova and Kh. Tarnev, *Plasma Sources Sci. Technol.* **12**, 608–618 (2003).
5. I. Pérès and M. Kushner, *Plasma Sources Sci. Technol.* **5**, 499–509 (1996).
6. B. Ramamurthi and D. J. Economou, *Plasma Sources Sci. Technol.* **11**, 324–332 (2002).
7. E. Speth et al., *Nucl. Fusion* **46**, S220–S238 (2006).
8. M. Bacal, *Nucl. Fusion* **46**, S250–S259 (2006).
9. St. Kolev, A. Shivarova, Kh. Tarnev and Ts. Tsankov, *Plasma Sources Sci. Technol.* **17**, 035017 (1–13) (2008).
10. Ts. Paunska, A. Shivarova, Kh. Tarnev and Ts. Tsankov, *"2D Model of a Tandem Plasma Source: The Role of the Transport Processes"*, AIP Conf. Proc. (1st Int. Conf. on Negative Ions, Beams and Sources, Aix en Provence, 2008), submitted.
11. H. Schlüter and A. Shivarova, *Phys. Rep.* **443**, 122–245 (2007).

Characteristics of the RF Negative Ion Source Using a Mesh Grid Bias Method

O. Fukumasa[a], J. Okada[a], Y. Tauchi[a], W. Oohara[a], K. Tsumori[b] and Y. Takeiri[b]

[a]*Graduate School of Science and Engineering, Yamaguchi University, Tokiwadai 2-16-1, Ube 755-8611, Japan*
[b]*National Institute for Fusion Science, Toki 509-5292, Japan*

Abstract. Using a mesh grid bias method for plasma parameter control, volume production of hydrogen negative ions H⁻ is studied in pure hydrogen rf plasmas. Relationship between the extracted H⁻ ion currents and plasma parameters is discussed. Both high and low electron temperature T_e plasmas are produced in the separated regions when the grid is negatively biased. In addition, with changing grid potential V_g, values of n_e increase while T_e decrease in their values. The negative ion production depends strongly on the grid potential and related plasma conditions. It is also confirmed that grid bias method is more effective than the so-called magnetic filter method to optimize plasma parameters for H⁻ production. Production of deuterium negative ions D⁻ is also studied briefly.

Keywords: Negative ion source, RF plasma, mesh grid bias method, magnetic filter method
PACS: 29.25.Ni, 52.20.-j, 52.27.Cm, 52.80.Pi

INTRODUCTION

Development of negative ion sources for neutral beam injection (NBI) system is required for large experimental fusion devices such as the ITER. In the present negative ion sources, source plasma is generated by dc arc discharge, so the lifetime of the ion source is limited to several hundred hours due to erosion and fatigue of the cathode filaments. Thus a long-lifetime ion source is required for future NBI systems. Microwave-discharge ion sources [1] and rf-driven ion sources [2, 3] are promising as long-lifetime ion sources because they have no filaments.

In pure hydrogen (H₂) plasmas, most of the H⁻ ions are generated by the dissociative attachment of slow plasma electrons (electron temperature $T_e \sim 1$ eV) to highly vibrationally excited hydrogen molecules H₂(v'')(effective vibrational level $v'' \geq$ 5-6) [4]. These molecules are mainly produced by collisional excitation of fast electrons with optimum energy of 40 eV. Therefore, spatial control of electron energy distribution (i.e., T_e) is necessary [5, 6]. With the use of a magnetic filter (MF), the electron energy distribution function in dc plasmas is well controlled for H⁻ formation. Unfortunately, in rf discharge plasmas, plasma parameter control with the MF is not so effective as one in dc plasmas [1, 2]. On negative ion sources, another important and interesting aspect is to develop the ion sources without filaments, MF and cesium injection. Therefore, using a mesh grid bias (MGB) method [7, 8] for plasma parameter control, volume production of hydrogen negative ions H⁻ is studied in pure hydrogen rf plasmas [9].

The purpose of the present study is as follows [9]: one is to investigate the possibility of controlling plasma parameters, in particular T_e, with the MGB method in rf-driven plasmas; and the other is to realize negative ion production in rf plasmas and to discuss the difference in T_e control and H⁻ production between the MGB method and the MF method. Finally, experimental results on isotope effect of H⁻/D⁻ production are presented and also briefly discussed.

EXPERIMENTAL SET-UP AND PROCEDURE

Figure 1 shows a schematic diagram of rf negative ion source. The source chamber (21 cm in diameter) made of stainless steel is divided by a mesh grid (MG) flange into two parts, i.e., a source region and an extraction region. The rf power (13.56 MHz) is supplied to a stainless circular disk antenna with 12 cm diameter. In the present experiment, both hydrogen (H_2) and deuterium (D_2) gases are used under a pressure of 1-6 mTorr., and the rf power is varied from 100 to 300 W. To produce both high and low electron temperature plasmas separately in the chamber, the mesh grid of 20 cm in diameter is placed (18.5 cm away from the rf antenna). Three different meshes are used, i.e., No. 1 (7 mesh/in.), No. 2 (30 mesh/in.), and No. 3 (50 mesh/in.). Details are shown in Table 1. Compared with the usual magnetically filtered multicusp ion source, in Fig. 1, the magnetic filter (MF) flange (i.e., the rod-type filter) is set instead of the MG flange. The present MF is composed of four rods where diameter of the rod is 10 mm and the distance between two rods is 54 mm. In the present experiment, magnetic field intensities of the MF are 60 and 100G.

The negative ion currents are directly detected by a magnetic deflection-type ion analyzer. The plasma grid has a single hole (10 mm in diameter) through which negative ions were extracted from the ion source. Plasma parameters (electron density n_e, electron temperature T_e, plasma space potential V_s, and floating potential V_f) are measured by three Langmuir probes. The one is movable along the axial direction from the mesh grid in the extraction region. The other two are set at $z = 3.7$ cm and $z = 11$ cm in the source region.

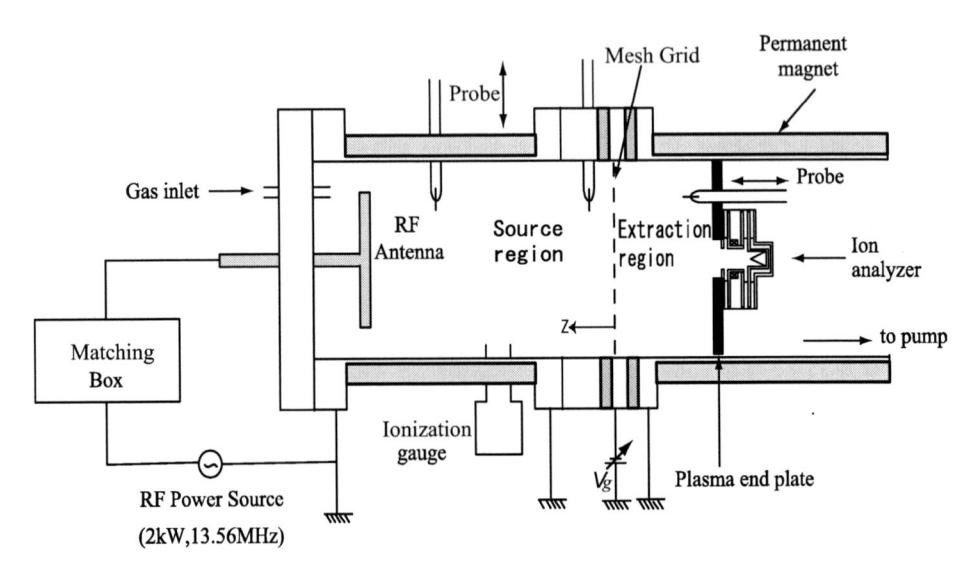

FIGURE 1. Schematic diagram of the experimental apparatus.

TABLE 1. Dimensions of three meshes used in the present experiment.

Mesh	Mesh size (mesh/in.)	Diameter of wire (mm)	Distance between two wires (mm)	Geometric transmittance (%)
No.1	7	1.03	2.36	48.5
No.2	30	0.25	0.597	49.7
No.3	50	0.05	0.458	81.3

EXPERIMENTAL RESULTS AND DISCUSSION

Plasma Production and Control

The aim of the present work is production and control of rf plasmas to enhance and to optimize H⁻ volume production in rf plasmas. So far, we have confirmed that the grid bias method is effective for H⁻ production in dc discharge H_2 plasmas [10-12]. Now, we test application feasibility of the grid bias method for rf discharge plasmas [9].

Figure 2 shows axial distributions of plasma parameters (i.e., n_e and T_e) for two different controlling methods. The one is the mesh grid bias (MGB) method and the other is the magnetic filter (MF) method.

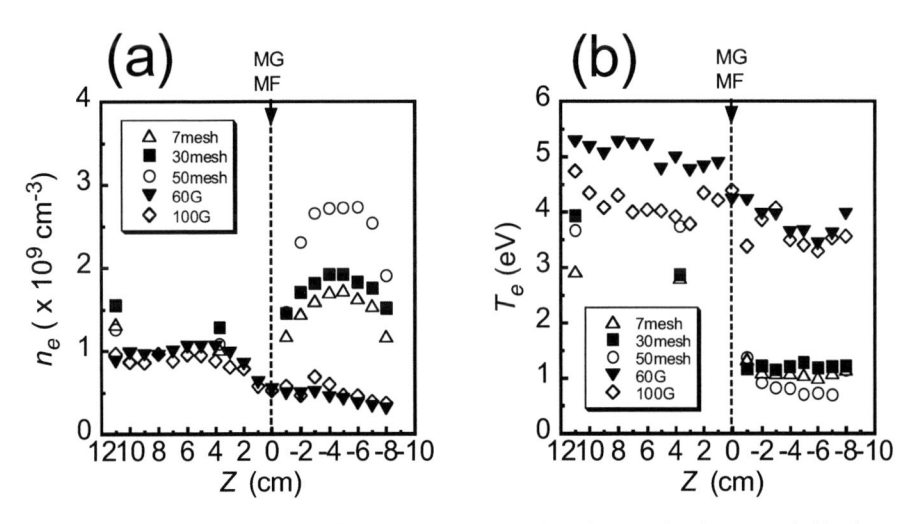

FIGURE 2. Axial distributions of plasma parameters: (a) Electron density n_e and (b) electron temperature T_e for three different mesh grids and two different MF fields. The end plate is set at z_{end} = -10.5 cm. Experimental conditions: input rf power P_{rf} = 200W, gas pressure $p(H_2)$ = 3 mTorr, and grid biasing voltage V_g = -50V.

Using the MF method, T_e is not well decreased in the extraction region: namely, $T_e \simeq$ 4 eV. The value of T_e is high for negative ion production. Although T_e is decreased by increasing the field intensity of the MF, B_{MF}, at the same time, n_e is also decreased more drastically. On negative ion volume production, a desired condition for plasma parameters is as follows: T_e in the extraction region should be reduced below 1 eV with n_e keeping higher. Therefore, in rf plasmas, plasma parameter control with using the MF method is not so effective.

Similar measurements have been made with using a mesh grid bias (MGB) method. High energy electrons pass the mesh (set at $z = 0$ cm) and enter into the extraction region. As is shown clearly, n_e increases in its value with z and reaches the maximum value and then decrease while T_e decreases in its value and keeps nearly equal to or lower than 1 eV. This is suitable condition for negative ion volume production. The mechanism is explained as follows: In the extraction region, the neutral particles are ionized by higher energy electrons flowing from the source region through the grid. The electrons produced in the extraction region cannot be accelerated by the external electric fields, because no additional heating power is fed into this region. Therefore, low energy electrons are generated in the extraction region. According to the results shown in Fig. 2, the MGB method is more suitable to optimize plasma conditions for negative ion volume production, compared with the MF method.

Figure 3 shows the dependence of plasma parameters on grid bias voltage for three different MGs in the extraction region. With decreasing V_g, values of n_e increase while T_e decrease in their values. Namely, there appears a clear decrease in T_e with a decrease in grid biasing voltage V_g. We also measure the T_e at $z = 3.7$ cm in the source region and obtain $T_e \simeq 3$ eV which is almost independent of V_g. Therefore, T_e in the extraction region has a lower value than that in the source region. Behaviors of plasma parameters are nearly the same as the ones in dc plasmas [10-12].

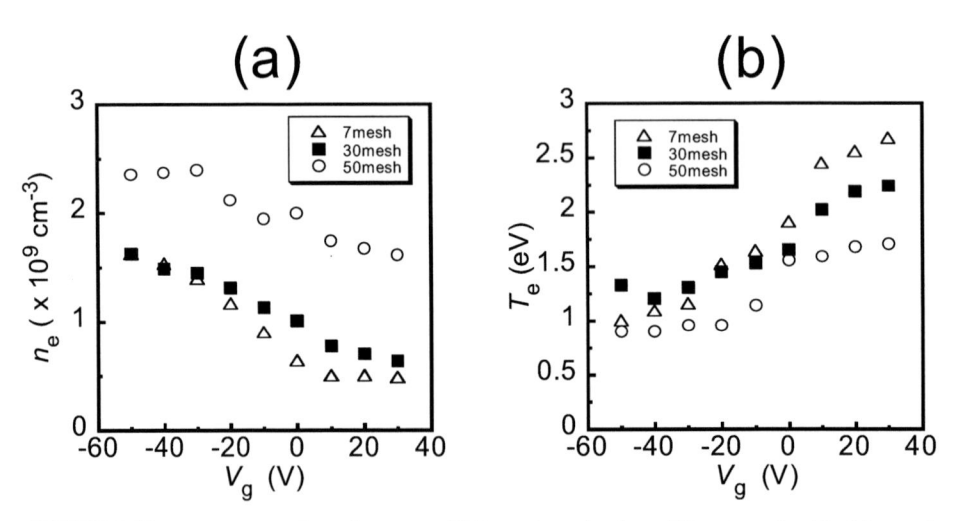

FIGURE 3. Plasma parameter dependence on grid bias voltage for three different mesh grids: (a) n_e and (b) T_e measured at $z = -5$ cm in the extraction region. The end plate is set at $z_{end} = -10.5$ cm. Experimental conditions: $P_{rf} = 200$W and $p(H_2) = 3$ mTorr.

The MGB method is the method to control electron transport electrostatically and is affected by the mesh size. Then, with using three different MGs, the effect of mesh size is also tested. We estimate the Debye length (λ_D) from the typical plasma parameters (i.e., $n_e \approx 1 \times 10^9 \text{cm}^{-3}$ and $T_e \approx 3.0$ eV) at $z = 3.7$ cm in the source region and obtain that $\lambda_D \approx 0.41$ mm. It is assumed that the sheath length is about several times of λ_D, then, distance between two wires of all meshes are within the sheath length. In the extraction region, n_e with No. 3 mesh is higher than other two meshes (i.e., No. 1 and No. 2). This is caused mainly by the difference of transmittance. It is easy for electrons to enter across the MG into the extraction region as No. 3 mesh has high transmittance.

We have confirmed numerically [13] that extraction probability of negative ions depends strongly on upstream distance from the extraction grid, i.e., the plasma end plate in the present case. At any rate, to increase the extraction of negative ion currents, production of negative ions near the plasma end plate should be enhanced by optimizing plasma conditions.

Negative Ion Production

We discuss H⁻ ion production in the present rf plasmas. According to the results shown in Fig. 2, plasma parameters in the extraction region and then production of negative ion depend on the position of the end plate. Figure 4 shows the effect of end plate position z_{end} on extracted negative ion currents I_{H^-} (i.e., H⁻ ion production). At first, I_{H^-} increases with z_{end}, reaches maximum at $z_{end} = -5$ to -7 cm, and then decreases. The changes in I_{H^-} are caused by the plasma parameters in the extraction region. For reference, two I_{H^-} from the plasmas with the MF methods (i.e. $B_{MF} = 60$ G and 100 G) are also plotted. These two I_{H^-} are much lower than those three I_{H^-} from the plasmas with the MGB method due to the difference of plasma parameters as shown in Fig. 2. Plasma production and then H⁻ ion production also depend on hydrogen gas pressure $p(H_2)$.

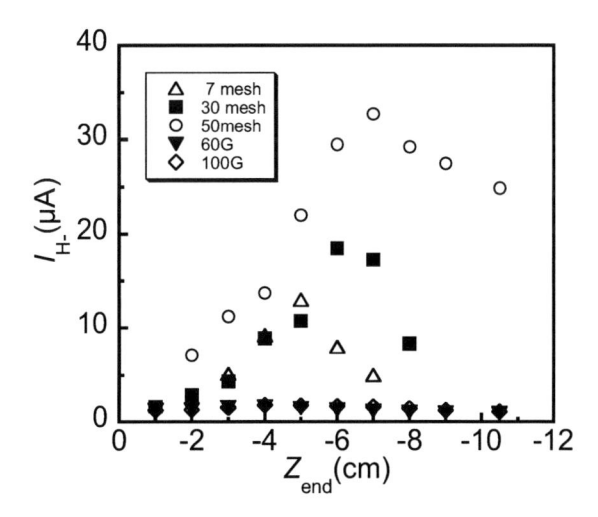

FIGURE 4. Extracted negative ion currents as a function of the end plate position for three different mesh grids and for two different magnetic filters. Experimental conditions: $P_{rf} = 200$W, $p(H_2) = 3$ mTorr, $V_g = -50$V, and extraction voltage $V_{ex} = 1$ kV.

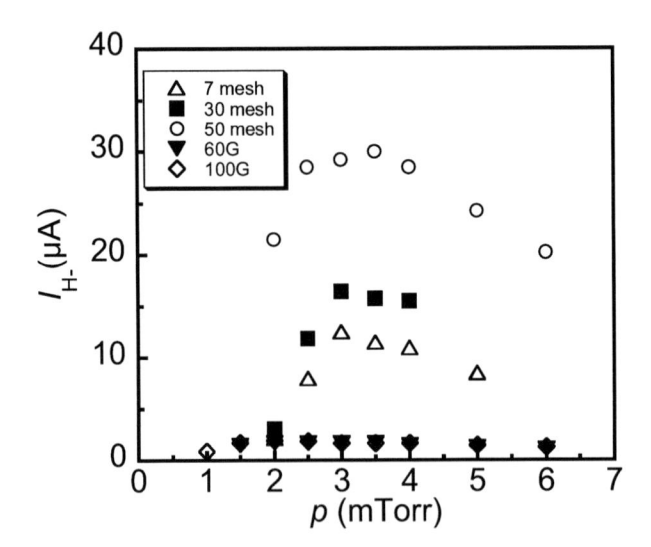

FIGURE 5. Extracted negative ion currents as a function of gas pressure for three different mesh grids and for two different magnetic filters. Experimental conditions: P_{rf} = 200W, V_g = -50V, and V_{ex} = 1 kV. However, z_{end} (7mesh) = -5 cm, z_{end} (30mesh) = -6 cm, z_{end} (50mesh) = -7 cm, and z_{end} (60G) = z_{end} (100G) = -4 cm respectively.

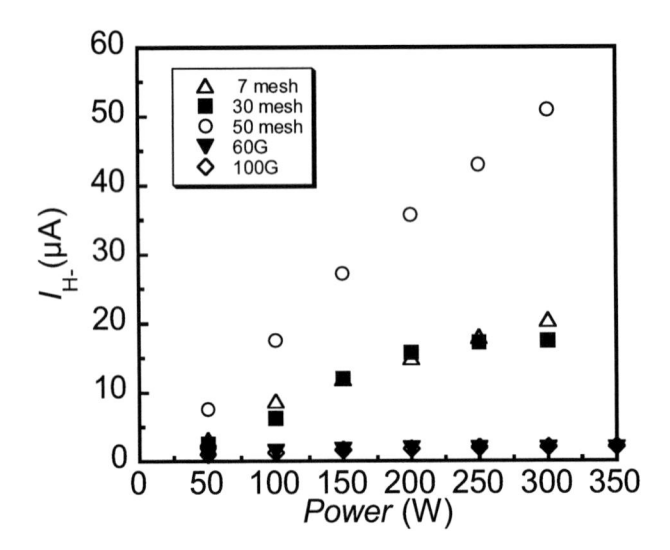

FIGURE 6. Extracted negative ion currents as a function of rf power for three different mesh grids and for two different magnetic filters. Experimental conditions: $p(H_2)$ = 3 mTorr, V_g = -50V, and z_{end} (7mesh) = -5 cm, z_{end} (30mesh) = -6 cm, z_{end} (50mesh) = -7 cm, respectively, for grid bias method. $p(H_2)$ = 2 mTorr and z_{end} = -4 cm for the magnetic filter method. In both cases, V_{ex} = 1 kV.

Typical example of pressure dependence on I_{H^-} is shown in Fig. 5. There are certain optimum pressures p_{opt}. Namely, the changes in I_{H^-} are caused by H⁻ ion production in the extraction region.

Figure 6 shows the rf power dependence of I_{H^-}. According to the results shown in Figs. 4 and 5, pressure and z_{end} are optimized. In addition, rf power dependence of I_{H^-} with the MF method is also shown. With increasing power, I_{H^-} increases linearly. I_{H^-} in the MGB method is much higher than that in the MF method. This difference is caused by plasma conditions. Within the present experimental conditions, plasma production and control in the extraction region are well done for negative ion production when No. 3 mesh is set due to its high transmittance.

As is shown clearly, extracted negative ion currents from plasmas with the MGB method is much higher than ones from plasmas with the MF method. These differences in extracted negative ion currents, i.e. negative ion production, are cased by difference in plasma conditions (i.e. mainly electron temperature) as shown in Fig. 2. Then, according to those results shown in Figs. 2, 4-6, we have confirmed that negative ions in the source are produced by the two-step volume process described in the introduction. However, in the MGB method, the grid is biased negatively and then bombarded by accelerated positive ions. So, a certain surface effect of negative ion production may be occurred, namely recycling H into $H_2(v")$ [14] and Cs-free surface production [15]. Recently, it is also pointed out by V. Dudnikov [16] that our recent results [9] on dependence of H⁻ current versus a grid voltage and gas pressure are very compatible with surface plasma generation of negative ions. We are now studying the effect of mesh grid materials (stainless steel, Ni and Mo) on negative ion production.

Isotope Effect of H⁻/D⁻ Production

Finally, we present the preliminaly results concerning H⁻/D⁻ production and discuss briefly isotope effect of H⁻/D⁻ production.

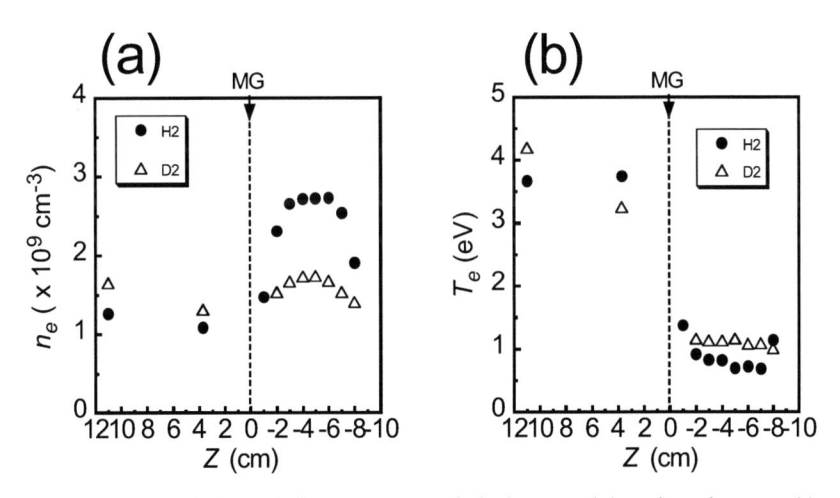

FIGURE 7. Axial distributions of plasma parameters in hydrogen and deuterium plasmas: (a) n_e and (b) T_e. Experimental conditions: P_{rf} = 200W, $p(H_2)$ = $p(D_2)$ = 3 mTorr, V_g = -50V, z_{end} = -10.5 cm, and mesh size is 50 mesh/inch.

Figure 7 shows axial distributions of plasma parameters (i.e. n_e and T_e). In the extraction region, n_e in H_2 plasmas is higher than one in D_2 plasmas. On the other hand, T_e in D_2 plasmas is slightly higher than one in H_2 plasmas.

Figure 8 shows the effect of z_{end} on the extracted negative ion currents I_{H^-} and I_{D^-}. Two currents have nearly the same dependence on z_{end}. Figure 9 shows the rf power dependence of I_{H^-} and I_{D^-}. With increasing rf power, the difference between I_{H^-} and I_{D^-} becomes remarkable although extraction voltage is set to be the same value in both cases.

So far, we have studied application feasibility of the MGB method for negative ion production in low density rf plasmas. In the future, we will study further this program in high density plasmas including the effect of grid materials.

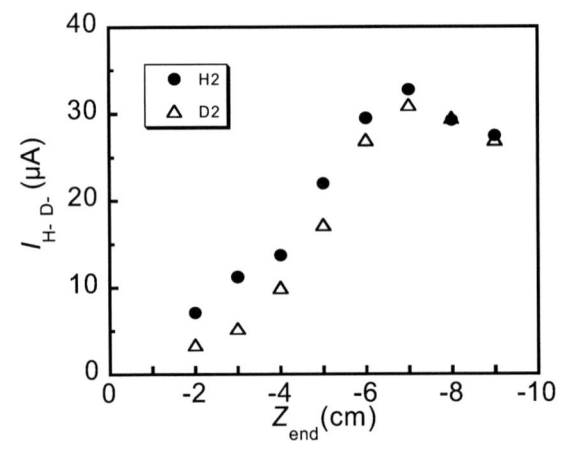

FIGURE 8. Extracted negative ion currents as a function of the end plate position. Experimental conditions: $P_{rf} = 200W$, $p(H_2) = p(D_2) = 3$ mTorr, $V_g = -50V$, $V_{ex} = 1$ kV, and mesh size is 50 mesh/inch.

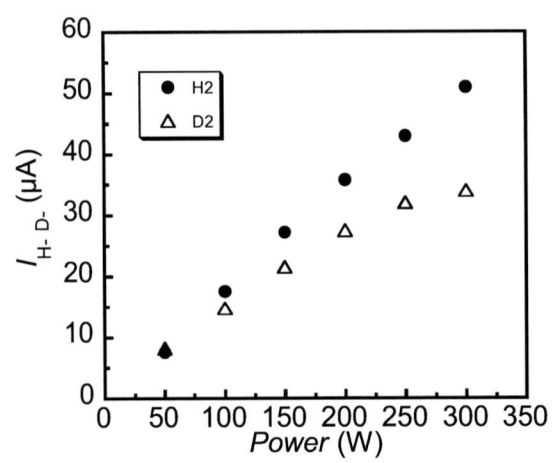

FIGURE 9. Extracted negative ion currents as a function of rf power. Experimental conditions: $p(H_2) = p(D_2) = 3$ mTorr, $V_g = -50V$, $z_{end} = -7$ cm, $V_{ex} = 1$ kV and mesh size is 50 mesh/inch.

SUMMARY

Control of plasma parameters and enhancement of extracted H⁻ ion currents in rf plasmas are studied by using the magnetic filter method and the mesh grid bias method. With using the magnetic filter method, electron temperature in the extraction region is not well controlled for optimizing H⁻ ion formation. However using the mesh grid bias method, it is confirmed that both high and low electron temperature plasmas are produced in the separated regions in the chamber, respectively, when grid potential V_g is negative. Extracted negative ion currents I_{H^-} depend on some experimental conditions and I_{H^-} in the grid bias method is much higher than that in the magnetic filter method. H⁻/D⁻ volume production is also discussed briefly. As, in the future, rf negative ion source is required for the NBI systems, the mesh grid bias method is quite useful to control and enhance negative ion volume production in rf plasmas.

ACKNOWLEDGMENTS

The authors would like to thank Professor emeritus N. Sato, Tohoku University, for his interest and encouragement to this work. They also thank T. Yoshioka and T. Nakano (Yamaguchi University) for their assistances. A part of this work was supported by the Grant-in-Aid for Scientific Research from Japan Society for the Promotion of Science. This work was also performed with the support of the NIFS Research Collaboration.

REFERENCES

1. O. Fukumasa and M. Matsumori, Rev. Sci. Instrum. **71**, 935(2000).
2. T. Takanashi, Y. Takeiri O. Kaneko, Y. Oka, K. Tsumori and T. Kuroda, Rev. Sci. Instrum. **67**, 1024(1996).
3. E. Speth et al., Nucl. Fusion **46**, S220(2006).
4. O. Fukumasa, J. Phys. **D22**, 1668 (1989).
5. O. Fukumasa, S. Mori, N. Nakada, Y. Tauchi, M. Hamabe, K. Tsumori and Y. Takeiri, Contrib. Plasma Phys. **44**, 516(2004).
6. O. Fukumasa and S. Mori, Nucl. Fusion **26**, S287 (2006).
7. K. Kato, S. Iizuka and N. Sato, Appl.Phy. Lett. **65**, 816 (1994).
8. S. Iizuka, K. Kato, A. Takahashi, K. Nakagomi N. Sato, Jpn. J. Appl. Phys., Part 1 **36**, 4551(1997).
9. J. Okada, Y. Nakao, Y. Tauchi and O. Fukumasa, Rev. Sci. Instrum. **79**, 02A502(2008).
10. Y. Nakao, D. Ito, J. Ono, Y. Tauchi and O. Fukumasa; Proceedings of the 6th International Conference on Reactive Plasmas and 23rd Symposium on Plasma Processing, (2006), p.185.
11. O. Fukumasa, D. Ito and Y. Jyobira, Proceedings of the 11th Int. Symp. PNNIB, AIP CP **925**, (2007), p31.
12. Y. Jyobira, D. Ito and O. Fukumasa, Rev. Sci. Instrum. **79**, 02A508(2008).
13. O. Fukumasa and R. Nishida, Nucl. Fusion **46**, S220(2006).
14. M. Bacal, Nucl. Fusion **46**, S250(2006).
15. A. Ueno, K. Ikegami and Y. Kondo, Rev. Sci. Instrum. **75**, 1714(2004).
16. V. Dudnikov, private communication.

Enhancement of D⁻ Negative Ion Volume Production in Pure Deuterium Plasmas

O. Fukumasa[a], T. Nakano[a], S. Mori[a], W. Oohara[a], K. Tsumori[b] and Y. Takeiri[b]

[a]*Graduate School of Science and Engineering, Yamaguchi University, Tokiwadai 2-16-1, Ube 755-8611, Japan*
[b]*National Institute for Fusion Science, Toki 509-5292, Japan*

Abstract. Pure volume production of D⁻ negative ions is studied in rectangular negative ion source equipped with an external magnetic filter (MF). Plasma parameters (n_e and T_e) in D_2 plasmas are varied mainly in the downstream region by changing the magnetic field intensity of the MF (i.e., B_{MF}). Production and control of D_2 plasma is nearly the same as that of H_2 plasmas, although the values of n_e and T_e in D_2 plasma are slightly higher than that of H_2 plasmas in both the source and the extraction regions. On D⁻/H⁻ production, however, it appears some different points. By varying the B_{MF}, H⁻ production in the extraction region is remarkably changed corresponding to the variation of n_e and T_e in the extraction region. On the other hand, D⁻ production is not so varied under the same discharge conditions in H_2 plasmas. This difference in D⁻ production is not well explained only variation of n_e and T_e in D_2 plasmas. Optimum B_{MF} and gas pressure for D⁻ production is slightly higher than that for H⁻ production.

Keywords: negative ion source, volume production, magnetic filter, isotope effect of D⁻/H⁻ production
PACS: 29.25.Ni, 52.20.-j, 52.27.Cm

INTRODUCTION

Sources of H⁻ and D⁻ negative ions are required for efficient generation of neutral beams with energies above ≈ 100 keV/nucleon. The magnetically filtered multicusp ion source has been shown to be a promising source of high-quality multiampere H⁻ ions. In pure hydrogen (H_2) discharge plasmas, most of the H⁻ ions are generated by the dissociative attachment of slow plasma electrons e_s (electron temperature $T_e \sim 1$ eV) to highly vibrationally excited hydrogen molecules H_2 (v'') (effective vibrational level $v'' \geq 5\text{-}6$). These $H_2(v'')$ are mainly produced by collisional excitation of fast electrons e_f with energies in excess of 15-20 eV. Namely, H⁻ ions are produced by the following two-step process [1, 2]:

$$H_2\ (X^1\Sigma_g,\ v'' = 0)\ +\ e_f\ \rightarrow\ H_2^*(B^1\Sigma_u,\ C^1\Pi_u)\ +\ e_f' \quad (1a)$$

$$H_2^*(B^1\Sigma_u,\ C^1\Pi_u)\ \rightarrow\ H_2\ (X^1\Sigma_g,\ v'')\ +\ h\nu \quad (1b)$$

$$H_2\ (v'')\ +\ e_s\ \rightarrow\ H^-\ +\ H. \quad (2)$$

where H_2 ($X^1\Sigma_g$) means the ground electronic state of the hydrogen molecule and $H_2^*(B^1\Sigma_u)$ and $H_2^*(C^1\Pi_u)$ mean the excited electronic states. The transitions from the $B^1\Sigma_u$ and $C^1\Pi_u$ levels to the ground state, $X^1\Sigma_g$, of H_2 are termed the Lyman and Werner Bands, respectively, and are found in the vacuum ultraviolet (VUV) region.

The reaction-producing D⁻ ions is believed to be the same as that for the production of H⁻ ions. To develop efficient D⁻ ion sources with high current density, it is important to clarify production and control of deuterium (D_2) plasmas and to understand difference in the two-step process of negative ion production between H_2 plasmas and D_2 plasmas. Caesium (Cs) seeding of this type of ion source is often used as it enhances the extracted negative ion currents and reduces the extracted electron currents. There are some studies on optimization of volume-produced D⁻ ion with or without Cs [3-5]. However, here we focus on understanding the negative ion production mechanisms in the "volume" ion source where negative ions are produced in low-pressure pure H_2 or D_2 discharge plasmas.

For this purpose, we are interested in estimating densities of highly vibrationally excited molecules and negative ions in the source. The production process of $H_2(v'')$ or $D_2(v'')$ is discussed [6] by observing the photon emission, i.e. VUV emission associated with the process (1b) [7, 8]. To clarify the relationship between plasma parameters and volume production of negative ions, the densities of H⁻ or D⁻ ions in the source are measured [9] by the laser photodetachment method [10]. The influence of plasma parameter distributions on H⁻ or D⁻ production is discussed using estimated rate coefficients and collision frequencies based on measured plasma parameters [9, 11, 12]. In this article [13], plasma parameters are controlled by using the usual magnetic filter method although we have shown recently that mesh grid bias method is also useful to optimize plasma conditions for negative ion volume production [14,15]. To study further the isotope effect of D⁻/H⁻ production, we discuss the relationship between negative ions in the source and extracted negative ion currents, including the results on VUV emission measurements.

EXPERIMENTAL SET-UP

Figure 1(a) shows a schematic diagram of the ion source [9, 11-13]. The rectangular arc chamber is 25 cm × 25 cm in cross section and 19 cm in height. Four tungsten filaments with 0.7 mm in diameter and 20 cm in length are installed in the source region

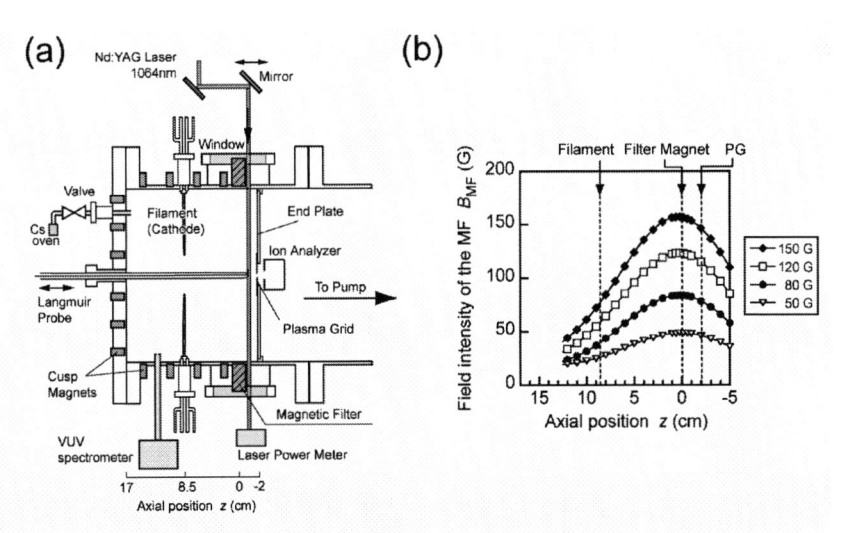

FIGURE 1. (a) Schematic diagram of the ion source. The probe, the laser path and power meter used in photodetachment experiments are also shown. (b) Axial distributions of field intensities for four different MFs.

from side walls of the chamber. The line cusp magnetic field is produced by permanent magnets which surrounded the chamber. The external magnetic filter (MF) is composed of a pair of permanent magnets in front of the plasma grid (PG). Figure 1(b) shows profiles of the field intensities for five different MFs along the axis of the ion source. The different MFs are created by changing the distance between the pair of the permanent magnets. These MFs gradually separate the extraction region from the source region with filaments. In the present experiment, using these MFs, production and control of H_2 and D_2 plasmas to enhance negative ion volume production are studied. The end plate is kept at floating potential and the PG potential is kept at ground potential throughout the present experiments for both H_2 and D_2 plasmas.

In the source region, the VUV emission measurements related to the $H_2(v")$ or $D_2(v")$ production, i.e. the process (1b), are carried out by using the VUV spectrometer. The spectrometer was normally operated at a resolution of 0.1 nm. The optical pipe is equipped with collimators such that a rectangular plasma volume with a cross section of about $6 \times 3 mm^2$ is imaged. Emission intensity yields results averaged over the line of sight.

The plasma parameters are measured by an axially movable cylindrical Langmuir probes, supported by a quartz glass pipe with diameter of 3 mm. This probe is also used to measure negative ion density. Negative currents are extracted through a single hole 5 mm in diameter on the PG. These currents are introduced into a magnetic deflection type ion analyzer for relative measurements of the extracted H⁻ or D⁻ current. On the other hand, H⁻ or D⁻ densities in the source are measured by the laser photodetachnment method [3, 10].

EXPERIMENTAL RESULTS AND DISCUSSION

Production and Control of D₂ Plasmas

On H⁻/D⁻ volume production, desired condition for plasma parameters is as follows: T_e in the extraction region should be reduced below 1 eV while keeping n_e higher. To realize this condition, namely to enhance H⁻/D⁻ production by dissociative attachment

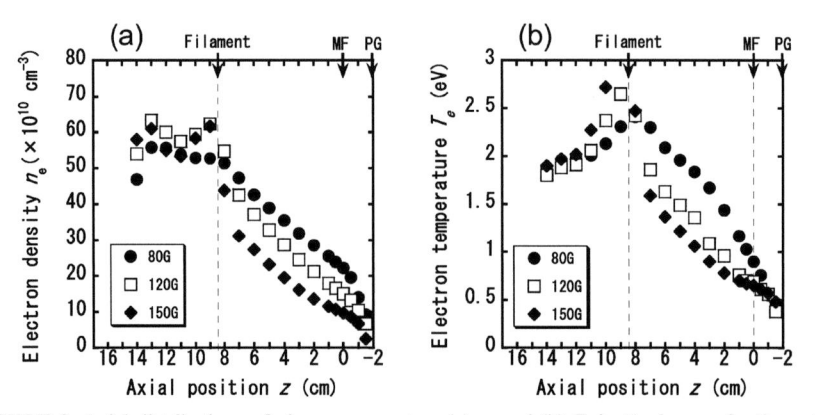

FIGURE 2. Axial distributions of plasma parameters (a) n_e and (b) T_e in D_2 plasmas for three different magnetic filters. The end plate is set at $z_{end} = -2$ cm. Experimental conditions are as follows: discharge voltage $V_d = 70$ V, discharge current $I_d = 10$ A, and gas pressure $p(D_2) = 2$ mTorr.

and to reduce H⁻/D⁻ destruction by electron detachment including collisions with energetic electrons, the MF is used. For this purpose, plasma parameter control is studied by varying the intensity of the MF.

As is shown previously [11, 12], as a whole, the axial profiles of n_e and T_e in hydrogen (H_2) and deuterium (D_2) plasmas have nearly the same patterns with each other. Figure 2 shows a typical example for axial distributions of n_e and T_e in D_2 plasmas for three different values of B_{MF}. The corresponding axial distributions of n_e and T_e in H_2 plasmas are also shown in Fig. 3. In general, for the same discharge conditions, both n_e and T_e in D_2 plasmas are higher than ones in H_2 plasmas. A stronger MF field is required for control of T_e in D_2 plasmas. This indicates that plasma production and transport are different in D_2 and H_2 plasmas, respectively.

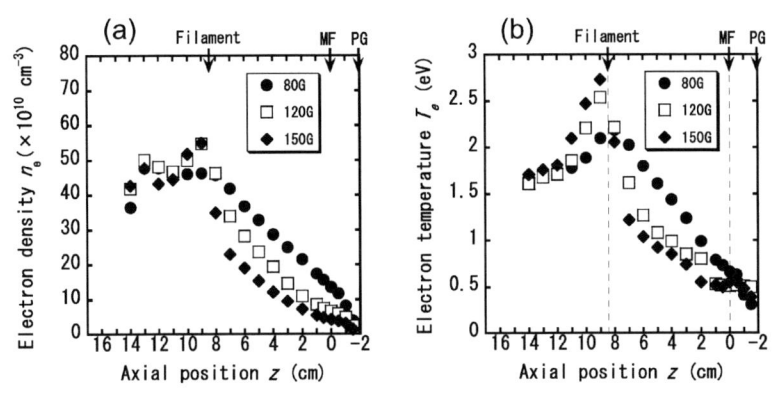

FIGURE 3. Axial distributions of plasma parameters (a) n_e and (b) T_e in H_2 plasmas for three different magnetic filters. Experimental conditions are the same as in Fig. 2.

However, these plasma conditions are well controlled by choosing the favorite combination of gas pressure and B_{MF}. To clarify the details in the extraction region, we show the enlarged axial distributions of n_e and T_e in D_2 and H_2 plasmas, respectively, in Figs. 4 and 5. Here, gas pressures are set to about optimum conditions, respectively (see Fig.7).

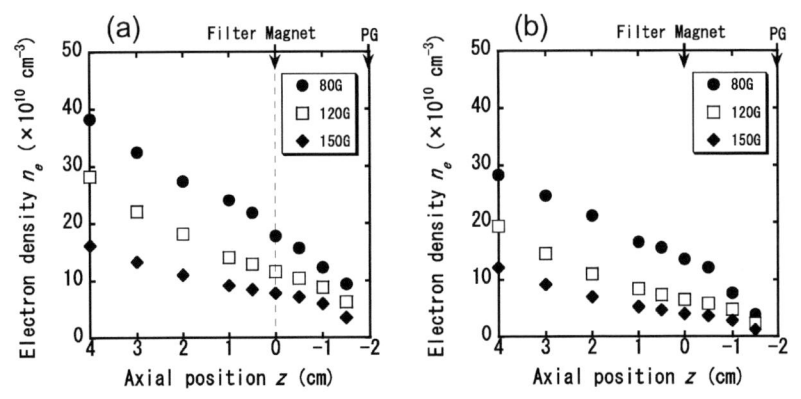

FIGURE 4. Axial distributions of electron densities in (a) D_2 plasmas and (b) H_2 plasmas for therr different magnetic filters. Experimental conditions are as follows: V_d = 70 V, I_d = 10 A, $p(D_2)$ = 4 mTorr, $p(H_2)$ = 2 mTorr and z_{end} = -2 cm.

Parameter is B_{MF}. When B_{MF} = 80 G, values of n_e and T_e in D_2 plasmas are higher than those in H_2 plasmas. T_e in the extraction region is decreased below 1 eV in both D_2 and H_2 plasmas. The plasma conditions are good for D^- and H^- volume productions.

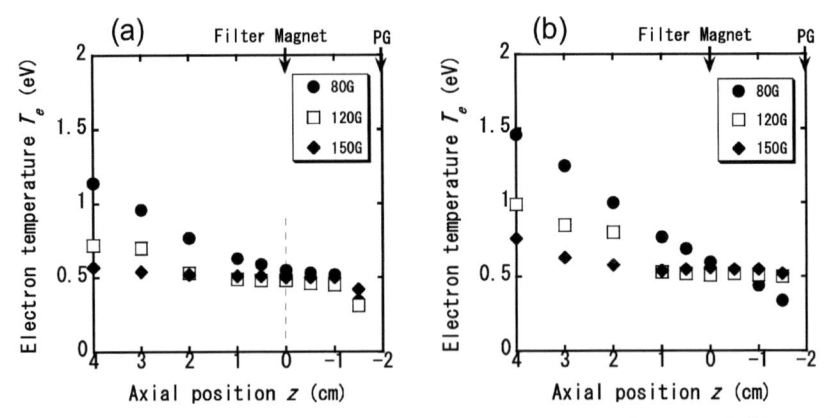

FIGURE 5. Axial distributions of electron temperature in (a) D_2 plasmas and (b) H_2 plasmas. Experimental conditions are the same as on Fig. 4.

Production and Extraction of Negative Ions

Plasma parameters in the extraction region depend strongly on the MF intensity, and therefore plasma conditions for negative ion volume production are also varied [11-13]. As a consequence, the extracted negative ion currents are found to be strongly dependent on B_{MF}. The extraction probability of negative ions depends strongly on the distance from the extraction electrode [16]. To increase the extraction of negative ion currents, the production of negative ions near the extraction electrode should be enhanced by optimizing the plasma conditions.

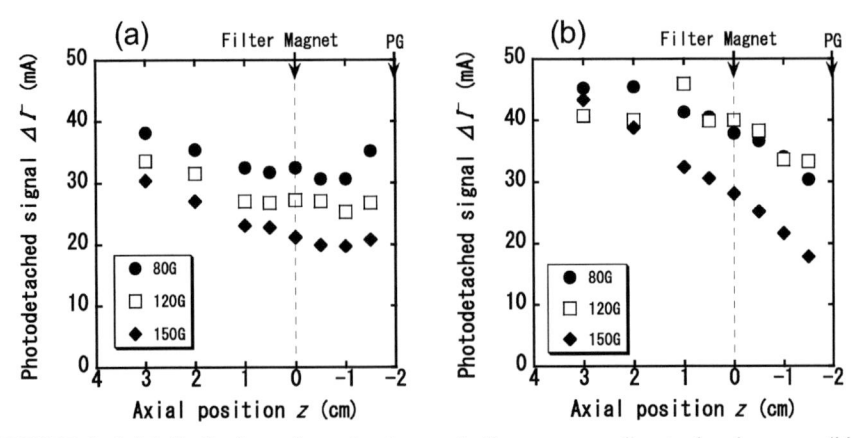

FIGURE 6. Axial distributions of negative ion productions corresponding to the plasma conditions in Figs. 4 and 5: (a) $p(D_2)$ = 4 mTorr and (b) $p(H_2)$ = 2 mTorr.

Figure 6 shows axial distribution of negative ion density in D_2 plasmas and in H_2 plasmas, respectively, corresponding to the plasma conditions shown in Figs. 4 and 5. Parameter is B_{MF}. In these figures, not a negative ion density but a photodetached electron current $\Delta\Gamma$ is plotted. For measurement of $\Delta\Gamma$, probe bias voltage is kept at 20 V.

By changing B_{MF}, plasma parameters in the downstream region (i.e., z = 4 to -2 cm) are varied. Then, negative ion production is varied and the measured $\Delta\Gamma$ is also varied in both D_2 and H_2 plasmas. However, the variation of $\Delta\Gamma$ in D_2 plasmas is lower than that in H_2 plasmas although variation of plasma parameters in both plasmas is nearly the same. At any rate, negative ion production in the vicinity of the end plate is nearly equal with each other.

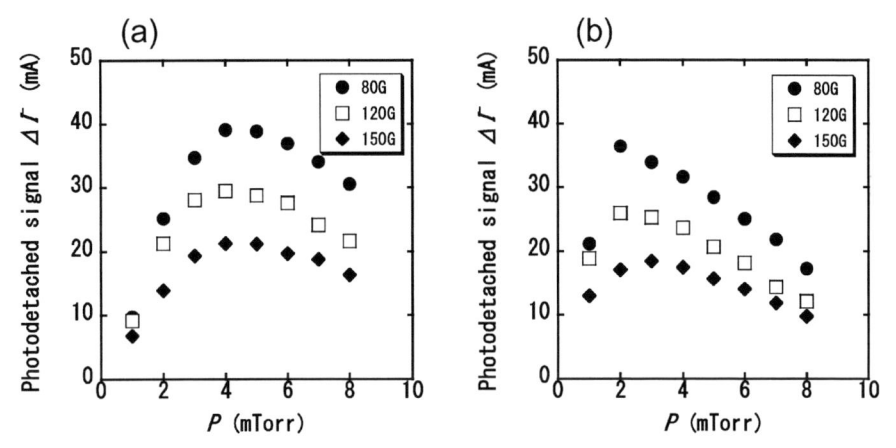

FIGURE 7. Pressure dependence of negative ion densities in the vicinity of the extraction electrode: (a) D⁻ density and (b) H⁻ density. Parameter is B_{MF}. Experimental conditions are as follows: V_d = 70 V and I_d = 10 A.

Figure 7 shows the pressure dependence of negative ion densities in (a) D_2 and (b) H_2 plasmas. In both cases, as described above, the negative ion densities are varied due to the change in plasma conditions with changing the magnetic field intensity B_{MF} of the MF. As shown clearly, there are some optimum pressures. With increasing gas pressure, negative ion densities increase in their magnitude, reach the maximum value, and then, decrease. Decreasing the B_{MF}, the optimum pressure p_{opt} shifts to higher pressure. For D⁻ production, p_{opt} is changed from 4 to 5 mTorr. On the other hand, for H⁻ production, p_{opt} is from 3 to 2 mTorr. Optimum pressure in D_2 plasmas is slightly higher than one in H_2 plasmas.

In Figs. 8 and 9, corresponding plasma parameters (n_e and T_e) are shown as a function of pressure. With increasing pressure, n_e in D_2 plasmas keeps nearly constant and n_e in H_2 plasmas has nearly the same manner although values of n_e is lower than ones in D_2 plasmas. Values of T_e are nearly equal to each other and are decreased gradually with gas pressure. According to variation of plasma parameters shown in Figs. 8 and 9, H⁻ production rate (i.e. disassociative attachment) keeps nearly the same value and then could not be well explained the pressure dependence of negative ion production shown in Fig. 7.

The corresponding extracted negative ion currents, I_{D^-} and I_{H^-}, are shown in Fig. 10.

As a whole, pressure dependences of the extracted currents have the same feature as ones of negative ion production shown in Fig. 7 although details are slightly changed. When B_{MF} = 150G, extracted negative ion currents at optimum pressure are nearly equal with each other. With decreasing B_{MF}, however, I_{D^-} is limited in low level compared with I_{H^-} although negative ion production in the source is increased. This is partly because n_e in the extraction region becomes high with decreasing B_{MF} (see Fig. 8(a)) and then extraction of I_{D^-} becomes inefficient compared with that of I_{H^-}. Details are now under study.

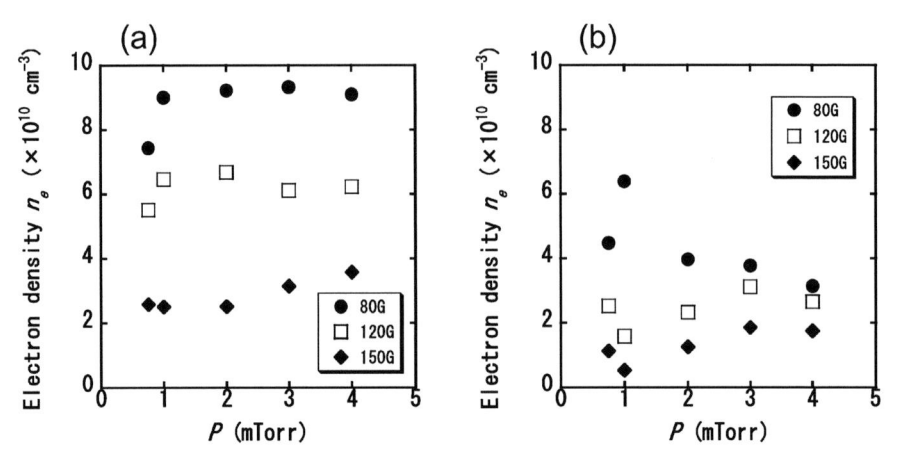

FIGURE 8. Pressure dependence of n_e in the vicinity of the extraction electrode, corresponding to the results in Fig. 7: (a) n_e in D_2 plasmas and (b) n_e in H_2 plasmas.

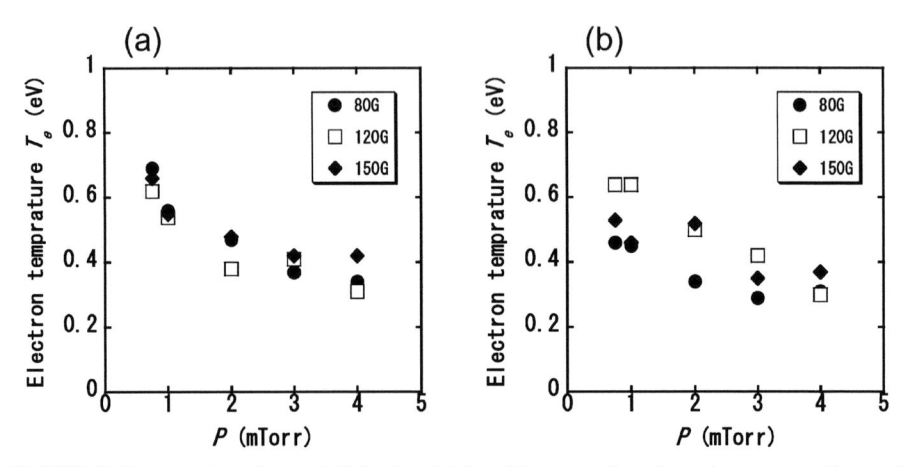

FIGURE 9. Pressure dependence of T_e in the vicinity of the extraction electrode, corresponding to the results in Fig. 7: (a) T_e in D_2 plasmas and (b) T_e in H_2 plasmas.

On negative ion production, intensity of the VUV emission caused by the process (1b) is measured [11, 13]. VUV emission intensity is measured in both the source region ($z = 8.5$ cm) and the extraction region ($z = 3$ cm). The values of integrated intensities in the source region are increased with increasing gas pressure and the B_{MF}. As shown in Fig. 6, the D⁻ and H⁻ densities vary with the B_{MF}. It is noted that the integrated intensity of the VUV emissions and the negative ion densities vary in opposite directions, respectively, when the B_{MF} is varied. Numerical calculations [17] show that the VUV emissions associated with the process (1b) are a function of fast primary electrons, i.e. its density and behavious. With increasing the B_{MF}, fast electron density in the source region is increased and then the collisions with process (1a) are also increased. Then, the intensity of VUV emission increases with the B_{MF} as ones observed in our experiments[11,13].

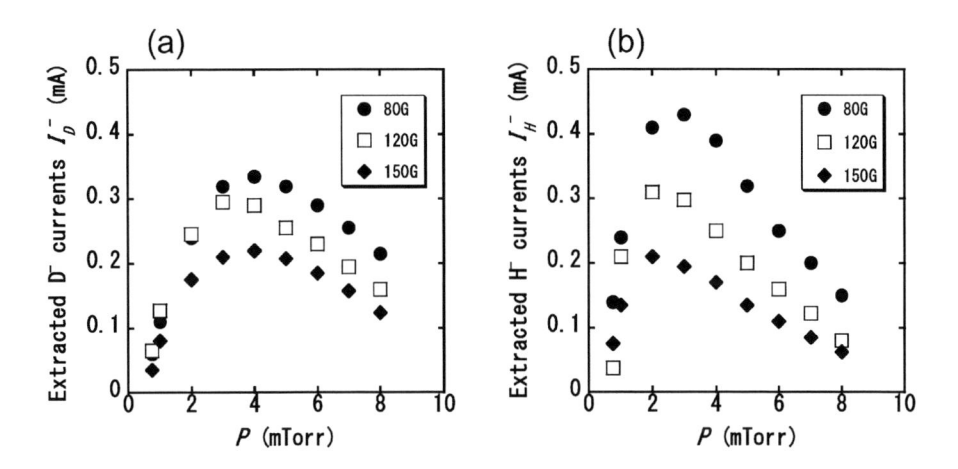

FIGURE 10. Pressure dependence of extracted (a) D⁻ and (b) H⁻ currents, corresponding to the negative ion densities shown in Fig. 7, where extraction voltage $V_{ex} = 1.5$ kV.

According to the results shown in Figs. 7 and 10 and related discussions, our present picture on negative ion production is as follows: In the present experimental conditions with low-pressure, electron-neutral collision mean free paths for destruction of the vibrationally excited molecules (i.e. ionization and dissociation collisions) are a few tens of centimeters. Therefore, sufficient amount of $D_2(v")$ and $H_2(v")$ are transported to the extraction region, although $D_2(v")$ and $H_2(v")$ are produced by the collisions between the ground state molecules and fast primary electrons in the source region. The negative ions are produced by the process (2) of slow plasma electrons to $D_2(v")$ and $H_2(v")$ in the extraction region. Namely, negative ion production is rate-determined by the plasma parameters in the extraction region.

SUMMARY

Production and control of D_2 plasmas are performed by varying the intensity of the MF. The values of T_e and n_e in D_2 plasmas are slightly higher than ones in H_2 plasmas. T_e in D_2 plasmas cannot be decreased and is kept above 1 eV in the extraction region with the same MF intensity for optimizing H_2 plasmas. A stronger MF field is required for control of T_e in D_2 plasmas. Therefore, plasma production and/or transport in D_2 plasmas are different from those in H_2 plasmas. Namely, an isotope effect of plasma production is observed. D^- and H^- densities have different spatial distributions corresponding to those different plasma conditions. The extracted D^- and H^- currents are mainly determined by D^- and H^- densities in the vicinity of the extraction hole, respectively. According to the discussions based on estimated rate coefficients and collision frequencies of main collision processes for production and destruction of negative ions, it is reconfirmed that T_e in the extraction region should be reduced below 1 eV while keeping n_e higher for enhancement of D^- production.

In the future, we will discuss further the isotope effect of D^- and H^- production including Cs injection and the atomic density. We will also try to control plasma parameters with using the mesh grid bias method.

ACKNOWLEDGMENTS

The authors thank Y. Tauchi (Yamaguchi University) for his support in the experiments. A part of this work was supported by the Grant-in-Aid for Scientific Research from Japan Society for the Promotion of Science. This work was also performed with the support of the NIFS LHD Project Research Collaboration.

REFERENCES

1. J. R Hiskes and A. M. Karo, J. Appl. Phys. **56**, 1927 (1984).
2. O. Fukumasa, J. Phys. D **22**, 1668(1989).
3. M. Pealat, J-P. E. Taran, M. Bacal and F. Hillion, J. Chem. Phys. **82**, 4943(1985).
4. T. Inoue, G. D. Ackerman, W. S. Cooper, M. Hanada, J. W. Kwan, Y. Ohra, Y. Okumura and M. Seki, Rev. Sci. Instum. **61**, 406 (1990).
5. U. Fantz, et al., Nuclear Fusion **46**, S297 (2006).
6. O. Fukumasa, Y. Tauchi, Y. Yabuki, S. Mori and Y. Takeiri, *Proceedings of the 9th International Symposium on the Production and Neutralization of Negative Ions and Beams, AIP CP **639**,* (2002), p28.
7. W. G. Graham, J. Phys. D **17**, 2225(1984).
8. O. Fukumasa, N. Mizuki and E. Niitani, Rev. Sci. Instrum. **69**, 995(1998).
9. O. Fukumasa, S. Mori, N. Nakada, Y. Tauchi, M. Hamabe, K. Tsumori and Y. Takeiri, Contrib. Plasma Phys. **44**, 516(2004).
10. Bacal, M. and Hamilton, G. W., Phys. Rev. Lett. **42**, 1538(1979).
11. O. Fukumasa and S. Mori, Nuclear Fusion **46**, S287(2006).
12. S. Mori and O. Fukumasa, Thin Solid Films **506-507**, 531 (2006).
13. S. Mori and O. Fukumasa, Rev. Sci. Instrum. **79**, 02A507(2008).
14. Y. Jyobira, D.Ito and O. Fukumasa, Rev. Sci. Instrum. **79**, 02A508 (2008).
15. J. Okada, Y. Nakao, Y.Tauchi and O. Fukumasa, Rev. Sci. Instrum. **79**, 02A502 (2008).
16. O. Fukumasa and R. Nishida, Nuclear Fusion **46**, S275(2006).
17. S. Mori and O. Fukumasa, "Behaviors of fast primary electrons with collisional effects including production of vibrationally excited hydrogen molecules in a hydrogen negative ion source for a NBI system", IEEJ Trans. FM **127**, 747 (2007). (in Japanese)

Experiments on the Detection of Negative Hydrogen Ions in a Small-size Tandem Plasma Source

Stiliyan St. Lishev, Antonia P. Shivarova and Tsanko V. Tsankov

Faculty of Physics, Sofia University, BG-1164 Sofia, Bulgaria

Abstract. A small-size inductively-driven tandem (two-chamber) plasma source, with a magnetic filter for electron cooling located in the second chamber of the source, is studied experimentally. The results are for the concentration of the negative hydrogen ions in the expanding plasma region of the source and its correlation with the axial variation of the electron density and temperature. The position of the magnetic filter has been varied in the measurements. The obtained spatial distribution of the electron temperature and density is in agreement with theoretical predictions for the operation of the filter. The results suggest that although the detected negative ions are locally produced, the locations of the regions of their effective production in the source – positions of the lowest electron temperature and/or of the highest electron density – are nonlocally formed, not only by the plasma expansion through the magnetic filter but also by plasma expansion from the driver into the second bigger-size chamber of the source.

Keywords: hydrogen discharges, sources of negative hydrogen ions
PACS: 52.70.-m, 52.50.Dg

INTRODUCTION

As it is known, the tandem plasma source has been introduced as a design for the plasma sources of negative hydrogen/deuterium ions [1] developed regarding fusion applications, in particular, for neutral-beam-injection plasma-heating in big fusion reactors. Space separation of regions of high- and low- energetic electrons that favours the two-step reaction of volume production of negative hydrogen ions by electron attachment to highly-vibrationally excited hydrogen molecules is in the basis of the idea for the design of the tandem plasma sources. The current research on these sources covers their two main modifications based on dc- [1-7] and rf- [8] discharges, respectively.

The work on the dc hot-cathode-discharge based sources, started in the 80's of the last century has provided plenty of results [1-7, 9-14] accumulated during the years, not only for the extracted current of the negative ion beams but also from discharge diagnostics. The research on the rf sources [8], started recently straight towards the ITER development, aims at proving their capabilities to achieve the ITER requirements.

Although in both cases – of arc- and rf- discharge based sources – the sources are of a tandem type, with spatial separation of the source in two regions with high- and low- electron temperatures, there are differences in their design. The arc-discharge based sources are one-chamber sources and the magnetic filter is that providing the separation of the two regions in the source. The rf sources are two-chamber sources. The first chamber where the driver – an inductive discharge – is located is of a smaller size. The second bigger-size chamber where the magnetic filter and the extraction are located is a volume for plasma expansion from the driver. Such a difference in the design of the arc- and rf- sources means different mechanisms of plasma maintenance in the sources. In the arc sources the discharge before the filter is under the conditions of locality. The design of the rf sources forces strong nonlocality of the discharge maintenance, as both recent experiments and modelling of the source show [15-17]. Particle- and electron-energy- fluxes from the driver ensure the plasma existence in the second chamber of the rf source. In the particular case of the operation of the magnetic filter for electron cooling considered here, differences in the mechanisms of the discharge maintenance in the arc and rf sources mean different gas discharge conditions before the filter region. Whereas in the arc discharges the filter acts on axially-homogeneous high-density plasma causing inhomogeneity of both electron temperature and density in the filter region, in the rf discharges – as it is shown here – its action is superimposed on the decrease of the electron temperature and density caused by the plasma expansion (from the driver) into the second chamber of the source.

This study presents experiments on diagnostics in the expansion plasma region of a small-size inductively driven tandem plasma source operating in a hydrogen gas. The measurements include determination of the plasma electronegativity and the concentration of the negative hydrogen ions by the laser photodetachment technique in its combination with probe measurements as well as determination of the axial distribution of the electron density and temperature by probe diagnostics. Test measurements with a Faraday-type extraction device are briefly commented on. The discussion stresses on the correlation between the negative ion concentration and the spatial distribution of the electron density and temperature for different positions of the magnetic filter for electron cooling.

EXPERIMENTAL ARRANGEMENTS AND METHODS

The experimental set-up (Fig. 1) constructed by analogy with the rf sources developed for ITER [8] is for small-scale experiments. It is a two-chamber inductively-driven plasma source. A quartz tube and a stainless steel cylinder are the two chambers of the discharge vessel. Their diameters (d_1 and d_2) and lengths (l_1 and l_2) are given in Fig. 1. The position $z = 0$ used further on in the presentation of the results is at the transition between the two chambers. Positive z-values correspond to positions on the axis in the second chamber. The inductive discharge is with cylindrical external 9-turn coil tightly wound around the quartz tube. In the experiments presented here the discharge is at 27 MHz, in flowing gas. The magnetic filter in the second chamber of the source, shown in Fig. 1, is movable in the axial direction. The magnetic field is perpendicular to the plasma flow from the driver and its strength can be varied.

The experimental arrangements for the discharge diagnostics are for probe diagnostics, photodetachment-technique measurements and measurements of the charged-particle currents extracted from the second chamber of the source.

The axial variation of the electron temperature and density is measured by a Smart-Probe™ probe system. The probe tip is a tungsten wire (10 mm in length and 0.38 mm in a diameter). The electron temperature is obtained from the transition region of the

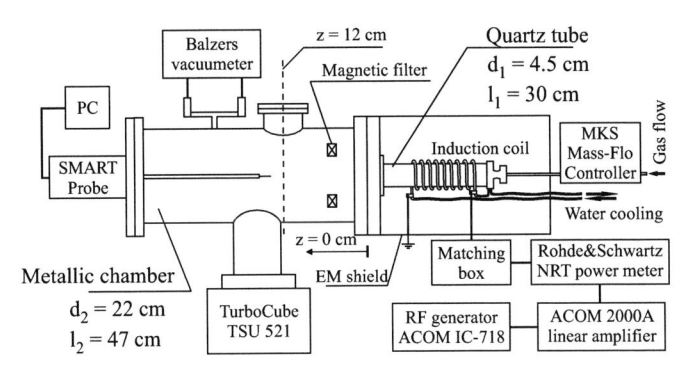

FIGURE 1. Experimental set-up: two-chamber inductively-driven tandem-type plasma source. The dashed line at $z = 12$ cm marks the position of the photodetachment measurements. The magnetic filter is also shown.

probe characteristics. The electron density is determined from the ion saturation current, according to the ABR theory [18].

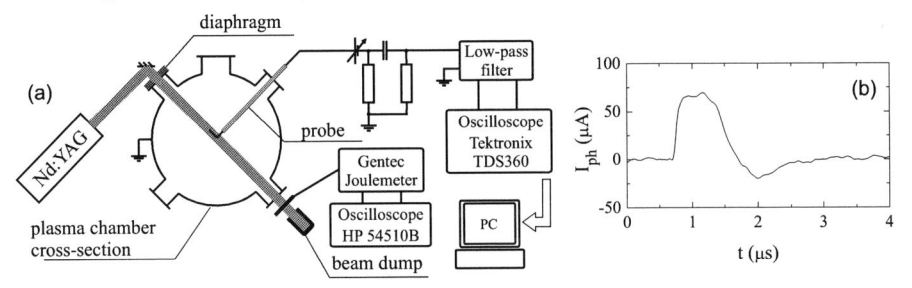

FIGURE 2. The set-up for the laser photodetachment technique measurements in (a) and measured shape of the photodetachment signal in (b).

A Surelite III-10 Nd:YAG laser operating at its second harmonic (wavelength of $\lambda = 532$ nm) is used in the laser photodetachment technique measurements. The method is employed in its combination with a probe. The experimental arrangements for these measurements are schematically given in Fig. 2(a). The beam diameter is fixed at 12 mm by a diaphragm. The probe is L-shaped Langmuir probe, with a tip (10 mm in length and a radius of 0.2 mm) oriented against the laser beam. It is biased at 30 V above the plasma potential. According to the laser photodetachment technique employed with a probe [13], the electronegativity $\alpha = n_-/n_e$ (where n_- and n_e are, respectively, the concentrations of the negative ions and of the electrons) is determined from the ratio of the plateau amplitude I_{ph} of the pulse (Fig. 2(b)) of the electron current to the probe (caused by detachment of the electrons from the negative ions by the laser beam photons) to its stationary value $I_{e(dc)}$:

$$\frac{I_{ph}}{I_{e(dc)}} = \alpha \frac{\delta n_-}{n_-}. \tag{1}$$

For determining α from (1), the theoretical dependence [13]

$$\frac{\delta n_-}{n_-} = 1 - \exp\left(-\frac{\sigma}{h\nu}W\right) \qquad (2)$$

for the photodetachment fraction (i.e. for the relative change ($\delta n_-/n_-$) in the concentration of the negative ions due to the detachment of electrons), checked to be valid in experiments [19] carried out at the same set-up, has been used. In (2), $h\nu$ is the photon energy, σ [13] is the cross section for photodetachment ($H^- + h\nu \rightarrow H + e$) and W is the laser pulse energy density.

The test experiments on the extraction are performed by a Faraday-cup based extraction device completed with a plasma electrode and an electron-extraction electrode with magnets installed on it. In these measurements, the extraction device is mounted on the back flange of the second chamber of the source (Fig. 1), replacing the Smart probe. It is movable in the axial direction.

RESULTS AND DISCUSSIONS

The experimental results presented here are for the electronegativity and the concentration of the negative ions, for the axial profiles of the electron temperature $T_e(z)$ and density $n_e(z)$ on the axis of the second chamber of the source, starting from the position $z = -2$ cm inside the first chamber, as well as for the currents to the electron extraction electrode and to the Faraday cup of the extraction device.

Results for the electronegativity and the concentration of the negative ions and their correlation with the local values of the electron temperature and density

All the results given here are at the position $z_{ph} = 12$ cm where the photodetachment technique measurements have been carried out. Results for two values of the gas pressure $p = 6$ and 8 mTorr and of the magnetic field $B_0 = 50$ and 100 G of the filter, at its maximum, are presented. The absorbed power is $P = 700$ W and the gas flow is 20 sccm. The position of the magnetic filter has been varied between $z_{MF} = 4$ cm and $z_{MF} = 10$ cm. Thus, the distance $\Delta z = z_{ph} - z_{MF}$ between the position of the photodetachment measurements and the position z_{MF} of the maximum value of the magnetic field of the filter varies between $\Delta z = 2$ cm and $\Delta z = 8$ cm.

The lower value of the gas pressure $p = 6$ mTorr shows up with lower electronegativity α (Fig. 3(a)) and lower concentration n_- of the negative ions (Fig. 3(b)). This goes together with higher electron temperatures T_e (Fig. 3(c)) and lower electron densities (Fig. 3(d)). Obtaining higher T_e and lower n_e for lower gas pressure is in accordance with the general trends of the gas discharge behaviour. Although α (Fig. 3(a)) is higher for higher B_0 ($B_0 = 100$ G), n_- remains almost the same (Fig. 3(b)) since n_e (Fig. 3(d)) is lower.

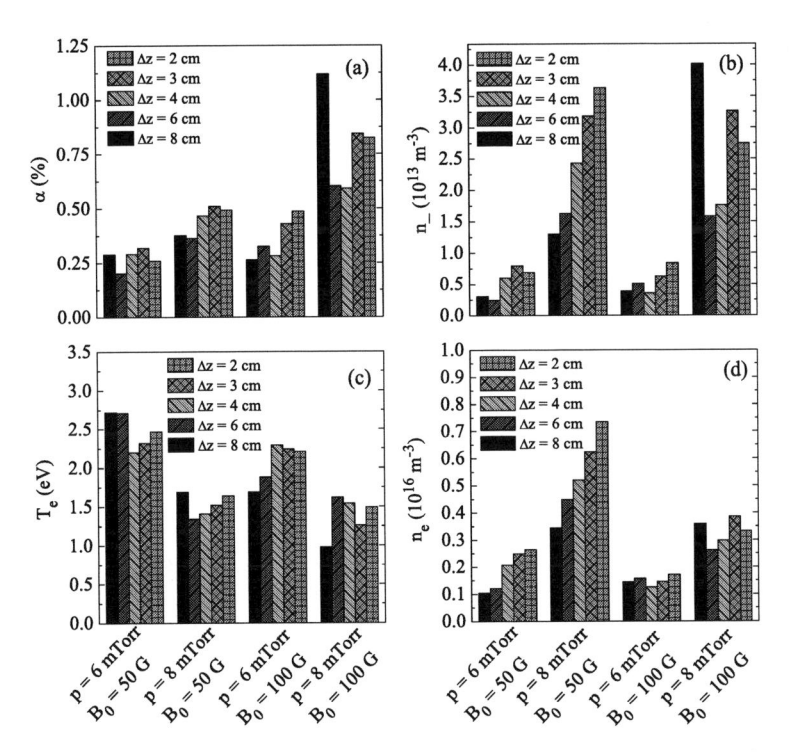

FIGURE 3. Electronegativity $\alpha = n_-/n_e$ (a), concentration of the negative ions n_- (b), electron temperature T_e (c) and electron density n_e (d) obtained at $z_{ph} = 12$ cm for different combinations of values of p and B_0 and different values of $\Delta z = z_{ph} - z_{MF}$. The location of the different cases in (a) and (b) is the same as in (c) and (d).

For $p = 8$ mTorr and $B_0 = 50$ G, the electronegativity α (Fig. 3(a)) is also comparatively low. However, this results from comparatively high values of both n_- (Fig. 3(b)) and n_e (Fig. 3(d)). For $p = 8$ mTorr and $B_0 = 100$ G, n_- is also comparatively high (Fig. 3(b)). Due to the lower values of n_e in this case (Fig. 3(d)), the electronegativity α is also high (Fig. 3(a)).

The highest values of n_- (Fig. 3(b)) are at $p = 8$ mTorr and $B_0 = 100$ G with a magnetic filter located far away from the position $z_{ph} = 12$ cm of the registration of the negative ions ($z_{MF} = 4$ cm, $\Delta z = 8$ cm) and at $p = 8$ mTorr, $B_0 = 50$ G and a magnetic filter close to the position of the registration ($z_{MF} = 10$ cm, $\Delta z = 2$ cm) of the negative ions. In the first case ($p = 8$ mTorr, $B_0 = 100$ G, $\Delta z = 8$ cm) the electron temperature T_e is the lowest one measured (Fig. 3(c)) and in the latter case ($p = 8$ mTorr, $B_0 = 50$ G, $\Delta z = 2$ cm) the electron concentration is the highest one measured (Fig. 3(d)). Having the highest concentration of negative ions for the highest n_e and the lowest T_e is in agreement with the requirements for efficiency of the reaction of local production of negative hydrogen ions via dissociative attachment of electrons to vibrationally excited molecules ($e + H_2(v) \rightarrow H^- + H$).

Axial profiles of the electron temperature and density

The axial profiles of T_e and n_e presented here are under the same conditions as in the previous subsection: $p = 6$ and 8 mTorr, $P = 700$ W, $B_0 = 50$ and 100 G and positions $z_{MF} = 4, 6, 8, 9$ and 10 cm of the magnetic filter corresponding, respectively, to $\Delta z = 8, 6, 4, 3$ and 2 cm. Figures 4 and 5 show results for given p and B_0 and different z_{MF}-values whereas in Fig. 6 axial variations of T_e and n_e for different B_0-values are compared. The theoretical results in Fig. 7 provide qualitative comparison with the predictions of the model [20] for the magnetic filter operation.

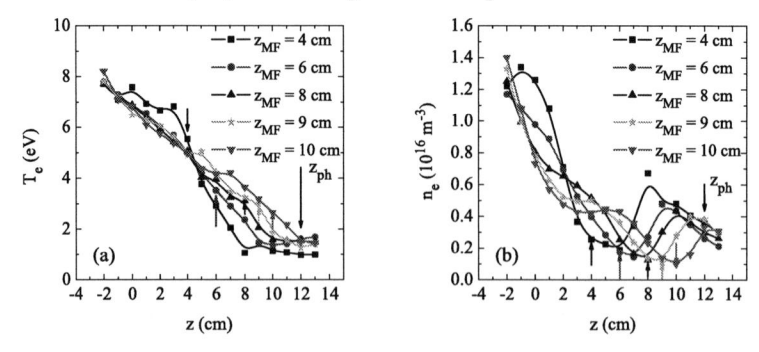

FIGURE 4. Axial profiles of the electron temperature (a) and density (b) for $p = 8$ mTorr, gas flow 20 sccm, $B_0 = 100$ G and different positions z_{MF} of the magnetic filter. The positions of the photodetachment measurements z_{ph} and of the magnetic filter are marked by arrows.

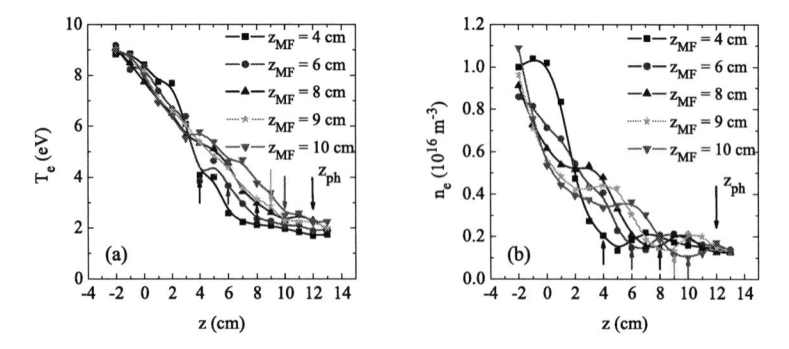

FIGURE 5. The same as in Fig. 4 but for $p = 6$ mTorr and $B_0 = 100$ G.

The main conclusion from the results in Figs. 4 and 5 is that the effect of the magnetic filter to reduce both T_e and n_e is superimposed on the effect of the plasma expansion from the driver into the bigger second chamber of the source, the latter acting in the same direction [15].

The farther from the transition ($z = 0$) between the two chambers the filter is located the more pronounced the display of the effect of the plasma expansion is. The axial profiles of T_e (Fig. 4(a)) and n_e (Fig. 4(b)) for $z_{MF} = 10$ cm show in an almost complete form the effect of the plasma expansion: fast axial drop of T_e followed by a region of an almost constant T_e (at 4 eV) and structuring of the axial profile of n_e composed by fast drop close to the driver (for $z \leq 2$ cm), a plateau region for 2 cm $\leq z \leq 6$ cm and a slower decrease further on. The magnetic filter affects the end part ($z \geq 7$ cm) of the profiles formed by the plasma expansion. Positions of the magnetic filter $z_{MF} = 8$ and 9 cm still keep well-pronounced effects of plasma expansion. In general, except for the ($z_{MF} = 4$ cm)-position (i.e., the position which is the closest one to $z = 0$) where the modifications in the profiles due to the magnetic filter start almost from the very beginning, all the profiles go together formed by the plasma expansion and each of them separates from the others at about 2 cm before the position of the magnetic filter.

Whereas the case of $z_{MF} = 10$ cm shows clearly pronounced effects of plasma expansion, the display of the effects of the magnetic filter is the best for the position $z_{MF} = 4$ cm, i.e. when the filter is close to the transition between the two chambers. In accordance with previous experiments [18], also carried out with a magnetic filter close to the transition between the two chambers, the axial profile of T_e for $z_{MF} = 4$ cm (Fig. 4(a)) shows up with a strong drop caused by the filter field, followed by a slight minimum and a constant T_e deeper in the second chamber. The axial profile of n_e (Fig. 4(b)) is formed by a strong drop of n_e, due to the filter field, a minimum in the filter region and a maximum behind the filter, followed by a slower decrease of n_e deep in the second chamber of the source. Shifting the position z_{MF} deeper into the second chamber shows the same behaviour of the axial profiles of T_e and n_e, with disappearance of the very end regions of their profiles when the magnetic filter is close to $z = 13$ cm, i.e. to the end position of the probe measurements. In general, Figure 4 displays very well the sequence of the appearance of the different regions in the axial profiles of T_e and n_e obtained for the different positions z_{MF} of the magnetic filter and their shifting away inside the second chamber with the shift of the filter position z_{MF}.

Reduction of the gas pressure (to $p = 6$ mTorr, Fig. 5) shows the same behaviour of the axial profiles of T_e and n_e, however, with less pronounced structuring of the profiles (e.g., the minimum in the filter region of n_e and its maximum behind the filter are not so well pronounced). Also the electron temperature becomes higher.

The comparison (Fig. 6) of the axial profiles of T_e and n_e for different values ($B_0 = 50$ and 100 G) of the magnetic field shows lower T_e and higher maximum of n_e behind the filter for the higher B_0-value. Results obtained for higher gas flow (30 sccm) show no influence of the gas flow.

Figure 7 presents theoretical profiles of n_e and T_e obtained from the fluid-plasma model [20] of plasma expansion through a magnetic filter in hydrogen discharges. The model stresses on that the transport processes govern the operation of the filter. The comparison with the experimental results (Figs. 4-6) could be only qualitative because of the simplified configuration of the source in the model (a rectangular single gas-discharge chamber and a high-frequency power deposition located in the beginning of the discharge vessel (at $0 \leq z \leq 6$ cm) before the filter). The theoretical results for the axial profiles of T_e and n_e in Fig. 7(a) and (b) obtained for $z_{MF} = (10 - 14)$ cm should be compared with the experimental T_e- and n_e- profiles for $z_{MF} = (4 - 8)$ cm in Fig. 4(a) and (b). (In Fig. 7 the ($z = 0$)-position marks the beginning of the driver whereas in Figs. 4-6 the ($z = 0$)-position is at the transition between the two chambers, i.e., between the driver and the expansion plasma volume.)

The conclusions from the comparison of the theoretical and experimental results can be summarized, as follows: (i) Both the theoretical and experimental results for

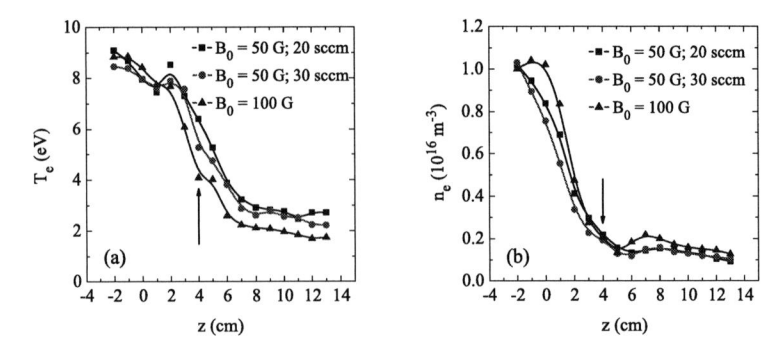

FIGURE 6. Axial profiles of the electron temperature (a) and density (b) for $p = 6$ mTorr, $z_{MF} = 4$ cm, gas flow of 20 sccm and two values of B_0. For $B_0 = 50$ G, results for different values (20 and 30 sccm) of the gas flow are shown. The arrows mark the position of the magnetic filter.

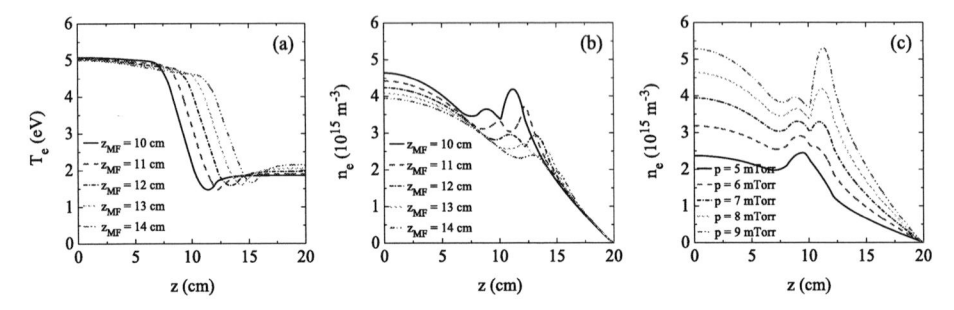

FIGURE 7. Theoretical results for the axial profiles of the electron temperature T_e (a) and density n_e (b) obtained from the model in [20] for $p = 8$ mTorr, $B_0 = 50$ G and different positions z_{MF} of the magnetic filter as well as for the electron density n_e (c) for $B_0 = 50$ G, $z_{MF} = 10$ cm and different values of the gas pressure.

$T_e(z)$ show strong drop of T_e caused by the filter, a minimum behind the filter (better pronounced in the theoretical results) and a lower T_e-value behind the filter when it is located closer to the driver; (ii) Both the theoretical and experimental gradients of T_e in the filter region are almost the same for the different positions of the filter, with a value of the order of 1 eV/cm; (iii) Both the theoretical and experimental results for n_e show up with a minimum in the filter region which in the theoretical results is located exactly in the center of the filter; (iv) Both in the theoretical and experimental profiles, the maximum of n_e behind the filter is higher when the filter is positioned closer to the driver; (v) The position of the maximum of n_e behind the filter coincides with the position of the minimum of T_e there or – when the latter does not shows up clearly in the experimental results – with the position where T_e reaches its lowest value. The theoretical results obtained for different gas pressure values (Fig. 7(c)) come to confirm the experimentally obtained suppression (Fig. 5) in the structuring of the n_e-profile at lower p.

Results from test experiments with a Faraday-type of extraction device

The test experiments with the extraction device (Fig. 8(a)) have been carried out at gas pressure $p = 9$ mTorr and rf power of $P = 100$ W applied for the discharge maintenance. The magnetic filter ($B_0 = 50$ G) is located at $z_{MF} = 4$ cm.

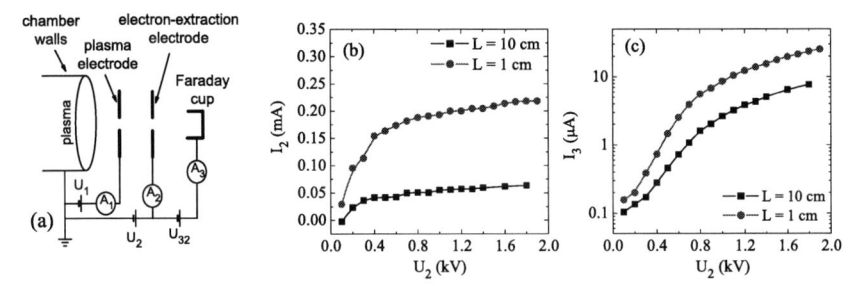

FIGURE 8. Schematical presentation of the electrodes of the extraction device, with the electrical circuits shown (A_{1-3} are ampermeters) in (a) and comparison of the currents to the electron extraction electrode (b) and to the Faraday cup (c) for two distances L between the extraction device and the magnetic filter.

Figure 8 shows results from measurement of the variation – with the voltage U_2 applied to the electron extraction electrode (at 250 G magnetic field induction of its magnets) – of its current I_2 (Fig. 8(b)) and of the current I_3 (Fig. 8(c)) to the Faraday cup (biased at $U_{32} = 1$ kV above U_2), for two positions of the extraction device with respect to the filter (at distances $L = 1$ and 10 cm between the center of the filter and the plasma electrode of the extraction device). Both I_2 and I_3 (Fig. 8) are higher for smaller L, i.e. when the extraction device is closer to the filter, at a position of higher n_e. However, whereas the saturation value of I_2 is 4 times higher for $L = 1$ cm than that for $L = 10$ cm, the ratio of the values of I_3 obtained at the two positions of the extraction device is about 3. Thus, although larger I_2 correlates with larger I_3, the increase of the currents is not the same. This discards the possibility of a Faraday-cup current being due solely to electrons and shows that this current may be attributed to negative ions.

CONCLUSIONS

Measurements of electronegativity and concentration of the negative hydrogen ions in the second chamber of a tandem two-chamber rf plasma source are combined with determination of the spatial distribution of electron density and temperature in the source. The position of the magnetic filter has been varied in the experiment. The conclusion is that the location in the source of the region of the most effective negative ion production – positions of the lowest electron temperature and of the highest electron density – is nonlocally formed based on transport processes of plasma expansion through the magnetic filter. Due to the design of the rf sources, this effect is superimposed on the effects of plasma expansion from the smaller size driver to the second bigger size chamber. Extraction just behind the magnetic filter for electron cooling is usually considered as a proper decision with a view to high efficiency of the source. The results presented here show that this should be specified based on detailed consideration of the spatial distribution of the plasma parameters. The choice of a proper position of the extraction is not too much sensitive with respect to the electron temperature: There is an extended region behind the filter where the electron temperature stays almost constant at its lowest value. However, this region starts at least few centimeters behind the filter. The choice of a proper position of the extraction with respect to the electron density is sensitive because the maximum of the electron density appearing few centimeters behind the filter should be "hit". Fortunately, the position of this maximum is in the region where the electron temperature has reached its lowest value. The results show that a magnetic filter close to the driver and an extraction at the maximum of the electron density behind the filter should be the proper decision.

ACKNOWLEDGMENTS

The authors thank Prof. Dr. Kiss'ovski for discussions. The work is within the programme of the Bulgarian Association EURATOM/INRNE (task P2) and DFG-project 436 BUL 113/144/0-1. Support to the experimental equipment by the National Science Fund (Bulgaria) and Sofia University through project D01-413 is highly acknowledged.

REFERENCES

1. M. Bacal, *Nucl. Fusion* **46**, S250–S259 (2006).
2. M. Bacal, A. M. Brunetau and M. Nachman, *J. Appl. Phys.* **55**, 15–24 (1984).
3. T. Inoue *et al.*, *Nucl. Instrum. Methods Phys. Res.* B **37/38**, 111–115 (1989).
4. K. N. Leung *et al.*, *Rev. Sci. Instrum.* **61**, 2378–2382 (1990).
5. F. A. Haas, L. M. Lea and A. J. T. Holmes, *J. Phys. D: Appl. Phys.* **24**, 1541–1550 (1991).
6. O. Fukumasa, M. Hosoda and H. Naitou, *Rev. Sci. Instrum.* **63**, 2696–2698 (1992).
7. G. I. Dimov, *Rev. Sci. Instrum.* **73**, 1–3 (2002).
8. E. Speth *et al.*, *Nucl. Fusion* **46**, S220–S238 (2006).
9. M. B. Hopkins, M. Bacal and W. G. Graham, *J. Phys. D: Appl. Phys.* **24**, 268–276 (1991).
10. M. Bacal, *Plasma Sources Sci. Technol.* **2**, 190–197 (1993).
11. O. Fukumasa, H. Naitou and S. Sakiyama, *J. Appl. Phys.* **74**, 848–852 (1993).
12. A. G. Nikitin, F. El Balghiti and M. Bacal, *Plasma Sources Sci. Technol.* **5**, 37–42 (1996).
13. M. Bacal, *Rev. Sci. Instrum.* **71**, 3981–4006 (2000).
14. B. Crowley, D. Homfray, S. J. Cox, D. Boilson, H. P. L. de Esch and R. S. Hemsworth, *Nucl. Fusion* **46**, S307–S312 (2006).
15. Zh. Kiss'ovski, St. Kolev, A. Shivarova and Ts. Tsankov, *IEEE Trans. Plasma Sci.* **35**, 1149–1155 (2007).
16. N. Djermanova, Zh. Kiss'ovski, St. Lishev and Ts. Tsankov, *J. Phys.: Conf. Series* **63**, 012013 (1–6) (2007).
17. St. Kolev, A. Shivarova, Kh. Tarnev and Ts. Tsankov, *Plasma Sources Sci. Technol.* **17**, 035017 (1–13) (2008).
18. I. Djermanov, St. Kolev, St. Lishev, A. Shivarova and Ts. Tsankov, *J. Phys.: Conf. Series* **63**, 012021 (1–6) (2007).
19. Zh. Kiss'ovski, St. Kolev, S. Müller, Ts. Paunska, A. Shivarova and Ts. Tsankov, *Plasma Phys. Control. Fusion* (2008), submitted.
20. St. Kolev, St. Lishev, A. Shivarova, Kh. Tarnev and R. Wilhelm, *Plasma Phys. Control. Fusion* **49**, 1349–1369 (2007).

Study of Multi-cusp Magnetic Field in Cylindrical Geometry for H^- Ion Source

Ajeet Kumar, V.K.Senecha and R.M.Vadjikar

Raja Rammana Centre For Advance Technology, Indore, India - 452013

Abstract. The computation of magnetic field generated by the permanent magnets (PM) in the multipole configuration in a cylindrical geometry has been carried out analytically and compared with the experimental measurements. It has been shown that the magnetic field intensity variation inside the cylindrical plasma chamber follows a power law functional form rather than the conventionally used exponential functional form. The measurement of magnetic field inside the cylindrical multicusp geometry has been performed using a Hall probe method with 3-D motorized movement bench which shows excellent agreement with the analytical calculations. A detailed analytical computations of multi-cusp magnetic field for the cylindrical geometry and its comparison with the conventional exponential functional form has been presented.

Keywords: Magnetic Field Computation, Multi-cusp magnetic field, H^- Ion Source, Multipole Configuration, Multi polar Field
PACS: 52.75.-d ; 29.25.Ni

INTRODUCTION

The multicusp magnetic field (MMF) has found wide applications in the development of ion sources due to their capability to confine large volume of high density uniform plasma [1]. Although the principle of plasma confinement by MMF originated with fusion research but MMF based negative ion source has demonstrated its capability for high current low emittance and stable H^- ion beams which is essential for the new generation of accelerators. There exist different cusp magnetic field geometries for plasma confinement but the full line cusp gives better confinement efficiency [2, 3]. Specifically in the design and development of H negative ion source for use in high energy accelerators, MMF confinement of plasma along with filter magnetic has become indispensable.

Two important parameters associated with the MMF design are the knowledge of field free region and line cusp loss width. Field free central region plays a crucial role in ion sources as it determines the volume available for the formation of uniform plasma as well as the working area used for the placement of filaments. For the proper emission of electrons from the filaments to generate the uniformly dense plasma filaments needs to be kept in the field free region. Libearmann et al. [4] has worked out in the approximation and showed that the MMF decays exponentially inside the plasma chamber. Similarly Koch et al. [5], has stated that in picket fence of multi polar arrangement using cylindrical magnets around the cylindrical geometry, the field decays exponentially inside the chamber. However, an exact analysis using the vector potential and Fourier decomposition has been carried, that shows that the magnetic field intensity varies as power law rather than decaying exponentially for the cylindrical geometry. This

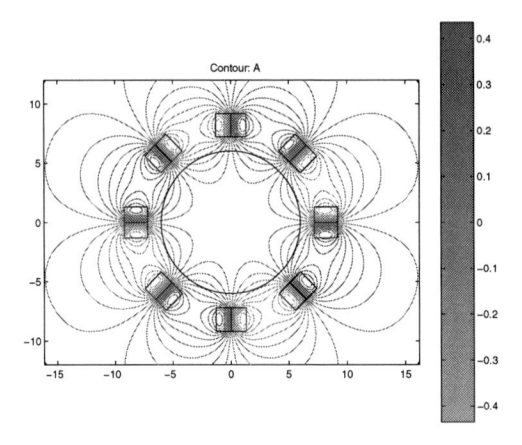

FIGURE 1. Permanent magnet arranged on the surface of cylindrical chamber along the length which is perpendicular to the plane of paper. The Plot shows the constant contour line of Vector potential, which is same as magnetic field lines in this case.

has further been cross examined by accurate measurement of MMF using Hall probe method in cylindrical geometry which proves that power law functional form is the best approximation. In fact Milne et al. [6] has fitted the MMF measurements using exponential curve fit, but had to use an artificial correcting factor of value < 1 which further reduces with the increase in the number of magnetic poles so as to obtain the best possible fit. This clearly indicates that field decays slowly as one moves from cylindrical wall surface toward the cylinder axis, in comparison with the prediction made using exponential functional form [4].

MAGNETIC FIELD CALCULATION

In a typical magnetic multipole configuration PM are placed with alternating polarity all along the surface of the plasma chamber. In this configuration the magnetic field strength is maximum near the surface of the plasma chamber which decreases as one moves toward the axis of the plasma chamber. Typically the negative ion sources employed in the accelerator are relatively smaller in size, cylindrical in shape and driven either by RF or discharge. We have considered here two configuration (i) When PM are placed around the surface of cylinder along its length and with alternate polarity (Figure 1) (ii) PM are arranged on a plane surface, with alternate polarity (Figure 2). Both arrangement generate full line cusp along the length of the magnet as magnetization is assumed to be normal to the length of the magnet.

There are numerous techniques for computing the magnetic field due to PM. Libermann et al. in [4] the magnetic field equivalent to case(ii) has been computed with the assumption that the permanent magnets are equivalent to point dipole. This assumption was justified on the basis that the separation between the adjacent magnet is large

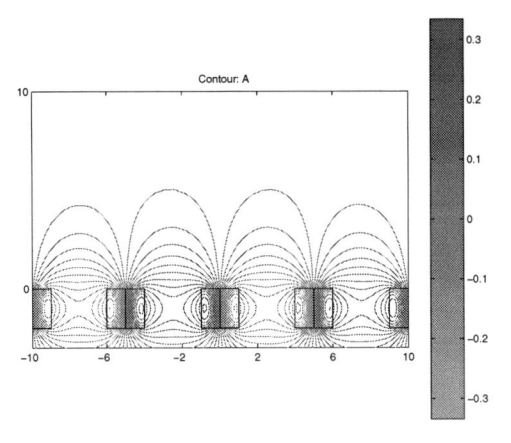

FIGURE 2. Permanent magnet arranged on a plane surface, perpendicular to the plane of paper, with alternate polarity. The Plot shows the constant contour line of Vector potential, which is same as magnetic field lines in this case.

compared to the width of the pole faces of the magnets. However in practical situation the width of the pole face of PM are not negligible compared to the distance between adjacent magnets used in an ion source, hence they cannot be considered as the point dipoles. Nevertheless the magnetic field generated by the multi pole configuration are periodic, hence the best way to compute the magnetic field would be to decompose it in terms of its various harmonics. We ignore the magnetic field dependence along the axis of the plasma chamber (z-axis). Thus the vector potential has only z component i.e. $\vec{A} = (0,0,A_z)$ and in the current free region $\nabla^2 \vec{A} = 0 \Rightarrow \nabla^2 A_z = 0$ and with the knowledge of A_z, $\vec{B}$ can be computed using $\vec{B} = \vec{\nabla} \times \vec{A}$. For the solution of $\nabla^2 A_z = 0$ and to determine A_z uniquely one should know the boundary value of A_z. A_z value can be explicitly computed on the boundary of plasma chamber using the method described in the Appendix. For two configuration considered in Figure(1 and 2), there can be two different arrangements for the placement of PM. (i) the PM are closely spaced -Case(A)- and (ii) PM are widely separated- Case(B). The Case(B) is somewhat extreme nevertheless, it will help in understanding the magnetic field in general, when PM are placed neither too far nor too close to each other. Figure(3) shows the plot of vector potential on the boundary surface in Case A and B. For clarity, the x-axis (along the boundary) has been scaled for Case(B) and plotted in the same figure as that of Case(A). For $2N$ pairs of PM in cylindrical multipole configuration (Figure 1) the result for radial and axial component of magnetic field(computed from A_z) are:

$$B_r = \frac{N}{R} \sum_{n=1}^{\infty} na_n \left(\frac{r}{R}\right)^{Nn-1} \cos(Nn\theta) \qquad (1)$$

$$B_\theta = -\frac{N}{R} \sum_{n=1}^{\infty} na_n \left(\frac{r}{R}\right)^{Nn-1} \sin(Nn\theta) \qquad (2)$$

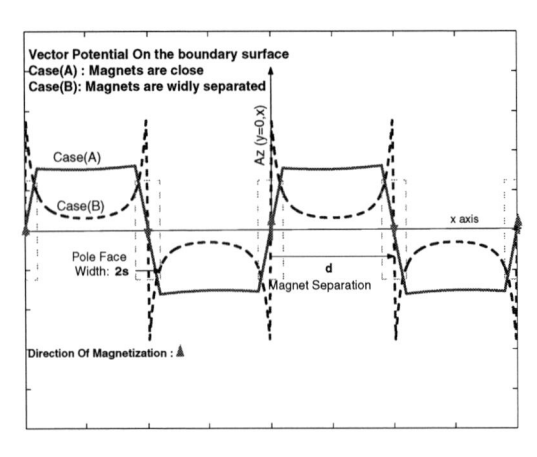

FIGURE 3. Magnetic Vector potential along the boundary

and planar multipole configuration (Figure 2), with the periodicity $2d$ the magnetic field along the x and y components are given by

$$B_x = \frac{-\pi}{d} \sum_{n=1}^{\infty} na_n exp\left(-\frac{n\pi y}{d}\right) \sin\left(\frac{n\pi x}{d}\right) \tag{3}$$

$$B_y = \frac{\pi}{d} \sum_{n=1}^{\infty} na_n exp\left(-\frac{n\pi y}{d}\right) \cos\left(\frac{n\pi x}{d}\right), \tag{4}$$

In essence, we have decomposed the actual magnetic field in its various harmonic component such that the boundary conditions are satisfied. The magnitude of the nth harmonic component of magnetic field for above two configuration (Figure 1 and 2) are given respectively as:

$$|B_n| = \frac{N}{R} na_n \left(\frac{r}{R}\right)^{Nn-1} \tag{5}$$

$$|B_n| = \frac{\pi}{d} na_n exp\left(-\frac{n\pi y}{d}\right) \tag{6}$$

From above it is evident that in both the configurations, $|B_n|$ depends only on R and y respectively hence they are independent of θ and x. But the absolute value of $|B|$ computed after summing over all the harmonics will be dependent on θ and x. For example, $|B|$ can be easily expressed for cylindrical multipole configuration (Figure 1) as

$$|B|^2 = \left(\frac{N}{R}\right)^2 \sum_{n=1}^{\infty} n^2 a_n^2 \left(\frac{r}{R}\right)^{2(Nn-1)} + 2\left(\frac{N}{R}\right)^2 \sum_{n>m} nm a_n a_m \left(\frac{r}{R}\right)^{N(n+m)-2} \cos N(n-m)\theta \tag{7}$$

All the above expressions obtained so far are quite general in the sense that for different magnet shape, or magnetization, only the coefficient a_n will change but the dependency

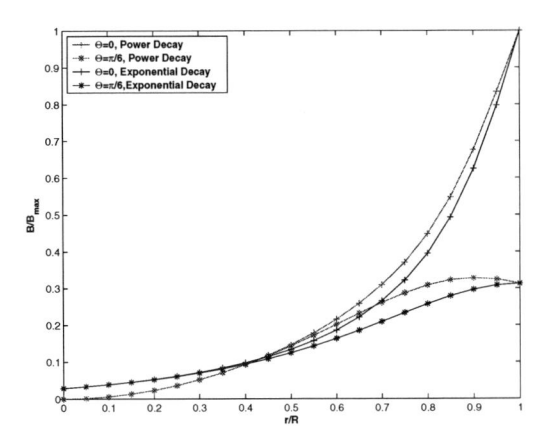

FIGURE 4. Magnetic Field for case (A)

on r and θ or x and y will be unaltered. Because of the nature of dependency of B_n on r and y, it is evident that the higher harmonics would vanish much rapidly when r and y increases, compared to the lower harmonics. Also because the coefficient (Ref. Appendix)) $a_n \alpha 1/n^2$ in most practical cases hence the contribution due to higher harmonics in $|B|$ is very small. Thus considering only first harmonic, an approximation for $|B|$ inside the plasma chamber surrounded by $2N$ permanent magnet forming line cusp would be

$$|B| = B_s \left(\frac{r}{R}\right)^{N-1} \tag{8}$$

But if we impose the solution of planar configuration (Figure 2) on the cylindrical configuration [1], and by considering only first harmonic, the field variation along the radius of the plasma chamber would be

$$|B| = B_s exp\left(-\frac{Ny}{R}\right) \tag{9}$$

Here $y = R - r$ and B_s is surface field (near pole face). It is Eqn.(9) which has been derived in [4] and used by [6] for fitting their magnetic field data in their study of carbon film deposition using plasma confined by magnetic-multipole system. The difference between the value of $|B|$ when computed using Eqn.(8) with that when computed using Eqn.(9) has been plotted and shown in Figure(4 and 5). The contribution due to higher harmonics has also been included in the plot.

Various anomalies creep in when Eqn(9) is used in place of Eqn(8). For example as mentioned by Libermann et al [4], $|B|$ will be zero on the plasma surface between the two adjacent magnets, but it is to be noted that, (as shown in Figure- 4), in cylindrical

---

[1] For this the $\frac{\pi}{d}$ in Eqs. 3 and 4 is replaced with $\frac{N}{R}$ where $2N$ is the total number of the permanent magnet and R is the radius of the chamber.

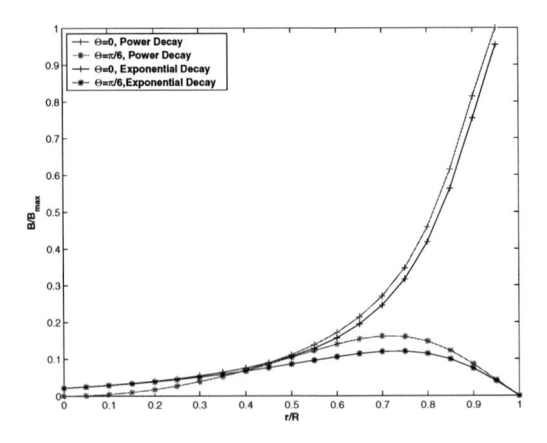

FIGURE 5. Magnetic Field for case (B)

geometry it is not necessary that $|B|$ will be zero any where inside or on the surface of plasma chamber. Further to estimate the radius of field free region, using the first harmonic of $|B|$ gives $r_{free} = R \left(\frac{B_{free}}{B_s} \right)^{1/(N-1)}$. Here field free region is defined as the region where $|B| < B_{free}$, and B_s is the magnetic field intensity at the plasma wall. But if we consider exponential decay form as per Eqn(9) then field free region would be $R \left(1 + \frac{1}{N} ln \frac{B_{free}}{B_s} \right)$. This can be quite intriguing, suppose B_{free} is chosen less than $\frac{B_s}{e^N}$, then it will predict complete absence of field free region. Further as per Eqn(9) it follows exponential decay, initially the field would decrease much faster compared to that predicted by Eqn(8), but toward the center the magnetic field predicted by Eqn(9)would decrease much slowly as compared to the prediction by Eqn(8). Hence there would be crossover in between the field obtained by Eqn(9)and Eqn(8), and if the B_{free} is larger than the crossover value then Eqn(9) would predict larger field free region. This will be valid when N is large and vice versa when B_{free} is less than the crossover value. In the central region close to the axis ($r \approx 0$), $|B|$ according to Eqn(9) varies linearly, given by $e^{-N}(1 + Nr/R)$, where as according to Eqn(8) it varies as $(r/R)^{N-1}$. On the other hand variation of $|B|$ near the surface is slightly faster ($B_s(1 - Ny/R)$) as predicted by Eqn.(9) compared to ($B_s(1 - (N-1)y/R)$) as predicted by Eqn.(8) , where y is measured from the surface. This explains the reason why the difference in the two curve is significant in the central region where as they are close near the surface.

RESULT AND DISCUSSION

The measurement of magnetic field inside cylindrical geometry has been performed using a Hall probe technique having measurement accuracy of one Gauss. The Hall probe is mounted on the measuring bench having x,y and z movement accurate up to $50 \mu m$ and controlled through stepper motor. The PM were arranged on the aluminum

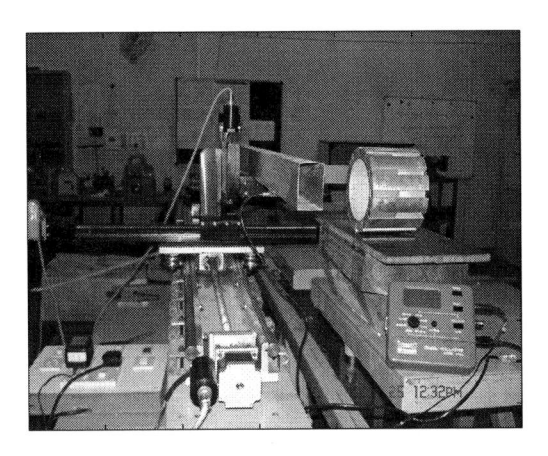

FIGURE 6. The experimental set-up to measure the multi-cusp magnetic field using the Hall probe system having 3-D stepper motor control movement for cylindrical plasma chamber used for the H- ion source.

cylindrical chamber, specially fabricated for holding them in proper slots. Two sets of 12 PM have been used. Ist set consist of 12 Wide magnet (12.5mmx25mmx50mm) and IInd set consist of 12 Square Magnet (12.5mmx12.5mmx50mm). Figure 6 shows the experimental setup used for the same. Figure 7(a-b) and Figure 8(a-b), shows that power law functional form (Eqn 8) in general predicts the variation in magnetic field more accurately compared to exponential functional form(Eqn 9). The discrepancy between the actual magnetic field |B| and those predicted by Eqn(9) increases in the central region where as Eqn(8) shows excellent agreement with the actual data even in the central region. Because of this in our arrangement the field free region predicted by Eqn(9) is nearly three times less as compared to experimental value, which is estimated correctly using the Eqn(8). In Figure 8(a-b), contributions up to two harmonics have been plotted for both power form and exponential form and again it is evident that the variation in magnetic field is more accurately predicted by power law dependence rather than exponential form. If we define the absolute error for the computed data with that of the experimental data by the expression:

$$\varepsilon_{rr} = \frac{1}{N} \sum_{i=1}^{N} |Bm_i - Bc_i| \qquad (10)$$

Where Bm_i is the measured magnetic field and Bc_i is the calculated magnetic field at the same location. The typical value of the above expression, using the power dependence is about 10 Gauss where as for the exponential dependence, it becomes 150 Gauss. The magnetic field,for the present experimental setup of 12 PM, near the chamber axis is about 5 G and near the surface of chamber wall it is about 2.3 kG . This clearly indicate the absolute error in using exponential form is much larger as compared to that of power law form.

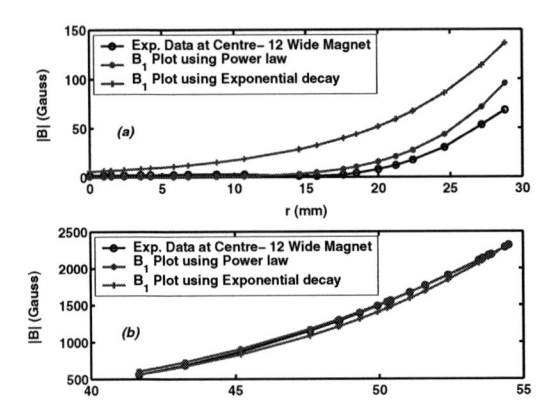

FIGURE 7. : The plot of magnetic field measurement data (blue curve) compared with exponential form (red curve) and power law form (green curve) for 12 wider permanent magnets surrounding the plasma chamber. In the central region (a) and near the chamber surface(b)

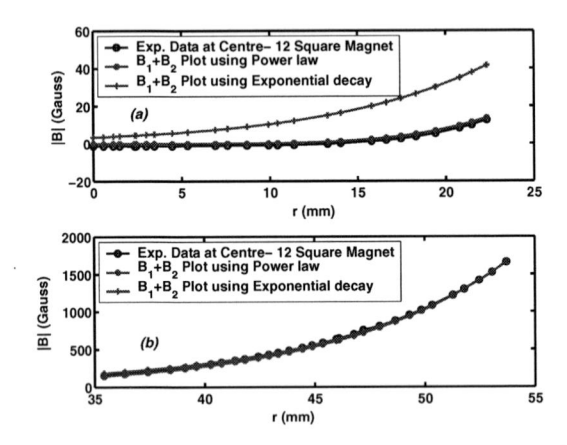

FIGURE 8. Same as above fig, but for square magnets considering the contributions up to two field harmonics

CONCLUSION

The result of the present work can be used in the plasma simulation which requires the knowledge of magnetic field inside the plasma chamber. The knowledge of field free region is crucial for placement of filament or RF antenna and electron filter magnet and for the proper design of extraction geometry. It also enable us to get an estimate of volume of uniformly dense plasma. Finally, power law functional form (Eqn- 8)represents a better functional form to compute multi cusp magnetic field inside the cylindrical plasma chamber than the conventional exponential form.

ACKNOWLEDGMENTS

Thanks are due to R. S. Shinde and Dr. Vinit Kumar for providing NdFeB Permanent Magnets. We thankfully acknowledge R. S. Shinde and his colleagues from ferrite lab. and R. K. Mishra and his colleagues for their help in Magnetic field measurement.

APPENDIX

MMF has been computed using vector potential ($\vec{A}$). For cylindrical geometry Figure(1) we need to use cylindrical coordinate and for planar geometry Figure(2) we use cartesian coordinate. In both the case z is perpendicular to the plane, along the length of the magnet. Since $\vec{A} = (0,0,A_z)$ which simplifies the expression for $\vec{B}$ as

$$\vec{B} = \frac{1}{r}\frac{\partial A_z}{\partial \theta}\hat{r} - \frac{\partial A_z}{\partial r}\hat{\theta} \tag{11}$$

$$\vec{B} = \frac{\partial A_z}{\partial y}\vec{i} - \frac{\partial A_z}{\partial x}\vec{j} \tag{12}$$

And the equation that governs the vector potential in the current free region simplifies to:

$$\frac{1}{r}\frac{\partial}{\partial r}\left(\frac{r\partial A_z}{\partial r}\right) + \frac{1}{r^2}\left(\frac{\partial^2 A_z}{\partial \theta^2}\right) = 0 \tag{13}$$

$$\frac{\partial^2 A_z}{\partial x^2} + \frac{\partial^2 A_z}{\partial y^2} = 0 \tag{14}$$

In cylindrical and cartesian coordinate respectively. The solution include origin in former case and vanishes at $y \to \infty$ in the latter case. Further as the permanent magnets are periodically arranged for plasma containment on the surface hence A_z is periodic on the boundary surface. The general solution for the above equation consistent with the boundary condition is written as

$$A_z = \sum_{m=0}^{\infty} r^m (C_m \sin m\theta + D_m \cos m\theta) \tag{15}$$

$$A_z = \sum_{m=0}^{\infty} e^{-my} (C_m \sin mx + D_m \cos mx) \tag{16}$$

in cylindrical and cartesian coordinate respectively. The C_m and D_m is to be determined from exact knowledge of A_z on the boundary. But the above form of solution can be exploited, if we can represent the A_z at the boundary in Fourier series and match it with the solution in the current free region. This will give the complete solution of the A_z. Hence the task is to compute A_z at the boundary and then decompose it in terms of Fourier series.

The $A_z(r = R, \theta)$ or $A_z(y = 0, x)$ can be easily computed using

$$\vec{A} = \frac{\mu_o}{4\pi}\int \frac{\vec{J_b}}{r}dv + \frac{\mu_o}{4\pi}\int \frac{\vec{K_s}}{r}da; \tag{17}$$

The knowledge of surface current ($\vec{K}_b$) and volume current ($\vec{J}_b$) can be obtained from the relation

$$\vec{J}_b = \vec{\nabla} \times \vec{M} \quad \vec{K}_b = \vec{M} \times \hat{n} \tag{18}$$

Where $\vec{M}$ is the magnetization of the permanent magnets. To demonstrate explicit expression for $\vec{A}$ let us assume $\vec{M}$ to be constant then

$$\vec{J}_b = \vec{\nabla} \times \vec{M} = 0; \vec{K}_s = \vec{M} \times \hat{n} = M\hat{k} \tag{19}$$

Hence, choosing the center of one magnet with positive magnetization as origin the expression for $A_z(y=0,x)$ is,

$$A_z(0,x) = \frac{M\mu_o}{4\pi} \int_{-b}^{b} \int_{-a}^{a} \sum_{i=-\infty}^{i=+\infty} \left(\frac{1}{\sqrt{(x-id\pm s)^2+y^2+z^2}} - \frac{1}{\sqrt{(x+id\mp s)^2+y^2+z^2}} \right) dydz \tag{20}$$

Where $2b$ is the length of magnet (along z axis) and $2a$ is width of magnet (along y axis) and $2s$ is the width of pole face (along x-axis) and $2d$ is the periodicity of the magnet (along x-axis). This can be numerically computed as a function of x. The typical plot of $A_z(0,x)$ has been shown in the Figure(3)Case(A) where the separation of the magnet is comparable to the pole width and in Figure(3)Case(B) is the typical plot of $A_z(0,x)$ where the magnets are widely separated. A similar expression can be written for $A_z(r=R,\theta)$ and can be computed as a function of θ. But as evident from the plot of $A_z(0,x)$ at the surface a convenient approximation for it can be made through the following periodic function $f(\theta)$ on the surface, where $\theta = \frac{x}{R}$ for planar case and $\theta = \frac{l}{R}$ for cylindrical case. l is arc length.
For Case(A)

$$f(\theta) = \begin{cases} -\frac{\theta+\theta_o}{\varphi}A & ; \quad -\theta_o < \theta \leq -\theta_o+\varphi \\ -A & ; \quad -\theta_o+\varphi < \theta \leq -\varphi \\ \frac{\theta}{\varphi} & ; \quad -\varphi < \theta \leq \varphi \\ A & ; \quad \varphi < \theta \leq \theta_o-\varphi \\ -\frac{\theta-\theta_o}{\varphi}A & ; \quad \theta_o-\varphi < \theta \leq \theta_o \end{cases} \tag{21}$$

And a convenient approximation of $f(\theta)$ for Case(B) is

$$f(\theta) = A(\delta(\theta-\varphi) + \delta(\theta-\theta_o+\varphi) - \delta(\theta+\varphi) - \delta(\theta+\theta_o+\varphi)) \tag{22}$$

Here $\theta_o = d/R$ where d is the separation between the neighbouring magnet and $\varphi = s/2R$, where s is the width of the pole face of the magnet. Note that $2\theta_o$ is the periodicity of the magnet. And since the function $f(\theta)$ is odd, it will have Fourier series expansion in terms of *sin* function. i.e.

$$f(-\theta) = -f(\theta) \Rightarrow f(\theta) = \sum_{n=1}^{\infty} a_n \sin\left(\frac{n\pi}{\theta_o}\theta\right) \tag{23}$$

the Fourier coefficient a_n is evaluated to
For Case(A)
$$a_n = (1 + -1^{n+1}) \frac{2A\theta_o}{n^2\pi^2\varphi} \sin\frac{n\pi\varphi}{\theta_o} \qquad (24)$$

For Case(B)
$$a_n = (1 + -1^{n+1}) \frac{2A}{n\pi} \sin\frac{n\pi\varphi}{\theta_o} \qquad (25)$$

When n is even $a_n = 0$. Hence only odd harmonics will contribute for magnetic field. And also in both the cases a_n decreases as n increases. Now matching the boundary value of A_z given by $f(\theta)$ to the general solution given Eqn. 15, i.e.
For Cylindrical configuration

$$\sum_{n=1}^{\infty} a_n \sin\left(\frac{n\pi}{\theta_o}\theta\right) = \sum_{m=0}^{\infty} R^m (C_m \sin m\theta + D_m \cos m\theta) \qquad (26)$$

Which implies $D_n = 0$, $m = \frac{n\pi}{\theta_o}$ and $C_m = \frac{a_n}{R^m}$ and if N pair of permanent magnets is surrounding the plasma chamber then $N = \frac{\pi}{\theta_o}$ hence the expression for $A_z(r,\theta)$ in cylindrical co-ordinate is

$$A_z = \sum_{n=1}^{\infty} a_n \left(\frac{r}{R}\right)^{Nn} \sin Nn\theta \qquad (27)$$

and for planar multipole configuration, matching the boundary value of A_z to the general solution given by Eqn. 16

$$\sum_{n=1}^{\infty} a_n \sin\left(\frac{n\pi}{\theta_o}\theta\right) = \sum_{m=0}^{\infty} (C_m \sin mx + D_m \cos mx) \qquad (28)$$

Since $\theta = \frac{x}{R}$, the above equation give $D_n = 0$, $m = \frac{n\pi}{\theta_o R}$ and $C_m = a_n$ and if periodicity of the permanent magnet along $y = 0$ is $2d$ then $\theta_o = \frac{d}{R}$ Hence the general solution for A_z in this case can be written as

$$A_z = \sum_{n=1}^{\infty} a_n \exp\left(-\frac{n\pi y}{d}\right) \sin\frac{n\pi x}{d} \qquad (29)$$

Further when $\vec{A} = (0,0,A_z)$ then the contours of constant A_z is same as magnetic field lines as $d\vec{s} \times \vec{B} = 0 \Rightarrow B_\theta dr - B_r r d\theta = 0 \Rightarrow \frac{\partial A_z}{\partial r} dr + \frac{\partial A_z}{\partial \theta} d\theta = 0 \Rightarrow dA_z = 0$. Some other useful properties of two dimensional multi polar magnetic field has been given in [8]

REFERENCES

1. R. Limpaecher and K.R. MacKenzie, Rev. Sci. Instrum. 44, 726(1973)
2. K.N.Leung, T.K.Samec and A.Lamm, Phys. Lett. A51,490(1975)
3. K.N.Leung, Noah Herskowitz and K.R.Mackenizie, Phys. Fluid 19,1045(1976)

4. Michael A. Liberman and Allan J. Lichtenberg,"Principal of Plasma Discharge and Material Processing",Publisher: John Wiley and sons, 1994
5. C. Koch and G. Matthieussent, Phys. Fluid 26,545(1983)
6. N.Tomozeiu, W.I.Milne, Journal of Non-Crystalline Solids 249(1999)180-188
7. C. Gauthereau and G. Matthieussent, Physics Letters 102A 5,6 (1984)
8. D.E.Lobb, Nuclear Instru. and Methods 64(1968) 251-267

Development of Small Multiaperture Negative Ion Beam Sources and Related Simulation Tools

M. Cavenago*, V. Antoni[†], T. Kulevoy**, S. Petrenko**, G. Serianni[†] and P. Veltri[†]

*INFN-LNL, Legnaro, Italy
[†]Consorzio RFX, Associazione Euratom-ENEA sulla fusione, Padova, Italy
**INFN-LNL, Legnaro, Italy and ITEP, Moscow, Russia

Abstract. In the design of extraction systems for negative ion sources several fundamental questions still deserve further investigation, as the distribution of particles near the extraction sheath, the optimal magnetic structure and the space charge compensation length after acceleration. Large (and undesired) deflection differences may develop between beamlets of a multiaperture source, so that equalization of the magnetic field effect is necessary. To guarantee an uniform strength of filter field at extraction and in the acceleration, several configuration of arrays of permanent magnets were studied and fast simulation tools were developed. As an example of optimized magnetic configuration and as a possible experimental tool, the design of NIO1 (Negative Ion Optimization try 1) is here discussed. This project consists of a 3 x 3 matrix of 8 mm extraction holes, aimed at a total H^- current about 130 mA with an extraction voltage $V_s = -60$ kV. A modular design is used, so several parts (the extraction grid, the acceleration grid, the filter assembly, the source multipoles) can be rotated by 90 degrees for versatility. Space charge compensation was included into a two dimensional self consistent code for negative beams, here used for NIO1 simulation.

INTRODUCTION

The development of Neutral Beam Injectors (NBI) for the ITER project [1] and the large beam test facilities planned to be built at Consorzio RFX (Padova, Italy) described elsewhere [2] are a strong motivation to further improve the negative ion sources (NIS) and the understanding of their beam extraction, also in the perspective of tokamaks beyond ITER. In this paper we describe two simulation codes under development for negative ion sources and the project of a small ion source (NIO1), useful for code validation and for testing of advanced beam diagnostic.

A scheme of a typical NIS is shown in Fig 1, with a full option accelerating column. The use of radiofrequency to heat plasma (as opposite to injection of 100 eV electrons by filaments) has the practical advantage of robustness; moreover an electron temperature about 4 eV is enough to produce the ionization required for plasma global balance [3, 4]. So considering radiofrequency ion source for fixing ideas, electrons heated to about $T_e = 4$ eV are confined by a magnetic filter field $\mathbf{B}^f$ in the rear of source (region 1), while a lower temperature plasma $T_e = 1$ eV diffuses in region 2 towards the source exit, named plasma grid electrode (PG). An intermediate and optional electrode named bias plate (BP) returns to the plasma a large flow of electrons, necessarily lost on PG. Ions H^- are formed on the PG rear wall (cesiated and bombarded by fast H^0), or in the region

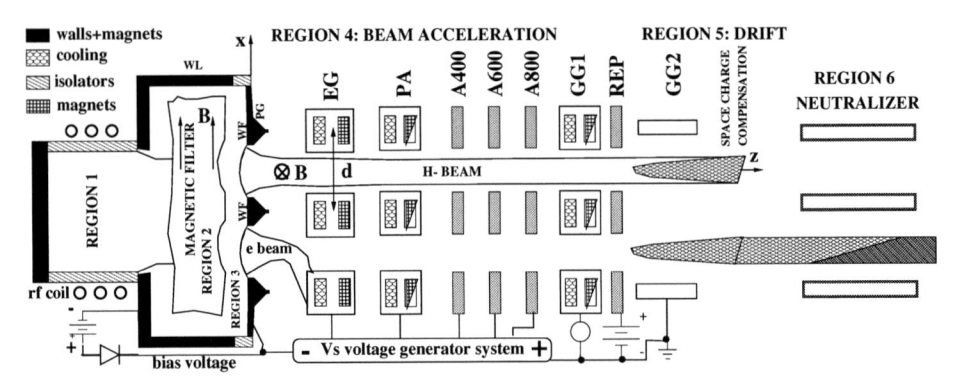

FIGURE 1. Scheme of an rf negative ion source and its accelerating column; for the sake of generality, we included several optional components: a repeller REP between the two nearly grounded grid GG1 and GG2 and the intermediate electrodes PA, A400, A600, A800. From GG2, the H⁻ beam is space charge compensated (dotted area) by the slow ions H_2^+ produced. Conversion (neutralization) from H⁻ to fast H⁰ (hatched green area) happens later.

2 volume.

A major parameter of the NIS operation is the ratio $R_j = j_e/j_{H^-}$ between the current density of electrons emitted from the plasma j_e and the ion current density j_{H^-}. The extraction grid electrode (EG) immediately follows the PG and typically houses permanent magnets (PM). Their magnetic field (called EG field for brevity) deflects most of extracted electrons, against the EG itself or back to plasma. The next electrode is called preacceleration grid (PA). The last electrode of the accelerator column is usually connected to ground and called grounded grid GG. In accelerators below 100 kV, including NIO1 (60 kV nominal), GG and PA typically coincide. Let z be the beam axis and y be the dominant direction of the EG field. The filter field $\mathbf{B}^f$ may be in the x or the y direction.

A quantity typically used in NIS plasmas [5] is the ratio $\alpha_R = n_{H^-}/n_e$ between particle densities n_e and n_{H^-}. These ratios can be roughly related at the plasma border by $R_j = R_m/\alpha_R$ with the constant $R_m = (m_{H^-}/m_e)^{1/2} \cong 43$. Since $R_j \leq 1$ is an ITER specification and advanced NIS easily maintain $R_j \leq 2$, we conclude that $\alpha_R \cong 20$ at the plasma border. On the contrary, the inner plasma region 1 must have $\alpha_R \cong 0$ (electrons dominates over H⁻). This spatial arrangement is the opposite of a well-known case[5], where negative ions are much colder than electrons and concentrate in the inner plasma.

So we have to assume that electron density reduces near PG by the effects of the space charge of H⁻ and of the filter field, for consistency with a coextracted current density j_e of electrons within reasonable limits ($R_j \leq 2$). Other open issues in the physics of NIS includes: the effect of crossing the filter field and the EG field, the uniformity between beamlets, the minimization/optimization of the cesium use, the simulation and the control of the plasma-beam interface (called meniscus in ion source physics).

Known models of the plasma-wall interface or the plasma-beam interface are typically very simplified systems, restricted to one space dimension and to positive ion plasmas[6, 7, 8]. Typical features are: a presheath where a small slope of the electrical potential ϕ

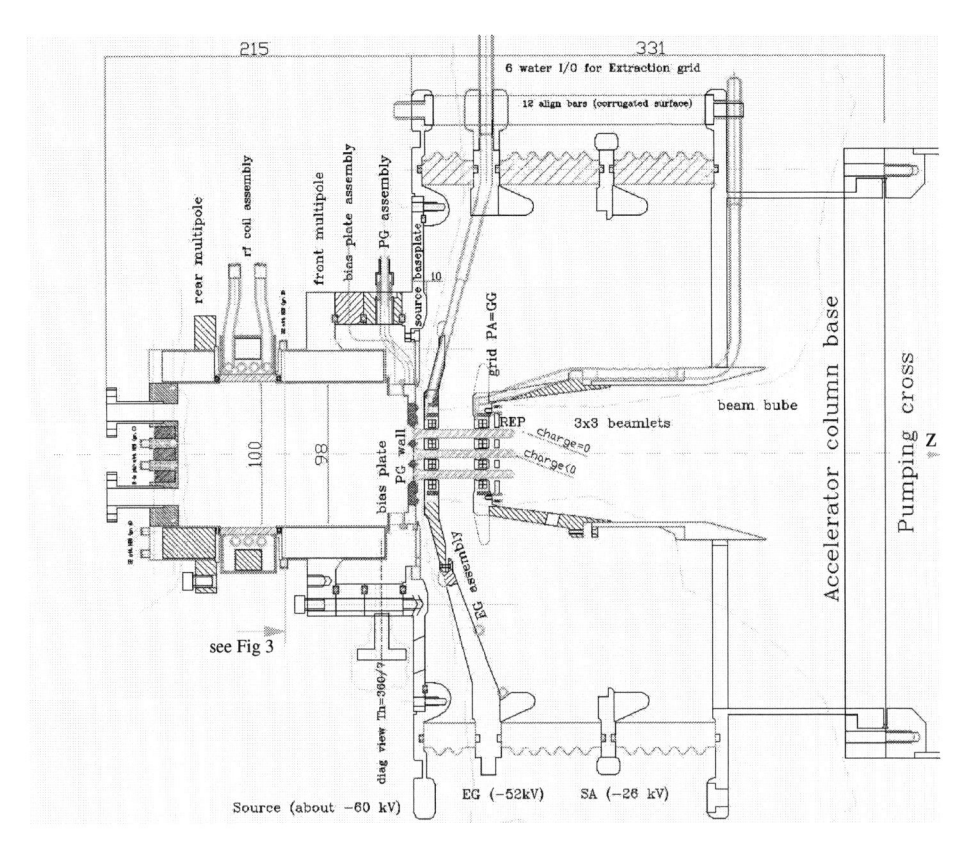

FIGURE 2. NIO1 conceptual design (with different cuts to show the yz section of the magnets inside EG and PA, and one of the optical diagnostic ports, here labeled 'diag view'): note the 3 x 3 beamlet matrix, the strong water cooling of the intermediate electrodes and the modularity of the ion source. Source baseplate is reentrant inside the accelerator column end flange.

guarantees the balance of charges and outwards particle flows; a thin sheath (with a net charge), where electric fields are strong; the need of large computing efforts[9, 10]. Sheath thickness is several times the Debye length λ_D.

The following sections will discuss a small test source design, a 3D tool for source magnet optimization and progresses in selfconsistent beam simulations.

THE NIO1 DESIGN

The small source conceptual design (called NIO1, for Negative Ion Optimization 1, see Fig. 2) emphasizes modularity, for quick repair of parts, so that source is a tower of disk assemblies (connected by O-rings). Rotation of parts of 90^0 is possible, to test the better direction of the source magnetic filter (crossed or parallel to the EG field).

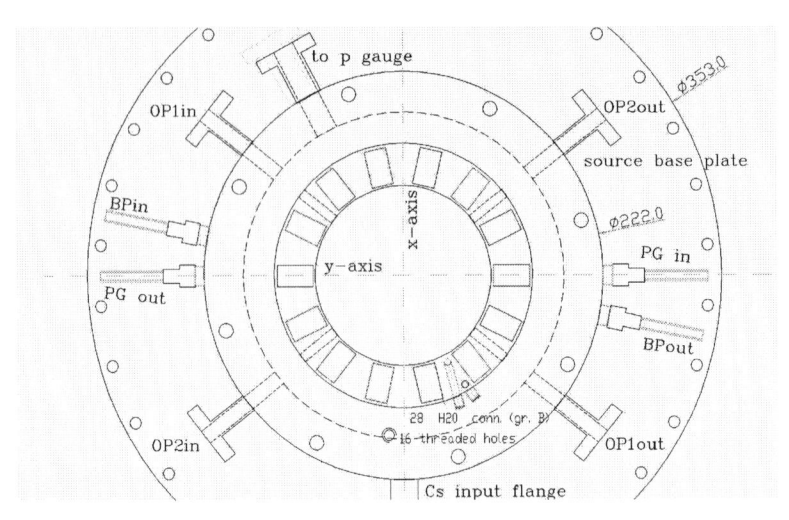

FIGURE 3. NIO1 conceptual design: *xy* section of the front multipole. The source baseplate and the outlines of the optical diagnostic ports OP, of the current inputs PGin and BPin, of the cesium port and of the pressure gauge port are also visible.

Consequently we have 9 beam holes in the PG (diameter about 8 mm) in a square pattern with $L_x = L_y = 14$ mm, where L_x is the spacing between centers in the x direction and L_y similarly for the y direction.

Requiring $j_{H^-} > 280$ A/m² (equivalent to 200 A/m² of D⁻, which is the ITER design value), total extracted ion current is in the order of 130 mA (an economic limit); nominal source voltage is $V_s = -60$ kV, with isolators up to $V_s = -100$ kV.

Manufacturing the EG electrode (movable along z) will be an important technical test. EG is a disk with 9 holes, supported by six arms based on a circular flange, sealed by O-rings between two accelerating column isolators; large windows between arms provide a strong lateral pumping. Up to 6 water feedthroughs can be installed. In Fig. 2 the four PM bar yz sections, centered at $y = \pm 7 = \pm L_y/2$ and ± 21 mm, are visible. Magnetization **M** directions are $\pm \hat{z}$, with a 180 degree turn from one bar to the other. Since each bar opposes the adjacent bar(s), the resulting B_y field is larger in the first and third beamlet in the yz section, as opposed to the central beamlet (the 2nd one). To equalize the magnetic EG field between holes, we use some iron shimming bars, centered at $y = \pm 27$ mm $\cong \pm 2L_y$; indeed each shim produces an image of the permanent magnets (with M_z reversed), so weakening the EG field at the outer beamlets. PM bars should be long enough in the x direction, so that their end effect lies away from the beamlets that are centered at $x = \pm L_x$. Since the EG outer envelope is circular, shims can be 56 mm long and PM bars 64 mm long; simulations proved the shim effectiveness (see next section). Anyway, for experimental flexibility, both shims and bars are planned to be removable (and thus adjustable) from rear of EG.

The PA design also includes removable PMs, to reduce the beam steering, if needed. As for EG, six pumping windows are provided. After acceleration, the beam travels inside a 84 mm ID (internal diameter) tube (labelled 'beam tube') to the pumping cross and then to the diagnostic chamber (a 350 mm ID tube, 1.5 m long). With reference to Fig 1, note that PA and GG1 coincide; moreover a repeller REP may be placed inside the 'beam tube', as shown in Fig 2. If compatible with mechanical stability of the accelerating column, we speculate to isolate the PA from the accelerator column base, with another O-ring (not shown in Fig 2). The accelerator column base will act as the true grounded grid GG2.

Accelerator column compression bars are made of PEEK (PolyEtherEtherKetone, a kind of thermoplastic). Alignment and stability of the accelerating column is complicated, since: 1) beam axis z is horizontal in source operation; 2) source will be often mounted and dismounted from the column in the experimental campaigns; 3) isolator raw tolerance is 1 mm (on a 369 mm nominal ID). Therefore alignment is guaranteed by external supports and suspensions (probably of PEEK). On the contrary the accelerator column base, the pumping cross and the diagnostic chamber are joined together by 16.5 inches Conflat flanges, making them a stable base of the system.

Source walls are adequately cooled and are covered with magnets (Fig 3), organized in the multipole assemblies and in the filter field plate; some bars may be removed.

The plasma grid assembly includes the plasma grid itself (a 90 mm diameter disk, with thickness from 5 mm to 8 mm at border) and a custom flange, with outer diameter (OD) 222 mm and O-ring grooves on both sides; major PG connections are: the optical ports OP, the Cs input and two tubular feedthroughs (named PGin and PGout) into which a current I_y and a fluid like hot air can be passed. This current can add a tunable B_x term to the filter field. Note that the I_y conductors have some bends in z direction for space reason.

The bias plate assembly is similar, in particular has two feed BPin and BPout into which the current I_y can return; this clear determination of the current loop seems one of the good points of NIO1 design. We also speculate to later add an iron ring, to close the B_x magnetic return path.

Operation is limited to clean H_2. Plasma is inductively coupled to an external rf coil[11], wound over a 40 mm long ceramic tube. Some effort was put to make operation without a Faraday shield possible: the rf coil is water cooled and a jacket around the ceramic allows for air cooling of it. Optional PM may be placed behind the coil to supplement plasma confinement. Main rf frequency is 2 ± 0.2 MHz, but low power experiments between 1 and 60 Mhz and several studies of matching techniques for rf circuitry are also planned. Modular design easily allows to use a longer rf coil (as usual) and a Faraday screen, if needed.

A quickly removable Cs oven is under development. A CF16 flange on the PG assembly allows a thermally isolated copper pipe for cesium transport (ID 4 mm, OD 8 mm) to reach the bias plate. Cesium oven is placed in a fiberglass pocket; by carefully regulating its temperature, cesium injection is controlled; on the contrary the copper pipe is overheated (at 470 K or more) to avoid Cs sticking. A plug (sealed and moved by bellows) or, if available, a suitable valve, allows to separate the Cs oven from the source vacuum, for maintenance purposes.

Source has an $m = 7$ multipole field (14 poles), so as to merge smoothly with the dipole of the filter field. Two lines of view pass between multipole bars (see Fig 3), at a $d = 19$ mm distance from the PG, as a compromise between the measuring needs ($d \rightarrow 0$) and the material space. Measurement of the H$^-$ density (near PG wall) is a

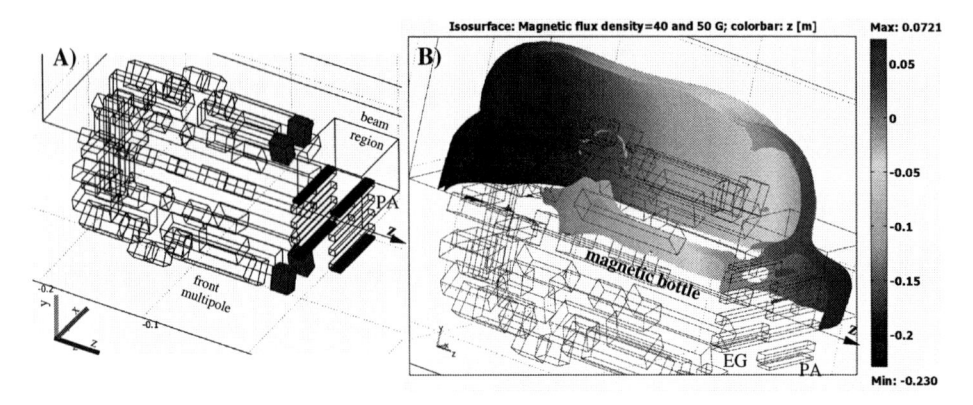

FIGURE 4. A) NIO1 conceptual design of the PM system: note the iron shims (blue) and the filter magnets (in brown) while other bars are shown only as a wireframe. B) isosurface of |B| plot; the levels of 40 and 50 G were chosen to put the 'magnetic bottle' in better evidence.

fundamental information for NIS physics[10].

Among other innovations to be tested we list: wall material effects; extended bias plate in the ion source; and many beam diagnostic systems, including calorimetric beam profile monitors (BPM) and a fast Alison scanners to measure emittance, under development. Emittance measure will be a much more informative test for code validation that any BPM is.

PERMANENT MAGNET ARRAYS

The permanent magnet system of NIO1 is shown in Fig 4, in the case when the filter and the EG fields are parallel. Then $y = 0$ is a symmetry plane of the system, with the boundary conditions $B_x = 0$ and $V^M = 0$; here V^M is the magnetic potential when no current is present, so that $\mathbf{H} = -\text{grad}\, V^M$ [12, 13]. Moreover $x = 0$ is another symmetry plane, with $V^M_{,x} = 0$, where $a_{,b}$ is the derivative of a with respect to b. Thanks to symmetry, only the quarter $0 \leq x$, $0 \leq y$ needs to be simulated.

Due the large number of bars, each one with a different magnetization axis, input files tend to be obscure. Thanks to the extended scripting capability of a solver software[13], it was possible to prepare some programs to build the more common structures of permanent magnet arrays, which in our case are: 1) the single bar; 2) the linear array of bars; 3) the multipole, that is the circular array of bars.

In fig 4.A for example, the PA magnets are a linear array, while the PA iron shims are two single bars. We here specify all bars (to obtain a complete plot), even if only the bars in $x > 0, y > 0$ enter the simulation.

Multipoles protect walls from electron bombardment[14], so they may help to save rf power. We assumed that all PM have a residual field density $B_R = 0.96$ T, as for SmCo, except for the optional magnets over the rf coil with $B_R = 0.33$ T, considering the possible difficulties of finding an adequate ferrite. In fig 4.B we plot the surface of

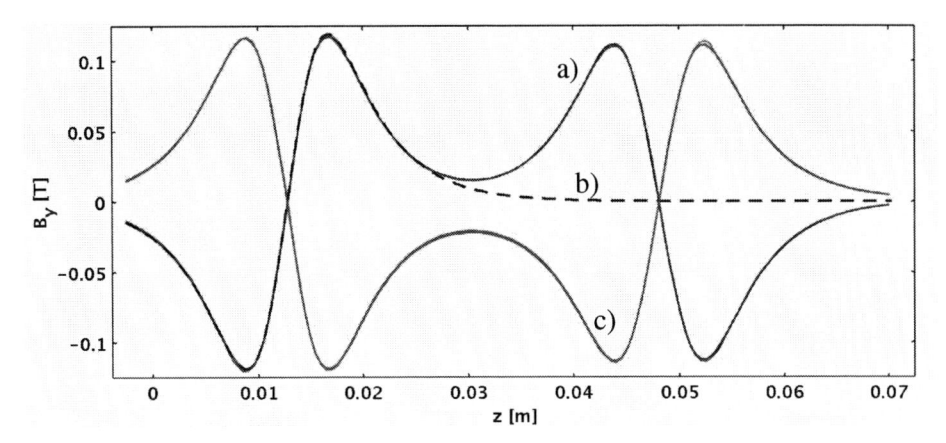

FIGURE 5. Profiles of B_y: a) simulation results for $y = 0$ and $x = 0, 2, \ldots, 18$ mm (solid lines); they are fully superposed after optimization; b) analytical result for the EG array only (centered at $z = 12.8$ mm); c) simulation results for $y = 14$ mm and $x = 0, 2, \ldots, 18$ mm (magenta lines). In the beam region, equivalent mesh size was less than 1 mm, so that whole simulation required 1100000 degrees of freedom.

constant $|B|$, to verify weak points, due to interference of different bars. It appears that an almost close surface with $|B| \cong 45$ G is formed inside the plasma chamber, which was the goal of having a so-called magnetic bottle. A second surface $|B| \cong 45$ G is visible outside the source. It must be noted that the efficiency of the 'magnetic bottle' concept for NIO1 is a matter of experimental verification, to be further studied.

Effect of the whole PM system on the beam region ($z > 0$) is shown in Fig 5.a), where we compute the B_y profiles for several lines of views with $y = 0$ (or $y = 14$ mm) and $x = 2k$ mm, with $k = 0, \ldots, 9$ to cover all the beamlets. After some trials and error, EG and PA iron shims were adjusted so the ten profiles with $y = 0$ superpose, showing a very good equalization. The analytical result[2] for an infinite linear array of PMs placed at the EG position is shown in curve b). Curves a) and b) agree for $z < 0.03$ m; thereinafter the PA magnetic field should be also considered.

In summary, magnetic optimization (based on symmetric arrays of PM and on image theory) can reach the design goals.

PROGRESSES IN SELFCONSISTENT BEAM SIMULATIONS

Magnetic field effect on the electron space charge can be represented in a 2D planar geometry, but not in 2D cylindrical one. As a possible and necessary step towards a 3D simulation of a real plasma meniscus, a family of codes was developed[15], named BYPO since a B_y perpendicular to a zx 2D planar geometry is assumed. The sheath is simulated by a space charge with a nonlinear dependency from the electrostatic potential ϕ, while the presheath is still represented by a nonlinear boundary condition for ϕ at the ion startline[16]. By convention, $\phi = 0$ at the PG, while V is the potential referred to

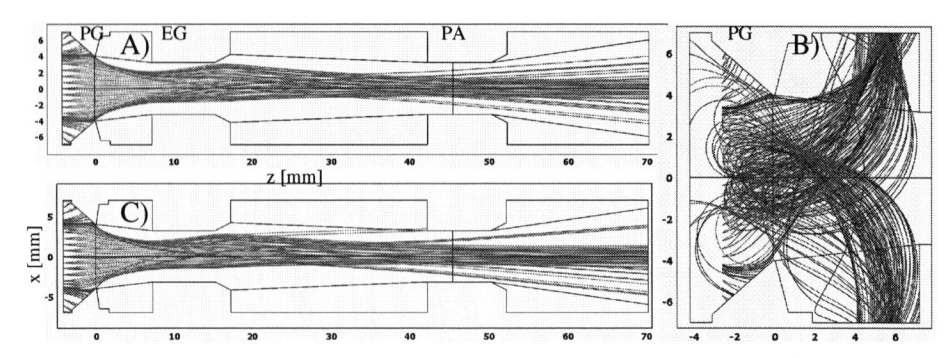

FIGURE 6. Beam simulations: A) Ion trajectories with a 75^0 focus angle, voltages $V_{EG} = -56.8$ kV, $V_{PG} = -65$ kV, current density j_{H^-} 341 A/m^2 and B_y as the 80 % of fig 5. Ion starting angles (respect to the $-\mathbf{E}$ direction) are $\alpha_s = -0.1$ rad (red), $\alpha_s = 0$ rad (brown), $\alpha_s = 0.1$ rad (green); B) Zoom on the electron trajectories, same simulation; electrons exiting from the upper side are reinjected from the lower side; C) As for panel 'A', but with a 82^0 focus angle.

ground GG.

Let N_x the number of beamlets in the x direction and N_y the number in the y direction. When $N_x = 1$, BYPO geometry is the zx section of the system (Fig 6) and boundary condition on open gaps is $\phi_{,x} = 0$. When $N_x > 1$, simulation geometry is the zx section of the central beamlet channel, and periodic conditions for ϕ on open gaps represent effects of possible x-asymmetries of the space charge; moreover electrons exiting from the upper open gap are reinjected from the lower open gap (up to $n \cong N_x/2$ times).

Several new physical effects are being added to the BYPO code, as the space charge compensation since BYPO16 and the gas flow since BYPO17; see next subsections. BYPO is implemented into a Comsol environment[13], with the following advantages: 1) use of state of art numerical solvers; 2) powerful graphics (in 2D and 3D); 3) a highly structured programming language. By some suitable routines, the geometry is fully parametric. Preparation of a structured atomic database is also well in advance.

The ray tracing routine in BYPO features an (iterative) refinement of the starting position x_s of the rays (trajectories) near the PG edge, to better define the beam envelope. Rays with different starting angles α_s are used, as visible in Fig. 6.A. Ray current is proportional to the difference of x_s with nearby rays, as taken in account by the BYPO integration of the space charge.

Comparisons of BYPO17 to SLACCAD[1] in the ISTF geometry were discussed elsewhere [2]. Some BYPO17 results for NIO1 are shown in Fig. 6.B: electrons are stopped before the EG, but this enhances the electron space charge in PG-EG gap. Ion beam also suffers some deflection inside the EG aperture, see Fig. 6.A, which is later corrected by the reversing B_y and the focusing effect of the acceleration field; we see a halo, mainly due the rays passing near the PG edge, and secondly to the imperfect optimization of the focus angle $\beta = 75^0$ of the PG. Indeed, in Fig. 6.C with $\beta = 82^0$ halo increases. Even if current carried in the halo is small, finer optimization of NIO1 electrodes seems needed.

Some 3D simulations of the NIO1 geometry were performed[17], using well known codes, as KOBRA[18] and, with a very limited resolution, SCALA[19]. Preliminary results are reasonably consistent with BYPO, in view of the differences of the detail of the meniscus physics used (in SCALA a fixed emitter, in KOBRA a volume region).

Compensation of beam space charge

After the GG, the negative ion beam (with speed v_b, number density n_b and radius r_0) reaches a radial equilibrium [20, 21] of the space charge compensation in an unknown distance z_2. Let σ_i the cross section to generate slow positive ions (mostly H_2^+) from the H_2 gas molecules with number density n_g; similarly let $\sigma_e \cong \sigma_1 + 2\sigma_2$ the cross section of producing electrons, where σ_2 (or σ_1) is the cross section of double (or single) detachment. Let the potential inside the beam be

$$V(r,z) = V_0(z) + V_2(r/r_0)^2 + V_4(r/r_0)^4 + \ldots \qquad (1)$$

where V_2 is the $-\phi_0/e$ of Ref [20]; at the GG2 we have $V = 0$. Radial transport (free fall for ions and collisional for electrons) gives

$$n_{i0} = m_i^{1/2} r_0 n_b n_g v_b \sigma_i (\tfrac{1}{2} T_i - 2eV_2)^{-1/2} \, , \quad n_{e0} = r_0 n_b n_g v_b \sigma_e (\pi m_e/2T_e)^{1/2} e^{-eV_2/T_e} \qquad (2)$$

where n_{i0} is the number density of positive ions on z-axis and n_{e0} the electron one; here T_i is the thermal energy of ions (due to beam collision) and T_e the electron one[20]. Poisson equation gives $n_b + n_{e0} = n_{i0} + (4\varepsilon_0 V_2/er_0^2)$. Note that if total charge happens to be positive (overcompensated beam) we get $V_2 < 0$ and thus $V_0 > 0$, which is risky: positive ions may be pulled back into the accelerator, making the radial equilibrium assumption invalid (and increasing the power load from backstreaming ions as well). In the undercompensated case we may have $V_2 > 0$ and $V_0 < 0$. Estimating T_e as in Ref [21] and setting $T_i = 0.8$ eV, the radial equilibrium can be solved. A tabulation was prepared for quick input in BYPO17. We see that for $n_g \leq 1.2 \times 10^{19}$ m^{-3} (gas pressure < 0.05 Pa), $n_b \leq 10^{15}$ m^{-3} and $r_0 \cong 4$ mm, the undercompensated case holds (with a 99.7 % compensation for beam voltages from 10 kV to 1 MV). BYPO17 extrapolates this radial equilibrium to an empirical 2D zx-model of the space charge compensation (ion axial transport is thermal, electrons are pulled to GG), showing that operation without a repeller may be possible.

Gas flow

The pressure difference is $\Delta p = F_z/C$ with F_z the flow and C the conductance. From the well known Knudsen formula $C = 4v_m/3 \int dz(P/A^2)$ for a long channel with area A and section perimeter P (smoothly changing with z), where v_m is the molecular average speed [22], we extrapolate that

$$p_{,z} = 3F_z b^{-3}/(2\pi v_m) \qquad (3)$$

for a circular channel of radius b. Similarly $p_{,z} \cong 3(F_z/a)b^{-2}/2v_m$ for a rectangular channel of width b and a much larger height a.

Since the actual radius (or width) $b(z)$ of our channel is rapidly changing, we empirically smoothed it; then eq 3 and the gas flow conservation div $\mathbf{F} = 0$ give an approximated equation for p, which can be easily solved numerically[13]. Gas stripping of the H$^-$ ion can be then computed as usual.

It must be noted that a large lateral pumping exists in NIO1 and that gas flow conservation neglect ionic pumping. BYPO17 thus corrects stripping losses for an user defined factor (typically 50 %).

CONCLUSIONS

Overall design of the NIO1 source has been completed, and several useful computer codes for source design were concurrently developed. Some insight on the halo formation and the optimal EG field for source operation was obtained. The flexibility of NIO1 promises to be extremely valuable for a full validation of codes and for experimental campaigns.

REFERENCES

1. R. S. Hemsworth, J. H. Feist, M. Hanada, B. Heinemann, T. Inoue, E. Kussel, A. Krylov, P. Lotte, K. Miyamoto, N. Miyamoto, D. Murdoch, A. Nagase, Y. Ohara, Y. Okumura, J. Paméla, A. Panasenkov, K. Shibata, M. Tanii, and M. Watson. *Rev. Sci. Instrum.* **67**, 1120 (1996).
2. P. Agostinetti et al., *this conference*.
3. P. N. Wainman et al., *J. Vac. Sci. Technol.*, **A 13**, 2464 (1995)
4. M. A. Lieberman and A. J. Lichtenberg, *Principles of Plasma Discharges and Material Processing*, John Wiley, New York, 1994
5. N. St J. Braithwaite and J. E. Allen, *J. Phys. D*, **21**, 1733 (2003)
6. L. Tonks, I. Langmuir, *Phys. Rev.*, **34**, 826 (1929).
7. K-U. Riemann, *J. Phys. D*, **24**, 493 (1991)
8. R. N. Franklin, *J. Phys. D*, **36**, R309 (2003)
9. J. H. Whealton et al., *J. Appl. Phys.*, **64**, 6210(1988)
10. D. Wunderlich, R. Gutser, U. Fantz, AIP Conf. Proc. **925**, 46 (2007).
11. R. F. Welton et al., *Rev. Sci. Instrum.*, **75**, 1789 (2004).
12. J.D. Jackson, *Classical Electrodynamics*, Wiley & Sons, New York, 1975
13. *Comsol Multiphysics 3.3a*, (2007), see http://www.comsol.eu
14. A. T. Forrester, *Large Ion Beams*, John Wiley, NY, 1996
15. M. Cavenago, "Meniscus formation and beam optics of negative ion sources", Report RFX_3.1.1_Part_1 of EFDA/06-1499 Task TW6-THHN-NBD1 (2008), unpublished
16. M. Cavenago, P. Veltri, F. Sattin, G. Serianni, V. Antoni, *IEEE Trans. on Plasma Science*, **36**, pp 1581-1588 (August 2008)
17. T. Kulevoy and S. Petrenko, private communications (2008).
18. P. Spadtke, *Rev. Sci. Instrum.*, **75**, 1643 (2004)
19. Vector Fields, http://www.vectorfields.co.uk
20. A. T. Holmes, *Beam Transport*, in *The Physics and Technology of Ion Sources*, (ed. I.G. Brown), J. Wiley, NY (1989)
21. E. Surrey, AIP Conf. Proc. **925**, 278 (2007).
22. B. Ferrario, *Introduzione alla tecnologia del vuoto*, (ed. A. Calcatelli), Patron editore, Bologna (1999)

ION SOURCES FOR ACCELERATORS

H⁻ Ion Source Development for the LANSCE Accelerator Systems*

R. Keller, O. Tarvainen, E. Chacon-Golcher, E. G. Geros,
K. F. Johnson, G. Rouleau, J. E. Stelzer and T. J. Zaugg

*AOT Division, Los Alamos National Laboratory,
P. O. Box 1663 - MS H817, Los Alamos, NM 87545, USA*

Abstract. Employment of H- ion sources for the LANSCE accelerator systems goes back about 20 years, to the construction of the Proton Storage Ring (PSR). The standard ion source consists of a filament driven multi-cusp discharge vessel with a biased converter electrode for negative-ion production and an 80-kV extraction system feeding into a 670-kV electrostatic pre-accelerator. The source typically delivers 18 mA pulsed beam current into the pre-accelerator column and reaches up to 35 days between services at 60 Hz pulse repetition rate. Recent development efforts with this source have been dedicated to improved filament material, improved cesium oven geometry and operating the source at elevated temperatures. A second line of development focuses on filament-less devices driven by a helicon discharge. Performance data obtained with the standard source as well as key results for the helicon experiments are given in this paper; the helicon work is discussed in a separate paper in much greater detail.

Keywords: Negative Hydrogen Ion Source, Multi-cusp, Filament, Converter, Elevated Temperature, Cesium Oven, Helicon discharge
PACS: 29.25.Ni

INTRODUCTION

Because of their high conversion efficiency to neutral atoms and protons at beam-energy levels around 1 GeV, negative hydrogen ions are employed as primary beams in many high-power accelerator facilities that include accumulator rings and rely on charge-exchange injection. The Los Alamos Neutron Science CEnter (LANSCE) facility at Los Alamos has been utilizing H- ion beams for about 20 years, to facilitate injection into the Proton Storage Ring (PSR) accumulator ring as well as to increase the overall available Linac beam power by simultaneously transporting protons and negative hydrogen ions in the same structure.

The H⁻ ion source exclusively used for beam production to serve the LANSCE accelerator systems, see Fig. 1, is of the multi-cusp type with filament driven discharge and biased converter where H⁻ ions are formed on a cesium covered surface and accelerated by the converter bias voltage of about 300 V towards the outlet aperture [1]. This standard source needs about 36 hours of preconditioning of the

---
This work was supported by the US Department of Energy under Contract Number DE-AC52-06NA25396

FIGURE 1. Inside view of the LANSCE converter ion source. The discharge chamber has a cylindrical shape of 192 mm aperture, and the created negative ions travel radially from the converter surface at the center of the chamber to the outlet aperture on the right. The converter electrode is supported by a water-cooled tube; its concave surface is hidden in this view. One of two cathode filaments is seen reaching in from a planar flange on the rear side; cesium vapor is injected into the discharge chamber from the bottom, and the electron repeller is shown on the right, hiding the outlet aperture from view.

cathode filaments, performed on a separate processing stand. After relocation to the high-voltage dome, the source is brought up to full power again at a pulse repetition rate of 120 Hz, and then cesium is fed into the discharge chamber, saturating the chamber walls. When the discharge has steadied after another 36 hours, the repetition frequency is reduced to 60 Hz, and the operational parameters are adjusted for this mode. This transition takes about 18 hours.

The extraction system includes a Pierce-type outlet electrode and two acceleration gaps that increase the beam energy to 80 keV. In the production set-up, the beam is then injected into a multi-gap 670-kV electrostatic column and achieves the Linac injection energy of 750 keV. Typical performance data are 18 mA beam current downstream of the 80-kV extraction system and a lifetime between services of up to 35 days when the source is operated at 60 Hz repetition rate and 835 µs pulse length. The lifetime is limited by filament breakdown, and considerable effort has gone into understanding the failure mode and possibly extending the lifetime as discussed in the following. Effects related to optimized H⁻ ion production on the cesiated converter surface, have been studied as well and will also be discussed below.

Apart from attempts to improve the filament-driven converter source, a source modification development effort is underway in parallel where the plasma is created by an rf driven helicon-type discharge [2, 3]. This work is being discussed in detail elsewhere [4], and only a description of the main features of this novel source type and some key results are included in this paper.

FILAMENT ISSUES

The two filaments serving the standard converter source are made from 250-mm long, 1.6-mm thick, high-purity tungsten [5]. The key to longevity is the generation of large grains [6] which is achieved during the pre-processing sequence. Voltage and heating current are being recorded daily during a production run, and the resistance values computed from these data show a very characteristic pattern that is typical for the behavior of metals subjected to significant stress and is termed 'creep' [7-8]. The evolution of filament resistances, see Fig. 2, starts with an initial shock period where the resistance increases significantly over about 5 - 10 days. After that, the resistance increase follows a simple exponential curve as long as the heating power is not significantly changed, and ultimately the increase becomes very rapid and turns into catastrophic failure within a few hours.

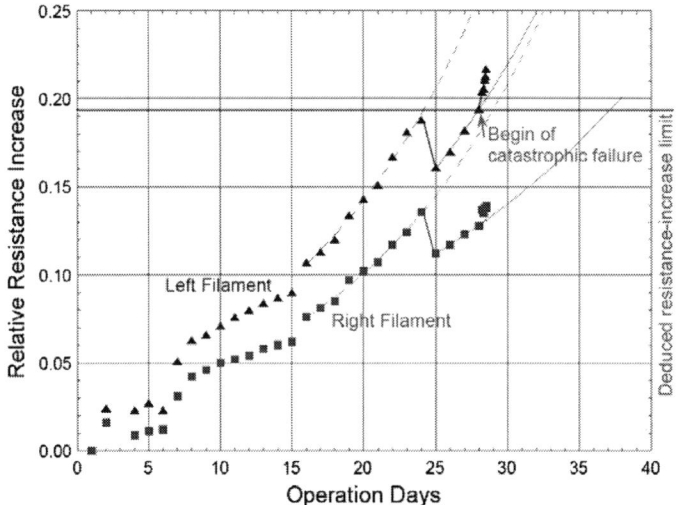

FIGURE 2. Evolution of the resistance of the two cathode filaments installed in a standard converter source during a test run in May 2008, showing the phases of initial shock Days 1-7), exponential growth, and catastrophic failure (Left filament on Day 28). Three phases of exponential growth are shown during Days 8-15, 16-24, and 25-28; the latter two are fitted with exponential curves and extrapolated. The filament heating power values were different for each of these three phases.

From the data plotted in Fig. 2 a practical resistance-increase limit of 19% can be deduced for the filaments under these conditions, but it is not yet assured that this limit applies universally for the LANSCE standard sources. Until more data can be collected an increase of 15% is deemed safe for reaching the mandated lifetime goal for any given run. In any case, these resistance-increase plots provide useful guidance how to adjust the filament heating power to achieve maximum beam-current output while safely reaching the desired filament lifetime for any given production run.

A dedicated analytic and computational modeling study [9] demonstrated that exponential resistance increase and catastrophic failure are well understood in principle, see Fig. 3. This model is based on the power balance between ohmic heating on one hand and heat conduction and radiation losses on the other. As main results, the formation of a hot spot as main cause for failure is identified, and the importance of very small defects ('nicks' with a thickness reduction in the order of 0.025 mm for 1.6 mm diameter that are commonly noticed on new filaments) becomes evident, see Fig. 4. To arrive at these key results within a short time, no attempt has been made to include metallurgical effects such as grain structure changes in this model, and this is the reason why the initial shock is not represented at all.

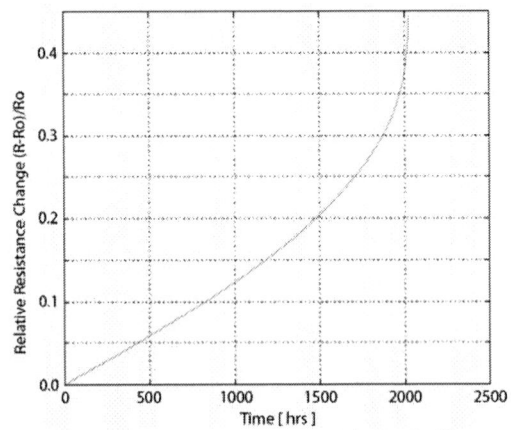

FIGURE 3. Simulated relative change in resistance $((R - R0)/R0)$ of a filament as a function of time for a perfectly uniform initial diameter profile.

To improve on the presently established filament lifetimes we have just begun testing tungsten/rhenium alloy filaments with 3% rhenium content. This material is widely being used in industrial plasma applications and promises to significantly extend the lifetimes [10]. Our own modeling [9] predicts a 30% increase in lifetime, mostly due to the slightly larger wire diameter of the material at hand.

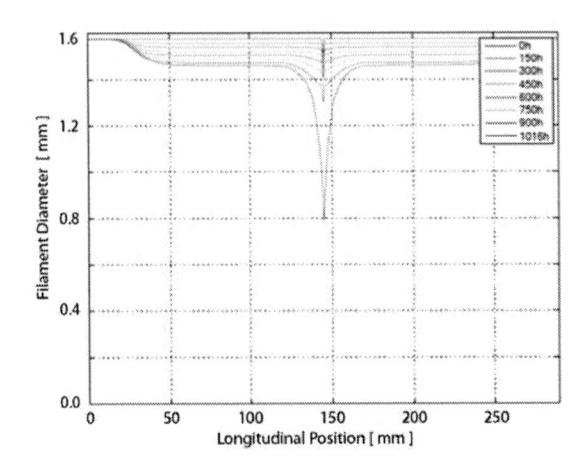

FIGURE 4. Simulated filament diameter versus position as a function of operating time, for a diameter profile with an initial 2% nick in the middle. For otherwise equal condition, the presence of the small imperfection accelerates failure by a factor of two (1016 hrs compared to just over 2000 hrs).

CESIUM-ENHANCED H⁻ ION PRODUCTION

Theoretical Considerations

The formation of H- ions in the standard LANSCE H⁻ source takes place on the surface of a negatively biased electrode (converter) exposed to a flux of positive ions (H^+, H_2^+, H_3^+ and Cs^+) created by the filament-driven discharge. Hydrogen atoms absorbed on the converter can be sputtered off from the surface by the incident ion flux. A fraction of the sputtered (and scattered) particles can acquire negative charge via resonant tunneling [11] charge exchange with the converter electrode surface. This fraction depends strongly on the work function of the surface material, which can be lowered by depositing an adequate layer of alkali metal (typically cesium) on the surface. In addition, the cesium ion influx has an instrumental role on the sputtering of hydrogen atoms during the discharge pulses and contributes significantly to the yield of negative ions [12-13].

Cesium is introduced into the ion source by a resistively heated oven that vaporizes the metal and injects it into the plasma chamber. Because the oven is being heated continuously, most of the vapor condenses on the cold surfaces of the plasma chamber and converter during the time between the discharge pulses, and it gets ablated from the converter surface during the discharge pulses.

It is well justified to assume that in the standard H⁻ ion source the cesium vapor is in dynamic equilibrium with the chamber walls because the cesium accumulation on the walls is significant, up to 20 grams over a 4-week running period. In fact, the start-up procedure of the ion source is based on severe over-cesiation. After the startup period, the cesium-oven temperature is reduced such as to just re-supply cesium that is lost through the outlet aperture or on parts of the chamber wall from where it cannot be recovered. Therefore, it can be expected that the temperature of the plasma chamber walls would directly affect the H⁻ yield of the ion source due to enhanced sputtering of H⁻ ions from the converter surface under the conditions of elevated cesium vapor pressure and the subsequent increase of Cs^+ influx. On the other hand, excessive neutral cesium density in the discharge chamber will adversely affect the extractable H⁻ beam current because an increasing fraction of the generated negative ions will be neutralized by collisions on their way to the outlet aperture.

A quantitative discussion of these counteracting effects is given elsewhere [14]. The optimum wall temperature for the standard H⁻ ion source resulting from the developed model lies between 115 and 122°C. The model was expanded to assess the effect of lowering the work function of the cesium covered molybdenum surface, based on a semi-empirical model [15]. As a main result, a work function of 1.6 eV can be reached when the converter surface is covered by 0.60 mono-layers of cesium. For comparison, the work function of pure molybdenum is 4.6 eV.

Unfortunately, the cesium coverage of the converter surface will not be constant during source operation: During the discharge pulses cesium is ablated, and between pulses it is accumulated. This fact leads to an operational limit for the cesium coverage when one considers that the delivered beam current must not vary by more than 10% over the pulse length. Expression (1) allows gauging the expected yield y of H⁻ ions formed in kinetic surface ionization (resonant tunneling) [11]:

$$y \propto \frac{2}{\pi} e^{-\frac{\pi(\phi - E_a)}{2av}} \qquad (1)$$

where E_a is the electron affinity of hydrogen, v is the velocity of the ions leaving the surface (corresponding to an assumed energy of 10 eV) and a is a constant derived from experimental data ($a = 2\text{-}5 \times 10^{-5}$ eV s/m). After inserting work function values calculated using a semi-empirical formula given in Ref. 15, we ultimately obtain that the converter coverage needs to be ranging from 0.47 to 0.78 mono-layers.

Fig. 5 shows the variation in cesium coverage as a function of the chamber-wall temperature for the two operational cases of 60 and 120 Hz pulse repetition rate.

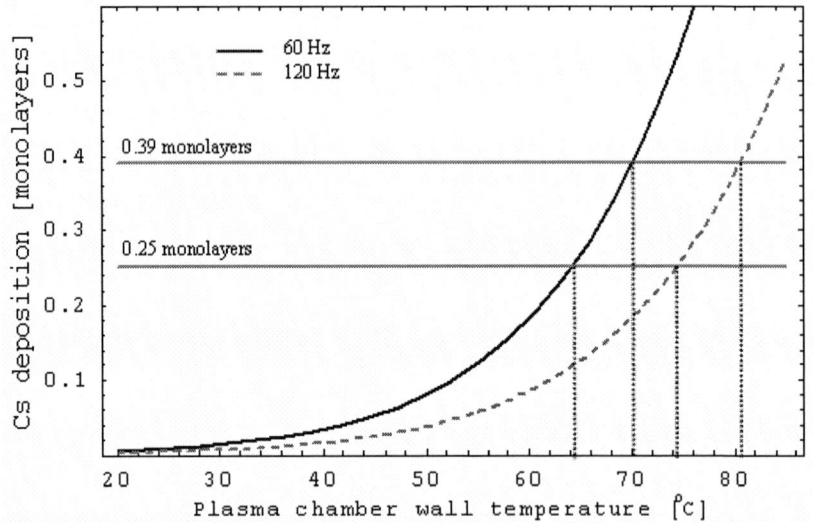

FIGURE 5. Estimated cesium accumulation rates on the converter surface at 60 and 120 Hz discharge pulse frequencies, based on the balance of deposition rates due to the cesium vapor pressure and evaporation rates due to the converter temperature [14]. The limits for cesium accumulation marked by horizontal lines correspond to the extreme values of 2 and 5, respectively, for the constant 'a' in Eqn. 1. The vertical dashed lines mark the two operational windows for the two relevant pulse repetition rates.

Our result suggests that the limitation for the plasma chamber wall temperature as indicated by the vertical dashed lines in Fig. 5 may originate from the rate of cesium accumulation on the converter surface. However, this limitation could be overcome by raising the evaporation rate of cesium by increasing the converter surface temperature beyond 120°C. The result implies that operating the source at 120 Hz would enable us to utilize elevated plasma chamber wall temperatures more efficiently than at 60 Hz. It also explains the fact that the cesium oven temperature is typically higher for 120 Hz operations than for 60 Hz operations (at equal pulse length).

A second effect can be exploited by raising the bias voltage applied to the converter because, when the H⁻ ions leave the surface at a higher velocity, stripping effects due to the built-up image charges are less pronounced [11].

Experiments at Elevated Chamber-wall Temperatures

During regular production runs beam current increases in the order of 10% had been occasionally noted when the air conditioning unit in the high-voltage dome of the injector failed, causing the temperature of the heat exchanger to rise that serves the source cooling circuit. Stimulated by these experiences and supported by the theory outlined above, an experimental study was performed where the chamber walls of a standard ion source were served by heated water. For this purpose, the supply water temperature was varied between the standard value of 22°C and 60°C. Again, a much more detailed account of these experiments is given elsewhere [14] and the main results are displayed in Fig. 6 where experimental data obtained from dedicated test runs using the research ion source 'Roxanne' are compared to the performance of standard production ion sources operated at the normal cooling temperature.

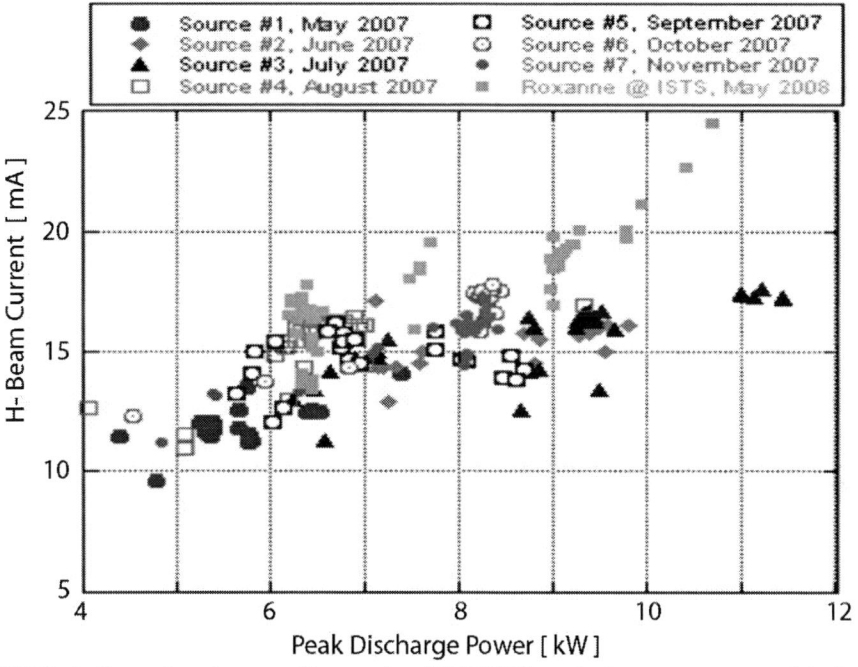

FIGURE 6. Comparison between 'Roxanne' and LANSCE production sources operated in beam production runs during the 2007 run cycle. There are essentially no differences between these two source types; the reason for choosing 'Roxanne' is its availability for such tests.

Fig. 6 demonstrates a 25% gain in beam current at given discharge power level when the source was operated at elevated temperatures. For 60 Hz as well as 120 Hz pulse rates, the beam currents increased linearly with increasing temperature within the range of adjustment (22-60 C). Typically, every data point for the 'Roxanne' runs was obtained during about 6-8 hours of operation at constant conditions. The scattered values for 'Roxanne' at several discharge power values represent variations of the water temperature.

For a complete evaluation, the effect of the elevated plasma chamber wall temperature on filament lifetimes and on the emittance of the extracted H⁻ ion beams needs to be studied in the future.

We are also considering the installation of a hot liner inside the discharge chamber as had been suggested some time ago [16]. At temperatures of several hundreds of degrees centigrade, the cesium vapor pressure in the discharge chamber would essentially be determined by the oven temperature, and the converter temperature could be independently adjusted to optimize cesium coverage during the discharge pulses.

HELICON-DRIVEN CONVERTER SOURCE

For our standard filament-driven H⁻ ion sources, the achievable beam current within the emittance of 0.2 π mm mrad is limited by the maximum plasma density that can be reached under the conditions where the source filaments last for 28-35 days. In this situation, a filament-less discharge appears very attractive, and in this study we are following an approach that utilizes helicon discharge technology [2-3, 17]. Many more details on this development effort are given elsewhere [4].

Helicon plasma generators are characterized by very high plasma density, up to 10^{13} cm⁻³, achieved with exceptionally good power efficiency. A development line combining a helicon plasma generator with an SNS-type H⁻ ion source has been pursued at Oak Ridge National Laboratory [18].

The focus of our efforts has been on developing a helicon-driven surface-conversion ion source. The source mainly consists of the outlet flange of a standard LANSCE H⁻ source, including the electron repeller, and is equipped with a biased converter electrode similar to the standard filament source. Upstream of the outlet flange, the helicon discharge chamber with permanent magnets and the external rf antenna is attached, see Fig. 7.

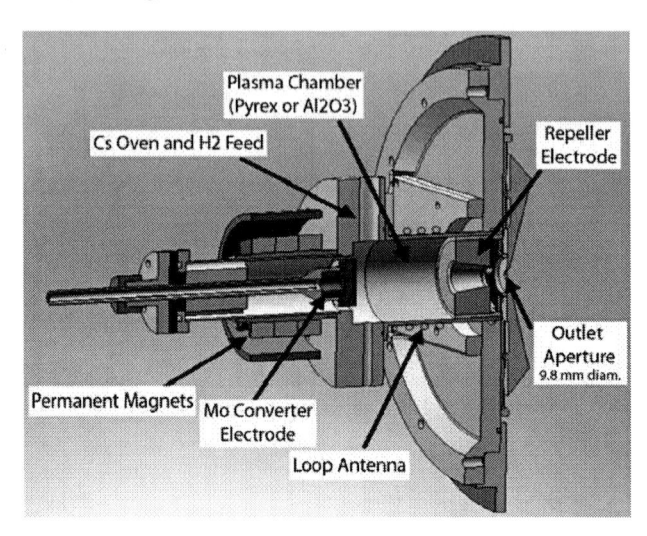

FIGURE 7. Schematic of our new helicon H- ion source.

Replacing the filaments by an rf-antenna is expected to yield higher beam currents, due to increased plasma density, as well as a longer source lifetime. A benefit of this approach (in contrast to the designs presented in References 3 and 4 is the "self-extraction" of the H⁻ ions from the converter electrode yielding a low emittance ion beam, typically less than 0.2 π mm mrad, (95 %, normalized, 1-rms) [1]. In addition, due to the "self-extraction" i.e. energetic H⁻ ions, the stripping losses of H⁻ in collisions with electrons are reduced significantly because of lower cross sections. Compared to purely inductive rf-coupling, helicon discharges provide better stability at low neutral gas pressures around 1-10 mTorr which helps to minimize the losses of H⁻ in collisions with neutrals.

After frequently observing cesium bursts with the standard oven that occasionally had compromised production runs in the past even with the standard source, a new oven was designed and successfully tested, see Fig. 8. The design follows a practice adopted be several other accelerator laboratories that use cesium ovens, namely including a second heater unit to actively keep the connecting shaft between oven and discharge chamber at a higher temperature than the oven itself. This ensures that whatever amount of vapor is released by the oven will not be subject to condensation in the pipe and thus lead to clogging. Such effects have occasionally been observed with our standard sources and lead to significant performance degradations when they occur.

A summary of results achieved by Aug. 2008 with the helicon-based H⁻ source is given in Table 1.

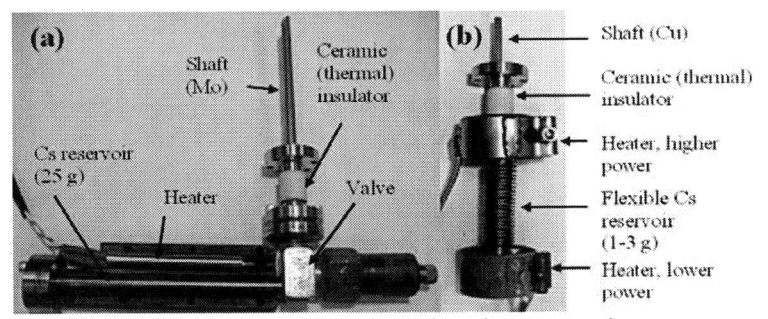

FIGURE 8. Standard (a) and new (b) cesium oven versions.

TABLE 1. Comparison of the LANSCE filament ion source to new helicon-based ion source.

	Filament Source	**Helicon source**
Neutral gas pressure	1-3 mTorr	4-7 mTorr (with Cs, without "ignition system)
Input power	4-11 kW discharge + 2-2.5 filament heating	1-5 (rf)
Plasma density	$5 \times 10^{11} - 1.2 \times 10^{12}$ cm⁻³ (probe measurement)	$5@10^{11} - 3.5 \times 10^{12}$ cm⁻³ (calculated from converter current)
Beam current, 9.8 mm aperture	14-24 mA (15-18 is typical)	7-12.5 mA
Emittance	< 0.2 π mm mrad	?
e/H⁻ ratio	3-5	5-8
Duty factor	6-12 %	1-3.7 % demonstrated

In summary, the helicon-source development has demonstrated very encouraging results, but to date they are not yet competitive as compared to the performance of the standard filament-driven source.

ACKNOWLEDGMENTS

We thankfully acknowledge the technical support given by H. Alvestad, N. Okamoto and G. Sanchez in preparing and conducting the development work.

REFERENCES

1. J. Sherman, E. Chacon-Golcher, E. G. Geros, E. Jacobson, P. Lara, B. J. Meyer, P. Naffziger, G. Rouleau, S. C. Schaller, R. R. Stevens and T. Zaugg, "Physical Insights and Test Stand Results for the LANSCE H- Surface Converter Source," Proc. 10th Intern. Symp. on Production and Neutralization of Negative Ions and Beams, AIP Conference Proceedings, 763, American Institute of Physics, Melville, NY, 2005, pp. 254-266.
2. F. F. Chen and R. W. Boswell, IEEE Transactions on Plasma Sci, Vol. 25, No. 6, (1997), p. 1245.
3. R. F. Welton, M. P. Stockli and S. N. Murray, Rev. Sci. Instrum. 77, 03A506 (2006).
4. O. Tarvainen, E. Geros, G. Rouleau, T. Zaugg and R. Keller, "First Results with a Surface Conversion H- Ion Source Based on Helicon Wave Mode-Driven Plasma Discharge," Proc. this conf., NIBS, Aix-en-Provence, France, Sept. 10-12, 2008; to be published in AIP Conference Proceedings Series, American Institute of Physics, Melville, NY.
5. KAMIS Incorporated,☐P.O. Box 67, ☐Mahopac Falls, N.Y. 10542, USA.
6. G. Rouleau, E. Geros, J. Stelzer, E. Chacon-Golcher, R. Keller, O. Tarvainen, and M. Borden, "Tungsten filament material and cesium dynamic equilibrium effects on a surface converter ion source," Rev. Sci. Instrum. 79, 02A514 (2008).
7. H. Carter and S. W. Gibbs, "Automated data acquisition and analysis in a mechanical test lab ," ASTM Special Technical Publication 1208, American Soc. for Testing and, Materials, Philadelphia, USA, 1993, pp. 28-39.
8. H. P. Gao and R. H. Zee, "Effects of rhenium on creep resistance in tungsten alloys," Journal of Materials Science Letters Vol. 20, (2001), pp. 885-887.
9. E. Chacon-Golcher, "Erosion and Failure of Tungsten Filaments: A Computational Perspective," Internal Report LA-UR-08-05251, Los Alamos Nat. Lab., 8/12/2008.
10. N. O. Moraga and D. L. Jacobsen, "High temperature emissivity measurements of tungsten-rhenium alloys," 22nd Thermophysics Conference, Honolulu, HI, June 8-10, 1987, p. 10.
11. B. Rasser, J.N.M van Wunnik and J. Los, Surf. Sci. 118, (1982), p. 697.
12. C.F.A. van Os, P.W. van Amersfoort and J. Los, J. Appl. Phys. 64, 8, (1998), p. 3863.
13. C.F.A. van Os and P.W. van Amersfoort, Appl. Phys. Lett. 50, 11, (1997), p. 662.
14. O. Tarvainen, "The Effects of Cesium Equilibrium on H⁻ Ion Beam Production with LANSCE Surface Converter Ion Sources," submitted to Nucl. Instruments and Methods A (2008).
15. G.D. Alton, Surf. Sci. 175, (1986), p. 226.
16. A.B. Wengrow et al., Proceedings of the PAC97, IEEE, p. 2764.
17. O. Tarvainen, M. Light, G. Rouleau and R. Keller, "Helicon Plasma-generator Assisted Negative Ion Source Project at Los Alamos Neutron Science Center," Proc. 11th Intern. Symp. on Production and Neutralization of Negative Ions and Beams, AIP Conference Proceedings, 925, American Institute of Physics, Melville, NY, 2007, pp. 171-179.
18. R. F. Welton et al., Proc. this conf., NIBS, Aix-en-Provence, France, Sept. 10-12, 2008; to be published in AIP Conference Proceedings Series, American Institute of Physics, Melville, NY.

The NEW DESY RF-Driven Multicusp H⁻ Ion Source

J. Peters

Deutsches Elektronen-Synchrotron DESY, Notkestraße 85, 22607 Hamburg, German

Abstract. The HERA RF-Volume source is the only source that can deliver a H⁻ current of 40 mA without Cs routinely and for periods of several years. Step by step improvements have led to a completely new design. The plasma is now contained in an Al_2O_3 ceramic chamber and has no contact to metal surfaces except for the collar. The sensitive antenna and also the pair of filter magnets are outside of the plasma. The plasma chamber is mounted in a ceramic disk for high voltage insulation. The ignition of the plasma is done with an ignition source utilizing the higher pressure at the gas input. Details about the setup, performance, H^+ containment, sparking suppression and experience with a new 10cm diameter LEBT will be given.

Keywords: Ion source, negative ion, accelerator
PACS: 29.25Ni,

DEVELOPMENT OF THE DESY RF SOURCE

The first RF-Driven Multicusp H⁻ Ion Source at DESY with internal antenna was a copy of the LBL RF source [1, 2]. With this source and the newly introduced negative bias of a tantalum collar a record beam of 80 mA was achieved without cesiation [3]. A tantalum collar was used and for the first time an increase of H⁻ current due to a negative collar bias was demonstrated [3]. It was later discovered, that the antenna coating was a source of potassium in the plasma [4]. The internal antenna had been an unreliable construction for six basic reasons (see [5]). At DESY a lifetime of more than six months is required. For this reason an external antenna was introduced [3]. The longest and highest H⁻ pulse ever seen in a RF-volume source was produced with this construction [6]. First only the antenna was shielded by ceramic from the plasma then the plasma chamber was completely insulated and a biased sandwich collar was introduced [6, 7].

THE NEW DESIGN

The characteristic parts of the new source are the fully insulated plasma chamber, the ignition source invented at DESY, a flat antenna coil design with ferrite field guidance backed by a capacitive foil, multi cusp magnets and a disc shaped HV insulator. Fig.1 shows a drawing of the cross section of the source. Fig.2 is a photograph of the assembled source before it is mounted into the vacuum vessel.

CP1097, *Negative Ions, Beams and Sources: 1st International Symposium*, edited by E. Surrey and A. Simonin
© 2009 American Institute of Physics 978-0-7354-0630-8/09/$25.00

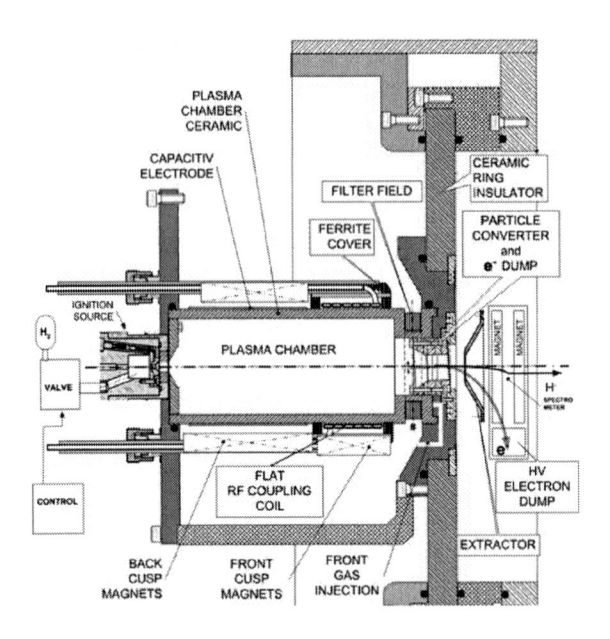

FIGURE 1. Cross section of the new DESY RF-Driven Multicusp H⁻ Ion Source

FIGURE 2. Picture of the new DESY RF-Driven Multicusp H⁻ Ion Source

Characteristic of the NEW DESY Source

Ignition Source

An ignition source for H⁻ sources was invented at DESY in 1999 (see [8]). A smaller sized version with a focusing as in our magnetron makes it possible to run a lower electrode voltage (see Fig. 3).

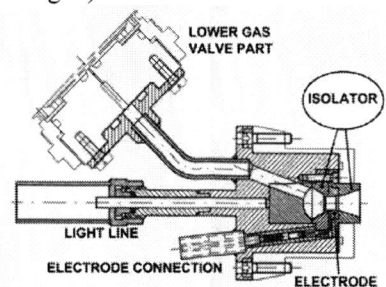

FIGURE 3. Picture of the small DESY ignition source for low electrode voltage

Plasma Chamber Size

With a ceramic insert the length of the chamber was changed (see[7]). The volume was 30% reduced. Advice was given to other institutes to try an increase in diameter [9].

Collar Choices and High Field Extraction Gap

Systematic studies of cylinder collars with different diameters and length were made [10]. In a next step cone and cone cylinder designs were tested with an external antenna. With the full ceramic plasma chamber a cone cylinder with ring (three electrodes) turned out to be the best.

Different aperture plate cones were tested [10,11]. Due to the magnetic field in this area a flat aperture plate gave the best result. Cone constructions (Pierce shape) cause a trapping of electrons. A high electric field is necessary up to the aperture opening.

The physics of the collar and electron dumping in the collar were studied with a 9 ring symmetric cylinder collar [11] and a cone-cylinder sandwich collar with vertical insulation slit (11 independent electrodes) [12].

Beam Alignment

In previous designs the beam leaves the source under an angle of 3° [1]. Instead of turning the whole source in a bellow [13] the collar was made horizontally adjustable up to 3°. The opposite extractor electrode is adjustable (see [14]). In addition the beam can be adjusted with the two dipole magnets which dump the electrons and correct the H⁻ beam. The filter field electromagnet (see PROJECTS IN PROGRESS) opens the possibility to adjust the beam without opening the source.

HV Source Insulator

The ceramic ring HV insulator (see Fig. 1) is a simple to manufacture Al_2O_3 ceramic disc with a center hole. It provides a long HV insulation path. The price is about 10% of the manufacturing costs of previous constructions.

Source Cooling for High Duty Cycle

Different faraday cages were tested with the first external antenna source [3]. No difference was found compared to the unshielded ceramic. For this reason it was not used in our low duty cycle chamber. In a high duty cycle source it can be built with water cooling channels and used as a shield for cooling (see Fig. 4).

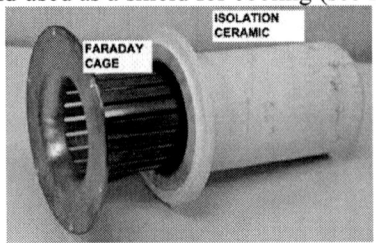

FIGURE 4. Picture of the antenna ceramic with a faraday cage

Getting Rid Of Positive Ions

Sparking is not a serious problem in the operation of the DESY RF source. There is typically one spark per hour, with periods up to eight hours between events. So far no damage has been found.

One reason for sparking is the formation of positive hydrogen ions. They can be formed by stripping of the accelerated H⁻ or by collisions of the electrons with the rest gas.

After some months of operation traces were found on the metal plate (1),(2) and on the ceramic (3) (see Fig.5). Traces (1) and (3) are due to positive ions captured out in the vacuum in front of the ceramic ring. They are accelerated by the HV between ground and the plasma aperture plate and produce trails at the outer plate edge and on the ceramic. Other positive particles are generated in collisions between beam and rest gas(2). Oscilloscopic measured extractor currents show a plasma discharge. However the sparks are so sharp that a clear analysis is impossible due to reflections on the wires.

Is there a way to avoid sparking by insulating the aperture plate?

This was first attempted with an Al_2O_3 plate with sealed screws and a mica plate underneath. The result was an almost constant sparking which seriously hampered the operation of the source. Obviously positive charges accumulate in front of the negatively polarized insulated plate and even find their way through sealed holes. A controlled discharge of the positive particle accumulations seems to be the best way to avoid sparking. In order to take the energy out of sparks which might damage the extractor (area (2)) it is helpful to insert a resistor between the extractor and ground.

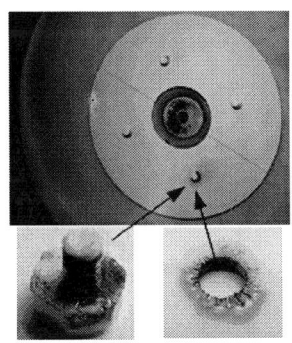

FIGURE 5. Traces of the bombardment with positive ions and the result of an insulation setup.

Enhancement by Potassium

In the step by step development of the present source the antenna was first covered with an Al$_2$O$_3$ cylinder. In a next step the filter magnets and the end of the source were covered with an additional macor ceramic which was also vacuum sealed. This simulated the completely insulated plasma chamber shown in Fig. 1.

In a material analysis it was discovered, that macor has a high potassium content, similar to that of enamel. With the enamel coated internal antennae it was possible to generate 80 mA due to the potassium contamination in our first set up in 1996 (see [2]). Tab.1 shows the percentage of alkali metals in enamel and in the two different ceramics.

TABLE 1. Alkali metals in enamel and ceramic of Al$_2$O$_3$ or macor

porcelain coating (enamel)[15]		Al$_2$O$_3$ ceramic [16]		Macor ceramic [17]	
Si	46.9	SiO$_2$	2.5	SiO$_2$	46
Ti	29.6	CaO	2.3	MgO	17
Al	5.7	Al$_2$O$_3$	92	Al$_2$O$_3$	16
Na	2.6	Na$_2$O	0.03	B$_2$O$_2$	7
K	15.2	K$_2$O	0.01	K$_2$O	10

The present plasma chamber consists completely of Al$_2$O$_3$ ceramic with only 0.01 % of K$_2$O. In order to check for an influence due to the potassium a macor cylinder was inserted into the chamber (see Fig. 6). It was seen out that this ceramic macor cylinder and also a shorter version of it gave no improvement in performance.

FIGURE 6. Macor cylinders (left) ready for installation into the plasma chamber

In discussions came the suspicion that the H⁻ were not produced by the volume effect but by surface reactions with alkali metals. Producing an alloy with alkali turned out to be very cost intensive and difficult [18]. However it is possible to sinter a 80% Al and 20% K_2MoO_4 material [19]. This material was used for the cone of our collar (see Fig.7). Such a material mix is similar to what we used for the enamel antennae. But this construction did not give an increase in H⁻ production.

It turned out that the material is not very stable. With cone voltages of more than -20V the beam is destabilized. From earlier experience it is known that this can be due to small particles which get into the extraction area.

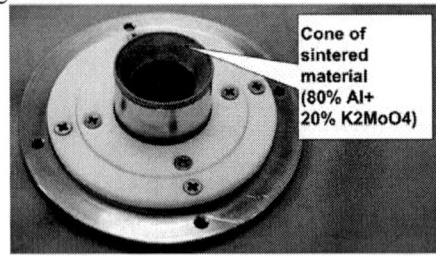

FIGURE 7. Three segment collar with a cone of a sintered material (80% Al +20% K_2MoO_4)

PRESSURE IN THE GAS PULSED SOURCE

The gas flow into the DESY source is pulsed with a piezo gas valve. This makes it possible to have a high pressure in the chamber during extraction (about 200 μsec) with lower average gas consumption and over all gas pressure. The reduced amount of rest gas decreases the production of H⁺ by $H_2 + e \rightarrow 2 H^+$. This reduces the destruction of H⁻ by $H^+ + H^- \rightarrow H^* + H$. Also less H⁺ accumulate and cause sparking. In general a reduced gas pressure reduces the risk of sparks. The achieved higher pressure in the plasma chamber of more than 4.5 mTorr by pulsing during extraction leads to a higher H⁻ current pulse.

Variation of the Pulse Pressure

During the tuning for the 3msec long pulse measurements and our beam and emittance studies (see [20]) it was discovered that a variation of the pulse pressure can improve beam performance. This was the motivation for further studies not just to vary the pulse pressure but also to independently measure it.

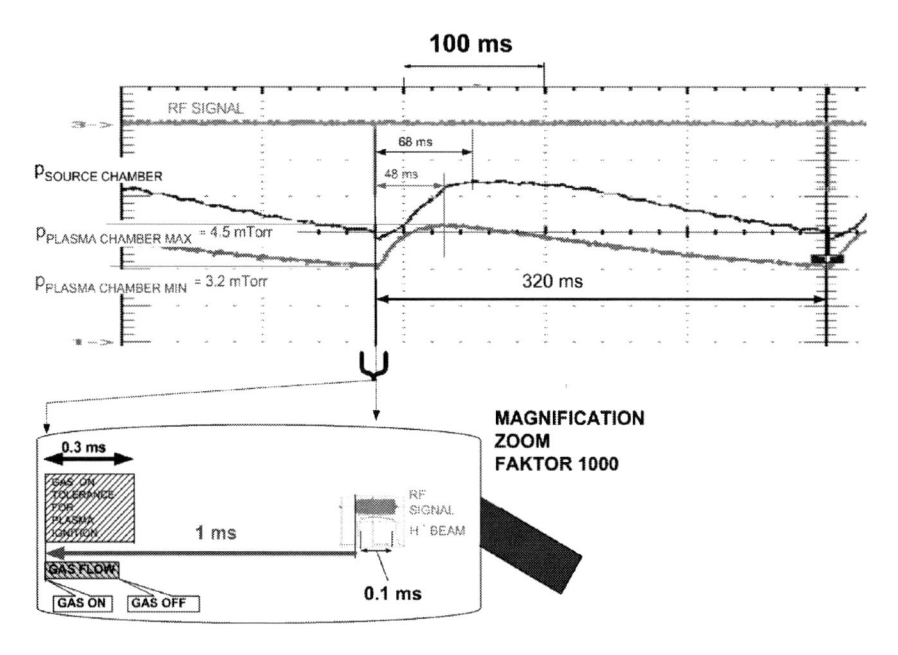

FIGURE 8. Measured pressure in the plasma and source chambers together with the RF signal. The timing of the gas flow together with the H⁻ beam and the RF signal are shown in the zoom.

Measuring the Pulsed Pressure

With the given chamber and feeding nozzle dimensions (d= 0.8 mm) one needs a gas flow during 0.28 msec with a start 1 msec before beam begin for optimum performance for a 100μsec beam pulse. It turned out that the timing tolerance of the gas flow start has to be below 0.3 ms . This was measured with a pulse period of 320 msec.

The pressure range of interest can be measured with a slow pirani gauge which depends on a filament. A calibrated penning gauge (which is faster) is not available for this range although it in principle works in this pressure region. By calibrating the penning with a pirani and utilization of a fast oscilloscope read out, calibrated measurements were possible.

The result of these measurements is that the pressure varies between 4.5 and 3.2 mTorr . There is a huge delay of 48 msec between the beam and the maximum pressure measured with the penning gauge. This is due to the pipe connection between the gauge and the chamber and the inertial of the penning itself. Due to that the fine structure of the pressure variation in the chamber can not be seen. As the timing of the source has been found to be critical within 0.3 msec, important information is lost.

Pressure Information via Beam Current, a Pressure Wave

Instead of using a penning source which is coupled to the plasma source as pressure meter one can also use the beam itself as an indication for the pressure. This makes sense because investigations showed that in the 40 mA region the current rises with pressure. By pulsing the pressure as shown in Fig.8 and moving the RF pulse together with the ignition source over the pressure curve one gets a sample representing to some extend the source pressure.

Fig. 9 gives an indication that with opening of the piezo valve a pressure wave is moving into the plasma chamber which adds pressure to the rest gas in the chamber. This way an optimum of pressure and beam can be reached. The critical timing of the source becomes also understandable.

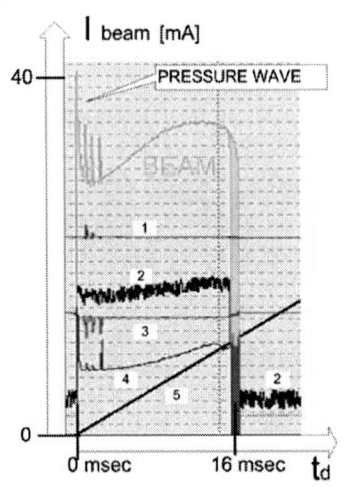

FIGURE 9. A pressure wave indicated by the BEAM current.

Shown is the BEAM while the start trigger of the RF and the ignition source is delayed by t_d relative to the gas valve start trigger. Curves 1,2,3 and 4 are part of the source histogram.

A NEW TRANSISTOR TRANSMITTER FOR THE DESY RF SOURCE

It was often necessary to change the expensive tube in the old transmitter after half a year. In order to reduce these maintenance costs a semiconductor transmitter was developed at DESY [21]. The semiconductor transmitter is 3 times smaller and nearly maintenance free. The latest design delivers 26 kW which gives at present about 40mA H⁻. The transmitter can be upgraded to deliver more power.

FIRST MEASUREMENTS WITH THE LEBT

The new DESY LEBT has an inner diameter of 10 cm and consists of two solenoids. Fields of 0.283 T are reached with pulsed currents of about 500 A. Fig. 10

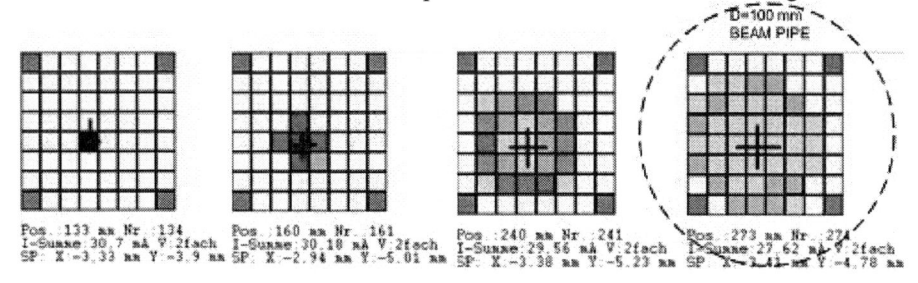

FIGURE 10. The transported H⁻ beam in the last part of the LEBT measured with a multi faraday cup. Source current is 40 mA.

shows the beam at the end of the second solenoid. It was measured with a 64 Faraday cup array. Each cup is 7.2 mm x 7.2 mm. The head is quadratic with a 60mm side length. So far 31 mA have been transported at a source current of 40 mA.

PROJECTS IN PROGRESS

At present there are preparations being made for testing a front gas injection (see Fig.1 and Fig.11) and a pulsed electrical filter field magnet (see Fig. 12).

Front Gas Injection

A pipe connection for front gas injection has been made and a second gas valve is installed. This gives the freedom to change the chamber pressure. During previous tests it turned out that the time dependent pressure profile is very important. As the front gas is not coupled to the injection source it provides more possibilities for performance optimization.

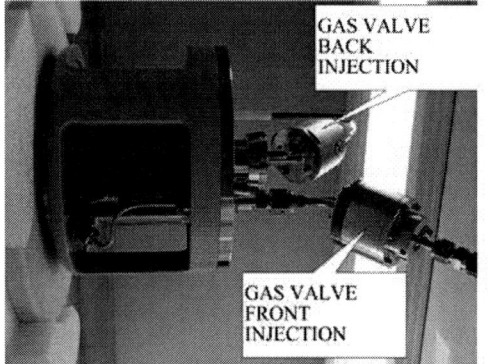

FIGURE 11. The front gas injection setup with the two gas valves mounted.

Variation of the Filter Field with an Electromagnet

Detailed measurements of the emittance were made with permanent filter magnets of different strengths [7]. With an electromagnet it is possible to study these effects without disassembling the source and thereby changing the conditioning of the source. Recent measurements demonstrated that the ratio of Ielectron/I H- influences the beam profile of the H⁻ beam. This effect is also filter field dependent. First measurements proofed that the electromagnet can replace the permanent magnets.

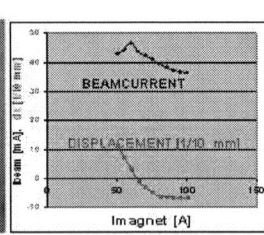

FIGURE 12. (LEFT) The filter field electromagnet in front of the plasma chamber. (MIDDLE) The filter field electromagnet mounted on the chamber.(RIGHT) First measurements

ACKNOWLEDGMENTS

The author is grateful for the contribution of the following colleagues at DESY: I.Hansen, M.Marx, K.Müller, H.Sahling, and R.Subke. The author also wishes to thank H.Weise and the technical groups at DESY for their support and M.Lomperski of DESY for helpful suggestions to the wording of the article.

REFERENCES

1 K.N. Leung, G.J.DeVries, W.F.DiVergilio and R.W.Hamm, Rev.Sci.Instrum.62(1),100(1991)
2 J. Peters, Rev.Sci. Instruments, 67, No. 3, March 1996, pp 1045-1047.
3 J. Peters, Rev.Sci.Instrum.69 (2) February 1998, pp 992-994
4 J. Peters, Proceedings of the XIX International Linear Accelerator Conference (August 1998) pp 1031-1035.
5 J. Peters, Proceedings of the EPAC 02, Paris June 2002
6 J. Peters, Proceedings of the 2006 Linear Accelerator Conference (August 21-25,2006).
7 J. Peters, AIP Conf. Proc. 925, 79 (2007).
8 J. Peters, Proceedings of the XX International Linear Accelerator Conference (August 2000).
9 Private communication with R. Welton, SNS
10 J. Peters, Rev. Sci. Instrum.77, 03A528 (2006).
11 J. Peters, Ph.D. thesis, Universität Frankfurt,2001.
12 J. Peters, Rev.Sci.Instrum.79, 02A515 (2008)
13 R.Keller R. Thomae, M. Stockli and R. Welton AIP Conf. Proc. 639, 47 (2002).
14 J. Peters, Proceedings of the 2005 PARTICLE ACCELERATOR CONFERENCE (PAC 05) May 16-20, 2005
15 Analysis Univ. Hamburg, Germany
16 Analysis Al$_2$O$_3$, TECHNISCHE KERAMIK FRÖMGEN, Germany
17 Analysis Macor , TB Takac GmbH, Germany
18 private communication with IME, RWTH Aachen, Germany
19 T. Büttner, Fraunhofer-Institut für Fertigungstechnik und Angewandte Materialforschung, Institutsteil Dresden
20 J.Peters, H.-H. Sahling and I. Hansen, Rev.Sci.Instrum.79, 02A523 (2008)
21 K.Müller, J.Hannemann, HERA report 2008

Next Generation H⁻ Ion Sources for the SNS

R.F. Welton[a], M.P. Stockli[a], S.N. Murray[a], D. Crisp[a], J. Carmichael[a], R.H. Goulding[a], B. Han[a], O. Tarvainen[b], T. Pennisi[a], M. Santana[a]

[a]*Spallation Neutron Source, Oak Ridge National Laboratory, P.O. Box 2008, Oak Ridge, TN, 37830, USA*
[b]*Los Alamos National Laboratory, Los Alamos, New Mexico, 87545, USA*

Abstract. The U.S. Spallation Neutron Source (SNS) is the leading accelerator-based, pulsed neutron-scattering facility, currently in the process of ramping up neutron production. In order to insure meeting operational requirements as well as providing for future facility beam power upgrades, a multifaceted H- ion source development program is ongoing. This work discusses several aspects of this program, specifically the design and first beam measurements of an RF-driven, external antenna H- ion source based on an AlN ceramic plasma chamber, elemental and chromate Cs-systems, and plasma ignition gun. Unanalyzed beam currents of up to ~100 mA (60Hz, 1ms) have been observed and sustained currents >60 mA (60Hz, 1ms) have been demonstrated on the test stand. Accelerated beam currents of ~40 mA have also been demonstrated into the SNS front end. Data are also presented describing the first H- beam extraction experiments from a helicon plasma generator based on the Variable Specific Impulse Magnetoplasma Rocket (VASIMR) engine design.

Keywords: negative ion sources, particle accelerators, ion formation
PACS: 29.25.Ni; 52.75.Di; 52.77.Dq; 52.80.Pi; 52.80.Vp

INTRODUCTION

High-brightness H⁻ ion sources are widely used in large accelerator facilities which utilize charge-exchange injection into accelerators or storage rings [e.g. 1]. One such facility, the U.S. Spallation Neutron Source (SNS) [1,2,3] currently employs a Radio-Frequency (RF), multicusp ion source developed at Lawrence Berkeley National Laboratory (LBNL) and is based on a porcelain-coated Cu antenna immersed in the plasma volume [4]. Advances in antenna coating technology [5] have contributed to a successful SNS commissioning period in which the source produced 20 - 50 mA with duty-factors of ~0.1% for periods of many weeks with availability approaching ~100%. The SNS is ramping-up neutron production (currently ~0.55 MW on target) and the source is now routinely delivering 32 mA (670us, 60Hz, ~4% duty-factor) to the accelerator for 2-3 week periods. Over the last 11 months, the baseline source performance has been dramatically improved by modifications of the LBNL Cs collar and by improving procedures of ion source preparation and start-up procedures [4]. Similar modifications have also been incorporated into the elemental Cs collar, and improvements noted. After each run the internal antennas are inspected and their condition is entered into a database. So far in 2008 ~30% of the antennas were found to exhibit some kind of defect after run periods.

Over the next several years the SNS will increase neutron production, providing between 1 and 1.4 MW of beam power to the target by the end of 2009 and possibly 2-3 MW a few years afterwards. A beam power of 1.4 MW requires the ion source to produce a linac H⁻ beam current of 38 mA, 1.0 ms in length with a repetition rate of 60 Hz (~6% duty factor) and an RMS emittance not exceeding 0.35 π·mm·mrad; while 3 MW would require 59 mA with identical other parameters. Although the impact of antenna defects on operations has been minimal, a more reliable RF coupling technique is desirable, given the facility plans to ramp beam current and pulse-length with the expectation of high availability.

In this work, we focus on meeting these requirements by continuing the development of the RF source [6] for the following reasons: (i) historically, there have been several examples of existing, short-pulse, low-rep-rate RF-driven H⁻ sources producing 60-100 mA with Cs with a 0.1% duty-factor, with reasonable emittance [7,8]; (ii) multi-year lifetimes have been observed in short-pulse, RF-driven H⁻ sources using external antennas, with lifetimes much greater than competing filament-driven or Penning sources [9, 10]; (iii) the SNS now has a considerable inventory of components, test equipment and experience related to the RF source; and (iv) the baseline SNS source is highly modular, allowing independent development, testing and evaluation of each subsystem.

Specifically, we focus on developing a ceramic plasma chamber capable of withstanding the thermal stress of ~100 kW RF power at ~7% duty-factor and a heavily electrically-insulated external antenna as well as boosting the beam current by improving the ionization and extraction efficiency of the source by employing improved Cs systems and optimized ionization surfaces as well as advanced extraction systems [11,12]. In order to minimize project risk, a second approach is being explored at ORNL [13] in collaboration with LANL [14], employing high-density plasma production using RF-helicon waves.

ORNL AlN EXTERNAL ANTENNA SOURCE

The ORNL Aluminum Nitride AlN external antenna source described here was based on the design of the Al_2O_3 external antenna source [6, 15]. As shown in Fig. 1, the source consists of a flanged, high-purity, AlN ceramic plasma chamber with inside dimensions: φ=6.8 cm; length: 18 cm; wall thickness: 0.7 cm. The outer surface of the AlN chamber is directly water-cooled by a Lexan serpentine jacket consisting of a single 3.3 x 9.6 mm water passage flowing 2.6 gal/min of de-ionized water (see Fig. 1). Computationally, the chamber was found to be capable of withstanding isotropic heat-loads of 100 kW at 7% duty factor while maintaining a thermal stress safety-margin of ~2x using coupled fluid dynamic, heat transfer, and thermal stress finite elemental analysis [12].

The RF-coupling antenna shown in Fig. 1 is located in the air space surrounding the Lexan cooling jacket and is constructed from φ=4.8 mm Cu tubing which is water-cooled and covered with three-layers of polyolefin shrink wrap for RF voltage isolation. The antenna is resonated in series with a variable capacitor driven by a 2 MHz (0-80kW) RF generator. The antenna is coiled in a 4 ½- or a 5 ½-turn double-layer 'stacked' geometry consisting of 2 or 3 inner and 2 outer turns which span an

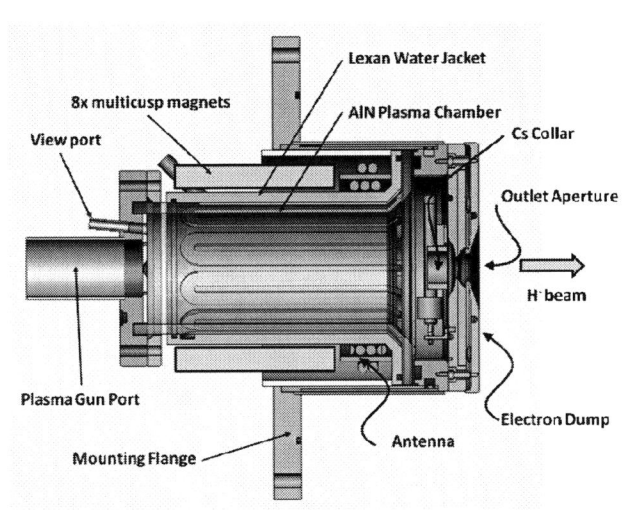

FIGURE 1. Cross-sectional view of the AlN source configured as tested. See Fig. 2 for details of the Cs collar. The AlN chamber is shown as transparent to allow viewing of the Lexan cooling jacket.

axial distance of ~2 cm on the plasma chamber. The inner and outer layers are also isolated from each other by a Teflon sheet. The antenna has been located as close as possible to the outlet aperture of the source to maximize extracted beam current as well as allow clearance for the 8 multicusp magnets. Each magnet has dimensions of 11.5 x 1 x 1 cm and is supported a distance of 1.9 cm from the inner wall of the plasma chamber by a Lexan holder.

The source body is also shown in Fig. 1 and functions to mechanically locate the plasma chamber assembly (described above) and the outlet aperture assembly as well as provide physical mounting to the accelerator. It is similar in design to the ORNL Al_2O_3 source body with the exception that the outer wall facing the antenna is now water-cooled Cu to reduce ohmic RF power losses [15]. The backflange of the source is also water-cooled and contains a port which accommodates our standard hollow-anode plasma gun (required for plasma ignition) behind a thick Ta aperture plate [6]. Plasma conditions are monitored through an optical spectroscopy view port also located on the backflange.

Two Cs collar configurations were tested: First, the modified LBNL Cs collar, which employs Cs chromate dispensers located in the collar itself, a conical Mo ionization surface located ~0.6 mm from the outlet aperture and 3 ceramic balls which radially locate the collar with respect to the outlet aperture [3].

The second collar configuration tested was the SNS elemental Cs collar, which was modified from the original design to be closer to the outlet aperture as well as support different ionization surface materials, as shown in Fig. 2. Here Cs is injected from an external reservoir and directed onto a conical ionization surface composed of either Mo, Ni or 304-stainless-steel [6,16]. Alignment as well as resistance to thermal-induced motion is achieved through the use of 4 small legs locked to the outlet aperture and two clamps rigidly affixed to the Cs air lines (not shown in Fig. 2). The collar-outlet aperture gap is then fixed to 0.13 mm.

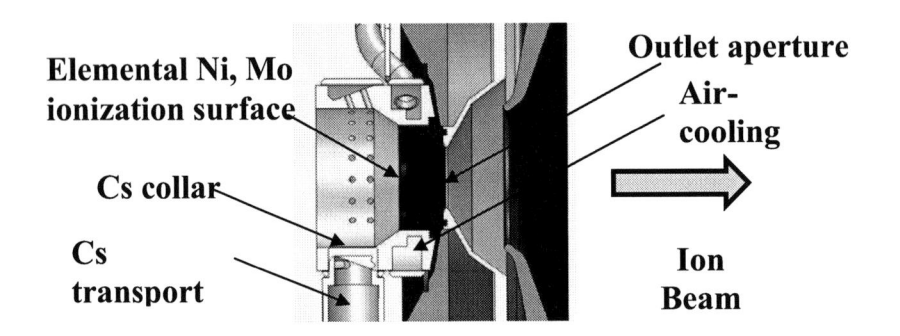

FIGURE 2. Cross-sectional view of the elemental Cs collar with Ni or Mo ionization surface.

Starting in 2008 the AlN external antenna source was tested on the SNS ion source test stand (7 experimental runs) and subsequently on the front-end of the SNS accelerator (3 experimental runs). Both test stand and front-end measurements employed an electrostatic Low Energy Beam Transport (LEBT). The unanalyzed beam current on the test stand was measured using a toroidal Beam Current Monitor (BCM) and a suppressed Faraday cup, both located near the exit of the LEBT [17]. The beam current on the SNS front end was measured using a calibrated toriodal BCM located near the exit of the radio frequency quadrupole (RFQ) accelerator at 2.5 MeV (BCM02) [3,4].

Experimental runs on the test stand were worked into the test facility schedule as time permitted, and varied in length from less than 1 day to ~3 weeks. Over this approximately year-long testing period, the maximum beam current attained from different source configurations on the test stand has increased from ~45 mA to ~95 mA and a sustained multi-day operation at >60 mA has also been demonstrated. Table I summarizes the best performance of the AlN external antenna source tested with each of the specified Cs-collar configurations. Figures 3 and 4 show the beam pulse shape and H⁻ beam current dependence on RF power, respectively. All current measurements are averaged over the entire pulse length, which varied from 800-1000 μs at a repetition rate of 60Hz. In test-stand run-6, the source took ~2.5 hours to start up from pump-down to the delivery of 60 mA (60Hz, 1ms). No emittances have been measured because the emittance scanner is not yet operational.

TABLE 1. Performance summary of the AlN source tested with different Cs collar configurations (see text) on the test stand

Run #/ Date	Collar Configuration	Ionization surface material	Maximum beam current / RF power	Sustained beam current / days of operation
TS-run-2 1/24/08	SNS elemental collar	304 stainless steel	60mA/ 50kW	>40mA/ ~3 weeks
TS-run-4c 6/17/08	Modified SNS elemental collar	Mo	81mA/ 48 kW	>50mA / ~3 days
TS-run-5 6/20/08	Modified LBNL collar	Mo	55mA / 56 kW	>40mA / ~3 days
TS-run-6 6/28/08	Modified SNS elemental collar	Ni	95mA/52 kW	>60mA / ~3 days

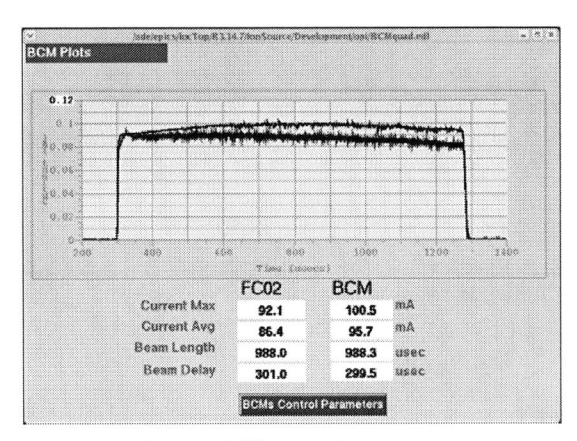

FIGURE 3. Beam current extracted from the AlN external antenna source with the elemental Cs collar and Ni ionization surface on test stand. Both BCM (upper) and Faraday cup (lower) traces are plotted.

It is important to note that on the test stand sources yield roughly twice the beam current that is measured near the RFQ exit when the source is operated on the front end with similar parameters. This test-stand to front-end current-ratio appears to vary not only from source to source, but also with operational conditions. Besides the uncertain RFQ transmission, there are likely many other factors contributing, including the lack of mass separation on the test stand.

Note the RF power specified in this work is the actual measured net power delivered by the 2 MHz RF generator as measured by a directional coupler located on its output, which is much lower than the normally reported requested value [e.g. 3] that the RF amplifier would deliver into a matched resistive load.

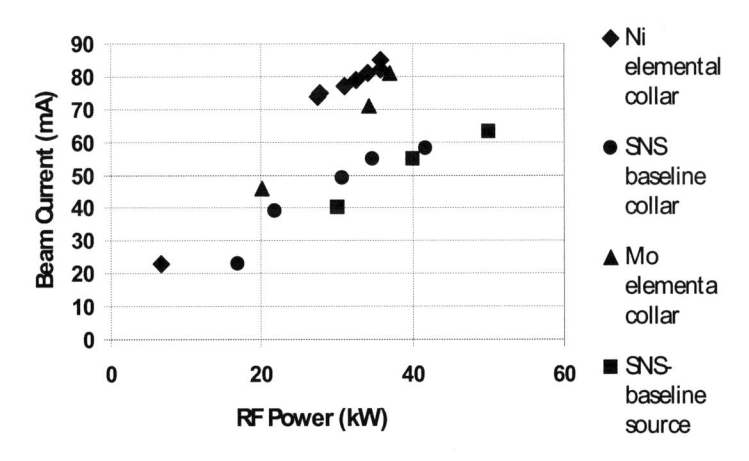

FIGURE 4. H⁻ beam current dependence on RF power for 3 AlN external antenna sources equipped with specified Cs-collar configurations. Also plotted is the SNS baseline source (internal antenna) [4]. All measurements were performed on the test stand.

TABLE 2. Performance of the AlN source tested with different Cs collars on the SNS front end

Run #/ Date	Collar	Ionization surface	Temperature	e-dump	LEBT ion energy	RF power	MEBT beam current
FE-run-2 7/10/08	Mod. LBNL collar	Mo	32 °C	2.3 kV	65 keV	30 kW	40 mA
FE-run-3 7/12/08	elemental collar	Ni	77 °C	4.5 kV	67 keV	30 kW	42 mA

To date, the AlN external antenna source has also been employed in 3 short experimental runs on the front end of the SNS accelerator. The first run was likely compromised by an air leak in the Cs collar and therefore only the latter 2 runs are listed in Table 2. During these experiments the H⁻ pulse width of the source was operated at 600 μs to match the current facility requirement.

As seen in Table 2, the second front end run used source and LEBT parameters, which were, at that time, considered to yield optimal performance with long-term stability and high reliability. The uncalibrated RF amplifier, likely, caused insufficient conditioning of the source, which led the beam current to gradually decay after establishing 40 mA.

In the third front end test run, the source and LEBT were tuned for maximum BCM02 beam current. It is well known that either increasing the Cs collar temperature and/or increasing the e-dump voltage can increase the beam current. It is also well known that increasing the ion energy to 67 kV increases the transmission through the RFQ while it compromises the beam bunching. The source in FE-run-3, however, was not well-aligned, which can slightly compromise the transmission through the RFQ. The run produced 42 mA for 6 hours with no decay noted. Considering all deficiencies and differences it is impossible to say whether the Ni or the Mo surface produced more ions on the front-end, although Ni appeared superior on the test stand. During the latter part of the 6 hours, the Cs reservoir temperature was slightly raised to increase the flow of Cs. While there were clear signs of increased Cs emission, the BCM02 current did not increase. This is similar to the modified LBNL baseline source where subsequent cesiations show no beam current increase on the front-end and contrasts with our test-stand experience where ~30% more current is typically observed for several hours [3].

Source emittance after the RFQ (2.5 MeV) was estimated using the beam profile from a combination of wire scanners and was found to be similar to that of the baseline SNS source. During FE-run-2 a leak developed in the Lexan water jacket and the source had to be replaced. These experiments identified several design issues which are now being corrected, e.g. the wall thickness of the Lexan jacket which leaked has been increased ~50%. Pending successful testing of these modifications, this source is scheduled for routine operation on the SNS-accelerator early in 2009.

FIRST BEAM MEASUREMENTS FROM A HELICON-DRIVEN H⁻ ION SOURCE

Plasmas produced by helicon wave excitation typically develop higher densities, particularly near the radial plasma core, at lower operating pressures than plasmas produced using traditional inductive RF coupling methods while using considerably less RF power [18]. Approximately two years' funding was received to develop an H⁻ ion source based on helicon wave coupling [19]. The idea was to combine the SNS H⁻ source with the existing high-density, hydrogen helicon plasma generator developed at ORNL for the VASIMR project [20, 21]. The scope of work spanned a two-year period with yearly milestones defined as follows: (1) demonstrate high-density plasma production in a prototype ion source in FY2007 and (2) to demonstrate the extraction of H⁻ in a 1.2 ms pulse with current levels in excess of the nominal beam current produced by the current SNS ion source in FY2008. The FY2007 milestone was achieved and reported in Ref. 6; densities >10^{13} e/cm³ were measured inside the ion source region of the plasma generator. This report discusses the initial H⁻ beam current extraction measurements from the source.

Two configurations of the ion source have been tested on the helicon test stand at ORNL to date: the original design, which is described in Ref. 6 and is shown in Fig. 5a; and a modified design shown in Fig. 5b. Also plotted in the figure are the magnetic flux tubes created by the solenoid configuration. Briefly, both configurations consist of a large solenoid magnet 65 cm in length that generates an axial magnetic field of 100-900 gauss (partially shown in the figure) and a mirror coil, both surrounding an evacuated quartz discharge chamber (ϕ=5 cm l=100cm). A half-turn helical RF antenna, oriented to excite the right-hand circularly-polarized helicon wave in the plasma, is located ~40 cm upstream from the center of the mirror coil, surrounding the discharge chamber located inside the solenoid magnet. Typically the mirror coil is operated at 1.5 to 5 times the axial B-field in the antenna region. As seen in Figure 5, the discharge chamber is terminated to the right by the SNS baseline outlet aperture assembly containing the modified LBNL Cs-collar.

Beam extraction is accomplished using a biased extraction/diagnostic module, shown in Fig. 6, which is located 7.2 mm downstream of the ion source. The entire module is supported by a standard 75 kV vacuum break and is biased to +65 kV which allows the ion source and associated power supplies to remain at ground potential

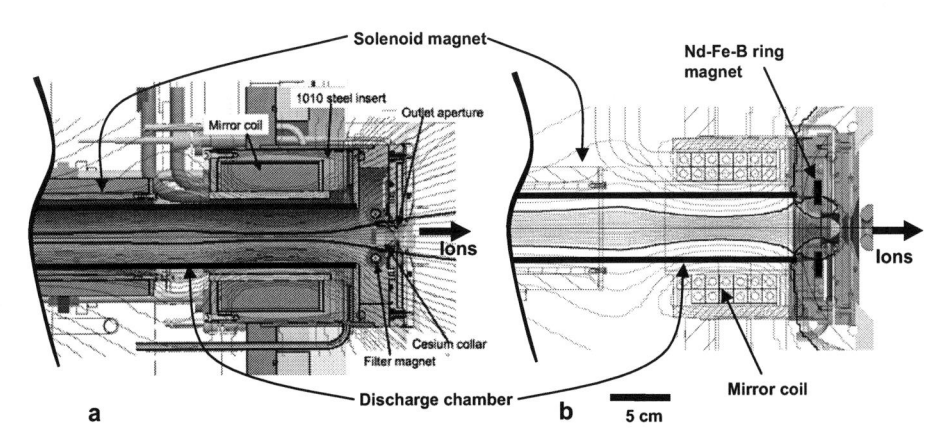

FIGURE 5. (a) Original version of the ORNL helicon ion source. (b) Modified version of ORNL source (see text). Magnetic flux mapping is also shown for both sources.

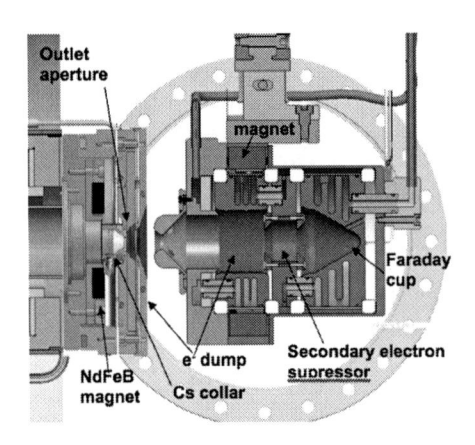

FIGURE 6. Beam diagnostic extraction module.

while extracting negative ions at 65 keV. The extraction module has been computer-designed to separate and independently measure any co-extracted electrons present in the beam. It consists of an ~300 Gauss transverse magnetic field, water-cooled electron dump, two water-cooled electron suppressor apertures, and a water-cooled Faraday cup. The three regions of the extraction/diagnostic module are electrically isolated and currents can be directly measured. The suppressor is normally biased to 140V below the Faraday cup and contains removable apertures which can be reduced in size to allow measurements of an emittance-filtered beam.

During initial operation of the helicon ion source shown in Fig. 5a, a peak plasma density comparable to that achieved in our original VASIMR helicon plasma generator was achieved. However, an axial ion density profile made using a Langmuir probe passing through the outlet aperture showed that a sharp density gradient existed in the region of expanding magnetic field, starting at ~ 2 cm from the opening in the cesium collar and extending through the collar. As a result, the plasma density at the entrance to the collar was reduced to < 2 x 10^{12} e/cm^{-3}, and much less at the exit. When beam extraction experiments were conducted using this configuration, it was found that only ~ 0.6 mA of H⁻ current could be extracted. The measured current on the electron dump electrode shown in Fig. 1 (biased to +5 kV) was also very low, reaching a maximum value of only 60 mA.

In order to increase plasma transport into the collar, reducing the observed density gradient, a Nd-Fe-B (MGO 46) permanent magnet was installed near the collar (see Fig. 5b). This resulted in an increase in the portion of the plasma volume mapped along the flux lines into the collar entrance from 6% → 75%. The highlighted (darkened) magnetic flux tubes shown in Figs. 5a and 5b delineate the outer boundary of flux which enters the collar region. Beam extraction was then performed on the source configuration shown in Fig. 5b and ~12 mA of H⁻ was obtained. The initial parametric study of beam current versus source operating parameters is shown in Fig. 7. During these measurements the magnetic field strength on axis was: ~830 gauss in the antenna region; ~1000 gauss at the mirror peak; and 1250 gauss at the peak under the Nd-Fe-B ring magnet. It is important to note that no Cs was injected into the

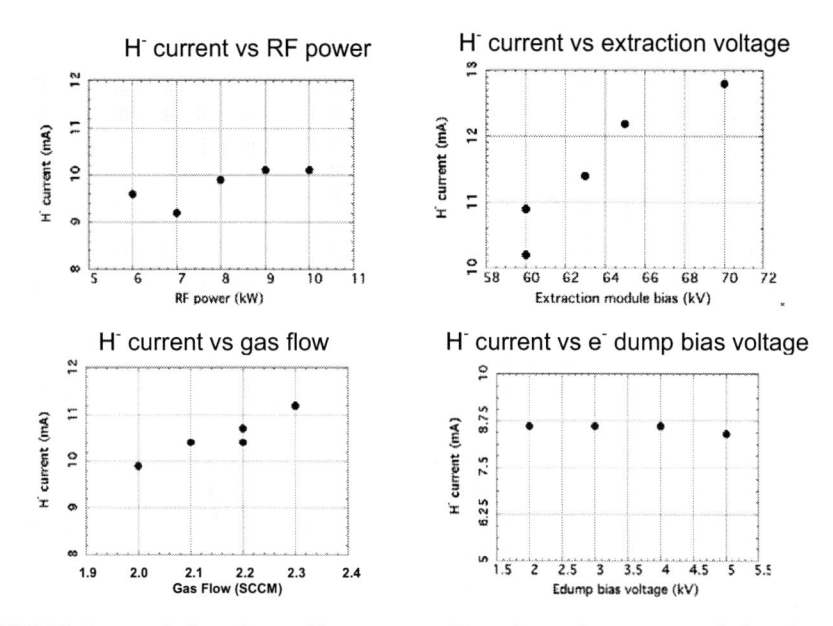

FIGURE 7. Parametric dependence of beam current. Normal operating parameters (unless changed above): RF Power: 7kW; extraction voltage: 65kV; H$_2$ gas flow: 2.2 SCCM; electron dump bias: 5kV.

source during these tests due to an inability of the plasma/RF to heat the Cs-collar. The extracted beam currents are less than the ≥20 mA produced by a un-cesiated modified LBNL baseline source. The use of Cs with the baseline source normally significantly increases the beam current. We are currently preparing for a cesiated experimental run using the elemental Cs system.

CONCLUSION

The first beam measurements of an RF-driven, inductively-coupled, H$^-$ ion source featuring an AlN ceramic plasma chamber and external antenna have been reported. Not a single ceramic chamber failure was observed over 7 experimental runs on the SNS test stand and 3 runs on the SNS front end, in contrast to earlier work with Al$_2$O$_3$ plasma chambers. Beam currents of up to ~100mA (60Hz, 1ms) have been observed and sustained currents >60 mA (60Hz, 1ms) have been demonstrated on the test stand. Beam currents of ~42 mA being transmitted through the RFQ have been demonstrated on the SNS front end.

The first uncesiated beam measurements of a helicon-coupled, H$^-$ ion source have also been reported. After improving the magnetic design of the source, preliminary data suggests that up to 12 mA of H$^-$ current could be extracted using only 7 kW of RF power and an H$_2$ gas feed-rate of ~2 SCCM. These power and gas consumption rates are considerably less than those required by conventional inductively-coupled, RF-driven H$^-$ ion sources. We expect much higher beam currents when cesiated experiments begin in the next fiscal year. The use of such low RF powers opens the possibility of extremely long-pulse or cw, high-current H^-/H^+ RF-driven ion sources as well as very high-reliability short-pulse devices.

ACKNOWLEDGEMENT

The work at Oak Ridge National Laboratory, which is managed by UT-Battelle, LLC, was performed under contract DE-AC05-00OR2275 for the US Department of Energy.

REFERENCES

1. http://www.sns.gov.
2. T. Mason, Physics Today, publisher: *AIP Press*, May 44 (2006).
3. M.P. Stockli, et al., these proceedings.
4. R. Keller, et al, *Rev. Sci. Instrum.* **73** 914 (2002).
5. R. F. Welton, M.P. Stockli, Y. Kang, M. Janney, R. Keller, R.W. Thomae, T. Schenkel and S. Shukla, *Rev. Sci. Instrum.* **73** 1008 (2002).
6. R.F. Welton, M. P. Stockli, S.N. Murray, T. R. Pennisi, B. Han, Y. Kang, R.H. Goulding, D.W. Crisp, D. O. Sparks, N.P. Luciano, J. R. Carmichael and J. Carr, *Rev. Sci. Instrum.* **79** 02C721 (2007)
7. K. Saadatmand, G. Arbique, J. Hebert, R. Valicenti and K.N. Leung., *Rev. Sci. Instrum.* **66** 3438 (1995)
8. G. Gammel, T.W. Debiak, S. Melnychuk and J. Sredniawski, *Rev. Sci. Instrum.* **65** 1201 (1994)
9. J. Peters, *Proceedings of the European Particle Accelerator Conference 2002*, Paris, France.
10. J. Peters, *Proceedings of the International Linear Accelerator Conference 1998*, Chicago, Il, USA.
11. R.F. Welton, M.P. Stockli, S.N. Murray, J. Carr, J. Carmichael, R.H. Goulding and F.W. Baity, *Eleventh International Symposium on the Production and Neutralization of Negative Ions and Beams 2006*, Santa Fe, NM, USA, AIP Conf. Proceedings **#925**
12. R.F. Welton et al., *Proceedings of the Particle Accelerator Conference 2007*, Albuquerque, NM
13. R.F. Welton, R. H. Goulding, F. W. Baity, M.P Stockli and Y. Kang, *Proceedings of the International Linear Accelerator Conference 2006*, Knoxville, Tenn., USA p. 367.
14. O. Tarvainen, G. Rouleau, R. Keller, E. Geros, J. Stelzer and J. Ferris, these proceedings.
15. R. F. Welton, M. P. Stockli and S.N. Murray, *Proceedings of the International Linear Accelerator Conference 2006*, Knoxville, Tenn., USA p. 373.
16. R. F. Welton, M. P. Stockli, S.N. Murray, Jr., J. Carr, Jr. and J.R. Carmichael, *Proceedings of the International Linear Accelerator Conference 2006*, Knoxville, Tenn., USA p. 364.
17. R.F. Welton, M. P. Stockli, S.N. Murray, *Rev. Sci. Instrum.* **75** 1793 (2004).
18. R. Boswell and F. Chen, *IEEE Trans. Plasma. Sci.* **25** (1997) 1229-1244
19. Laboratory Directed Research and Development Program of Oak Ridge National Laboratory, managed by UT-Battelle, LLC, for the U. S. Department of Energy under Contract No. DE-AC05-00OR22725.
20. Y. Mori, H. Nakashima, F.W. Baity, R.H. Goulding, M.D. Carter, D.O. Sparks, *Plasma Sources, Sci. Tehnol.* **13** 424 (2004).
21. R. H. Goulding, R.F. Welton, F.W. Baity, D.O Sparks, *Proceedings of the Topical Conference on Radio Frequency Power in Plasmas 2007*, Clearwater, Florida, USA.

A Proposal for a Novel H⁻ Ion Source Based on Electron Cyclotron Resonance Plasma Heating and Surface Ionization

O. Tarvainen and S. Kurennoy

Los Alamos National Laboratory, Los Alamos, New Mexico, 87545, USA

Abstract. A design for a novel H⁻ ion source based on electron cyclotron resonance plasma heating and surface ionization is presented. The plasma chamber of the source is an rf-cavity designed for TE_{111} eigenmode at 2.45 GHz. The desired mode is excited with a loop antenna. The ionization process takes place on a cesiated surface of a biased converter electrode. The H⁻ ion beam is further "self-extracted" through the plasma region. The magnetic field of the source is optimized for plasma generation by electron cyclotron resonance heating, and beam extraction. The design features of the source are discussed in detail and the attainable H⁻ ion current, beam emittance and duty factor of the novel source are estimated.

Keywords: Negative Ion Source, Electron Cyclotron Resonance Plasma Heating.
PACS: 29.25.Ni, 52.25.Jm

INTRODUCTION

The focus of the H⁻ ion source development program at Los Alamos Neutron Science Center (LANSCE) has recently been on gradually improving the performance of the filament-driven surface conversion ion source (see e.g. Ref. 1) and developing an rf-driven surface conversion ion source operated in helicon wave mode [2]. The benefit of the helicon-driven H⁻ ion source over the filament-driven source is higher plasma density and longer lifetime. However, the neutral gas pressure required for plasma ignition causes significant H⁻ losses, as the ions travel through the plasma from the converter electrode to the outlet aperture, which limits the performance of the ion source.

In this article we present a novel design for an H⁻ ion source based on electron cyclotron resonance plasma heating and surface ionization. The source is expected to operate within a neutral gas pressure range of 0.1-1 mTorr i.e. one order of magnitude lower than the helicon discharge. This helps to mitigate the H⁻ losses due to collisions with neutrals and reduces the volume formation of H⁻ preventing undesired increase of emittance, which is sometimes observed with surface converter ion sources [1].

Physical processes relevant for the H⁻ ECR ion source are discussed. The source concept is briefly compared to both filament-driven surface conversion source (used for beam production at LANSCE) and the prototype helicon-driven source. This is done in order to estimate the attainable H⁻ ion current, beam emittance and duty factor of the novel source. The design features and options for the new source concept are discussed and compared to earlier work by other authors.

PHYSICAL PROCESSES AFFECTING THE H⁻ ION BEAM PRODUCTION WITH SURFACE CONVERSION ION SOURCES

In this chapter physical processes affecting the formation of H⁻ ions on the conversion surface and the loss processes of H⁻ in the plasma are discussed in detail together with a brief description on the beam formation. The purpose of the discussion is to highlight the advantages of the new source design and to help estimate the presumably attainable H⁻ beam currents.

Surface Properties of the Conversion Surface Affecting the H⁻ Yield

The surface production of H⁻ ions via resonant electron tunneling on the surface of the converter electrode depends on several factors such as work function, Fermi energy and shifting/broadening of the electron affinity level. The most straightforward way of improving the conversion efficiency (H⁻ ions/bombarding ions) of the surface is to lower its work function by depositing a fractional monolayer of alkali metal (cesium) on the surface. In filament-driven surface conversion sources the metal surface of the converter electrode is subject to tungsten (or tantalum) deposition due to evaporation of the filaments. The inherent benefit of microwave-driven ion source is the lack of consumable parts in the plasma volume. Therefore, the "tungsten poisoning" of the conversion surface can be avoided meaning that this type of ion source could eventually provide better conversion efficiencies than the state-of-the-art filament sources.

Loss Processes of the H⁻ Ions

The H⁻ ions ejected from the conversion surface must travel through the plasma in order to become extracted. In LANSCE filament- and helicon-driven sources the distance from the converter electrode to the outlet is 12 cm and the typical energy of the ions is 225-325 eV corresponding to the negative voltage of the converter electrode (plasma potential is on the order of few volts and therefore negligible). Three types of loss processes of H⁻ due to collisions with other particles within this distance need to be considered:

(1) $H^- + e \rightarrow H + 2e$: The H⁻ mean free path λ_{H^-} can be calculated from the following equation: $\lambda_{H^-} = \sqrt{\frac{m_e E_{H^-}}{m_{H^-} E_e}} (\sigma n_e)^{-1}$ where m_x and E_x are the mass and energy of the particles, σ is the reported cross section [3] for $e + H^- \rightarrow H + 2e$ (note that the projectile and target are different than in reaction of interest) and n_e is the electron density. It can be estimated that the plasma density of the novel source is on the order of 10^{12} cm⁻³ and the temperature of the main part of the electron population on the order of 3-10 eV (with a significant tail to high energies similar to the LANSCE filament source [4]). The resulting calculation yields that in the case of the ECR-

driven source almost 90 % of the H⁻ ions should survive as they travel through the plasma (d = 90 mm). This fraction is estimated to be similar for the filament- and helicon-driven sources.

(2) $H^- + H_2 \rightarrow H^0$ (it is assumed that the neutral hydrogen is in molecular form): A typical neutral gas pressure in an ECR-heated plasma (at 2.45 GHz) is 0.1-1 mTorr. Based on cross section data from Ref. 3 it can be calculated that the H⁻ losses in this pressure range are less than 10% within a propagated distance of < 10 cm. The ability to reduce the neutral gas pressure from 5-10 mTorr (corresponding to H⁻ losses of tens of percent), which is typical for the LANSCE helicon source, producing 12-13 mA of H⁻, is the greatest motivation for the novel source design.

(3) $H^- + Cs^0 \rightarrow H^0$: The calculation for H⁻ losses in collisions with neutral cesium atoms, based on the reported cross section [5], for an H⁻ energy of 300 eV suggests that if the neutral pressure of cesium exceeds 0.7 mTorr the loss rate of H⁻ due to collisions with neutral cesium atoms exceeds the loss rate due to collisions with neutral hydrogen molecules at pressure of about 1 mTorr (of H_2). The cesium vapor pressure in the surface converter ion sources is typically much less than 0.7 mTorr.

Based on these considerations we estimate that > 70 % of the H- formed on the converter surface of the ECR-driven source could survive through the plasma to the outlet.

Brief Discussion on Beam Formation

Due to the negative bias of the converter electrode the H⁻ beam is formed in the plasma sheath adjacent to the converter surface and is "self-extracted" from the ion source. The beam is focused to the outlet aperture by shaping the surface of the converter. The formation mechanism of the beam explains the relatively small emittance values typically obtained with a LANSCE-type surface conversion ion sources (filament sources). The trajectories of the H⁻ ions are affected by the magnetic field in the region between the converter and the outlet aperture. In contrast to the helicon source, for which the magnetic field has to be a compromise between optimized coupling of rf-power (maximum plasma density) and beam propagation, the magnetic field of the ECR-driven ion source can be optimized for both functions simultaneously.

CONCEPTUAL DESIGN OF THE ECR SURFACE CONVERSION ION SOURCE

The main design objectives of the new ion source concept were high ionization efficiency (low neutral gas pressure) and compactness. Therefore, an ionization mechanism relying on a resonant process, i.e. electron cyclotron wave – plasma interaction, was an obvious choice. Negative ion sources (for H⁻) based on ECR plasma heating have been designed and built earlier. The approach of Tuske et al. [6] has been to separate the main plasma from the H⁻ production region by electric filter reducing the electron temperature and, consequently, stripping losses of (slow) negative ions near the outlet aperture. H⁻ ion beam currents on the order of 5 mA have been obtained with that source type. The main drawback of this approach is the drop of plasma density, imposed by the filter field, between the two stages of the source. This limitation can be overcome with a converter-type ion source. Takagi *et al.* [7] have designed and tested an ECR-driven plasma sputter ion source equipped with a converter electrode. In that source the converter electrode was used also as an antenna coupling the microwave power with the plasma. Negative ion beam currents of 7 mA have been obtained with the source [7]. However, the extracted ion beam contained almost 30 % of impurities (O⁻ and OH⁻) due to required microwave power level (3-4 kW) and subsequent heating of vacuum seals. In order to understand the origin of the problems encountered with this source design we used MicroWave Studio [8] to simulate the mode structure excited into the plasma chamber. According to our simulations it seems likely that the plasma chamber of the source described in Ref. 7 is, in fact, a multimode cavity. In addition, the resonant frequencies closest to 2.45 GHz deposit significant amounts of energy at the microwave window, which could explain the observed heating of vacuum seals near this location. The microwave-driven ion source design presented in this article has some similarities with the source described in Ref. 7. In order to optimize the ionization process we designed the plasma chamber to be a resonant cavity for TE_{111} eigenmode at 2.45 GHz, which should significantly improve the coupling efficiency of the microwaves. Variation of this design is used at CERN (ISOLDE) for ionization of noble gas radioisotopes (positive ions) [9].

A schematic drawing of the ion source is presented in Figure 1. Details such as vacuum seals and water cooling channels or subsystems such as cesium oven are not presented.

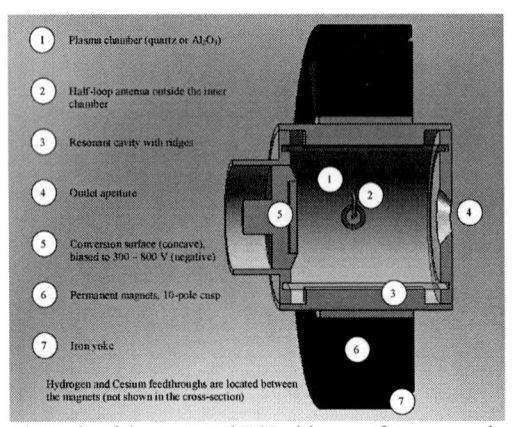

FIGURE 1. A schematic of the proposed ECR-driven surface conversion ion source.

The plasma chamber of the source is a quartz tube (or aluminum nitride tube for high power, high duty factor operation) located inside a resonant cavity. The quartz tube prevents the plasma to be directly in contact with the antenna and allows the main plasma volume and the remaining cavity volume to be in different vacuum conditions. MicroWave Studio (MWS eigensolver) was used to design the resonant cavity. Two metal ridges with gaps near the end walls are inserted 180 degrees apart on the inner wall of the cavity to allow the magnetic field of the TE$_{111}$ eigenmode at 2.45 GHz, corresponding to a frequency of a cheap commercial magnetron, to complete its loop and to separate it from the unwanted modes, including TM$_{010}$. The cavity dimensions are listed in Table 1.

TABLE 1. RF-cavity dimensions

Parameter	Value[mm]	Comment
Cavity inner radius, r_{cav}	45	
Cavity length, L_{cav}	80	
Quartz-tube outer radius, r_{out}	37.5	
Quartz-tube wall thickness, r_t	3.175	
Ridge width	20	
Ridge height	$r_{cav} - r_{out}$	Fills the space between the cavity and inner chamber
Ridge-end-wall gap length	8	

The desired eigenmode can be exited with a simple loop antenna inserted in the cavity mid-plane, with the loop plane oriented vertically, and connected to a 50-Ω coaxial cable. The magnetic field of the mode is well coupled to the antenna. Figure 2 shows the electric and magnetic fields of the TE$_{111}$ eigenmode excited in the cavity. The presented field normalization is the MWS default, i.e. the field total energy is 1J, which means that the presented values do not correspond to the cavity fields of the operating source.

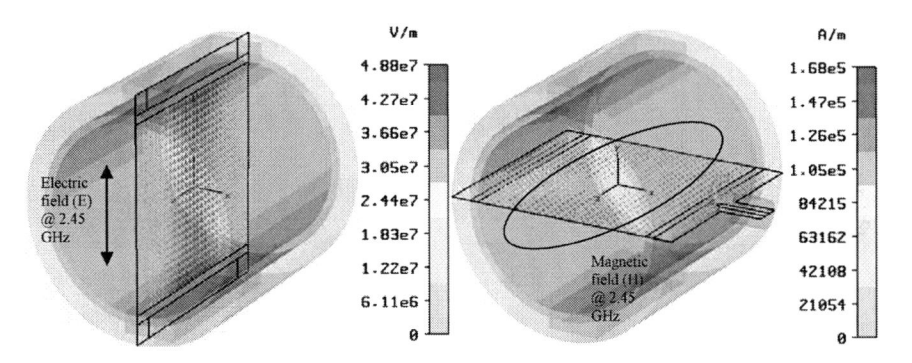

FIGURE 2. Electric field (left) and magnetic field (right) of the TE$_{111}$ eigenmode (at 2.45 GHz) excited in the cavity of the proposed ECR-driven surface conversion ion source.

The effects of the end wall (outlet and biased converter electrode) holes, gaps and shapes were also studied and taken into account in the preliminary design. It was observed that these features had a small effect on the resonant frequency. Also plasma loading of the cavity will slightly affect the resonant frequency. Effects of both kinds can be compensated by tuning the cavity length by moving the converter electrode acting as a tuner.

The electrons in the plasma are accelerated in a resonance when their Larmor frequency in the external magnetic field equals to the microwave frequency. These energetic electrons ionize neutral atoms and maintain the plasma. For 2.45 GHz the corresponding resonance magnetic field is 0.0875 T (875 G). The required magnetic field can be generated either with solenoids or permanent magnets. Magnetic field design based on permanent magnets is favorable for two reasons: it makes the source more compact and helps to maximize the extraction efficiency of the H⁻ ions. Ten rows of permanent magnets (NdFeB, grade 50 MGO, 1 inch by 1.5 inch cross section, and length of 2 inches) are placed around the cavity forming a typical 10 pole cusp structure. The resulting resonance surface (B = 0.875 T) is illustrated in Figure 3. The magnetization direction of the permanent magnets is indicated by arrows. The spatial location of the resonance can be varied by moving the magnets radially. The Figure also shows that the strength of the magnetic field falls to zero on the source axis.

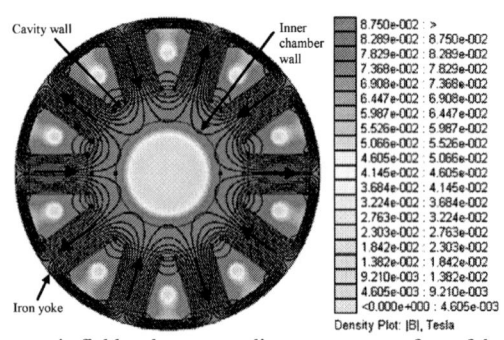

FIGURE 3. The cusp magnetic field and corresponding resonance surface of the proposed ECR-driven surface conversion ion source. Simulation with FEMM [10].

The magnetic field being zero on the source axis is favorable for the beam extraction and for attaining a more uniform plasma density due to $F = -\mu \nabla B$ force "pushing" the electrons (and ions) towards the axis (cusp confinement). The extraction efficiency was studied with an ion tracking code written with Mathematica. The magnetic field for the ion tracking was simulated with Radia3D [11]. The calculated beam spot at the outlet electrode is presented in Figure 4. The parameters used in this example of ion tracking calculation are: distance from the converter to the outlet 90 mm, converter voltage -500 V, converter radius of curvature 127 mm (concave), converter radius 19 mm. Reducing the number of magnetic poles would make the extraction of the H⁻ more problematic due to increasing magnetic field near the source axis. However, the maximum number of poles that can be used is 10 since for higher number of poles the resonance surface is outside the quartz chamber. Also the magnet length affects the size of the beam spot at the outlet aperture. The geometry of the magnetic field is reflected into the shape of the beam spot i.e. number of cusps on the beam spot is half of the number of magnetic poles.

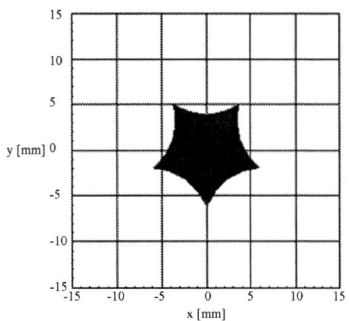

FIGURE 4. The results of the ion tracking calculation. The "star" represents the predicted beam spot at the outlet aperture.

The deposition of cesium and sputtered material (from the converter) on the quartz tube can have an adverse effect on the coupling of the microwave power with the plasma. If the thickness of the conductive layer exceeds the skin depth of 2.45 GHz microwaves, the coupling of the power to the electrons is prevented (strictly speaking, the field amplitude falls to 1/e from the original). The skin depth δ depends on the electrical conductivity of the metal. This process can be affected by choosing the right converter material. Rhenium seems to be favorable due to low sputtering yield and high skin depth (compared to other metals). In the worst case if the metal deposition on the quartz tube presents a problem, the tube can be omitted. In this case the antenna needs to be covered by an insulator preventing a direct contact with the plasma. Furthermore, the antenna-insulator assembly needs to be shielded from metal deposition. This can be realized with a metal (or dielectric) shield. It is important to isolate the metal shield from the cavity wall in order to prevent the formation of a surface current loop canceling the magnetic field in the antenna region and therefore preventing the coupling of the microwave power to the plasma. Our simulations have demonstrated that with a proper design the shielding does not affect the mode structure.

Based on the experience gained with both the LANSCE filament- and helicon-driven surface conversion ion sources we can estimate the expected H⁻ output for the microwave-driven source presented in this article. Taking into account differences in plasma density, electron temperature and beam dynamics (propagation through the plasma) we estimate that the H⁻ ion beam current extracted from the microwave source can be 25-30 mA.

The emittance of the ion beam is mainly defined on the converter surface (sputtering energy of the ions) and, therefore, it can be expected that the emittance of the H⁻ ion beams extracted from the ECR-driven surface conversion ion source does not differ significantly from the emittance of the LANSCE filament-source (typically 0.15-0.25π mmmrad, 95 % norm.-rms). The expected duty factor of the ECR-driven source is at least 12 %.

The source design can be used for producing other negative ions as well. The converter can be transformed into a cathode manufactured from a material to be ionized. The heavy ion (such as xenon or cesium) induced sputtering of the cesiated surface will result into emission of negative ions. In addition the source concept could be used for the production of intense proton ion beams. Ion beam (H^+) currents of >100 mA have been produced with 2.45 GHz ECR ion sources [12].

ACKNOWLEDGMENTS

This work has been supported by the US Department of Energy under Contract Number DE-AC52-06NA25396.

REFERENCES

1. J. Sherman et al., Proceedings of the 10th International Symposium on Production and Neutralization of Negative Ions and Beams, AIP Conference Proceedings, 763, (2005), p. 254.
2. O. Tarvainen, G. Rouleau, R. Keller, E. Geros, J. Stelzer and J. Ferris, Rev. Sci. Instrum. 79, 02A501, (2008).
3. H. Tawara, Y. Itikawa, Y. Itoh, T. Kato, H. Nishimura, S. Ohtani, H. Takagi, K. Takayanagi and M. Yoshino, Nagoya University Report No. IPPJ-AM-46, (1986).
4. O. Tarvainen, R. Keller and G. Rouleau, Proceedings of PAC07, JACoW, (2007), p. 1802.
5. F.W. Meyer, J. Phys. B: Atom. Molec. Phys. 13, p. 3823, (1980).
6. O. Tuske et al., Proceedings of the 11th International Symposium on Production and Neutralization of Negative Ions and Beams. AIP Conference Proceedings, Volume 925, p. 114, (2007).
7. A. Takagi and Y. Mori, Rev. Sci. Instrum., 71, (2000), p. 1042.
8. Microwave Studio, v.2006B, CST GmbH, 2007. www.cst-world.com
9. F. Wenander and J. Lettry, Rev. Sci. Instrum., 75, 5, p. 1627, (2004).
10. D. Meeker, http://femm.foster-miller.net/
11. http://www.esrf.eu/Accelerators/Groups/InsertionDevices/Software/Radia
12. J. Sherman et al. Rev. Sci. Instrum., 69, p. 1003, (1998).

First Results with a Surface Conversion H⁻ Ion Source Based on Helicon Wave Mode-Driven Plasma Discharge

O. Tarvainen, E.G. Geros, R. Keller, G. Rouleau and T. Zaugg

Los Alamos National Laboratory, Los Alamos, New Mexico, 87545, USA

Abstract. The currently employed converter-type negative ion source at Los Alamos Neutron Science Center (LANSCE) is based on cesium enhanced surface production of H⁻ ion beams in a filament-driven discharge. The extracted H⁻ beam current is limited by the achievable plasma density, which depends primarily on the electron emission current from the filaments. The emission current can be increased by increasing the filament temperature but, unfortunately, this leads not only to shorter filament lifetime but also to an increase in metal evaporation from the filament, which degrades the performance of the H⁻ conversion surface. In order to overcome these limitations we have designed and tested a prototype of a surface conversion H⁻ ion source, based on excitation of helicon plasma wave mode with an external antenna. The source has been operated with and without cesium injection. An H⁻ beam current of over 12 mA has been transported through the low energy beam transport of the LANSCE ion source test stand. The results of these experiments and the effects of different source parameters on the extracted beam current are presented. The limitations of the source prototype are discussed and future improvements are proposed based on the experimental observations.

Keywords: Negative ion source, Helicon plasma discharge.
PACS: 29.25.Ni, 52.25.Jm

INTRODUCTION

The accelerator facilities at Los Alamos Neutron Science Center (LANSCE) consist of two injector systems (H⁺ and H⁻), 800 MeV linear accelerator, proton storage ring and experimental areas. The H⁻ ion beam, required for charge exchange injection into the proton storage ring, is currently produced with a filament-driven surface conversion ion source (see e.g. Ref. 1). The H⁻ ion source regularly produces beam currents of 16-18 mA at 60 Hz discharge pulse repetition rate (835 μs pulses). The future scenarios for the LANSCE accelerator facility require improvements on the performance of the H⁻ ion source. The development strategy is comprised of several paths, namely 1) gradual improvement of the filament-driven surface conversion ion source discussed in Ref[2], 2) development of helicon-driven (*rf*) surface conversion ion source, 3) collaboration with SNS ion source R&D team and 4) feasibility studies of other ion source concepts (e.g. Ref. 3). This article focuses on the status and future prospects of the helicon-driven surface conversion ion source development at LANSCE.

Helicon plasma generators [4] are characterized by high plasma density, up to 10^{13} cm^{-3}, achieved with exceptionally good power efficiency. The use of helicon plasmas for negative ion production was proposed by Welton [5]. Efforts of combining a helicon plasma generator with an SNS-type H⁻ ion source have been carried out at Oak Ridge National Laboratory [6].

The focus of our efforts has been on developing a helicon-driven surface conversion ion source. The source is equipped with a biased converter electrode similar to the LANSCE filament-source. Replacing the filaments by an rf-antenna is expected to yield higher beam currents, due to increased plasma density, and longer source lifetime. A benefit of this approach, in contrast to the designs presented in References 5 and 7, is the "self-extraction" of the H⁻ ions from the converter electrode yielding a low emittance ion beam, typically less than 0.2 π mmmrad, 95 % norm.-rms [1]. In addition, the stripping losses of energetic (about 300 eV) H⁻ ions in collisions with electrons are reduced significantly, due to lower cross sections. Compared to purely inductive rf-coupling, a helicon discharge provides better stability at low neutral gas pressures around 1-10 mTorr (0.13-1.3 Pa), which helps to minimize the losses of H⁻ by collisions with neutrals.

HELICON-DRIVEN SURFACE CONVERSION ION SOURCE PROTOTYPE AND LANSCE ION SOURCE TEST STAND

A schematic of the helicon ion source prototype is presented in Figure 1. The mounting flange, extraction system, converter electrode and repeller electrode (magnetic + electrostatic) are identical to the parts used in the LANSCE filament source. The magnetic field required for the helicon wave mode excitation is created by a permanent magnet array in which the individual magnets are magnetized parallel to the source axis. The array consists of 18 rectangular magnets arranged in 6 rows. A 3-loop external rf-antenna, wrapped around the plasma chamber (OD 75 mm, both PYREX and Al$_2$O$_3$ have been used), is placed in the fringe field of the permanent magnets. The strength of the magnetic field on the source axis, at the location of the antenna, is about 0.01 T. This configuration is suitable for excitation of the m = 0 helicon wave mode [8]. The 13.56 MHz RF-power is coupled into the plasma with the aid of a capacitive matching circuit consisting of shunt- and series-capacitors. The reflected/forward power ratio is typically 1-20 %. The axial position of the converter electrode, whose concave surface (127 mm radius of curvature) is designed to focus the ion beam through the outlet aperture, can be varied with respect to the antenna and the magnet array. Applying the negative converter bias voltage of 250-350 V facilitates ignition of the discharge and thus tuning of the rf-matching for the plasma impedance. The permanent magnet ring embedded in the repeller electrode creates a magnetic field hump of 0.02 T near the outlet aperture of the source and reduces the amount of cold electrons leaking from the plasma into the extraction through the gradient-B force, which improves the e⁻/H⁻ ratio of the extracted beam.

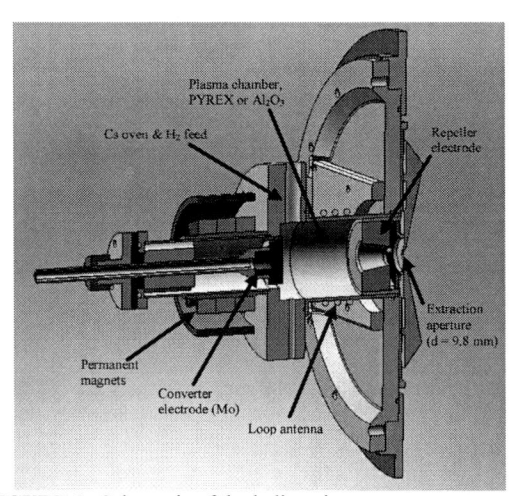

FIGURE 1. Schematic of the helicon ion source prototype.

The first cesiation experiments with the helicon source were performed using the cesium oven designed for the filament-driven sources (see Figure 2a). For the filament source the oven is typically loaded with 25 grams of cesium lasting up to 40 days, exceeding the filament lifetime. The drawback of the oven design is that the hottest part of the assembly is the cesium reservoir instead of the shaft conducting cesium vapor into the plasma chamber. This frequently causes uncontrolled bursts of cesium, as a result of cesium condensation on the inner surface of the shaft clogging the tube, and subsequent sudden release of cesium as the oven is brought up to operational temperatures (around 200 °C). In the case of the filament source this does not present a problem in most cases because the excess of cesium is deposited on the walls of the plasma chamber and affects the cesium balance of the source in a favorable manner since the source performance can be affected by controlling the wall temperature [9,2]. For the helicon source, excessive amounts of cesium in the plasma chamber present a major problem due to adverse effects on the rf-matching followed by a dramatic increase of reflected power. In order to improve the cesiation process of the helicon source, a new cesium oven, see Figure 2b, was employed. In this oven design, utilizing two resistive heaters, the cesium reservoir is the coldest spot of the assembly. Thus, the consumption of cesium during the initial cesiation of the ion source is reduced, and uncontrolled bursts of cesium are avoided. The oven was loaded with only 1 gram of cesium for each experimental run described in this article. The Cs ampule is cracked by gently bending the stainless steel cesium reservoir. The duration of each run was 4-7 days at a working rate of 8-10 hours a day.

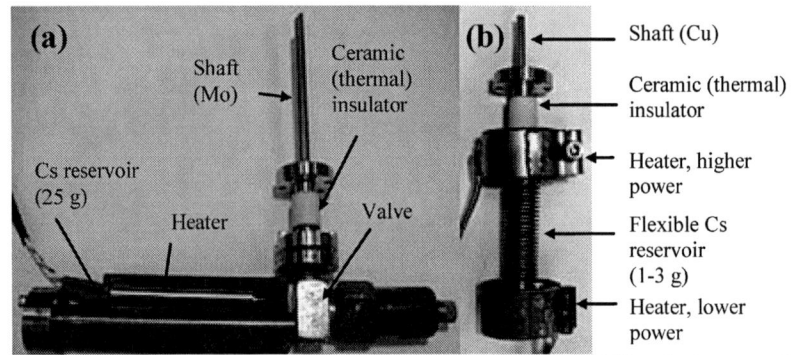

FIGURE 2. (a) Standard Cesium oven for the filament-driven source and (b) New Cesium oven for the helicon-driven source. The main parts of both oven designs are shown.

All measurements described in this article were performed at 80 keV extraction voltage on the LANSCE ion source test stand (ISTS) illustrated in Figure 3. The ISTS [10] is a reproduction of the LANSCE H⁻ injector system. The reported beam currents were measured with a Faraday cup at the end of the LEBT corresponding to the injection point of the 670 kV column of the LANSCE H⁻ injector. The beam current values are not affected by parasitic electrons since the beam current was measured after the 4.5 degree bend. The diameter of the extraction aperture used in the measurements with the helicon source is 9.8 mm, being identical to LANSCE filament sources.

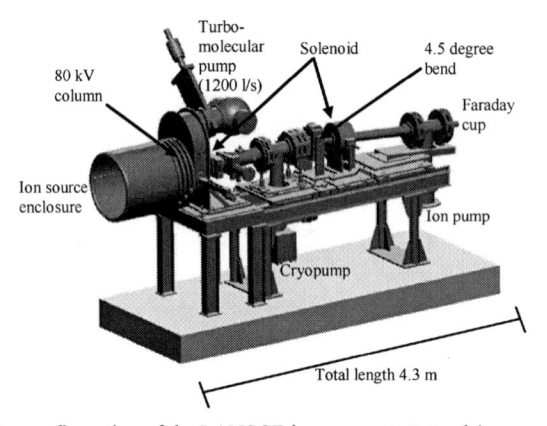

FIGURE 3. The configuration of the LANSCE ion source test stand (components used for the measurements in this article are highlighted).

EXPERIMENTAL RESULTS

The first beam extraction experiments with the pre-prototype of the helicon-driven surface conversion ion source were performed without cesium injection [11]. An H⁻ beam current of 3.4 mA was achieved at 1 % duty factor. The LANSCE filament sources typically produce 2-2.5 mA of H⁻ beam current in uncesiated conditions at 60 – 120 Hz repetition rate.

Recently the experiments with the helicon source concentrated on operating the source with cesium. The effects of different source parameters such as neutral gas pressure, rf-power and -matching have been explored to gain experience on the cesium equilibrium in this source. The duty factor was gradually increased up to 3.7 % at constant pulse length of 865 μs without any signs of droop in the extracted beam current. Five experimental runs were performed at ISTS with the improved cesium oven. The highest H⁻ beam currents achieved during each of these experiments are presented in Figure 4. So far, the highest obtained H⁻ beam current is 12.3 mA. The beam current was stable at the level of over 12 mA for 2 hours (the source parameters were deliberately changed after this). The highest transient current obtained so far is 16.1 mA, lasting for a few minutes. The e/H⁻ ratio of the extracted beam depends on the cesium conditions. Values of as low as e/H⁻ = 5 have been achieved.

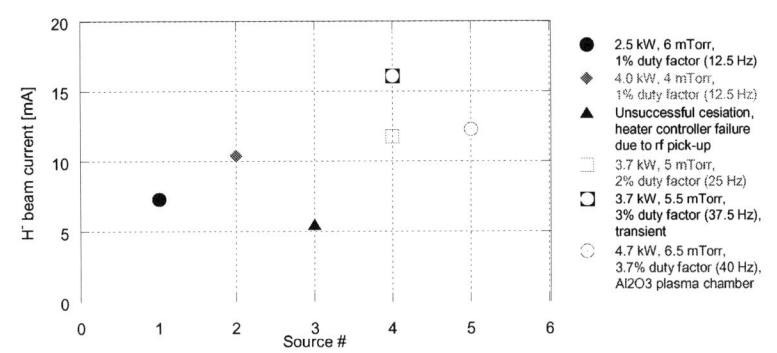

FIGURE 4. Performance of the LANSCE helicon ion source as of August 2008.

During the experiments we have observed the following trends affecting the H⁻ output of the helicon source:

1. Increasing the rf-power at constant neutral gas pressure results in an increase of the H⁻ beam current. A clear "jump" is typically observed at 1.0-1.8 kW, corresponding to the shift from purely inductive coupling into the m = 0 helicon mode. However, increasing the rf-power typically forces us to increase the neutral gas pressure in order to keep the reflected power under the limiting value set by the rf-amplifier (max 5 kW forward / 1 kW reflected). The cesium feed rate needs to be increased with increasing rf-power in order to maintain optimal cesium coverage on the converter electrode. This is due increasing plasma density and converter current resulting to higher sputtering rate of cesium from the converter. Converter currents up to 10 A have been measured at 5 kW of forward power (for the filament source the maximum is typically 2.5-3.0 A).

2. Decreasing the neutral gas pressure at constant rf-power results in an increase of the H⁻ beam current. This is due to reduced stripping losses of the H⁻ beam traveling through the plasma. But reducing the neutral gas pressure causes the converter current to decrease, and consequently, cesium feed rate needs to be lowered. Reducing the neutral pressure below 4 mTorr causes the reflected rf-power to exceed the limit of the

amplifier because ignition of the discharge in the beginning of each rf-pulse becomes unreliable. Signs of severe neutral starvation have not been observed when the source has been operated at pressures above 4 mTorr. In practice this means that the droop of the H⁻ beam current within the discharge pulse (865 µs in these experiments) is less than 10 %.

3. The feed rate of cesium affects the matching of the rf-power. Injecting cesium at a high rate helps to ignite the plasma at low neutral H_2 pressures, but it also causes the reflected power to increase and does not correspond to optimized surface conditions.

DISCUSSION AND FUTURE IMPROVEMENTS

The results obtained with the helicon-driven surface conversion ion source are very promising. However, in its present form the source is not reliable enough for production-grade operations. The extracted H⁻ beam current is limited mainly by the neutral gas pressure of 4-7 mTorr at sufficient rf-power levels, causing stripping losses. At pressures above 1 mTorr the dominating reaction causing H⁻ losses is neutralization in collisions with hydrogen molecules [12]. This is due to the rather low ionization degree of the plasma, calculated to be 1-3 %. Figure 5 shows the calculated fraction of surviving H⁻ ions as a function of traveled distance under different neutral gas pressures. The calculation demonstrates that it is more desirable to decrease neutral gas pressure than reduce the distance between the converter electrode and the outlet of the source. The distance of 12 cm corresponds to the source geometry used in the measurements. In the calculation it was assumed that the energy of the H⁻ ions emitted from the converter is 275 eV, and the neutral gas is entirely molecular hydrogen (H_2). The corresponding cross section for the neutralization was taken from Ref. 13. The losses due to other collision mechanisms (H⁻ with electrons, positive ions, neutral cesium etc.) can be estimated to be independent of the neutral hydrogen pressure and less important at pressure range of > 1 mTorr.

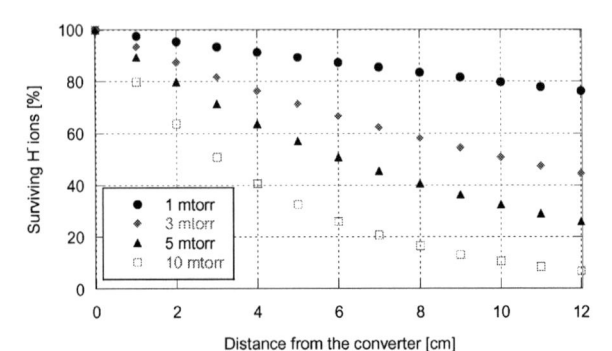

FIGURE 5. Calculated fractions of surviving H⁻ ions as a function of traveled distance at different neutral gas pressures. The dominating loss mechanism is collisions with neutrals (H_2) [12]. An H⁻ energy of 275 eV was assumed.

The results plotted in Figure 5 suggest that the extracted beam current could be increased by a factor of 2-3 if the neutral pressure was reduced from 4-6 mTorr to 1-2 mTorr and other loss processes would remain unchanged. Therefore, the focus of our efforts has recently been in developing techniques to assure source operation at very low neutral gas pressures, on the order of 1 mTorr. We have identified three possible methods to accomplish this:

1. *Increasing the volume of the plasma chamber*: Our preliminary studies with different sizes of plasma chambers have indicated that increasing the neutral gas volume helps to ignite the plasma at lower pressures. However, increasing the diameter of the plasma chamber reduces the power density and subsequently the plasma density, which is not desirable.

2. *Pulsed gas feed:* The purpose of the pulsed gas feed is to provide higher neutral gas pressure at the beginning of the rf-pulse. The neutral gas pressure is then allowed to decay towards the end of the rf-pulse. This method would require longer rf-pulses (on the order of 2-3 ms) due to the fact that stripping losses would still reduce the beam current during the 1-2 ms measured from the leading edge of the rf-pulse. A gas feed system capable of providing 150 µs gas pulses at 60 -120 Hz has been developed for the purpose, similar to the one reported in Ref. 14. Initial tests of the gas pulsing system were conducted at an ion source processing stand normally used for filament pre-processing. The neutral gas pressure variation within 3 ms intervals corresponding to the rf-pulse was observed to be several mTorr. The variation of the neutral gas pressure can be changed by varying the delay between gas pulses and rf-pulses as the neutral gas pressure decays exponentially. This method will be tested at ISTS at a later time.

3. *Ignition of the discharge in the gas feed line by a Tesla coil transformer*: The most promising option to facilitate plasma ignition at low neutral gas pressures, and consequently increase the extracted beam current, is to utilize a Tesla coil transformer. The purpose of the Tesla coil is to ignite the discharge in the gas feed line where the neutral gas pressure is significantly higher. The low-density plasma diffuses from the gas feed line into the main discharge chamber and makes it possible run the discharge at lower pressures. This method has enabled us to shift the threshold pressure for plasma ignition from 8 mTorr to 4 mTorr in pure hydrogen discharges. Experiments with a cesiated source have not yet been conducted. Ignition systems for inductively coupled rf ion sources have been used before by Peters[15] and Welton[16].

Reliable plasma ignition will most probably enable us to reduce the amount of cesium in the plasma chamber. This is desirable to obtain stable conditions for rf-matching.

Another issue that needs to be addressed in order to improve the production of H⁻ with the helicon ion source is the optimization of the magnetic field for beam extraction. So far our experiments have concentrated on maximizing the beam production through optimized plasma density. The effect of the magnetic field on the beam formation was studied with ion tracking simulations after it was observed that the surface of the repeller electrode had no signs of cesium accumulation near the extraction aperture while other parts of the source facing the plasma chamber were covered with a visible layer of cesium. This observation indicates that the repeller electrode intercepts a part of the ion beam. Figure 6 shows the calculated beam envelope in the plasma chamber for three different permanent magnet strengths within the range of 0.1 – 0.35 T (1 – 3.5 kG). The strengths of the magnets are measured on the surface of the each PM block. For the ion tracking calculations it was assumed that the ions are sputtered perpendicularly from the converter electrode (concave surface with 127 mm radius of curvature) and reach the energy corresponding to the converter voltage within a thin plasma sheath adjacent to the surface. In the example presented in the Figure the converter voltage was set to -275 V. The static magnetic field was calculated with Radia3D [17], and the magnetic field components at each point along the ion trajectories were imported to an ion tracking code written with Mathematica® [18]. Possible effects of electric fields on the beam trajectory were not taken into account.

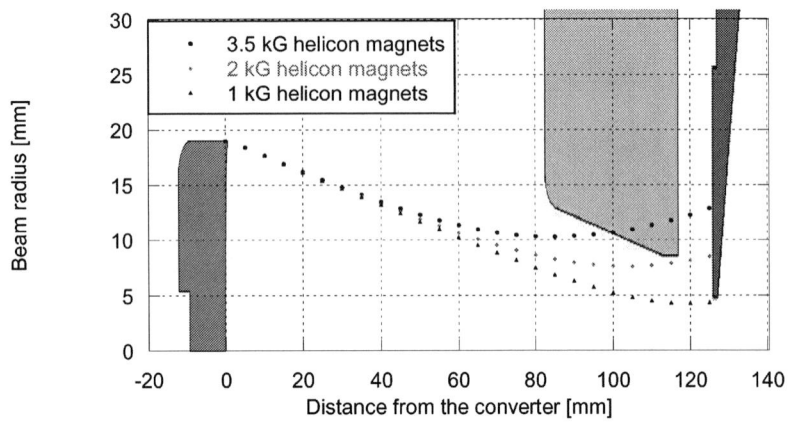

FIGURE 6. Calculated beam radius as a function of distance from the converter with different permanent magnet strengths. The solid surfaces are from left to right: converter electrode, repeller electrode and outlet aperture.

In the case of 3.5 kG magnets which are currently used the H⁻ beam envelope on the outlet plane is much larger than the extraction aperture. The difference between the footprints obtained as simulation result (point where the repeller intercepts the beam) and the observed cesium tracks on the repeller surface is less than 2 mm. The simulation result suggests that using weaker magnets would be beneficial to the beam extraction. Changing the converter curvature and/or voltage seems not to be as effective as reducing the magnetic field. However, in the helicon arrangement weaker magnetic field yields lower plasma density at a given rf-power and requires higher neutral gas pressure for plasma ignition. This was confirmed in an experiment without beam extraction. Therefore, reducing the neutral gas pressure remains the most important item of future developments.

Our near-term plans include emittance measurements. However, the source reliability has to be improved prior to this. Because of the similar source geometry, it can be expected that the emittance of the H⁻ ion beams extracted from the helicon source does not differ significantly from the emittance of the filament source (typically < 0.2 π mm mrad, 95 % norm.-rms).

As a summary of the status of the helicon ion source development at LANSCE, Table 1 compares the filament source to the helicon source (as of August 2008).

TABLE 1. Comparison of the LANSCE filament ion source to LANSCE helicon ion source.

	Filament Source	Helicon source
Neutral gas pressure	1-3 mTorr	4-7 mTorr (with Cs, without "ignition system)
Input power	4-11 kW discharge + 2-2.5 filament heating	1-5 (rf)
Plasma density	$5 \times 10^{11} - 1.2 \times 10^{12}$ cm^{-3} (probe measurement)	$5 \times 10^{11} - 3.5 \times 10^{12}$ cm^{-3} (calculated from converter current)
Beam current, 9.8 mm aperture	14-24 mA (15-18 mA is typical)	7-12.5 mA
Emittance	< 0.2 π mm mrad	?
e/H$^-$ ratio	3-5	5-8
Duty factor	6-12 %	1-3.7 % demonstrated

ACKNOWLEDGMENTS

This work has been supported by the US Department of Energy under Contract Number DE-AC52-06NA25396.

REFERENCES

1. J. Sherman et al., Proceedings of the 10th International Symposium on Production and Neutralization of Negative Ions and Beams, AIP Conference Proceedings, 763, (2005), p. 254.
2. R. Keller et al. in these proceedings.
3. O. Tarvainen and S. Kurennoy in these proceedings.
4. F. F. Chen and R. W. Boswell, IEEE Transactions on Plasma Science, Vol. 25, No. 6, (1997), p. 1245.
5. R. F. Welton, M. P. Stockli and S. N. Murray, Rev. Sci. Instrum. 77, 03A506 (2006).
6. R.F. Welton et al. in these proceedings.
7. O. Tarvainen, M. Light, G. Rouleau and R. Keller, Proceedings of the 11th International Symposium on Production and Neutralization of Negative Ions and Beams, AIP Conference Proceedings, 925, (2007), p. 171.
8. F.F. Chen and H. Torreblanca, Plasma Phys. Control. Fusion 49, A81-A93 (2007).
9. O. Tarvainen, submitted to Nucl. Instrum. and Meth. in Phys. Res. A.
10. W.B. Ingalls, M.W. Hardy, B.A. Prichard, O.R. Sander, J. Stelzer, R.R. Stevens, K.N. Leung and M.D. Williams, Proc. of the XIX Linac Conference (Chicago IL), Argonne National Lab Report ANL-98/28, (1998), p. 887.
11. O. Tarvainen, G. Rouleau, R. Keller, E. Geros, J. Stelzer and J. Ferris, Rev. Sci. Instrum. 79, 02A501 (2008).
12. C.F.A. van Os, A.W. Kleyn, L.M. Lea, J.T. Holmes and P.W. van Amersfoort, Rev. Sci. Instrum 60, 4, (1989), p. 539.
13. H. Tawara, Y. Itikawa, Y. Itoh, T. Kato, H. Nishimura, S. Ohtani, H. Takagi, K. Takayanagi and M. Yoshino, Nagoya University Report No. IPPJ-AM-46, (1986).
14. Y. Huang and M. Sulkes, Rev. Sci. Instrum. 65, (1994), p. 3868.
15. J. Peters, Proceedings of the 11th International Symposium on Production and Neutralization of Negative Ions and Beams, AIP Conference Proceedings, 925, (2007), p. 79.
16. R.F. Welton, M.P. Stockli, S.N. Murray, J. Carr, J. Carmichael, R. Goulding and F.W. Baity, Proceedings of the 11th International Symposium on Production and Neutralization of Negative Ions and Beams, AIP Conference Proceedings, 925, (2007), p. 87.
17. http://www.esrf.eu/Accelerators/Groups/InsertionDevices/Software/Radia
18. Mathematica 6, Wolfram Research, http://www.wolfram.com/products/mathematica/index.html

Negative Hydrogen Ion Source with Inverse Gas Magnetron Geometry

V.A. Baturin, P.A. Litvinov, S.A. Pustovoitov

*Institute of Applied Physics, National Academy of Sciences of Ukraine,
58 Petropavlovskaya St. Sumy, 40030 Ukraine*

Abstract. The work is dedicated to the experimental investigation of the intense volume-plasma H⁻ ion source. Formation of ions occurs in a volume of hydrogen plasma (without additives of cesium) due to two-step dissociative attachment of thermal electrons to vibrationally excited molecules H_2. Preliminary experimental researches of two parameters of the upgraded source - emission density of H⁻ ions and gas flow are represented below. The advancing of electrode system of the emission chamber of a source has allowed receiving the value of an emission density of H⁻ ions equal 440 mA/cm².

Keywords: Ion Source, Negative Hydrogen Ion, Plasma, Inverse Gas Magnetron.
PACS: 52.27.Cm, 52.50.Dg.

INTRODUCTION

The plasma dual-chamber of H⁻ ion source having a rapid starting, reliable and long-time operation and also simplicity in service was designed for the injector of a high energy accelerator. In more details design of this source can be found in reference [1]. It is a non-cesium ion source working on the basis of tubular discharge. For plasma generation the inverse gas magnetron is used. The dual-chamber design of a source allows receiving rather easily necessary vacuum conditions, using pumps with average speeds of pumping. At high-power of discharge it is capable to supply an emission density of H⁻ ions current exceeding 200 mA/ cm².

The record values of emission density of H⁻ ions were obtained from the surface - plasma sources. However in some cases the using of these sources creates problems connected with existing of cesium vapors in their gas-discharge chambers. This fact, and a number of other circumstances, stimulates the development of non-cesium H⁻ ion sources. At the Institute of Applied Physics NAS of Ukraine the activities on advancing of earlier version of an axial - symmetrical source described in reference [1] are carried out.

Preliminary experimental researches of two parameters of the upgraded source - emission density of H⁻ ions and flow of working gas are adduced below. The increase of value of the first parameter and the decrease of value of the second one, are actual for using of ion sources in a structure of injectors of modern accelerators. These purposes in a described source are obtained by means of increase of a flow of slow electrons into the emission area and optimization of pulse supply of hydrogen into its discharge chamber.

DESCRIPTION OF THE SETUP

The scheme of the upgraded source is shown in Figure 1. The activity of this source is grounded on the most effective for today mechanism of formation of negative ions in a volume of hydrogen plasma (without cesium vapors). It is two-step process of dissociative attachment of low electrons by vibrationally excited molecules H_2 [2]. The cross-section of this process quickly grows up to a significant size ($\geq 10^{-17}$см^2) with growth of oscillatory quantum number at electron energy of several electronvolt. The main contribution to generation of H⁻ ions is introduced by molecules, exited on levels v = 5 - 11. The formation of vibrationally excited molecules realizes basically by fast electrons (> 10 eV). The optimization of conditions for vibratory excitation of molecules and for the subsequent formation of negative ions in this source is realized due to the creation of a gas discharge system generating in the emission chamber two areas of plasma - peripheral, with a rather large fraction of fast electrons and internal paraxial with slow electrons.

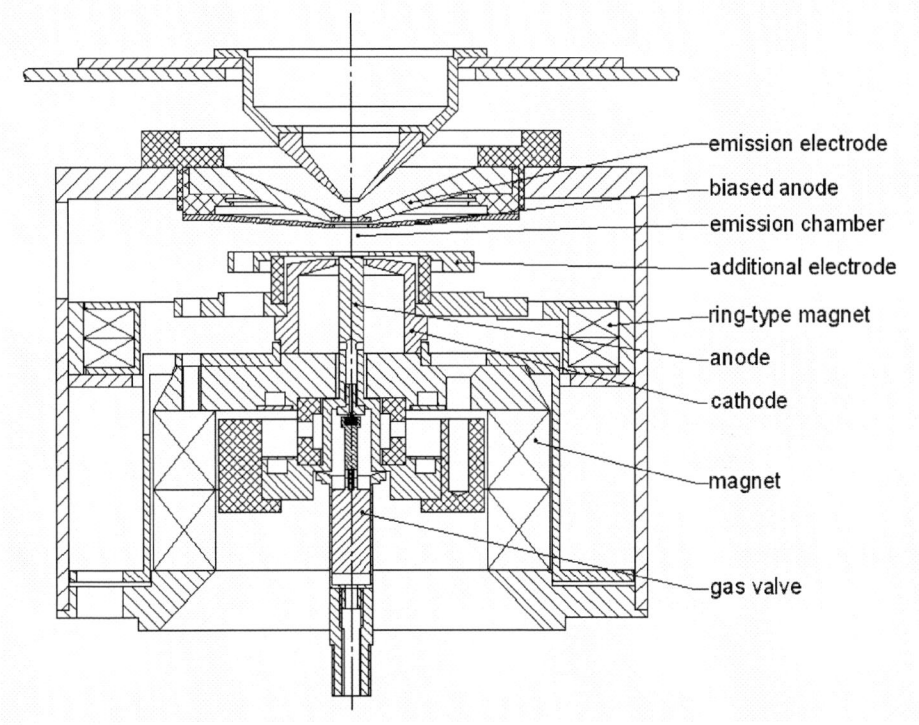

FIGURE 1. Schematic of ion source with a inverse gas magnetron geometry.

The discharge chamber consisting of the cathode and anode represents inverse gas magnetron geometry, which works on the basic of glow discharge in crossed ExH fields. In the inverse magnetron of the given design it is possible to realize both magnetic and electrostatic retention of fast electrons. The superposition of a longitudinal magnetic field excited by magnets Sm-Co$_5$, increases a life time of fast electrons, which start from a cylindrical surface of the cathode to a central anode. The cathode walls of the gas magnetron provide the retention of electrons along a magnetic field. As a result a life time of fast electrons in the chamber of the magnetron increases. They experience numerous collisions with atoms of gas and have time to make sufficient number of ionizations before they get on a surface of an anode.

The ion source operates as follows. Under supply of a potential pulse on electrodes, the discharge is excited in the chamber. The generated plasma at sufficient width of an annular slot penetrates along a magnetic field into the emission chamber and extends up to an emission electrode. Additional electrode, which is under the positive potential relative to the anode of the magnetron, provides reliable plasma drawing from the magnetron into the emission chamber of the source.

Due to formation of double layer before a narrow annular slot at magnetron output, fast electrons are delivered to the area of its volume. In peripheral plasma conditions favorable for vibrational excitation of molecules are created. Internal paraxial plasma, formed by diffusion of peripheral tubular plasma across a magnetic field, will contain the vibrationally excited molecules and the enriched fraction of slow electrons, while the fraction of fast electrons does not penetrate here because of the action of magnetic filter. Thus, in the internal plasma there are necessary conditions for effective realization of a finishing phase of two-step process of formation of negative ions.

It is necessary to mark, that in a considered source, the magnetic field of the filter by a natural mode coincides with a magnetic field of gas discharge and, accordingly, is formed by a general magnetic system. The magnetic system is designed on the basis of permanent Sm-Co$_5$ magnets which create in an interpolar gap magnetic field B_z =0,09 – 0,12 T. For correction of a magnetic field in emission area of an ion source the ring-type magnet was used (Sm-Co$_5$) with a radial magnetization.

One of the important problems at optimization of H$^-$ source activity is the optimization of gas flow. On the one hand for achievement of maximum density of negatively ions it is necessary to increase a gas pressure in the discharge chamber of a source, and on the other hand it is necessary to reduce a gas pressure in the field of formation of ion beam to reduce the destruction of H$^-$ ions.

For reduce of gas flow in the upgraded source, we have applied a quick-operating valve designed by us [3]. In this valve, for maintenance of a short gas pulse with a high rate of pulse rise, a hammer device is applied. Besides in the upgraded source a locking device of the valve is made as a unified structural member together with an anode of the magnetron. It has allowed to avoid practically completely the integrating of a gas pulse, and to reduce gas flow during a pulse. The power supply of the valve allows executing the adjustable delay between supply of gas into the discharge cell and discharge initiation in the source.

The design of a magnet system of a source allows supplying the necessary gas pressure difference between discharge and emission chambers.

Extraction of ions from a source is made from paraxial zone of the emission chamber. Suppression of accompanying electrons occurs due to their moving along a magnetic field onto the emission electrode serving as the source anode. Negative ions are practically not affected by the influence of magnetic field and at observance of a condition $\lambda_i > d$, they participate in emission of ion beam. Here λ_i – mean free path of a negative ion, d – distance from a place of its formation to the emission aperture.

EXPERIMENTAL RESULTS

Emission Characteristics

The ion source was placed in the vacuum chamber. The vacuum chamber was pumped out up to pressure 10^{-4} Pa by a diffusion pump with pump speed 2500 l/s. Electrodes of the ion source were cooled by water. The discharge current was varied within the limits of 30 -130 A with pulse duration within the limits 0,1 - 1,2 ms, pulses repetition rate - 1-5 Hz. H⁻ ion beam and accompanying electrons were extracted from the emission aperture 3,4 mm in diameter ($S_{em} = 0,1$ cm^2). Extraction voltage is 20 kV. In these experiments the formation of an ion beam was not made. To clean the ion beam from accompanying electrons the permanent magnets creating a cross-sectional magnetic field with magnetic-field strength ~ 0,01 T were established on an extraction electrode.

In a basic design of the ion source the energy distribution of electrons in internal paraxial diffuse plasma was determined only by parameters of peripheral plasma. It was not possible to change their energy and density in the area of extraction.

In modified version of the ion source we tried to create the mechanism permitting to operate slow electrons density near the emission aperture. Such mechanism can be realized by creation of potential difference between an emission electrode and internal paraxial diffuse plasma of a source across magnetic field. To realize this mechanism the design of the gas-discharge chamber of the source was changed as follows:

1. Near the emission aperture within a few mm the radial component of magnetic field Br was built. It was reached by introducing of non-magnetic (graphitic) insert into an emission electrode.
2. At a distance of 0,2 mm from the emission electrode one more electrode (biased anode) was inserted so that it was in the area of a radial component of magnetic field. This electrode is an anode of the emission chamber of a source.

Applying to the emission electrode controlled negative voltage offset concerning an biased anode of the sources it will be possible to accelerate unmagnetized positive ions into the emission area. These ions, in turn, will capture slow electrons into the emission area. Besides the slow electrons in this area under operating of crossed ExH fields will commit a closed drift that will cause the increase of their life time in the emission area. If in same area of plasma there will be a sufficient quantity of an vibrationally excited molecules of hydrogen, it is possible to expect increase of density of H⁻ ions generating and accordingly their emission density.

Figure 2 shows the dependence of negative ion beam current as a function of discharge current at voltage between emission electrode and biased anode equal 10V. From the graph it is visible, that at a discharge current 130 A the emission density of H⁻ ions comes to 440 mA/cm^2. Thus the modernization of a discharge chamber has allowed doubling the emission density of H⁻ ions current in comparison with basic design of a source.

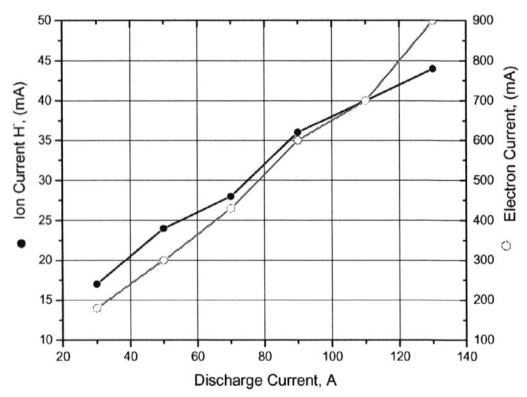

FIGURE 2. Dependence of H- ion current and accompanying electrons current on discharge current.

Dependence of Ie/Ii ratio as a function of discharge current is represented in Figure 3. The optimization of this ratio was not conducted. The voltage extraction was 20 kV.

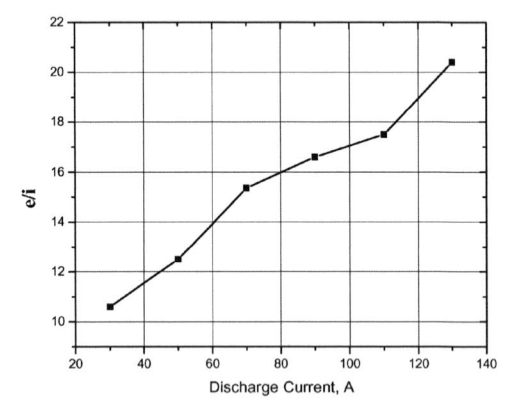

FIGURE 3. Dependence of I_e/I_i ratio as a function of discharge current.

Typical oscillograms of discharge current and H⁻ ions current at several values of a discharge current are represented in Figure 4. Pulse duration of discharge current - 0,4 ms.

We suppose, that is possible to reach further increase of an emission density of H⁻ ions and decrease of value I_e/I_i by further optimization of topology of a magnetic field and geometry of anode of ion source in the emission area, extraction voltage and performance parameters of a source.

Measurements of ion beam emittance were conducted in earlier version of H⁻ source with slot-hole geometry. With the noiseless discharge for the current being 50 mA, the normalized emittance is 0,26π mm mrad across the emission slit and 0,64π mm mrad along it [4]. For this version of a source the emittance of a beam was not measured.

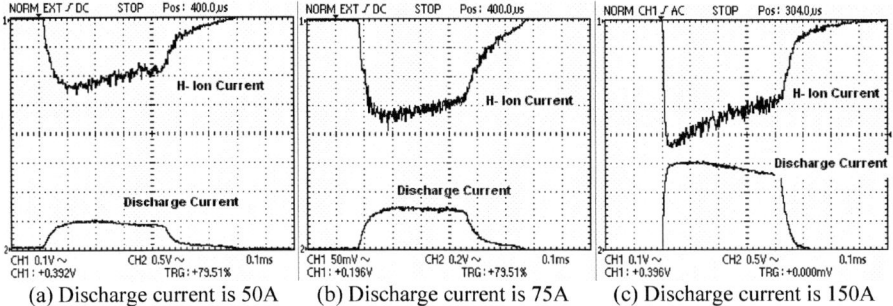

(a) Discharge current is 50A   (b) Discharge current is 75A   (c) Discharge current is 150A
FIGURE 4. Oscillograms of discharge current and H⁻ ions current.

Gas flow

The gas flow was measured by a flow meter with differential pressure gage. At duration of a gas pulse 1,2 ms in the upgraded version of the ion source the hydrogen flow has compounded 0,020 cm^3/pulse. Usage of the new gas valve in a source has allowed reducing total gas flow. As a result the vacuum conditions in the field of primary acceleration of ion beam were improved, and consequently in this area the loss of H⁻ ions has decreased.

CONCLUSION

In plasma volume of non-cesium H⁻ ion source the conditions for obtaining of increased density of H⁻ ions in the field of adjoining to the emission aperture were realized. It was made due to increase of flow of slow electrons to plasma adjoining to the emission aperture and their retention in this volume, and also due to decrease of gas pressure in an accelerating interval of a source.

The value of density of H⁻ emission current - 440 mA/cm^2 is obtained.

The designed source has high operating characteristics. It has rapid starting. The current of H⁻ ions with nominal parameters is usually reached in 1-2 minutes after achievement of necessary vacuum conditions and supply of high voltage.

ACKNOWLEDGEMENTS

The authors would like to thank M. P. Stockli and I.A. Soloshenko for the support. This work was supported partly by ORNL Subcontract No. 4000044353

REFERENCES

1. Yu. V. Kursanov, P. A. Litvinov, V.A. Baturin, "*H⁻ Source with the Volume-Plasma Formation of Ions*" AIP Conference Proceedings 763, American Institute of Physics, Melville, NY, 2005, pp. 229-234.
2. M. Bacal, G. W. Hamilton, *Phys.Rev. Letters* ,**42**, 1538 (1979).
3. V. A. Baturin, A. Yu. Karpenko, P. A. Litvinov, S. A. Pustovoitov, and I. I. Chemeris, *Instruments and Experimental Techniques,* **47**, 417-422, (2004).
4. P. Litvinov, V. Baturin, *Nuclear Instruments and Methods in Phys. Research B*, **171**, 573-576 (2000).

A 15 mA CW H- Source for Accelerators

Yu. Belchenko, A. Sanin, and A. Ivanov

Budker Institute of Nuclear Physics, Novosibirsk, 630090 Russia

Abstract. A cw surface-plasma type negative ion source has been developed. H- beam with current 15 mA and energy of 32 keV is regularly produced in the runs with duration >10^2 hours. Unattended operation of the source under computer control is realized. The source has a simplified maintenance and an easy access to consumable parts.

INTRODUCTION

Two versions of cw surface-plasma sources are under regular operation at BINP: an experimental one with the beam current 15 mA and energy 25 kV [1], and a source for tandem accelerator with beam current up to 10 mA and energy 25 kV [2]. An advanced source version for the prolonged operation in accelerators with beam current 15 mA and energy 32 kV was designed and put into operation recently. The source design, properties and long-term runs experience are described below.

DESIGN

The ion source geometry is similar to that of the experimental source, reported earlier [1, 3]. The source cross-section along magnetic field lines (in X direction) is shown in Fig.1. It uses a combined discharge with the Penning glow, driven by plasma injection from the hollow cathode arc. The discharge chamber consists of the massive cylindrical anode jacket with the cathode body enclosed. The high-current Penning glow in hydrogen is supported by electron oscillations along the magnetic field between the massive cathode bottom protrusions. In the directions perpendicular to the magnetic field, the Penning discharge volume is limited by the anode bottom cover and by the anode insert (top triangle in exploded view of Fig.1). The distance between the cathode protrusions is 8 mm, discharge area at the cathode, limited by the anode window, is 7 x 10 mm^2, discharge volume is of about 0.5 cm^3.

Penning discharge is driven by the plasma injection from the small hollow cathode units, made in the cathode protrusions. Each hollow cathode unit has the tip with the small aperture, as it is shown in Fig.1. Hydrogen and cesium enter to the Penning discharge region through the plasma of hollow cathode inserts. Cesium is needed to support the hollow cathode arc and to activate the electrodes of the Penning discharge. Discharge with the orificed hollow cathode driver operates under reduced hydrogen pressure ~30 mTor and has a decreased level of discharge and beam fluctuations as compared with that in the pure Penning discharge. The hollow cathode orifice contracts the discharge plasma in the area adjacent to the source emission aperture.

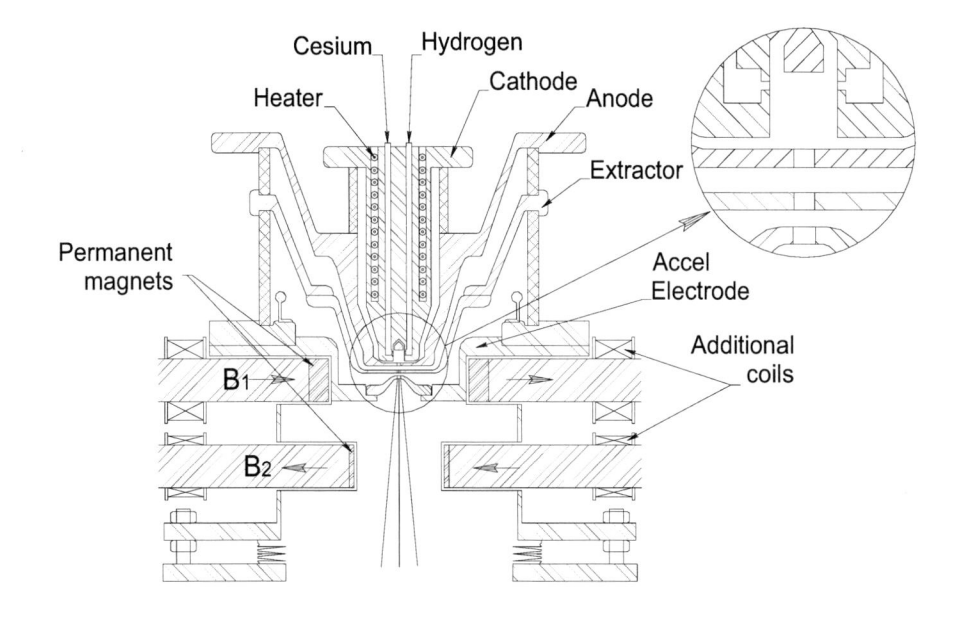

FIGURE 1. Source cross section along magnetic field lines.

Negative ions, produced on the cesiated discharge electrodes via surface-plasma mechanism, are extracted through the emission aperture, drilled at the center of the anode bottom cover. A triode ion-optical system is used for beam extraction to 4-5 keV and post-acceleration up to 32 keV. The anode body is biased negatively, while the conical accelerating electrode is grounded. Extraction and acceleration voltages are connected in-series between the anode and accelerating electrode.

Magnetic field in the discharge and extraction regions is produced by permanent NdFeB magnets. Additional coils are used for the magnetic field enhancement and adjustment. Since H- ions are deflected by the magnetic field in the discharge and extraction region, the second magnet system with the opposite field polarity is installed in the beam drift space in order to direct the beam back to the vertical axis.

Hydrogen feed is controlled by manual leak and by the electromagnetic valve. Cesium seed is supplied by heating of the external oven loaded with Cs pellets (Cs_2CrO_4+Ti). The change of oven heater power provides the cesium feed control. Cesium and hydrogen are delivered through the channels in the cathode body, preliminary heated by the built-in ohmic heater.

Power supply and control systems are assembled in a standard rack. High voltage rectifiers utilize the oil isolation. All power supply subsystems are equipped with the computer modules, which provide the source data adjustment and recording. The fiber optic links are used for data exchange between the source, rack and the main computer. The low capacitive multiple-core HV cable with 22 m length was designed and used for source connection with rack.

PARAMETERS

H- beam with current 15 mA is produced at discharge voltage 75-85 V, discharge current 6-7 A, hydrogen feed 0.15 L·Tor/s, magnetic field of about 950 Gs. An optimal anode temperature for H- beam production is about 250-350°C. Figures 2, 3 show the dependencies of source output data (of discharge voltage, of H- beam and of currents in the extractor and accelerating electrode circuit) vs basic input discharge parameters. As it is shown in Fig.2, H- beam and electrode circuit currents grow proportionally to the discharge current. The total current in the extracted circuit (mainly consisted of H- ions and co-extracted electrons) is about 40 mA for the 15 mA beam and it is about 2 times lower, than that in the experimental source, described earlier [1]. The total current in the post-accelerating circuit is about 25 mA for the 15 mA beam.

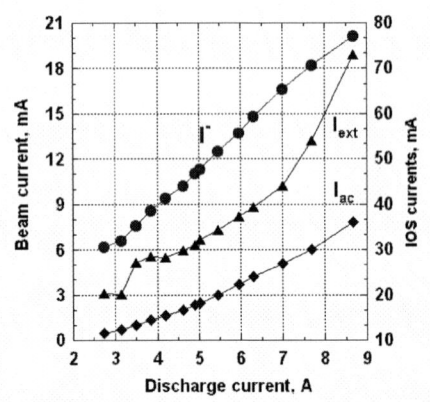

FIGURE 2. H- beam current I^-, and currents in the extractor I_{ext} and accelerating electrode circuit I_{ac} vs discharge current. Magnetic field 950 Gs, Hydrogen feed 0.15 LTor/s. Standard mode.

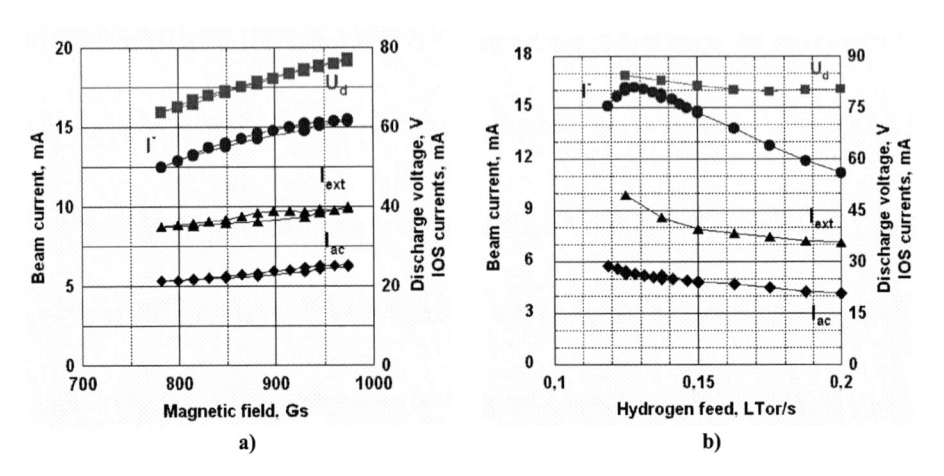

FIGURE 3. a) Discharge voltage U_d, beam current I^-, currents in extractor I_{ext} and in accelerating I_{ac} electrode circuit vs magnetic field. Discharge current 7 A. Hydrogen feed 0.15 LTor/s. Standard mode.
b) The same parameters vs hydrogen feed. Discharge current 7 A. Magnetic field 950 Gs.

Fig.3a illustrates the dependencies of output parameters on magnetic field, controlled in the range 780-980 Gs by an additional electromagnet coils. The magnetic field values for the discharge central area are indicated. Discharge voltage, H- beam and electrode currents are gradually increased with the magnetic field growth within the tested range.

The dependencies of source output parameters on the hydrogen feed are shown in Fig.3b. An optimal hydrogen density provides the maximal H- beam production and transport to the Faraday cup. Further increase of hydrogen flux in the range 0.13-0.2 L·Tor/s decreases the beam and the electrodes circuit currents.

EMITTANCE MEASUREMENT

Direct measurement of H- current density and of beamlets local divergence was carried out by an electric sweep scanner. Circular collimator with an aperture of 0.4 mm is used to decrease the beamlet distortion in scanner drift space. Angular distribution of beamlet current density is recorded by electric sweep across the entrance slits of the Faraday cup. A wide Faraday cup with seven parallel entrance slits is used to provide the registration of beamlets for the whole beam area, including the divergent parts at the beam periphery. Beamlet scanning with the large amplitude excursion, overlapping two neighboring slits, permits to calibrate the beamlet angular shift directly by measuring the distance between the 2 peaks in the recorded scan.

Digitally recorded scans were numerically processed [2]. The signal filtering, interpolation and data integration was done by the special written code. 3D current density distribution over the beam X-X' or Y-Y' space is measured directly and computed with the spline interpolation between the measured points. The contours of equal current density are plotted over the beam phase planes. An internal area of the contours as well as an integral beam current within the contours is calculated.

Compact size of sweeper parts, an open area of beamlet drift space and a suppression of secondary electrons during the beamlet current measurements decrease the input of parasitic and ghost signals, produced by secondary particles. An additional digital subtraction of signal RF pickup was realized. The residual ghost signal at the beam extended periphery does not contribute to the emittance of the beam core measured and computed inside the defined contours of beamlet current density and beam intensity.

Measured beam emittance data for the experimental source are following. Normalized 90% XX' emittance of the 7.8 mA beam is $\varepsilon_{xx'}$ = 0.75 π·mm·mrad, while normalized 1RMS emittance value is $\varepsilon_{xx'1RMS}$ = 0.18 π·mm·mrad. Normalized 90% YY' emittance of 9 mA beam is about $\varepsilon_{yy'}$ = 0.67 π·mm·mrad with the normalized 1RMS emittance value $\varepsilon_{yy'1RMS}$ = 0.15 π·mm·mrad [2].

LONG-TERM RUNS

Long-term runs with duration >10^2 hours were regularly produced with the source manual or computer control. The sequence of source parameters control at the run start is shown in Fig.4. Traces of Cs oven heater power, discharge voltage, extraction and acceleration voltage and of H- beam current are shown. CW hydrogen-cesium Penning discharge is ignited (at 23 min of Fig.4) after the cathode preliminary heating and cesium deposition to discharge chamber. The extraction and acceleration voltage are gradually increased to their nominal values 5 kV and 27 kV (within 25 - 36 minutes interval in Fig. 4). With the further increase in cesium seed (at 45 min in Fig. 4) the discharge goes to the standard mode, while the beam current goes to 17 mA (at 55-60 min in Fig. 4). All the procedures during discharge start and long-term run (switching the magnet coils, hydrogen and cesium feed, discharge voltage, high voltages) are produced by computer according to the specified scenario.

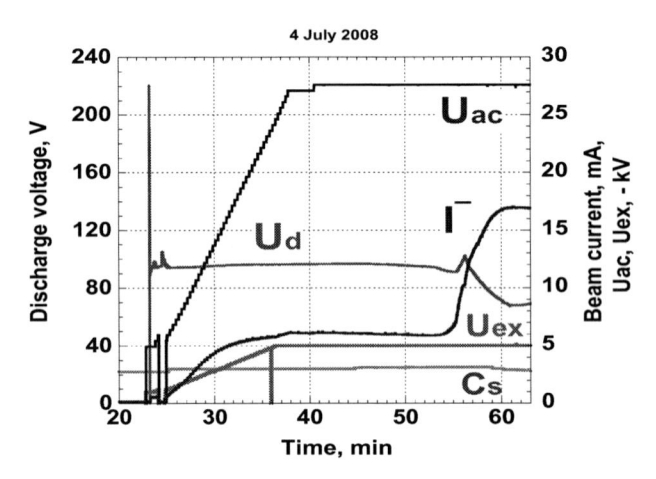

FIGURE 4. Control of source parameters during the source start. U_d- discharge voltage, I^- - beam current, U_{ac}- accelerating voltage, U_{ex} –extraction voltage, Cs- power of cesium oven heater

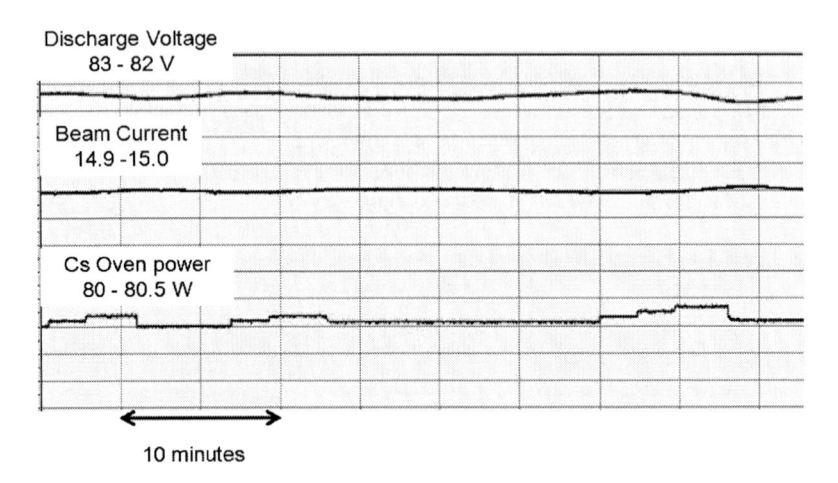

FIGURE 5. Monitoring of discharge and beam production with computer control.

An example of discharge voltage and of beam production monitoring during the long-term run by the computer control is shown in Fig.5. The feedback between the Cs amount and discharge voltage is explored by the specified scenario of the computer code. The power of Cs oven heater is increased (or decreased), if the discharge voltage and its increment become larger (or lower) than the values, prescribed in scenario. As it is shown in Fig 5, discharge with voltage in the range 82 - 83 V is supported, when the power of Cs oven heater is controlled in the range 80 - 80.5 W. H- beam current (central line in Fig.5) varied in the range 14.9 -15 mA for this run.

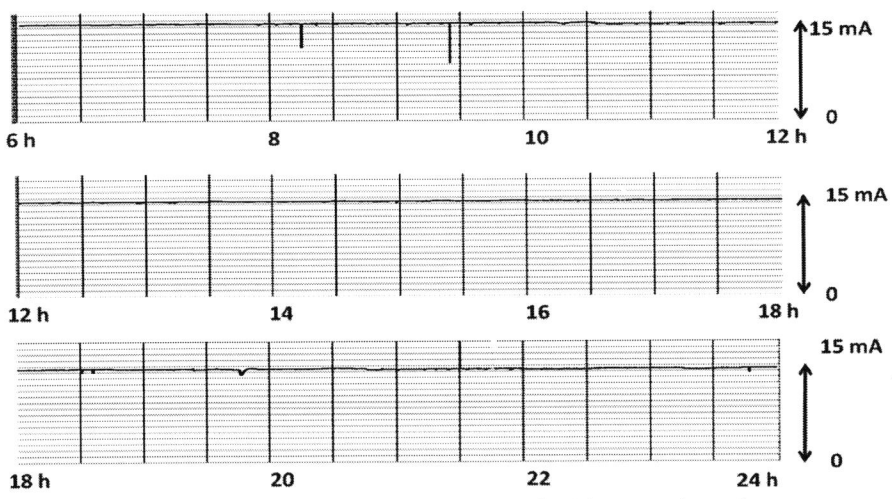

FIGURE 6. Beam current stability during 18 h of 160 hour experimental run.

An example of source stability during the long-term run is shown in Fig. 6. Trace of filtered H- beam current for 18 hours interval of 160-hour experimental run is presented. The computer control of the Cs heater and of the discharge voltage provides the long-term operation with variation of filtered beam current in the range 14.7-15.3 mA. Occasional drops of beam current are caused by breakdowns of accelerator or/and extraction voltage, or by discharge voltage breaks.

LOW CESIUM MODE

Less effective low cesium discharge mode (L mode) is realized with the decrease of cesium feed to the source. Discharge transition into the L mode after cesium feed decrease is illustrated by traces in Fig. 7a. After decrease of cesium oven power the discharge voltage increased, while the beam current decreased from 15 mA to 8 mA level. Comparison of H- production for low cesium and standard mode is presented in Fig. 7b. H- yield in the low cesium mode is about half of that for the standard mode.

The transition to L mode is apparently caused by the discharge rearrangement from the hollow cathode driving mode with the contracted plasma column to a pure Penning mode with the distributed plasma. Discharge voltage fluctuations are about 3 times larger for the L mode as compared with that in the standard H mode.

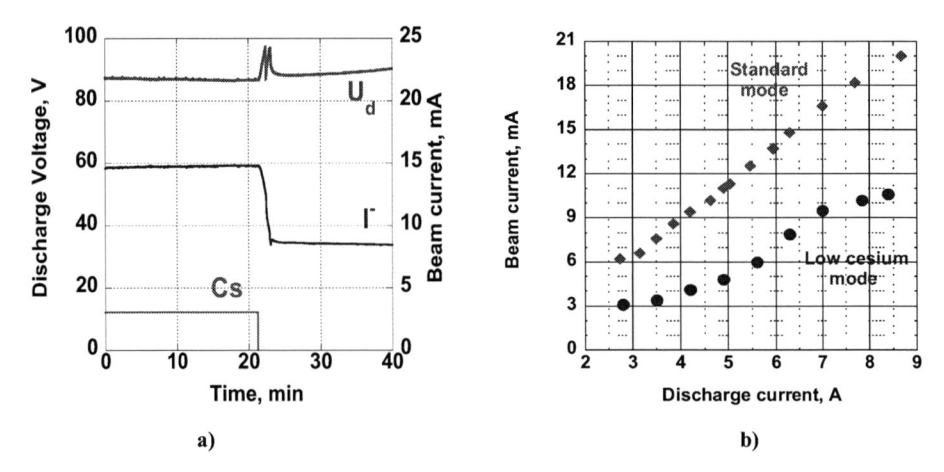

FIGURE 7. a) Discharge transition to L mode and beam drop after cesium oven heating decrease. U_d- discharge voltage, I⁻ - beam current, Cs- cesium oven heating power.
b) H- yield in standard mode (squares) and in L mode (circles).

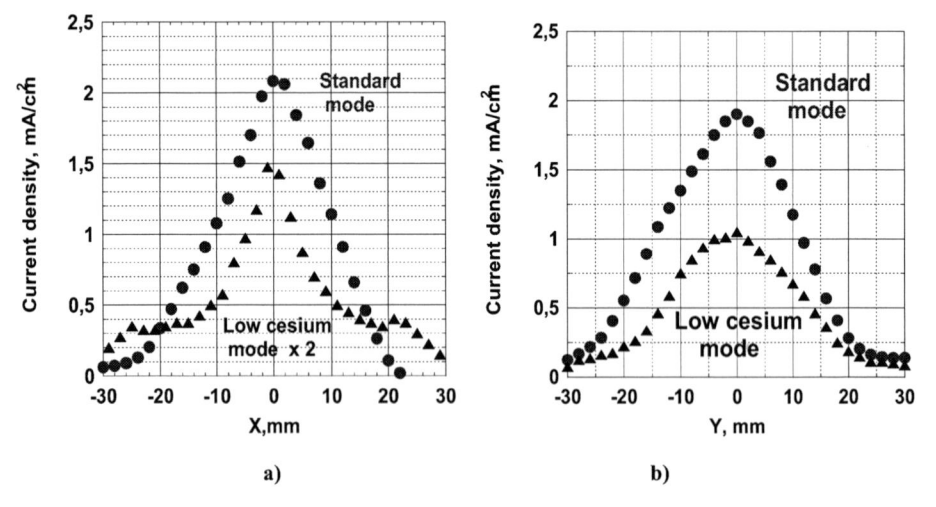

FIGURE 8. Beam cross section in X and Y directions for standard (circles) and for L mode (triangles).

Beam cross sections in X and Y directions were recorded by small movable Faraday cup. The beam cross sections measured at distance 205 mm from the source for standard 15 mA mode and for low cesium 8 mA mode are shown in Fig. 8. The smooth symmetric bell-shaped distributions were recorded. The shoulders are displayed for the distribution in X direction (along magnetic field lines) for the L discharge mode. There are no shoulders in the Y distribution for L mode and for the both X and Y distributions in the standard mode.

These shoulders in X distribution on the beam peripheral parts for the L mode presumably correspond to H- ions, produced on the cathode surfaces of the source [4]. Part of these ions survives during transport through the cw discharge with the decreased plasma and hydrogen density, especially through the low density plasma of the distributed L discharge. The plasma and hydrogen density in the described cw source is several times lower, than that in the high-current pulsed Penning surface-plasma sources [5], where the cathode-produced H- ions group is suppressed due to charge exchange of H- ions in the anode discharge region.

SOURCE MAINTENANCE

The mountable construction of the source promotes a simplified maintenance: an easy access for electrode wiring and cooling, a direct access to heaters and to hydrogen/ cesium feeding systems without source dismount. Detachable discharge and HV ceramic insulators are used. The simplified change of consumable parts is specified.

No essential problems with electrode sputtering, erosion and flakes formation were recorded during and after several $>10^2$ hours runs. Spikes of electrode circuit currents for the occasional breakdowns are limited by electronics and by serial resistors of power supplies. Breakdown currents do not damage the electrodes and do not disturb the source high-voltage stability. "Magnetron" sputtering of the cathode material by cesium ions is decreased due to cesium coverage on the cathode. The wear of replaceable cathode head after 11 months use is shown in Fig.9a. The working planes of cathode protrusions are sputtered, while the hollow cathode apertures are plugged with the sputtered molybdenum.

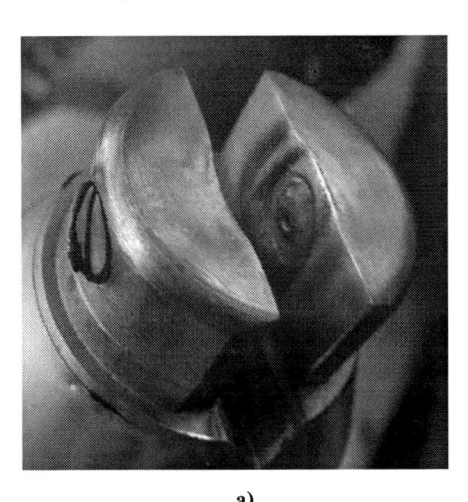

a) b)

FIGURE 9. a) Photo of replaceable cathode head after 11 months of operation.
b) Photo of replaceable anode cover with emission aperture.

The wear of the replaceable anode cover with the emission aperture after 160 hour run is shown in Fig.9b. Solid molybdenum film is deposited onto H- emission zone of the anode cover. No molybdenum flakes formation on the anode is achieved.

An arrow groove with 2-3 mm depth, 1 mm width is melted on the replaceable extractor by co-extracted electrons after ~200 hours work. This groove does not prevent the source nominal duty. A detachable accelerating electrode is gradually sputtered by the peripheral part of the ion beam, but it needs no repairing after ~700 hour integral use. The leak of accelerator voltage is revealed after about 500 hour operation due to cesium migration to the ceramic insulator. The HV resistance of insulator is recovered after the insulator surface washing.

REFERENCES

1. Yu. Belchenko, I. Gusev, A. Khilchenko et al, Rev. Sci. Instrum. **77**, 03A527 (2006).
2. Yu. Belchenko, A. Sanin, I. Gusev et al, Rev. Sci. Instrum. **79,** 02A521 (2008).
3. Yu.Belchenko and V.Savkin, Rev. Sci. Instrum. **75,** 1704-1708 (2004).
4. Yu.Belchenko and A.Kupriyanov, Rev. Sci. Instrum. **64**, 1179-1181 (1994).
5. H.Vernon Smit, P.Allison, and J.Sherman, Rev. Sci. Instrum. **65**, 123-128 (1994).

Ramping Up the SNS Beam Power with the LBNL Baseline H⁻ Source

Martin P. Stockli, B. X. Han, S. N. Murray, D. Newland,
T.R. Pennisi, M. Santana, and R. F. Welton

Spallation Neutron Source, Oak Ridge National Lab Oak Ridge, TN 37831, U.S.A

Abstract. LBNL designed and built the Frontend for the Spallation Neutron Source, including its H⁻ source and Low-Energy Beam Transport (LEBT). This paper discusses the performance of the H⁻ source and LEBT during the commissioning of the accelerator, as well as their performance while ramping up the SNS beam power to 540 kW. Detailed discussions of major shortcomings and their mitigations are presented to illustrate the effort needed to take even a well-designed R&D ion source into operation. With these modifications, at 4% duty factor the LBNL H⁻ source meets the essential requirements that were set at the beginning of the project.

Keywords: H⁻ source, RF ion source, multicusp ion source, Cesium, low-energy beam transport
PACS: 07.77.Ka, 29.25.Ni, 52.80.Pi

INTRODUCTION

Lawrence Berkeley National Laboratory (LBNL) designed and built the Frontend for the Spallation Neutron Source in Berkeley, CA [1]. Briefly, plasma is generated by driving radio frequency (RF) through a 2.5-turn antenna inside a multicusp ion source. Matching networks connect the antenna to a continuous, 600-W, 13.56-MHz supply, typically operating at 250 W, and to a pulsed, 6%-duty-factor, 80-kW, 2-MHz supply, typically operating between 40 and 70 kW. As shown in Fig. 1, a ~300 G filter-field cools the plasma that drifts towards the source outlet. The Cs collar contains less than 30 mg of Cs in Cs_2CrO_4 cartridges [2]. Most of the negative ions form when bouncing from the Cs-collar outlet aperture, which is next to the source outlet. A 1.6 kG dipole field, peaking ~7 mm outside of the source outlet, steers the co-extracted electrons towards one side of the e-dump, which is kept between +2 and +7 kV with respect to the -65 kV source potential. Some of the electrons hit the e-dump, while a significant

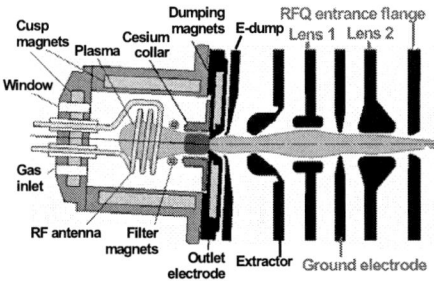

FIGURE 1. Schematic of the LBNL ion source and LEBT.

CP1097, *Negative Ions, Beams and Sources: 1st International Symposium*, edited by E. Surrey and A. Simonin
© 2009 American Institute of Physics 978-0-7354-0630-8/09/$25.00

fraction hit the extractor, where they generate thermal and radiation problems, especially when the extractor is run near its maximum +20 kV potential.

A 12-cm long, two-lens electrostatic low-energy beam transport (LEBT) focuses the beam into the radio frequency quadrupole (RFQ) with the required Twiss parameters. The second lens is divided into four electrically-isolated quadrants, which allow for steering, chopping, and blanking the beam. The compactness of the LEBT prohibits any characterisation of the low-energy H⁻ beam before it is accelerated by the RFQ to 2.5 MeV. Only after the H⁻ beam enters the medium-energy beam transport (MEBT) can the H⁻ beam be measured with a current torroid. The MEBT contains a travelling-wave chopper that improves the rise and fall time of the chopped beam. Clean chopped beam gaps are needed for minimizing the activation of the extraction switchyard in the accumulator ring.

EARLY ION SOURCE AND LEBT PERFORMANCE

Using a function generator to discard the initial beam during the meniscus formation, the Frontend commissioning at LBNL culminated with 85-µs long, 50 mA MEBT beam pulses at low repetition rate [3]. Shortly after, the Frontend was disassembled and shipped to Oak Ridge National Laboratory (ORNL), where it was reassembled. Starting late in 2002, the recommissioning used the new SNS timing system, which starts the ion source and RFQ simultaneously. Although the recommissioning culminated with a few hours of MEBT beam pulses with ~50 mA peak current, the timing system limited this current level to ~20 µs as seen in Fig. 2a. Many commissioning tasks preferred short, but square-shaped pulses that could be tuned at the cost of the peak current, as shown in Fig. 2b.

As listed in Table 1, throughout the commissioning similarly high peak currents were normally produced after installing a newly refurbished source. The highest beam currents, however, typically lasted only a few hours and gradually decayed to more modest currents between 10 and 25 mA, which were sufficient for most commissioning tasks [4].

The Frontend recommissioning was impeded by the availability: in the average week over 24 hours were spent troubleshooting, refurbishing, and repairing the ion source and LEBT. The dominating problem was the LEBT, which kept failing because some of the glued metal-ceramic joints overheated. After the recommissioning was completed, the glued, 20-kV extractor insulators were replaced with threaded ceramic standoffs, and the more complex, glued, 60-kV, lens-2 standoffs were replaced with custom-made ceramic standoffs featuring threaded studs brazed to both ends.

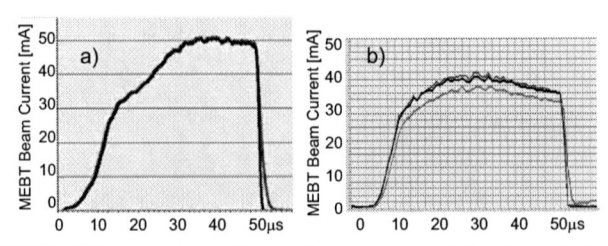

FIGURE 2. H⁻ beam pulses optimized for peak current (a) and constant current (b).

TABLE 1: Pulse length, MEBT beam current requirement and achievements, and the source and LEBT availability for all commissioning phases at ORNL.

Commissioning	Year	Pulse length	mA required	mA in MEBT	% Availability
FE @ORNL	02/03	~.05ms	Varies	51 pk	85.6
DTL-1	2003	~.05ms	Varies	45 pk	92.4
DTL-2/3	2004	~.05ms	Varies	48 pk	97.8
CCL1-3	2004	~.05ms	Varies	46 pk	98.6
SCL	2005	~.05ms	Varies	38 pk	99.8
Ring	05/06	~.05ms	Varies	41 pk	99.7
Target	2006	~.05ms	Varies	43 pk	100

These improvements cut the downtime roughly in half for the commissioning of the first drift tube linear accelerator. At that time the downtime was dominated by failures of the 2 MHz amplifier and its matching network, both of which were improved. Many additional modifications were implemented over time to reach the 100% availability during target commissioning in 2006.

During the first neutron production run the beam pulse length was doubled, and the average, unchopped MEBT beam current normally exceeded the requested unchopped 20 mA, requiring what was believed to be <50 kW of 2 MHz peak power, as shown in Fig 3a. However, when the pulse length was extended to 0.25 ms during the second neutron production run, it became difficult at times to produce the required 20 mA, as shown in Table 2, despite increasing the requested 2 MHz peak power as seen in Fig. 3b. Realizing the need for more RF power, a 24-hour, 6%-duty-factor test was conducted with what was believed to be 80 kW. The 2 MHz amplifier survived this test, but a later inspection found signs of serious overheating. A subsequent calibration of the 2 MHz amplifier revealed an extensive hysteresis in the delivered versus requested 2 MHz power caused by artifacts in the lookup table. The delivered power was up to 130% higher than requested low power levels, while never delivering more than 70 kW at higher power levels. The hysteresis makes it impossible to reconstruct the delivered power from the archived requested power. However, the hysteresis curve and experience suggest that most of the time the 2 MHz amplifier delivered ~70 kW. A new lookup table was generated from a power calibration with a matched load.

THE LEBT CHOPPER ARCING PROBLEM

During the 2006 neutron production runs, the LEBT chopper high voltage (HV) switches [5] experienced multiple failures. After one failure clearly occurred around the start-up of an ion source, the chopper HV switches were disconnected for

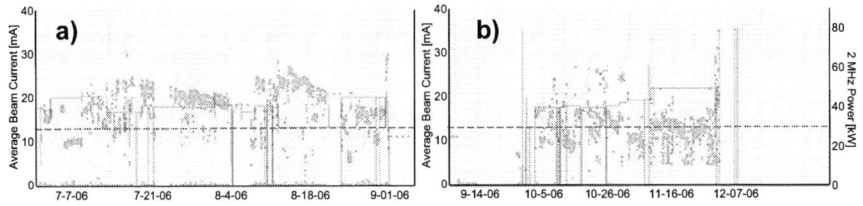

FIGURE 3. Requested 2 MHz peak power (solid line) and average chopped MEBT beam current (dots) compared with the required 13 mA (dashed line) for the 1st(a) and 2nd(b) production runs.

TABLE 2: Duty factor, pulse length, unchopped MEBT beam current requirement and achievements, and the source and LEBT availability for all neutron production runs. The run numbers were adjusted to reflect the calendar year rather than the fiscal year used at SNS.

Production Run	Duty Factor	Pulse length	mA required	mA in MEBT	%Availability
Run 2006-1		~.1 ms	20	20-28	99.9
Run 2006-2	0.2	~.25ms	20	14-30	99.98
Run 2007-1	0.8	~0.4ms	20	10-20	70.6
Run 2007-2	1.8	~0.5ms	20	11-20	97.2
Run 2007-3	3.0	~0.6ms	25	25-30	99.65
Run 2008-1	~4	~0.6ms	25/30	25-37	94.9
Run 2008-2	~4	~0.65ms	32	32-40	

every conditioning as well as for every cesiation of the ion source. Early in the first 2007 neutron production run, which increased the duty factor from 0.2 to 0.8%, LEBT chopper HV switch failures became frequent and caused significant downtime. The HV feedbacks and load current measurements of the 65 kV supply and the two lens HV supplies indicated infrequent arcing. A reliable lens-2 arc detector was implemented by recording all large signals obtained from pickups capacitively coupled to each of the four lens-2 segments. The first three days of recorded arc rates are shown in Fig. 4. It started with 30 arc/h (arcs per hour) generated during a cesiation after which it rapidly subsided to ~1 arc/h. However, the following afternoon the arc rate suddenly increased to 38 arc/h before subsiding to ~10 arc/h. The following morning it rose again to 25 arc/h without an apparent cause. Much later was the 4-pm peak arc rate correlated with increasing the pulse length from 0.25 to 0.4 ms.

The system was shut down to install a video camera, which revealed arcs occurring in many different locations at varying intervals. Subsequent visual inspections revealed discolorations on the insulators, which connect the four segments, but no obvious cause could be identified. Numerous implemented additional arc counters, the implemented residual gas analyzer, nor additional video studies were able to pinpoint the root cause of the arcing. Many mitigating ideas were tried without success, disproving numerous hypotheses. As many as 50 arcs could occur within 15 minutes without apparent cause. Chopper failures were minimized by disconnecting the chopper at the onset of every high frequency arc event, which interrupted neutron production until the arc rate subsided to an acceptable level. These and many other mitigating and investigative efforts accumulated a downtime of 16 days or 30% of the planned neutron production time.

At the same time the ion source test stand started to be used to test numerous LEBT mitigations, but none of them showed a significantly lower arc rate. After six weeks of 24/7 testing, a lens-2, which featured metal clamps in lieu of the inter-segment ceramics, showed a 90% reduced arc rate, bringing the focus to those joints.

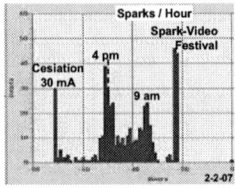

FIGURE 4. Lens-2 arcs per hour recorded early in run 2007-1.

The four lens-2 segments are held together by 4 alumina flats glued to the segments. These used Epon 826 and Versamide 140 mixed at a 1:1 weight ratio, which it was later learned only "resists 200 °F temperature" [6]. A lens-2 assembled with 3000 °F glue [7], baked and conditioned, did not yield a drastic reduction in arc rate.

Another concern was the unshielded inter-segment alumina insulators that were exposed to the high electric fields surrounding lens-2. Adding shields, however, did not drastically reduce the arc rate. Only when the shields were combined with the 3000°F glue [7] did the arc rate show drastic reductions similar to the all-metal lens-2.

These and other findings suggest the following issues with the original lens-2: The lens-2 fields accelerated low energy ions, some of which bombarded the unshielded inter-segment insulators, which charged up and occasionally discharged. Uncontrolled beam losses as well as secondary ions from the beam and corona heat up the electrostatic LEBT, which is found hot when probed after removing the ion source. Increasing the duty factor increases the LEBT temperatures, which in turn increases the outgassing of the low temperature glue, causing occasional discharges. Because the heating is uncontrolled, the lens-2 temperature is likely to slowly rise and fall with the small uncontrolled changes in the losses. When the temperature reached the boiling point of the low-temperature glue, bursting bubbles caused local pressure spikes that discharged the adjacent unshielded insulator, resulting in high frequency arc events. Switching off the beam allowed the lens to cool slightly before beam production was resumed, restarting the somewhat random clock.

As final mitigation the glue was eliminated by clamping the inter-segment insulators with screws. As shown in Fig. 5, two types of brackets provide the threads needed for clamping. In addition the brackets almost fully shield the insulators and form spark gaps with the screw heads attached to the neighboring segment. This produces 32 inter-segment spark gaps breaking down near 10 kV, compared to the four original spark gaps that would hold voltages in excess of 20 kV. After a major arc to ground, the numerous small spark gaps allow a voltage equilibration between the four segments with many small sparks dividing up the stored HV energy, rather than the few, much larger secondary sparks triggered in the original >20-kV spark gaps.

This lens-2 modification was tested in the test stand and then implemented for run 2007-2. Despite raising the duty factor to 1.8 %, the arc rate normally stayed below 1 arc/h. There was not enough time to replace the low-temperature glued lens-1 standoffs, which were mounted on lens-2. These caused a few high frequency arc events, which did not damage the chopper, likely due to the reduced transients.

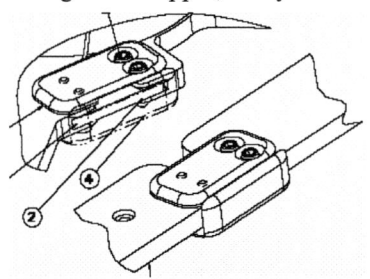

FIGURE 5. The new lens-2 inter-segment joints shield the ceramic plates, one of which (2) can be seen because one shield (4) was removed. The shields form spark gaps with the screw heads attached to the other segment.

Under pressure to improve the system, the LEBT was redesigned before the arcing was understood. The redesigned LEBT is mounted from the water-cooled RFQ flange rather than from the ion source flange, which keeps it cooler and eliminates the electrically-stressed 65 kV LEBT insulators. Most gaps were increased to reduce the electric fields. The redesigned LEBT was implemented for run 2007-3 without showing a significant decrease in arc rate. Surprisingly, the 65 kV arc rate appeared to increase, while chopper HV switch failures remained rare.

LEARNING EFFECTIVE CESIATIONS

Cs boosts the H⁻ current of long (>>100 μs) beam pulses, at least for high repetition rates (>>10 Hz). Accordingly SNS needs Cs to produce the high H⁻ beam currents for the beam power ramp up. To mitigate the risk of Cs-induced arcing in the ultra-compact LEBT and in the nearby RFQ, LBNL implemented Cs cartridges, which contain Cs_2CrO_4 and St101, a getter made of 16% Al and 84% Zr [8]. When heated to temperatures in excess of 550°C, the Cs_2CrO_4 reacts with the activated getter and releases Cs while forming Cr_2O_3, Al_2O_3, and ZrO_2 [9]. The eight cartridges loaded into the baseline LBNL Cs collar together contain less than 30 mg Cs [2], much less than the many grams of Cs typically used in external Cs reservoirs.

The standard cartridges shown in Fig 6b are heated with an electric current and therefore feature electric terminals at both ends [9]. The LBNL Cs collar features slots that tightly fit the Cs cartridge cross-section shown in Fig. 6a. Accordingly, it requires cartridges without terminals, as shown in Fig. 6c. There the Cs escapes through both ends rather than through the wire-covered slot along the top side of the cartridges.

The LBNL Cs collar is cooled or heated with cold or heated compressed air up to ~400°C. Shutting off the compressed air and adjusting the 2 MHz duty factor allows for controlling the higher temperatures required for the release of Cs. Naturally no manual existed for this novel method of heating the cartridges. After initially exploring a wide range of parameters, the newly refurbished sources were conditioned mostly with low duty-factor plasma before heating the collar for 20-30 minutes to 500 or 550°C. A 2004 study believed the residual gas was excessively deactivating the getter [10]. A follow-up paper [11] recommended maintaining the collar temperature below 250°C, ideally as cold as possible before cesiation.

Despite standardizing the conditioning and cesiation procedures, keeping the collar temperature as cold as possible before cesiating at 550°C for 30 minutes yielded inconsistent results. Frequently the cesiation process had to be repeated, and while the initial beam current could be quite high, it would decay to more modest levels. The decay happened within days at low duty factor as seen in Fig. 3, and within hours at high duty factors.

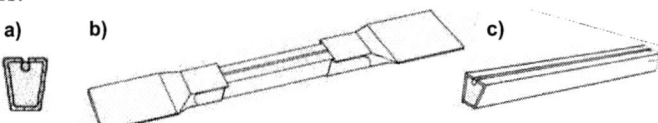

FIGURE 6. Cross section (a) and Cs-cartridges with (b) and without (c) terminals.

After a challenge in fall 2007 to routinely meet the 25 mA requirement, relevant manuals and records were reviewed and compared with new observations of the partial pressures and HV load currents observed during the conditioning and cesiation processes. The new data revealed that the cartridges emit gas when being heated, which is not surprising considering the large surface area of the Cs_2CrO_4 and St101 powder. It suggests that the poorly and inconsistently degassed powder caused the inconsistent release of Cs, rather than an excessive deactivation of the getter by the residual gas [10, 11].

The effort led eventually to the following procedure: Shortly after the initial pump-down, the cartridges are heated to ~80°C, which starts degassing the Cs cartridges. Shortly after, the RF generators are switched on, and the 2 MHz is tuned to 50 kW. Over the next 30 minutes the duty factor is raised to 6%, which gradually raises the collar temperature to ~120°C. Most important are the following two hours of 6%-duty factor, 50kW source conditioning, likely to sputter-clean the Cs collar outlet aperture. During the last 50 minutes of this conditioning period, the temperature of the Cs cartridges is raised to 350°C, which completes the degassing and increases the getter action of St 101 [8]. The subsequent 30-minute period at 550°C starts activating the getter by dissolving its surface oxide film, which enables the St 101 to readily react with the Cs_2CrO_4. This releases enough Cs to yield close to optimal beam performance, normally with a close-to-perfect persistence.

When a cesiation does not yield optimal performance, or the performance drops, the system is normally recesiated by heating the Cs collar to 500°C for 2-3 minutes. The lower temperature and shorter duration are likely sufficient because the getter is already activated. Up to 14 successful recesiations have been demonstrated, many more than the 2-3 found with the former procedure [10]. Normally, the first recesiation increases the beam current by a few % at best, after which the optimal performance is reached. This optimal performance cannot be further increased by adding more Cs. Normally the source and LEBT are tuned for a higher beam current than required before lowering the RF power until the required beam current is achieved. This allows the operators to adjust the RF power for raising or maintaining the beam current.

Figure 7 shows a not-untypical 2-week run: During the first day the 2 MHz (solid curve, out of calibration) is turned to zero while a newly refurbished source is installed. The initial cesiation and recesiation yield only about 22 mA, far below expectations. Next day, after fixing a faulty connection for a lens-2 segment, the unchopped MEBT peak current was ~40 mA peak current or ~37 mA when averaged over the 640 µs long, unchopped beam pulses. Over the next two days, the beam

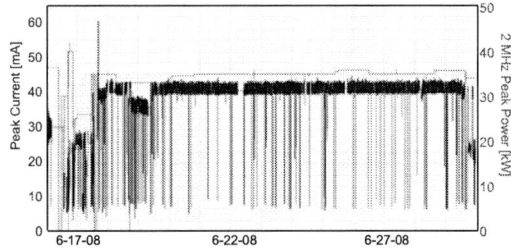

FIGURE 7. 2 weeks of MEBT Peak Beam Current (fluctuating) and RF power.

current was lowered for beam studies before tuning up the 480 kW beam for neutron production, which started on day 4. Over the 10-day production period, operators adjusted the RF in an effort to maintain the desired beam power. However, no significant changes are seen in the peak current because the RF-power is near its optimum value. The current is reduced during the last day for beam studies, before the source is replaced.

As seen in Fig. 7, the loss in beam current over a 2-week period is normally too small to be accurately quantified. Once the optimum performance level is reached, the current appears to remain constant. No drastic rise and decline of the beam current are observed when the last monolayer of Cs is gradually sputtered away, as predicted by calculations [12].

This leaves only two explanations: a) There are many monolayers of Cs, which sputter too slowly to reach a fractional monolayer or 2) on a clean surface the Cs atoms that bond to three neighboring surface atoms adhere well and resist sputtering, whereas all other Cs atoms are rapidly sputtered away, forming quickly a very stable, fractional mono-layer.

Cs deposited on insufficiently sputter-cleaned surfaces appears to have a dwell time in the range of hours with a 50 kW plasma at a few % duty factor, which explains the poor persistence of the beam current before the 50 kW conditioning at 6% duty factor was increased to 2 hours.

20 mA, THE BEST PERFORMANCE OF THE LBNL SOURCE

Initially the ion source was started together with the RFQ, accelerating only the first-50-μs slice of the entire beam pulse, to limit the heat load deposited on the MEBT beam stop. When the pulse length was extended using higher-power beam dumps, the rest of the beam pulse showed a much lower current, even when tuning the source and LEBT for maximum beam current at the end of the first-50-μs pulse, as seen in Fig. 8. Increasing the average beam current required retuning of the source; especially the RF match, the LEBT and the entire accelerator. This problem was solved on October 31, 2007 by implementing a delay that allows for the RFQ to accelerate a beam slice from any 50 μs window out of the entire pulse.

In addition, Fig. 8 shows an important feature of the LBNL source: while it can produce high beam currents for short pulses, the beam current is always significantly lower for pulses that are significantly longer than 100 μs. To get the highest beam current for short beam pulses the RF needs to be tuned to give a rapid discharge ignition, which requires high antenna currents in the absence of plasma.

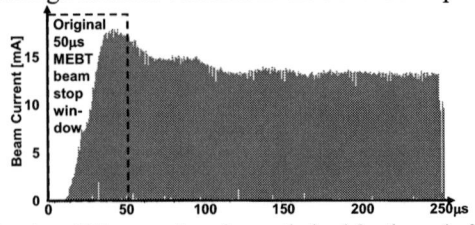

FIGURE 8. 250 μs long H⁻ beam pulse when optimized for the end of a 50 μs pulse.

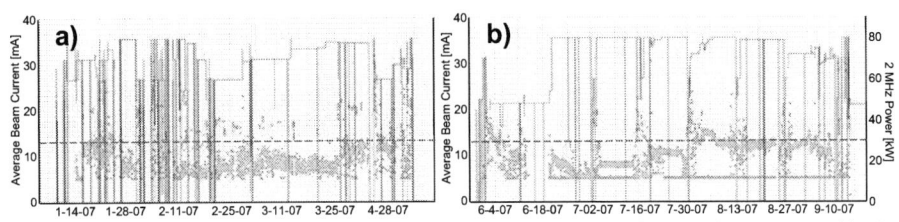

FIGURE 9. Requested 2 MHz power (solid line) and average chopped MEBT beam current (dots) compared with the required 13 mA (dashed line) for the 1st(a) and 2nd(b) 2007 neutron production runs.

To get the highest beam current for long pulses the RF needs to be tuned to the highest antenna current with plasma, requiring about a 20% change in the tuning capacitor. This in turn reduces the antenna current at the beginning of the pulse, which can cause a slow beam rise time, large fluctuations of the beam rise time, or even missing pulses. As countermeasures one can compromise the tune, or increase the 13 MHz power, the 2 MHz power and/or the Hydrogen flow in various combinations. Increasing the Hydrogen pressure lowers the beam current, especially with the original LBNL source, where the maximum beam current was obtained with the lowest pressure that would still ignite the discharge. Most of this learning curve is seen in Fig. 9b, showing the struggle to reach the required 20 mA (=13 mA chopped) MEBT beam current. Run 2007-1 in Fig. 9a focused on the LEBT problems rather than the 20 mA requirement. The 20 mA requirement was met by the end of run 2007-2 with all possible adjustments optimized. These 20 mA are the highest persistent MEBT beam current that the LBNL source ever produced for 0.5 ms at 30 Hz, rather than the "higher than 50 mA at 60 Hz and 1 ms with a normalized emittance of <0.2 π·mm·mrad" quoted in a major ion source book [13].

MODIFICATIONS REDUCING PERFORMANCE VARIATIONS

Source-to-source performance variations are an issue that has been addressed since starting up in 2002. Despite eliminating variations in the source materials and minimizing variations in mechanical dimensions and magnetic fields, significant source performance variation remained. Standardizing conditioning and cesiation procedures yielded no drastic reduction in the variations.

After identifying starkly different responses to standard cesiations, it was discovered that the Cs cartridges were 1.8 mm shorter than the slots in the Cs collar. With most of the Cs escaping through the ends, this allowed the greater part of the Cs to be delivered either near the Cs collar outlet aperture or at the other end, depending on the actual location of each cartridge. As final mitigation, springs push the cartridges away from the outlet, which creates large openings near the Cs collar outlet aperture. This lets most of the Cs escape near the outlet aperture and minimizes the Cs being lost into the plasma chamber.

Another problem was the Cs collar's radial legs, which were attached to the source outlet flange with tight-fitting screws to keep the collar on axis. When being heated to 200°C or 500°C, the thermal expansion would cause the legs to buckle and move the collar axially between 1.5 and 2.4 mm. Depending on the initial, slight, unintentional

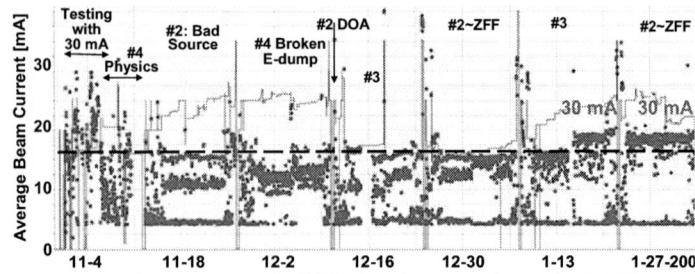

FIGURE 10. Requested 2 MHz power (solid line) and average chopped MEBT beam current (dots) compared with the required 13 mA (dashed line) for the neutron production run 2007-3.

bends, the Cs collar would move away from the source outlet, lowering the beam current output, or it would move towards the source outlet, enhancing the beam current. As final mitigation, the legs were made springy and moved backwards to consistently yield a forward force. This is the likely cause for source #2 to barely make 25 mA with 50 kW (Fig. 10) in November 2007, and at the end of December make 25 mA with 35 kW.

The original Delrin e-dump insulators featured inadequate standoff distances, which led to frequent breakdowns. Sometimes these breakdowns caused permanent shorts, which significantly lowered the source output, as seen in early December 2007 in Fig. 10. Ceramic insulators featuring increased standoff distances eliminated this problem.

With these and other improvements, source-to-source variations were reduced to <10 %. There is no clear pattern that links the variations to the actual source body (#). A significant fraction of the remaining variations appears to originate from the lack of time and other conditions that constrain the expert tuning when starting up a source.

MODIFICATIONS INCREASING PERFORMANCE TO >40 mA

Unable to deliver more than 20 mA and having to deliver 25 mA at the start of run 2007-3 [14], a test at the end of run 2007-2 with a 3rd 30-minute 550°C cesiation briefly delivered 32 mA in the MEBT. The subsequent test with the integrated Cs-collar [15] yielded only 28 mA MEBT beam current, after showing 60 mA on the test stand [10].

Shortly after, it was discovered that the gap between the 1 mm thick stainless steel Cs-collar outlet aperture and the source outlet (Fig. 11) was 3.2 mm, much larger than previously reported [15]. A rapidly-designed 4 mm thick stainless steel outlet aperture on the test stand yielded unprecedentedly high currents for the LBNL ion source. Ten configuration tests over 30 days led to a 4 mm thick Cs-collar outlet, made from Mo and tapered at 40°, featuring ceramic balls to maintain alignment and a 0.5 mm gap with respect to the source outlet aperture as shown in Fig 11. When tested on the

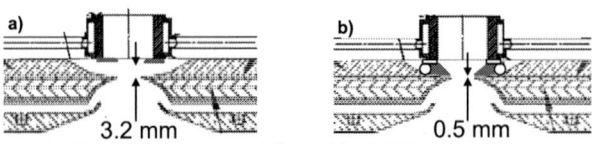

FIGURE 11. Cs collar configuration as delivered by LBNL (a) and the most sucessful (b).

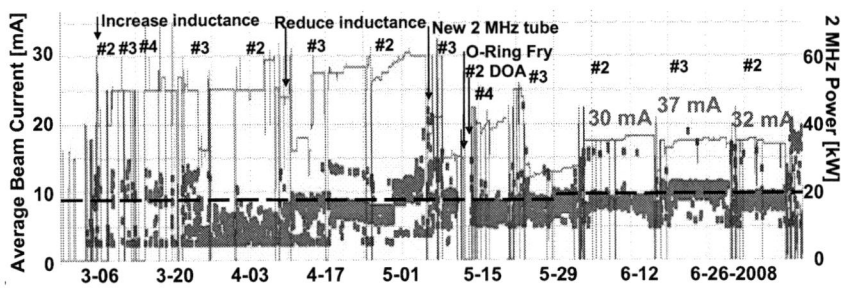

FIGURE 12. Requested 2 MHz power (solid line) and average chopped MEBT beam current (dots) compared with the requirement (dashed line) for the neutron production run 2008-1.

Frontend in the week before run 2007-3, this ball-centered, 40°, Mo collar outlet aperture yielded more promising results than the external Cs reservoir [16] and a modification of the source outlet.

After improving the Cs-collar mount discussed in the previous section, and producing 25 mA MEBT beam current with 35 kW RF, an increase to 50 kW allowed for 30 mA during the last three weeks of run 2007-3, 5 mA more than required, as shown in Fig. 10.

At the end of run 2007-3 the external antenna source was tested on the Frontend [17]. Its high inductance damaged the 2 MHz matching network, which was rebuilt with less stray capacitance. This required the inductance to be increased for matching the modified LBNL H⁻ source at the beginning of run 2008-1. Silent, likely-invisible discharges limited the antenna current which could be used for a stable plasma production, yielding less than 25 mA MEBT beam current as seen in Fig. 12. The subsequent inductance reduction and many other improvements re-enabled the production of 25 mA with 50-60 kW. As last resort, without time for a proper calibration, a new 2 MHz tube was installed, which re-enabled high antenna currents, but without knowing the actual RF power. At the end of May the MEBT beam current was raised to 30 mA, meeting the June 1 requirement. The last two sources produced 37 mA and 32 mA in the MEBT when operated at maximum performance.

Run 2008-2 required 32 mA, which was met by measuring 34, 40, and 38 mA MEBT beam current when using ~65 kW RF, although the RF power was reduced for delivering the 30-32 mA MEBT currents desired for the production runs. The second source was cesiated only once before delivering 40 mA. After delivering 32 mA for 13 days without another cesiation, the source was fine-tuned to deliver 46 mA (Fig. 13), which corresponds to an H⁻ current density in excess of 120 mA/cm² and a Cs consumption of less than 1 mg/day.

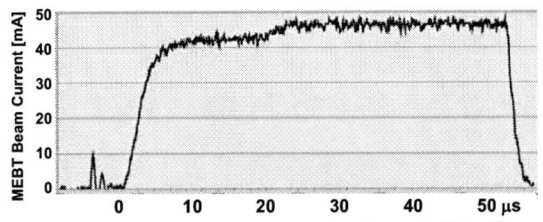

FIGURE 13. 46 mA MEBT beam current are measured at the end of the 50 μs pulse accelerated by the RFQ 0.35 ms after the start of the 0.65 ion source pulses at 60 Hz.

SUMMARY AND OUTLOOK

We are grateful to LBNL for supplying an H⁻ source and LEBT that contributed to the successful commissioning of the SNS accelerator. Numerous LEBT improvements were implemented to achieve an acceptable availability. High-power beams stress electrostatic LEBTs, and at least low-temperature glues lead to unacceptably poor availability numbers. Anticipating more problems at higher duty factors, a prototype 2-solenoid LEBT is being acquired for increased robustness.

The original LBNL H⁻ source and modified LEBT have routinely produced 20 mA MEBT beam currents for 0.6 ms pulses at 30 Hz with an unknown LEBT output emittance. This is below the 38 mA required for 1.23 ms at 60 Hz with a LEBT output normalized rms-emittance of ≤0.2 π·mm·mrad.

The Cs-collar mount needed to be modified to disable the uncontrollable axial movement of the Cs collar. The position of the Cs cartridges needed to be controlled to achieve uniform responses to cesiations. And most importantly, the 1-mm thick, stainless steel Cs-collar outlet aperture had to be replaced with a 4-mm thick, Mo aperture tapered at 40°, which is ~0.5 mm from the source outlet, to produce the 38 mA MEBT beam current required for the 1.4 MW beam power goal.

By September 2008 32 mA MEBT beam currents are routinely exceeded and the learning curve suggests that 38 mA will be routine by August 2009, when required by the ramp-up plan. 46 mA MEBT beam currents have been demonstrated.

Over the last year, procedures have been developed that yield predictable, high-performance results for cesiations and recesiations using the LBNL-introduced Cs_2CrO_4 cartridges. A single set appears capable of supporting ion sources lifetimes in the range of many years despite having <30 mg of Cs.

The only drawback of the modified LBNL H⁻ source is the internal antennas, which show a defect in ~50% of recent source uses after only two weeks. This has caused neither significant performance deficiencies nor downtime since a 2-week source cycle was implemented. It is, however, a potential problem as we gradually increase the duty factor and have recently increased the source usage to a 3-week cycle. Accordingly, we plan on implementing the external antenna source for production run 2009-1 [17]. Its four-times-larger inductance, however, increases the risk of RF failures.

ACKNOWLEDGMENTS

The commissioning of the SNS accelerator and early neutron production would not have been possible without LBNL delivering a first-class Frontend on a very tight schedule and on budget. In addition we are indebted to the SNS Research Accelerator team, which readily provided the needed support. Division Director Stuart Henderson, provided the challenge and the needed support for the recent successes. The proofreading by J. Green and P. Kite was invaluable. Work was performed at Oak Ridge National Laboratory, which is managed by UT-Battelle, LLC, under contract DE-AC05-00OR22725 for the U.S. Department of Energy.

REFERENCES

1. R. Keller et al, "Design, Operational Experiences and Beam Results Obtained with the SNS H⁻ Ion Source and LEBT at Berkeley Lab" in *Production and Neutralization of Negative Ions and Beams*, edited by M. P. Stockli, AIP Conference Proceedings CP639, American Institute of Physics, Melville, NY, 2002, pp.47-60.
2. CS/NF/3.6/11 from SAES Getters S. p. A., Via Gallarate 215, 20151 Milano, Italy.
3. R. Keller et al, "Commissioning of the SNS Front-end Systems at Berkeley Lab" in *Proceedings of EPAC 2002, Paris, France*, 2002, pp. 1025-1027.
4. A. Aleksandrov, "Commissioning of the Spallation Neutron Source Front-end Systems" in *Proceedings of the 2003 Particle Accelerator Conference*, 2003, pp.65-69.
5. Model PVX-4130-Q02-0017A from Directed Energy, Inc., Fort Collins, CO 80526, USA.
6. Patsy Munoz, Cirrus Enterprises, LLC dba E.V. Roberts, private communication 2007.
7. Ceramabond 552 from AREMCO PRODUCTS, INC., Valley Cottage, NY 10989, USA.
8. "St 101 non-evaporable getters" manual, SAES Getters, Via Gallarate 215, 20151 Milano, Italy.
9. "Alkali Metal Dispensers" manual, SAES Getters, Via Gallarate 215, 20151 Milano, Italy.
10. R.F. Welton, M. P. Stockli, S. N. Murray, and R. Keller, "Recent Advances in the Performance and Understanding of the SNS Ion Source" in *Production and Neutralization of Negative Ions and Beams*, edited by J.D. Sherman and Y.I. Belchenko, AIP Conference Proceedings CP763, American Institute of Physics, Melville, NY, 2005, pp.296-314.
11. R. F. Welton M. P. Stockli, S. N. Murray, and R. Keller, "Advances in the Performance of the SNS Ion Source," *Proceedings of 2005 Particle Accelerator Conference*, Knoxville, TN , 2005, pp. 472-474.
12. R. F. Welton, "Analysis of the Cs System for the SNS Ion Source", SNS Note SNS-NOTE-ION-0014, 2002.
13. K.N. Leung "Radio-Frequency Driven Ion Sources" in *The Physics and Technology of Ion Sources, 2nd edition*, edited by I.G. Brown, Wiley-VCH, 2004, pp 163-175.
14. S. Henderson, "Spallation Neutron Source Progress, Challenges, and Upgrade Options" in *Proceedings of EPAC 2008, Genoa, Italy*, 2008, pp. 2892-2896.
15. R. F. Welton and R. Keller, "Design considerations for a new Cs collar/outlet aperture," SNS Note SNS-NOTE-ION-0021, 2002, and R.F. Welton et al, "Enhancing Surface Ionization and Beam Formation in Volume-type H⁻ Ion Sources," in *Proceedings of EPAC 2002, Paris, France*, 2002, pp. 635-637.
16. R. F. Welton et al., "Ion Source Development at the SNS" in *Production and Neutralization of Negative Ions and Beams*, edited by M. P. Stockli, AIP Conference Proceedings CP925, American Institute of Physics, Melville, NY, 2007, pp.87-104.
17. R. F. Welton et al, these proceedings.

The HERA Magnetron: 24 Years Of Experience, A World Record Run And A New Design

Jens Peters

Deutsches Elektronen-Synchrotron DESY, Notkestraße 85, 22607 Hamburg, Germany

Abstract. The HERA magnetron is based on a design from FNAL. This technology was developed at the INP in the 70s, transferred to BNL and from there to FNAL. The DESY magnetron was modified according to the requirements of the HERA collider. During the first years it was the only source connected to the H⁻ LINAC. It was necessary to run the source for long uninterrupted periods with a high reliability and fortunately with a low duty cycle. Many improvements in the design were made, including the development of a vacuum module which could be changed in minutes, the addition of internal cathode heating, and cathode grooving. To study the influence of the cathode temperature a set-up with air temperature stabilization of the cathode was developed. The cesium consumption could be significantly reduced. The source has now been running without cleaning for more than two and a half years after a fine tuning of all parameters.

Keywords: ion source, magnetron, accelerator, negative ion
PACS: 29.25Ni

INTRODUCTION

The magnetron was the main H⁻ source for HERA even after a second source beam line for the RF source was installed in winter 1997/1998 [1].

As a proper start up of the magnetron takes more time than that for the RF source it was permanently operated during the HERA runs. At the beginning it was necessary to have a reliable magnetron for long uninterrupted periods. With a second test magnetron set up new developments were investigated simultaneously before they were installed in the HERA source. Under these conditions it was also possible to do important developments for the RF source.

ADAPTING TO HERA CONDITIONS

The HERA magnetron source is coupled to an RFQ via a 35mm ⌀ LEBT. This has the advantage of being operated on ground potential [2]. This cesium loaded surface plasma source has an exit slit of 1x10 mm² and an 18 kV extraction. The cathode is grooved parallel and opposite to the exit slit for an increase of H⁻ intensity [3]. To protect the RFQ and Alvarez accelerator against Cs contamination a cold box is installed (see Fig.1). For the cold box a bend in the horizontal plane is installed with n= 0.9 to get a round beam. It was possible to change from freon to water cooling for the (0.23 T) cold box bending magnet coil. After the new installation in winter 1997/1998 all sources were outside the radiation interlock zone.

The HERA magnetron has a lower repetition than the FNAL source [4]. At first the magnetron was run at 6.25 Hz for increasing the temperature. The pulse length can also be increased in the non extraction cycles for this purpose. The extraction was done synchronized with the 50 Hz of the synchrotron at 1/4 Hz. For tests 1 Hz was used. For proper operation of the magnetron a half-monolayer of Cs on the cathode is necessary. The build up could take days.

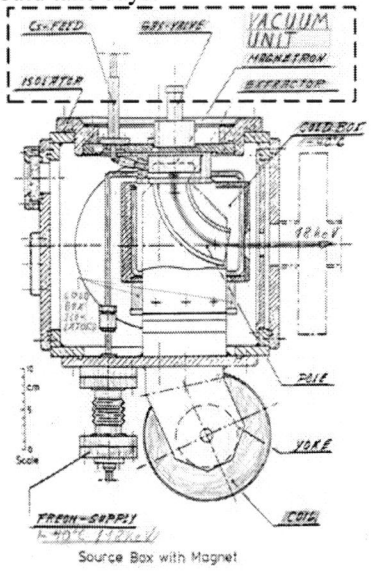

FIGURE. 1 Top view of the H⁻ source.

For details of the operation of the magnetron see [5] and Table 1.

TABLE 1. Data of the HERA magnetron H⁻ source.

beam energy	18 keV	arc voltage	140 V
H⁻ beam current	60 mA	arc current	47 A
emittance		arc pulse width	75 μsec
$\varepsilon_{x\,rms,norm}$ ($\varepsilon_{x\,90\%,norm}$) (35mA beam)	0.28(1.35) π mm mrad	extraction repetition rate	1/4 Hz -1Hz
$\varepsilon_{y\,rms,norm}$ ($\varepsilon_{y\,90\%,norm}$) (35mA beam)	0.25(0.81) π mm mrad	magnetron repetition rate	1/4 Hz / 6.25 Hz
cathode temperature	249 °C	Cs boiler temperature	70 °C
anode temperature	147 °C	Cs consumption 6 Hz magnetron repetition	3mg /day-0.5mg/day

IMPROVEMENTS

During the first years the source did not work reliably. A spare magnetron was built for improving the equipment. A module concept was designed. There were three

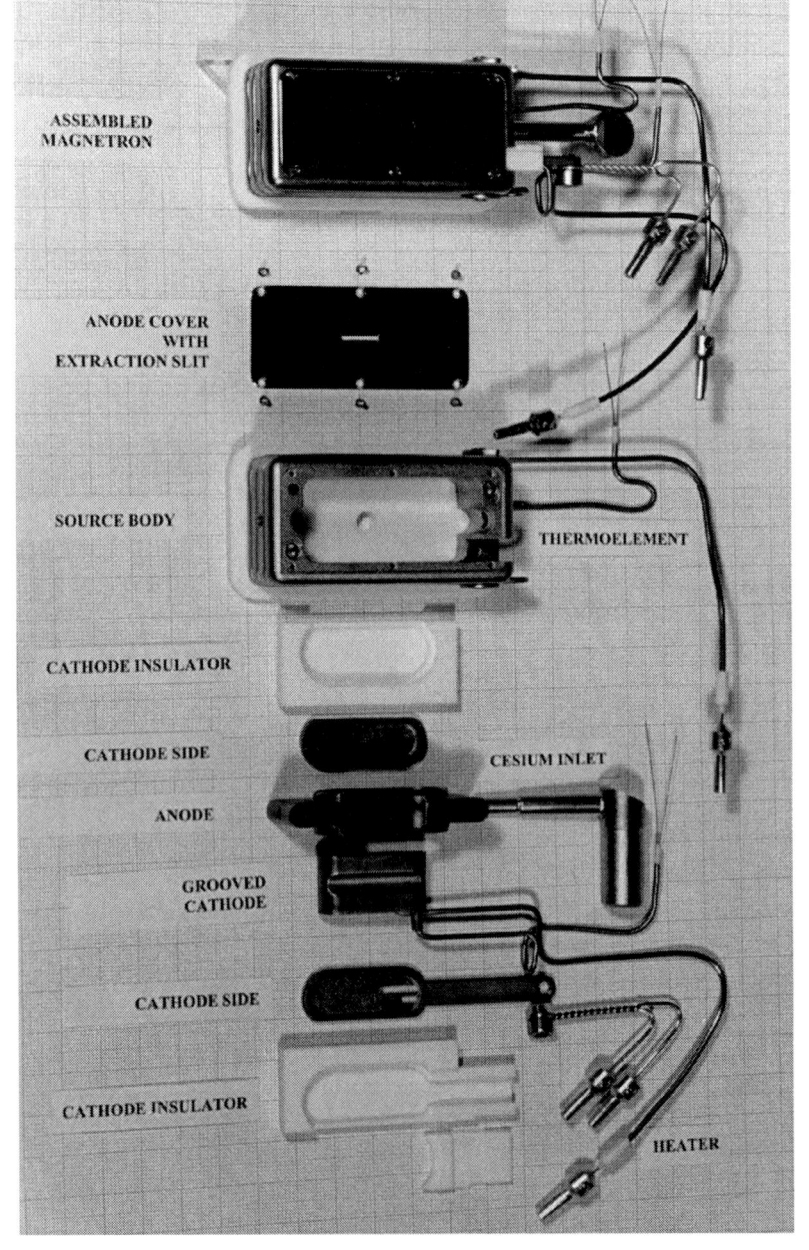

FIGURE. 2 Explosion view of the magnetron.

modules for all important components. As far as possible they were plugged. the vacuum module consisting out of magnetron, gas valve, extractor, Cs feed and isolator (see Fig.1) was important. With its plugged connections it was replaceable in minutes and inspect able. The main operational delay has been achieving the required vacuum and the build - up of the critical Cs layer.

A new magnetron design was crucial in order to avoid short circuits between anode and cathode. The critical areas were detected by routine inspections of the magnetron under dry argon after run periods. The short circuits build by Cs and Mo reduced the run time in the old design to several weeks.

A cathode heating was introduced to reaching the proper operation temperature in order to avoid an increased pulse length for heating. This reduced the start-up time and reduced H_2 and Cs consumption (see [6] and Fig.2)

Freon for cooling the cold box and HV for the extractor were fed through the same point. Due to the sometimes high humidity in Hamburg condensation problems developed. For this reason the freon was fed into two HV plastic insulators which were kept in an isolating plastic box under dry nitrogen (see Fig.3, Right).

A plug-in Piezo H_2 gas valve was developed with a mechanical fine adjust.

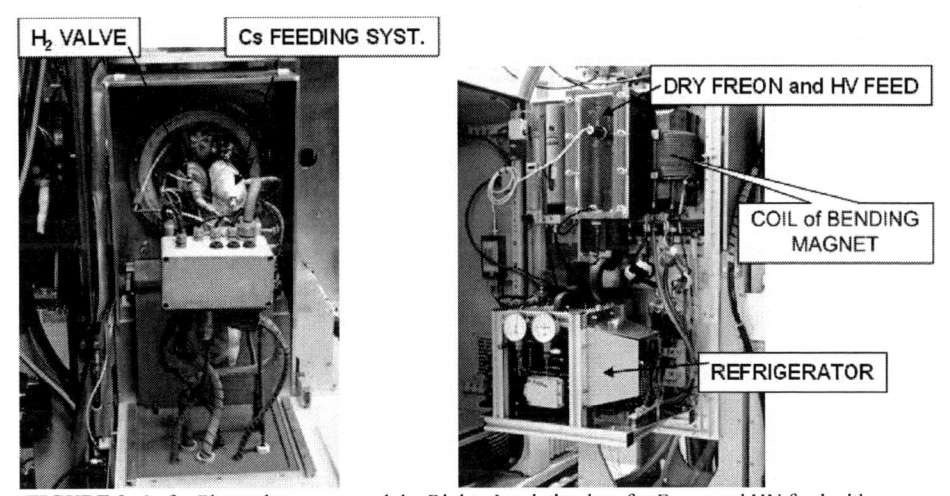

FIGURE 3. Left : Plugged vacuum module. Right : Insulation box for Freon and HV feed with refrigerator and coil for the bending magnet.

IS CESIUM A PROBLEM ?

No Cs was found with a quadrupole spectrometer for masses up to 200 in the RFQ, the Alvarez or in the source box. However there was clearly Cs in the box.

There was no problem with operating the RFQ during the last decade. It is not clear if the problems which emerged during the early years were due to Cs. The first gap of the Alvarez accelerator is very critical concerning multipactoring. It might be that it was spoilt during the beginning when Cs was used very generously (see Fig. 4).

Swipe samples analyzed with x-ray fluorescence and x-ray diffraction showed Cs in all locations of the source box. Mo was found at the pole tip of the bending magnet opposite of the magnetron and in all locations of the magnetron. The covered areas were in some parts clearly shaped. The vapors were probably guided by electrical fields in the box. The Cs/Mo aura is an important limit and risk for the operation of the magnetron. These problems come in addition to the operational delay for the build up of half a mono layer on the cathode and the sensitivity to moisture (see next paragraph).

LIFETIME LIMITS, RISKS AND SOLUTIONS

Measurements at DESY [7] showed that about 0.7µm per day are removed from the Mo surface of the magnetron cathode. This reduces the efficiency of the groove focusing and leads to a decrease of the extracted current. Even worse is that it forms together with Cs layers on the positive anode. These layers are rather stable and stick to the anode wall. But after some time they fall down and can cause short circuits between anode and cathode. This is the most important lifetime limit for the magnetron. There is no final solution for this limit. However one can extend the lifetime by reducing the magnetron current, the Cs supply and building a proper insulation wall between anode and cathode.

A Mo/Cs layer destabilization is often caused by changes of the operating parameters. The layers are hygroscopic. When exposed to moisture by contaminated hydrogen or a vacuum leak the magnetron usually becomes inoperable.

A periodic rise and fall of arc and extractor current was often noticed during the first weeks of magnetron operation. This is probably due to a destabilized layer of Cs/Mo on the anode which becomes positive ionized and moves to the cathode.

Two important risks are Cs vapors on the HV insulators and clogging of the Cs feeding. Insulators can be built out of ceramic with a curvy shape in order to increase the electrical length and to have different exposition angles. When the conductive layer of Cs becomes hot it will burn away. Plastic insulators burned away by these layer currents. At critical places like the cold box it is helpful to introduce a shielding of the insulator with capton foils.

The clogging risk of the Cs feeding can be reduced by heating the feeding system and using less Cs. However in the standard magnetron the inlet is on the anode side. Here the Cs/Mo layers mentioned above can clog the nozzle.

PERFORMANCE: A WORLD RECORD RUN

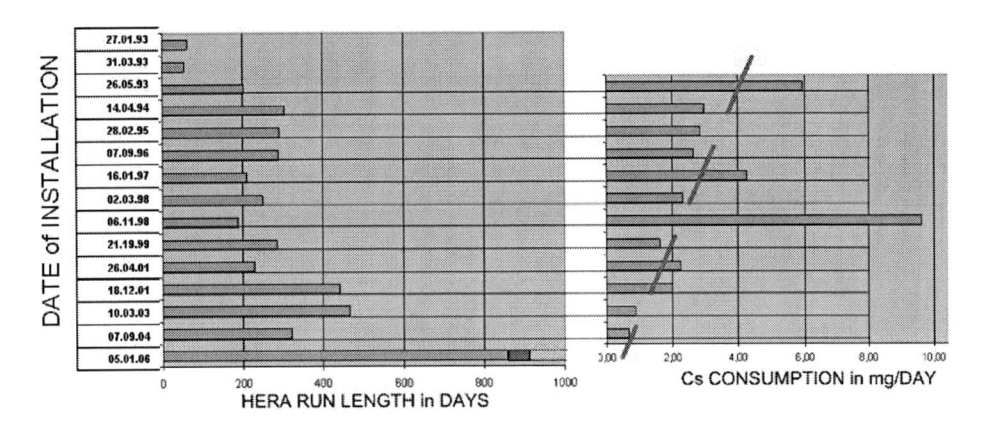

FIGURE 4. HERA run length and Cs consumption per day 1993 - 2008.

Fig. 4 shows the years when the magnetron was in operation for HERA. 1993 with three and 1994 with one termination of operation by the magnetron were bad years. Since those years the maintenance period of HERA determined the magnetron run. No breaking of the source vacuum was necessary. In order to minimize the risk a new vacuum unit was always installed before a new HERA run. There were only a few short interruptions due to electronics failures which are not shown.

The last run for HERA started on 05/01/06 was not terminated with the end of HERA operation. The source ran until September 08, for a total of 32 months. There were three interruptions due to short circuits between anode and cathode which were burned away after several days.

The Cs consumption has been very much reduced from 6 mg per day to less than 0.5 mg per day. This became possible due to a careful tuning of the source.

A NEW IMPROVED MAGNETRON

Not only was the magnetron invented by INP but also a further development the planotron [8, 9]. Unfortunately there has not been much long term experience in either Russia or in other countries with this type of source. A magnetron becomes similar to the planotron when the back side is closed with an insulator. Such a device was tested at DESY. It turned out that a short developed due to the transport of Mo/Cs by the ExB drift. Also the extracted current was reduced as the contribution from the back side is missing. For these reasons further planotron developments were stopped at DESY.

Recent accelerator plans require H⁻ sources with long pulses and high duty cycles. This makes cooling of the cathode and a complete new construction necessary. The Cs inlet should be at the cathode to avoid clogging. The cathode should have a dimple instead of a groove in order to avoid beam on the anode plate. Fig.5 shows such a magnetron which was tested successfully at DESY. The cathode feeding arms are made of steel for better machinability. They are covered by Mo shields. The Cs inlets should be only to the sides of the dimple.

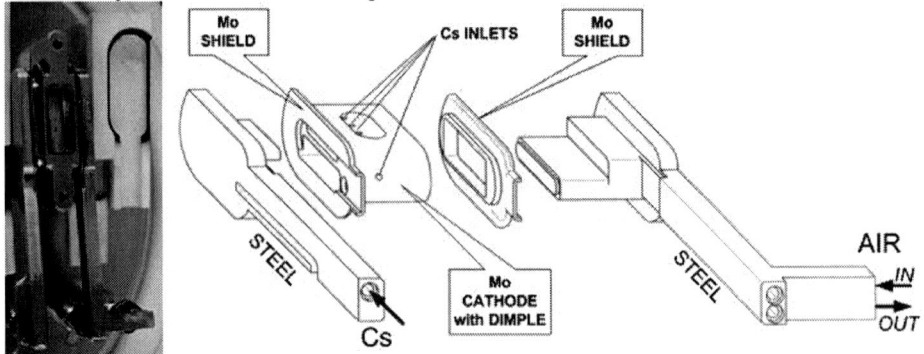

FIGURE 5. Magnetron with cathode Cs injection and cathode air cooling.

Magnetron versus Penning Source

The magnetron has a longer lifetime than the penning source as there is less deterioration of the walls by ion bombardment. The penning source is optimized for charge exchange reactions in order to reduce the emittance. In earlier publications very low emittance values were reported for the penning source [10]. A comparison of recent emittance measurements with $\varepsilon_{rms\ 90\%} = 0.31\ \pi$ mm mrad at 53 mA for the magnetron and recent publications of the ISIS penning source show no advantage of the penning [11].

ACKNOWLEDGMENTS

The author is grateful for the contribution of the following colleagues at DESY: I.Hansen, H.Runz, H.Sahling, and R.Subke. The author also wishes to thank H.Weise and the technical groups at DESY for their support and M.Lomperski of DESY for helpful suggestions to the wording of the article.

REFERENCES

1. C.M. Kleffner et al., Proceedings of the XIX International Linear Accelerator Conference (August 1998).
2. L.Criegee et al., Rev. Sci. Instrum. 62(4), April 1991.
3. R.L. Witkover, AIP Conf. Proc. **111**, 398 (1983).
4. Ch. W. Schmidt et al., IEEE Trans. Nucl. Sci. NS 26, 4120 (1979).
5. J.Peters, Proceedings of the XIX International Linear Accelerator Conference (August 1998)
6. J.Peters, Rev.Sci. Instruments, 65, No. 4, April 1994, pp 1237-1239.
7. J. Peters, Rev.Sci.Instrum.69(2) February 1998
8. Yu.I.Belchenko, G.I.Dimov, V.G.Dudnikov, NUCLEAR FUSION 14 (1974)
9. V.G.Dudnikov, Proc. IV All-Union Conf. on Charged Particle Accelerators, Moscow, 1974,Nauka 1975, Vol 1, p. 323
10. J.D. Sherman et al., Rev.Sci. Instruments, 62 (10)l 1994, pp 1237-1239.
11. D.C.Faircloth et al., Rev.Sci. Instruments, 79, 02B717 (2008).

Commissioning the Front End Test Stand High Performance H⁻ Ion Source at RAL

D. C. Faircloth[a], S. Lawrie[a], A. P. Letchford[a], C. Gabor[a], P. Wise[a], M. Whitehead[a], T. Wood[a], M. Perkins[a], M. Bates[a], P. J. Savage[b], D. A. Lee[b], J. K. Pozimski [a+b]

[a]*STFC, Rutherford Appleton Laboratory, Chilton, Didcot, Oxfordshire OX14 0QX, United Kingdom*
[b]*Department of Physics, Imperial College, London SW7 2AZ, United Kingdom*

Abstract. The RAL Front End Test Stand (FETS) is being constructed to demonstrate a chopped H- beam of up to 60 mA at 3 MeV with 50 p.p.s. and sufficiently high beam quality for future high-power proton accelerators (HPPA). High power proton accelerators with beam powers in the several megawatt range have many applications including drivers for spallation neutron sources, neutrino factories, waste transmuters and tritium production facilities. The aim of the FETS project is to demonstrate that chopped low energy beams of high quality can be produced and is intended to allow generic experiments exploring a variety of operational conditions. This paper details the first stage of construction- the installation and commissioning of the ion source. Initial performance figures are reported.

Keywords: Negative H⁻ Ion Sources, Test Stand
PACS: 29.25.Ni, 52.59.-f

INTRODUCTION

High power proton particle accelerators in the MW range have many applications including drivers for spallation neutron sources, neutrino factories, transmuters (for transmuting long-lived nuclear waste products) and energy amplifiers. In order to contribute to the development of HPPAs, to prepare the way for an ISIS upgrade and to contribute to the UK design effort on neutrino factories, a Front End Test Stand (FETS) is being constructed at the Rutherford Appleton Laboratory (RAL) in the UK. The aim of the FETS is to demonstrate the production of a 60 mA, 2 ms, 50 p.p.s. chopped H⁻ beam at 3 MeV with sufficient beam quality.

FETS consists of a high power ion source, a 3 solenoid magnetic LEBT, a 324 MHz, 3 MeV, 4-vane RFQ, a fast electrostatic chopper and a comprehensive suite of diagnostics.

ION SOURCE OVERVIEW

The basic design of the ISIS H⁻ source has previously been described in detail[1]. The source is of the Penning type[2], comprising a molybdenum anode and cathode between which a 55 A low pressure hydrogen discharge is produced. A transverse magnetic Penning field is applied across the discharge. Hydrogen and Caesium vapour are fed asymmetrically into the discharge via holes in the anode. The anode and cathode are housed in a stainless steel source body.

The beam is extracted through an aperture plate (plasma electrode) using an extraction electrode. On the ISIS operational source the aperture is a 0.6 mm by 10 mm slit and the extraction electrode is of an open ended jaw design, with a jaw spacing of 2.1 mm and a separation from the aperture plate of 2.3 mm. A +17 kV extraction voltage is used operationally. For the FETS high performance source a +25 kV extraction voltage is used, the aperture widened and the extract electrode terminated.

The source is pulsed at 50 Hz, the operational source runs with a 250 µs pulse length. The FETS source is modified to run with a 1.8 ms pulse length by improving the cooling system[3].

After extraction the beam is bent through a 90° sector magnet mounted in a refrigerated coldbox (Figure 1). The sector magnet has two main purposes; to analyze out the electrons extracted with the H- ions, and to allow the coldbox to trap Caesium vapour escaping from the source.

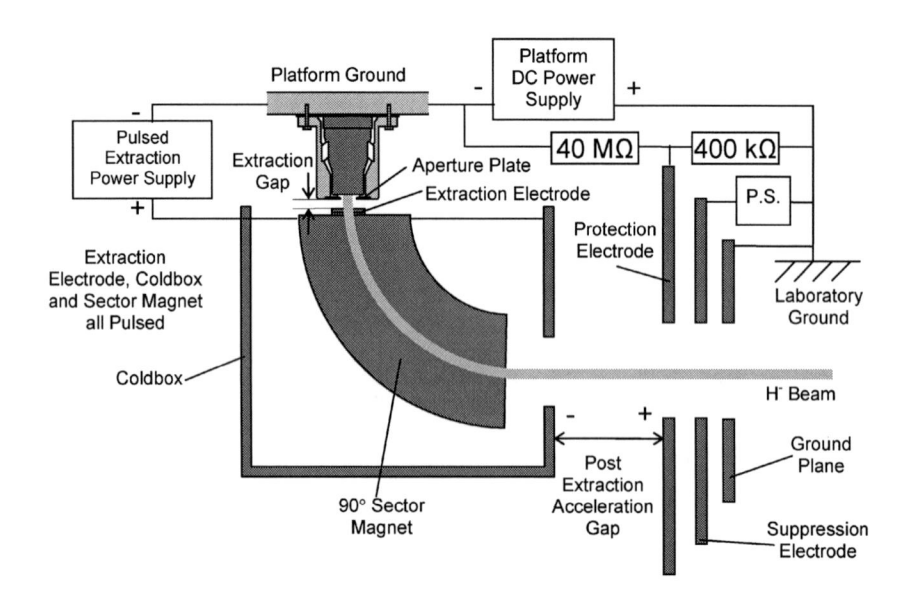

FIGURE 1. Schematic of the FETS ion source extraction and post acceleration system.

The H⁻ beam emerges through a hole in the coldbox and is further accelerated by a post extraction acceleration gap. On the ISIS operational source this is an 18 kV post acceleration voltage giving a total beam energy of 35 keV. For FETS this is a 40 kV voltage giving a total beam energy of 65 keV.

FIGURE 2. The FETS ion source.

SECTOR MAGNET MODIFICATIONS

A detailed study of beam transport[1,4] in the sector magnet has shown the poles used in the ISIS operational source can be modified to improve beam transport. The angle of the sector magnet pole faces can provide weak dipole focusing. The field gradient index, n of the sector magnet can be calculated as follows:

$$n = -\frac{R_e}{B_e}\left(\frac{dB}{dR}\right) \quad (1)$$

where, R_e is the radius the centre of the beam follows as it goes through the sector magnet and B_e is the magnetic flux density at that radius. When n > 1 the beam is defocused vertically (radially), when n < 1 the beam is focused vertically (radially).

Historically the field gradient index in the ISIS sector magnet was always quoted as being n = 1. Recent detailed finite element modeling[4] has shown that the field gradient index of the operational source actually closer to n = 1.4. By integrating the field around different radii the overall field gradient index can be calculated including the fringe field effects at the entrance and exit of the dipole. Figure 3 shows how n varies across the magnet aperture, the top of the plot is the inside radius and the bottom of the plot the outside radius of the sector magnet. Figure 3(a) shows field gradient index variation for the ISIS operational source; n varies from 1.25 to 2.25 across the magnet aperture. This explains why previous studies[1] have shown severe defocusing in the vertical plane. Figure 3 also gives an indication of what proportion of the magnet aperture is good field region. For the sector dipole magnet, the good field region is the area where the value of n is within a certain range. In Figure 3(a) when particles near the top and bottom edges of the good field region they see higher field gradient indexes and are therefore defocused out of the beam.

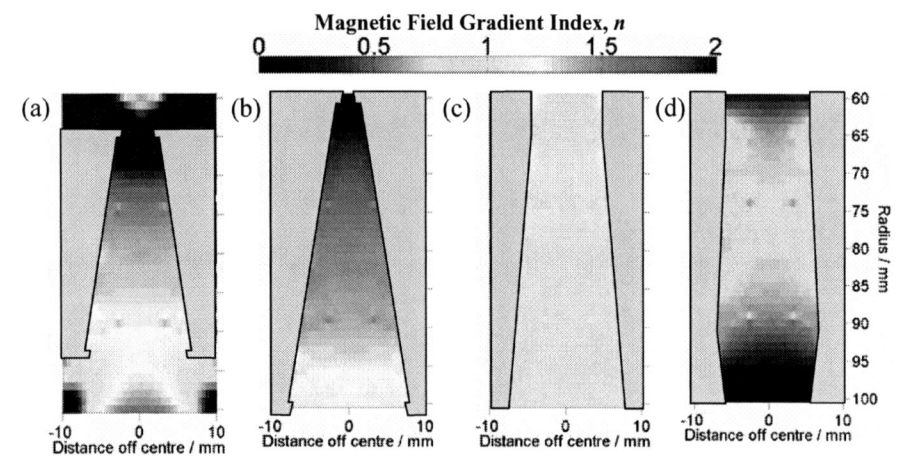

FIGURE 3. Contour plots of n for (a) the standard ISIS pole pieces; (b) ISIS pole pieces with wider radial extent; (c) n = 1 poles; (d) n = 0.75 poles with enlarged shims at the outer radius.

Figure 3(b) shows how n varies across the aperture for wider poles. Increasing the radial width of the poles increases the size of the good field region which means that particles that would have previously been lost are now transported through the dipole.

By changing the angle of the poles the field gradient index can be altered, Figure 4(c) shows the variation in n for an n = 1 average index. To provide focusing in the vertical (radial) direction a set of poles was also designed to give an n = 0.8 average field gradient index.

All three of these designs have been manufactured and tested[4], some results are shown in the final section of this paper.

POST EXTRACTION ACCELERATION

As mentioned in the overview, after being transported though the sector magnet the beam is then further accelerated by a high voltage gap. The focussing properties of this gap have been investigated in previous studies[5] and a field gradient of 9 kVm^{-1} found to provide the smallest emittance growth, some results are shown in the final section of this paper.

Ion Source 70 kV Insulator

To implement the post extraction accelerating voltage the ion source is mounted on a flange which is supported by an insulator. On ISIS this insulator must hold off 35 kV. For FETS it must hold off 70 kV: to do this its length must be doubled. The insulator must support the full 150 kg load of the ion source. The insulator is manufactured out of Noryl-GFN3 (a machinable, fibre-glass reinforced plastic). This material is used because of its high strength. Unfortunately Noryl-GFN3 is only manufactured in slabs of a certain thickness and so the longer insulator had to be constructed from two adhered sections.

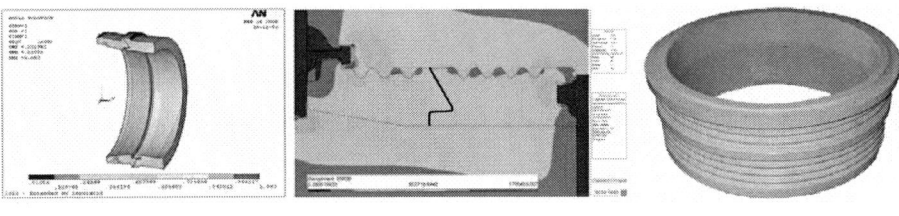

(a) Mechanical Model   (b) Electrostatic Model   (c) Completed Insulator

FIGURE 4: The 70 kV insulator.

Mechanical and electrical finite element modelling is used to confirm the design of the new 70 kV insulator.

Post Extraction Acceleration Electrode Assembly

The 70 kV insulator is mounted on the post extraction acceleration electrode assembly shown in Figure 5. This assembly has several purposes: It is a spacer to separate the insulator from the ion source vessel discussed in the next section. It supports the electrode to protect the extraction power supply from main platform supply in the event of a flashover (circuit is shown in Figure 1).

It supports the suppression electrode to prevent ions travelling back across the post acceleration gap. It supports the beam current toriod and ground electrode. It also employs mu-metal sheets to magnetically shield the diagnostics from the sector magnet's stray field. The post acceleration gap is variable between 4 and 10 mm.

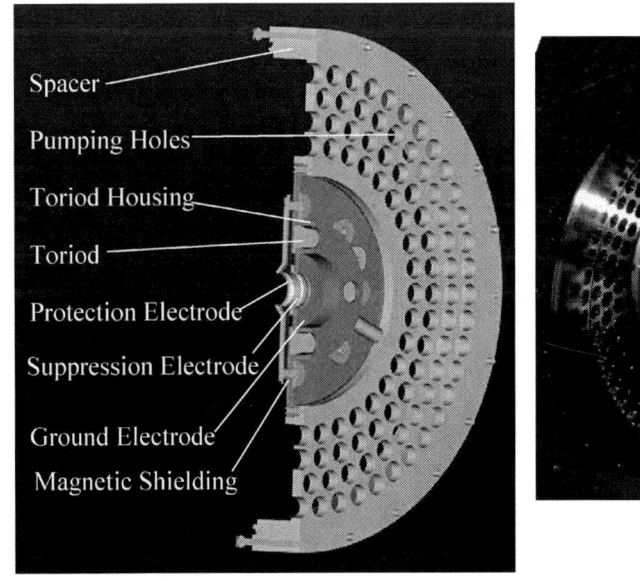

FIGURE 5: The post extraction acceleration electrode assembly.

Ion Source Vessel

The post extraction acceleration electrode assembly mounts on the ion source vessel shown in Figures 6 and 7.

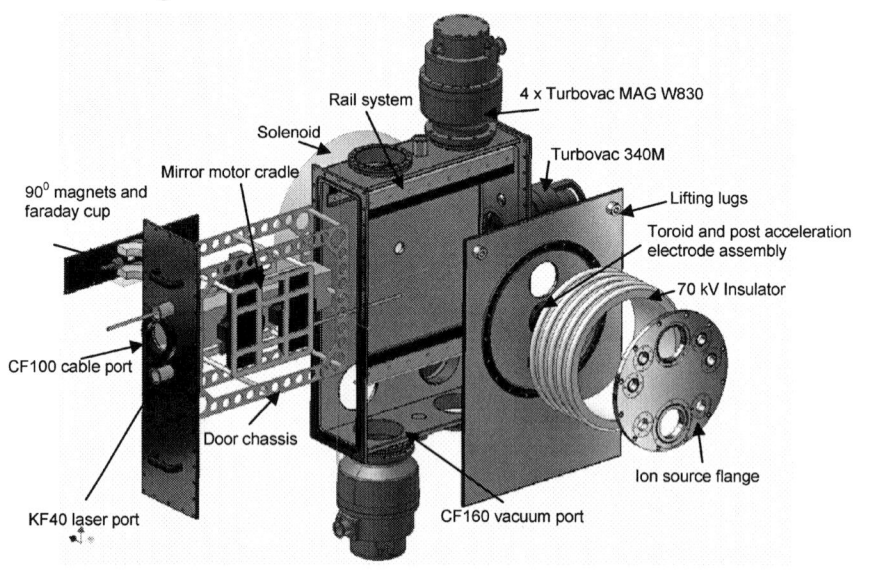

FIGURE 6: Exploded view of the beam profile measurement system inside the ion source vessel.

The vessel supports 4 turbo molecular pumps and pressure measurement heads. Several ports are included for electrical feedthroughs for connection to the post acceleration electrode assembly.

FIGURE 7: Ion source vessel in situ.

Diagnostics

The ion source vessel contains a laser wire beam profile measurement system[6], based on the photo-detachment of the outer electron of the H⁻ ions with a laser. The detached electrons are detected in the faraday cup arrangement shown in Figure 8. The laser wire system allows the transverse beam density distribution to be determined at full beam power without affecting the beam. This is achieved by stepping the laser beam through the ion beam at a variety of different angles to collect many different projections and then combining these using either the Algebraic Reconstruction Technique[6] or the Maximum Entropy algorithm[7].

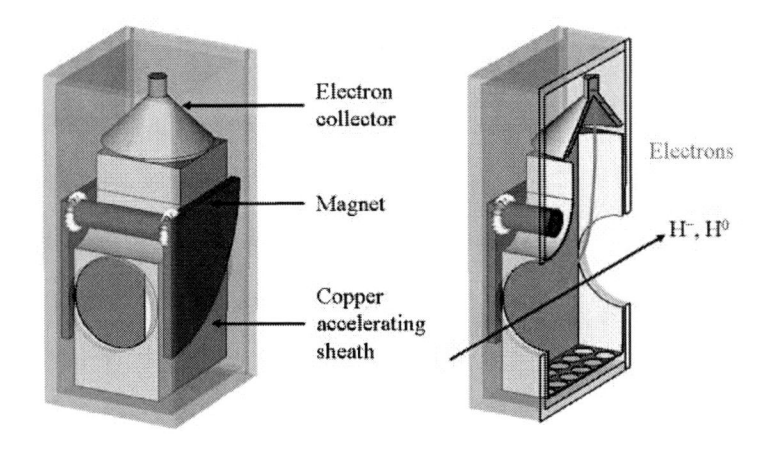

FIGURE 8: The electron collection system for the laser wire profile measurement system.

A temporary diagnostics vessel containing a pair of X and Y slit-slit emittance scanners and a scintillator profile measurement system is installed after the ion source vessel to allow commissioning of the laser wire system. The temporary diagnostic chamber will be moved along the beam line as it is constructed.

ANCILLARY EQUIPMENT

High Voltage Platform and Cage

A 70 kV a high voltage platform is required to support the ancillary equipment required to operate the ion source. This must be surrounded by an interlocked high voltage cage for personnel protection (Figure 9). The platform is supported using commercially available post insulators and has a handrail to allow safe working.

FIGURE 9: The high voltage cage and platform and the 70 kV DC power supply.

An in-house built 70 kV DC power supply is used to energise the platform and a 1 µF capacitor is used to minimise droop during the beam pulse. Platform voltage is monitored using a voltage divider and an automatic dumping system is used to earth the platform.

Two oil-filled 70 kV isolating transformers are used to provide single and three phase power to the ancillary equipment on the platform.

Other Equipment

The ion source requires numerous ancillary equipment to operate. This is all mounted in four racks on the high voltage platform. The layout of these racks is shown in Figure 10.

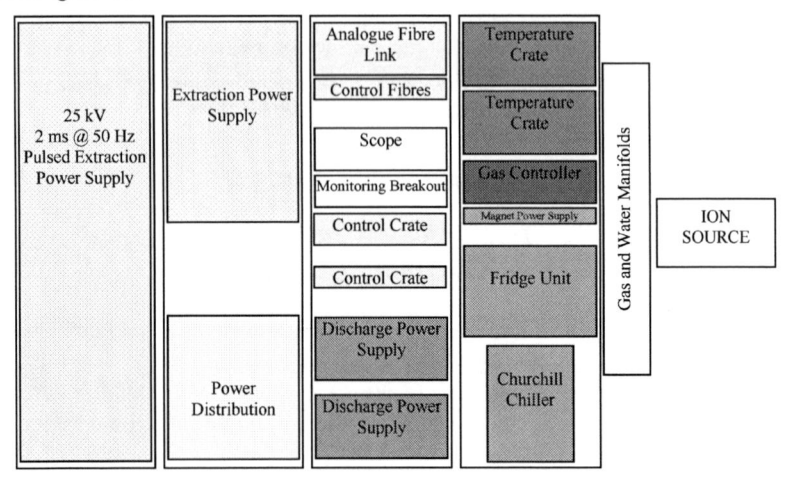

FIGURE 10: The layout ancillary equipment racks for FETS.

A new 25 kV (10% d.f. at 50 Hz) extraction voltage power supply has been developed, based on the standard ISIS design but using a larger TRITON 8960 tetrode tube. Danfysik DC and pulsed power supplies are used to power the source plasma discharge. New temperature controller crates have been built including extra temperature channels for additional monitoring. The rest of the equipment (H_2 controller, fridge unit, water chiller, monitoring and control) are standard ISIS items. Manifolds are used to distribute the cooling water, compressed air and hydrogen.

INITIAL PERFORMANCE

At the time of writing the installation is nearly complete. Alignment of the rail systems and completion of the upgraded pulsed extraction power supply are the main outstanding items. First beam should be produced in autumn 2008. It is possible to predict performance based on results[3] obtained from the Ion Source Development Rig (ISDR) at ISIS.

Extracted pulse lengths have been limited to 500 μs at 50 p.p.s. by the available pulsed extraction power supplies, however discharge lengths of up to 1.8 ms at 50 p.p.s. have been demonstrated. Extraction voltages have also been limited to 20 kV, but even at this voltage beam currents of 76 mA have been achieved when using the 0.8 mm wide aperture plate. Figures 11 and 12 show emittance and profile measurements taken 615 mm downstream from the ground plane of the post extraction acceleration gap for a 17 kV extraction voltage and a 2 mm, 18 kV post extraction acceleration gap. The beam current in both cases is 55 mA.

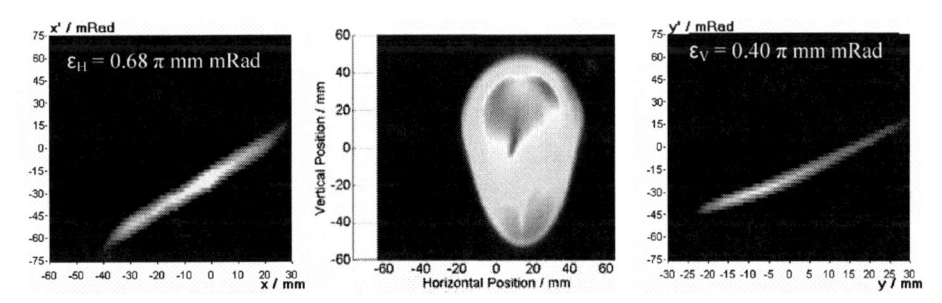

FIGURE 11. Beam emittance and profile from widened ISIS pole pieces with 17kV extraction voltage.

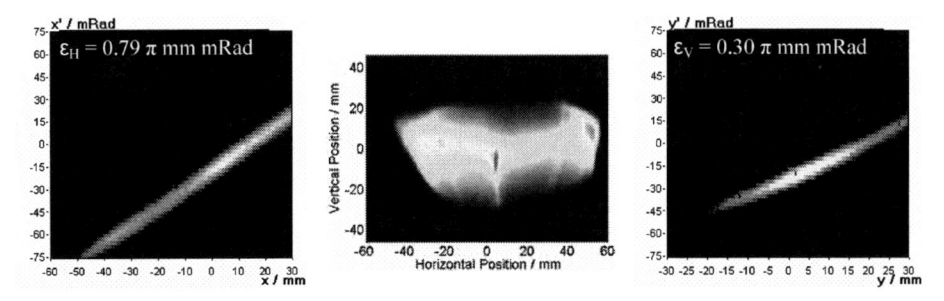

FIGURE 12. Beam emittance and profile from n = 1.0 pole pieces with 17kV extraction voltage.

DISCUSSION AND CONCLUSIONS

The installation is almost complete and although no beam has yet been measured on FETS, extensive testing on the ISDR indicate that the predicted performance will be close to the 60 mA, 2 ms, 50 p.p.s. design values. The only significant unknown is the amount of droop in beam current during the beam pulse, this can only be measured when the upgraded extraction power supply is completed.

REFERENCES

1. D.C. Faircloth, et al, "Understanding Extraction and Beam Transport in the ISIS H− Penning Surface Plasma Ion Source", Review Of Scientific Instruments 79, 02b717 2008.
2. V G Dudnikov, "Surface Plasma Source of Penning Geometry", IV USSR National Conference on Particle Accelerators, 1974.
3. D.C. Faircloth et al, ."Thermal Modelling of the ISIS H⁻ Ion Source", Review of Scientific Instruments, Volume 75, Number 5, May 2004.
4. S.R. Lawrie, et al, "Redesign of the Analysing Magnet In The ISIS H- Penning Ion Source", AIP this conference.
5. D.C. Faircloth et al, "Study of the Post Extraction Acceleration Gap in the ISIS H Penning Ion Source", Proceedings of EPAC08, MOPC142.
6. D.A. Lee, et al, "Laser-based Ion Beam Diagnostics for the Front End Test Stand at RAL", Proceedings of EPAC08, TUPC058.
7. J. J. Scheins, TESLA Report TESLA 2004-08, http://flash.desy.de/reports publications/ tesla reports.

Redesign of the Analysing Magnet in the ISIS H⁻ Penning Ion Source

S. R. Lawrie, D. C. Faircloth, A. P. Letchford, M. Westall,
M. O. Whitehead, T. Wood[a] and J. Pozimski[a, b]

[a]STFC, Rutherford Appleton Laboratory, Chilton, Didcot, Oxfordshire OX14 0QX, United Kingdom
[b]Department of Physics, Imperial College, London SW7 2AZ, United Kingdom

Abstract. A full 3D electromagnetic finite element analysis and particle tracking study is undertaken of the ISIS Penning surface plasma H⁻ ion source. The extraction electrode, 90° analysing magnet, post-extraction acceleration gap and 700 mm of drift space have been modelled in CST Particle Studio 2008 to study the beam acceleration and transport at all points in the system. The analyzing magnet is found to have a sub-optimal field index, causing beam divergence and contributing the beam loss. Different magnet pole piece geometries are modelled and the effects of space charge investigated. The best design for the analysing magnet involves a shallower intersection angle and larger separation of the pole faces. This provides radial focusing to the beam, leading to less collimation. Three new sets of magnet poles are manufactured and tested on the Ion Source Development Rig to compare with predictions.

Keywords: H⁻ Ion Source, Low Energy Beam Transport, Particle Tracking, Dipole Magnet, Weak Focusing, Extraction Electrode
PACS: 29.95.Ni, 41.75.Cn, 41.85.Lc

APPARATUS SETUP

The H⁻ Penning ion source [1] used on the ISIS pulsed spallation neutron source at the Rutherford Appleton laboratory (RAL) is one of the most successful in the world, due to its long service and gradual development over the years. The continual advancement of technology in the source has led to a high average beam current of 55 mA and a pulse length of up to 1.5 ms.

A Front End Test Stand (FETS) is being constructed at RAL to demonstrate that a high power H⁻ beam can be produced for future proton accelerators with sufficient beam quality [2]. The design criteria for the 65 keV beam produced from the FETS ion source is to have 60 mA current over a 2 ms pulse length, with horizontal and vertical normalised RMS emittances below 0.3 π mm mrad. The ion source needs to be upgraded to meet these requirements, with the majority of effort focusing on increasing the pulse length and reducing the emittance. To generate the necessary modifications, experiments are performed on the Ion Source Development Rig [3] (ISDR) at ISIS. This apparatus, shown in Fig. 1, allows experiments to be performed on an identical copy of the ISIS source, so as not to interrupt the user schedule for ISIS neutrons. Improvements can then be implemented on the ISIS and FETS sources.

Ion Source Development Rig

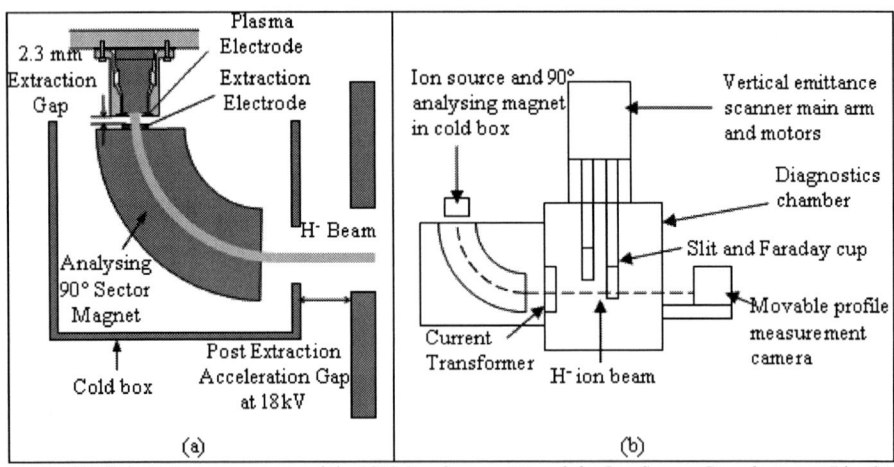

FIGURE 1. Main components of the ISIS Ion Source (a) and the Ion Source Development Rig (b)

On the ISDR, a suite of diagnostic tools is used to fully study the beam. A fast toroid is used to measure the beam current. The amount of electron stripping can be measured using a dipole magnet to separate neutrals from the H⁻ beam. The energy spread of the beam can be measured to an accuracy of a few eV. Horizontal and vertical slit-slit scanners are used to measure the beam emittance. The beam's profile can be studied at different positions along its path using a quartz scintillator and fast CCD camera. Finally, a pepperpot head can be mounted in front of the scintillator in order to measure the emittance and profile simultaneously.

The ISIS Ion Source

The ISIS and ISDR H⁻ ion sources are of the Penning (PIG) surface plasma type. The H⁻ beam is formed from a slit-shaped plasma electrode aperture using an extraction electrode held at 17 kV. To analyse out extracted electrons, the beam then passes through a 90° sector dipole magnet. The sector magnet is housed in a cold box, used to condense excess caesium vapour and prevent it propagating downstream. After leaving the cold box, the beam crosses the Post-Extraction Acceleration Gap (PEAG), gaining a further 18 keV of energy as it is accelerated to laboratory ground potential, resulting in a total beam energy of 35 keV. The PEAG on ISIS is presently rather large at 55 mm. Its 0.327 kV mm^{-1} electric field is too low to produce an einzel lens of sufficient strength to focus the beam sufficiently; resulting in large beam losses, due to collimation, upon entry into the Low Energy Beam Transport (LEBT) solenoids.

Simulations and experiments [4] have shown that a PEAG electric field of 9 kV mm^{-1} is optimal in terms of reducing the beam emittance. On the ISDR, therefore, a PEAG of 2 mm, with an applied potential of 18 kV, is used to fulfil this criterion. This small gap has successfully focussed the beam both horizontally and vertically, significantly reducing the emittance, as shown in Figs. 2 and 3; nevertheless the emittance is still much too large for the FETS requirements.

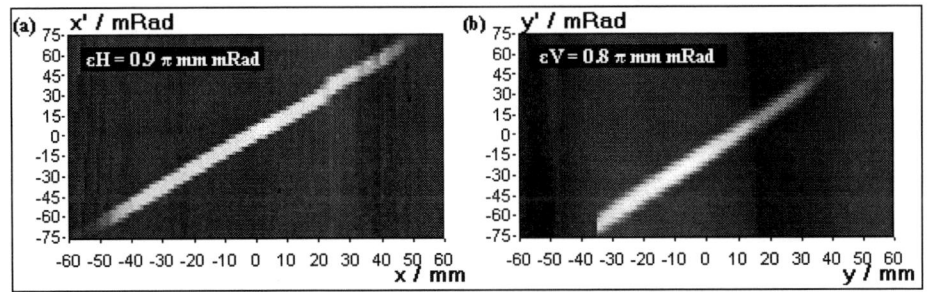

FIGURE 2. Horizontal (a) and vertical (b) emittance plots for a 55 mm PEAG.

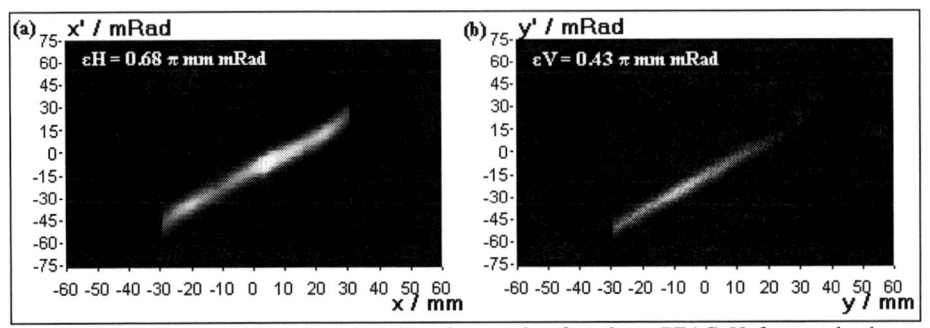

FIGURE 3. Horizontal (a) and vertical (b) emittance plots for a 2 mm PEAG. Unfortunately, the minimum scanning range in (b) was limited to -30 mm so some beam is missing, leading to a smaller emittance. If the entire beam were covered, the emittance is likely to be closer to 0.5 π mm mRad.

Beam Characteristics

Using the smaller post extraction acceleration gap of 2 mm, studies at low extraction energies showed [5] that the beam has an asymmetrical 'cobra-head' shape, which is wide at the top and long and tapered at the bottom. The degree of vertical defocusing is extreme, even with an optimal PEAG electric field. To see why, a set of five plasma electrode plates were manufactured with circular extraction apertures placed at positions along the standard slit aperture. It was found that – despite being extracted from a circular aperture – each beamlet still had a distinctive 'cobra-head' shape [6]. The only other component in the ion source which may focus the beam is the sector magnet, so a divergent beam indicates that the sector magnet is imperfect.

WEAK FOCUSING FORCES IN THE SECTOR MAGNET

To ensure parallel beam transport around a dipole magnet, the magnetic field must be inversely proportional to the radius. Hence the magnetic field

$$B = B_e \left(\frac{R_e}{R}\right)^n \quad (1)$$

where $n = 1$ and R_e is the optimum radius at which the beam is steered around.

In the ISIS ion source, R_e = 80 mm. For a 17 keV beam, this leads to the optimum magnetic field at R_e of B_e = 0.235 T. If $n \neq 1$, it can be directly calculated by rearranging (1) that

$$n = -\frac{R_e}{B_e}\left(\frac{dB}{dR}\right) \quad (2)$$

where n is the magnetic field index, and is a measure of the quadrupole component of the field inside a dipole magnet caused by the curvature of the field at the edges of the pole faces [7]. A value of the magnetic field index other than unity leads to non-parallel beam transport round a dipole magnet. This effect is Weak Focussing.

Specifically, $n < 1$ means the magnetic field is too weak at small radii and too strong at large radii, resulting in radial focussing. Conversely, $n > 1$ leads to defocusing. This being the case, it was suspected that the ISIS ion source sector magnet had a magnetic field index greater than unity, and hence was radially defocusing the beam, causing the stretched out 'cobra-head' shape.

An electromagnetic finite element model of the ion source was created in CST Particle Studio [8] to study how weak focusing affected beam transport.

Studying the Magnetic Field Index

Having solved the magnetostatic field, n was calculated using Eq. 2 in a Visual Basic macro. Figure 4(a) shows n as a function of radius in the dipole at the mid-plane between the pole faces. At R_e, n = 1.35, therefore the sector magnet does have a slightly sub-optimal geometry, causing the radial defocusing of the beam.

The area where the field index is approximately constant – the "good field region" – only extends about 10 mm either side of R_e. This is due to the small 55 mm radial width of the pole pieces. 5 mm wide, 1 mm tall shims on the inner and outer radii are used in an attempt to mitigate fringe field effects but these actually make the field index worse. The obvious solution to this problem is to increase the radial width of the pole pieces. The maximum width allowable by the internal dimensions of the cold box

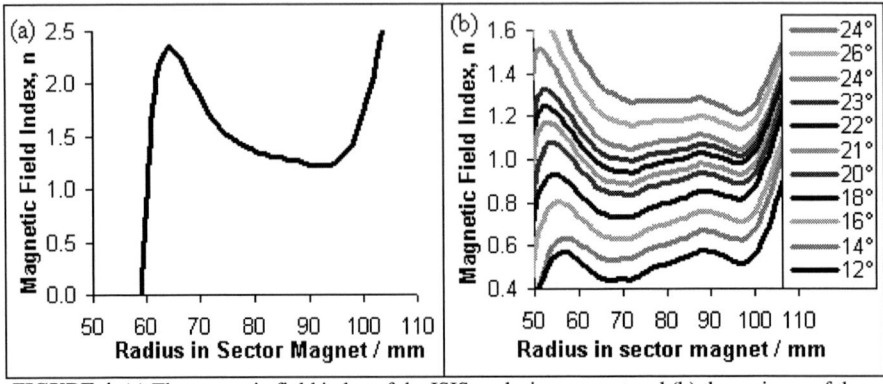

FIGURE 4. (a) The magnetic field index of the ISIS analysing magnet and (b) the variance of the n with the intersection angle of the pole faces, for a fixed gap at R_e of 32 mm.

is 74 mm. In order to vary n, either the intersection angle or the separation of the pole faces must be changed. For the ISIS poles – which have an intersection angle of 17.76° – it was found that to achieve a field index $n = 1$, the gap between the faces would have to increase from 25 mm to 28.6 mm. In fact, $n = 1$ can be achieved for any intersection angle, providing the gap is varied to suit. Varying the intersection angle for a fixed gap of 32 mm in Fig. 4(b), one can see not only how the value of n at R_e changes, but also the overall shape of n in the good field region. Note that in Fig. 4(b) and all successive plots, the magnetic fields were calculated for pole pieces with the wider 74 mm radial width in order to maximise the good field region.

It is not only the field index at the mid-plane of the gap that is important, though. If the magnetic field, and the subsequent n, is not sufficiently uniform across the whole of the gap, this limits the size of the good field region in the transverse direction.

Figure 5 shows how the shape of n changes as the field is sampled at three axial positions in the gap. If one wants to study this effect in more detail, then a line graph, as in Fig. 5(a), is insufficient. Therefore, in Fig. 5(b) the same magnetic field is sampled at a higher resolution across the gap, with the value of n represented as colours in a contour plot. This gives a much more intuitive feel of how n changes.

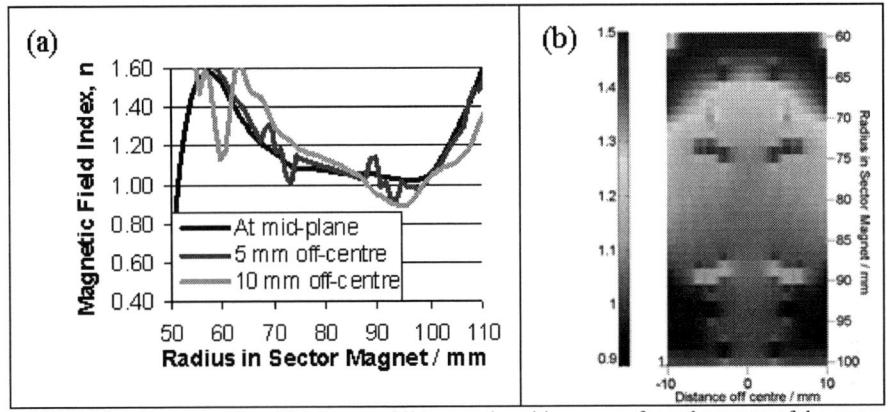

FIGURE 5. Variation of n at (a) three and (b) several positions away from the centre of the gap.

With this new technique to study how the pole piece geometry affects the magnetic field index, it is now simple to accurately design new pole pieces for any desired n.

PARTICLE TRACKING THROUGH THE ISIS ION SOURCE

To test the accuracy of the ion source CAD model and the solved electro- and magnetostatic fields in CST, particle tracking of the H⁻ beam was performed. A drift space of 700 mm was included in the model in order to compare the beam profiles at a position level with the slit-slit emittance scanners.

Previous attempts to simulate the shape of the beam have been unsuccessful; giving a rectangular profile rather than the distinctive 'cobra-head' seen on the ISDR. Indeed, performing particle tracking in CST produces this result when no space charge is present in the model. However there are reasons why space charge should be included:

1. Space charge compensating particles exist in drift sections of an accelerator, due to residual gas interactions, but not in accelerating regions. Therefore space charge must be taken into account in the extraction gap and PEAG.
2. At the standard operating conditions of +17 keV extraction and 35 keV total beam energy, the beam profile is round. However lowering the extraction energy reveals the cobra-head structure, indicating that the beam is collimating at higher energies as it passes through the PEAG.
3. Simulations with no space charge show a small beam totally dissimilar in shape to the real beam and which does not come close to collimating.

Experiments including space charge in the entire beam path caused the beam to be enormously divergent due to its low energy and large areas of drift relative to the size of the acceleration gaps. This confirmed that space charge should only be considered in the extraction and post-extraction acceleration gaps. When this arrangement was simulated, the resulting beam had the profile shown in Fig. 6(a); to be compared with the real measured beam in Fig. 6(b). The impressive agreement between the two indicates both that the electromagnetic and particle tracking solvers are accurate and that space charge is necessary in order to correctly simulate the beam.

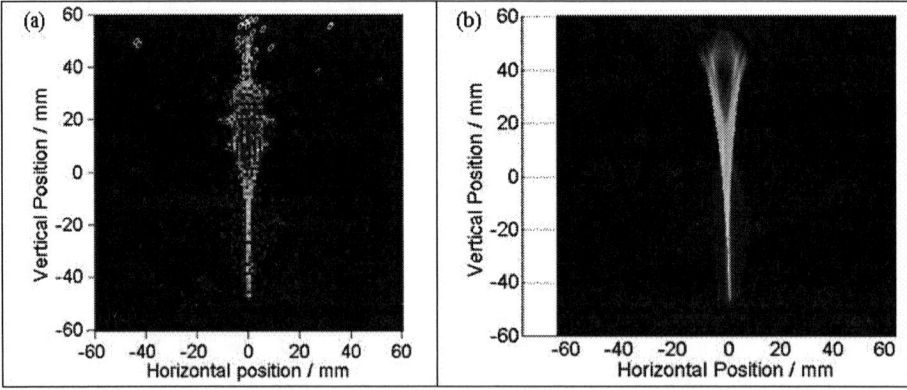

FIGURE 6. Beam profile using (a) particle tracking in the CST ion source simulation and (b) measured scintillation on quartz glass of the real beam on the ISDR.

With the standard ISIS H⁻ beam successfully re-created for the first time, improvements could be made to the sector magnet with confidence that the simulated beams from new pole pieces would be seen in the lab.

Design of the New Sector Magnet Pole Pieces

In order to improve the beam transport through the ion source, the good field region of the sector magnet was enlarged by widening the pole pieces radially to 74 mm. To see how this widening affects the beam, a set of poles with the same intersection angle, centre gap and shims as the old poles was created. Next, a set of poles with a field index $n = 1$ was designed, with the shims modified in order to enlarge the good field region still further, so as to keep $n = 1$ across the entire gap. Finally, in order to add radial focussing to the beam, pole pieces were made which have $n = 0.75$ as well as shims on the outer radius shaped so as to increase the local magnetic field and hence guide errant particles back in to R_e. The contour plots of n for these new poles are shown in Fig. 7 along with the plot for the old ISIS poles.

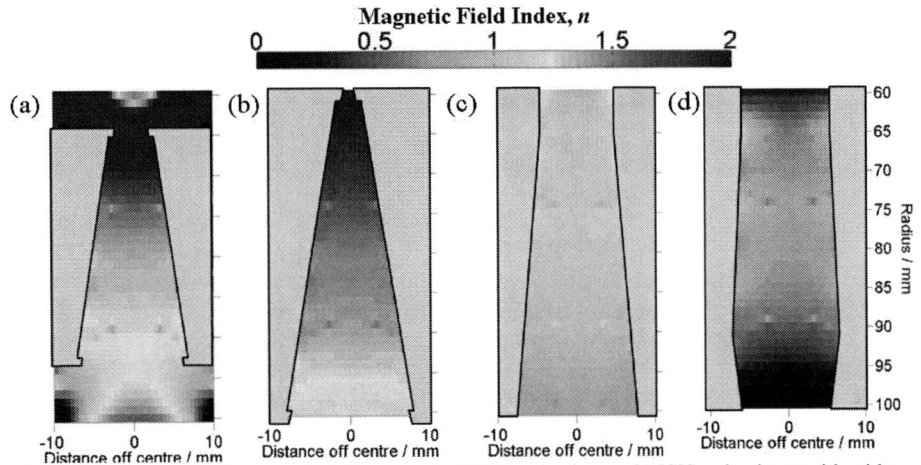

FIGURE 7. Contour plots of n for (a) the standard ISIS pole pieces; (b) ISIS pole pieces with wider radial extent; (c) $n = 1$ poles; (d) $n = 0.75$ poles with enlarged shims at the outer radius. Note the enlarged good field region in (b) compared to that in (a). Representations of the pole piece are overlaid on the contour plots for comparison, but are not to scale.

One further modification to the pole pieces was the cut-off from 90° of the sector magnet closest to the cold box exit. A cut-off is used in order to prematurely stop the magnetic field; otherwise the fringe field extent is such as to over-steer the beam vertically. Previous modelling work [9] has shown that the beam emerges parallel and on axis from the cold box when the cut-off is 3 mm and a tube of high permeability "maximag" steel is inserted to the cold box exit hole to remove fringe fields. However these models were performed with very little or no drift space after the cold box hole.

The present simulation has 700 mm of drift space, and it is clear that the beam does indeed emerge from the cold box on axis, but thereafter veers upward due to cumulative influence of the tiny fringe field remaining a substantial distance downstream. To remove this effect, the cut-off has been increased to 14 mm, which allows the residual field to fine tune the beam's vertical alignment. In this manner, the beam is correctly aligned to enter the first solenoid of the LEBT parallel and on axis.

With these modifications implemented to the pole pieces, particle tracking was performed. The predicted beam profiles from the new poles are shown in Fig. 8.

Explanation For The Predicted Beam Shapes

It can be seen that for smaller n, the beam becomes smaller in the radial (vertical) direction as predicted, but spreads in the axial (horizontal) direction. It may be tempting to attribute the asymmetrical focusing to the beam's space charge redistributing itself, but space charge is fully compensated in most of the model.

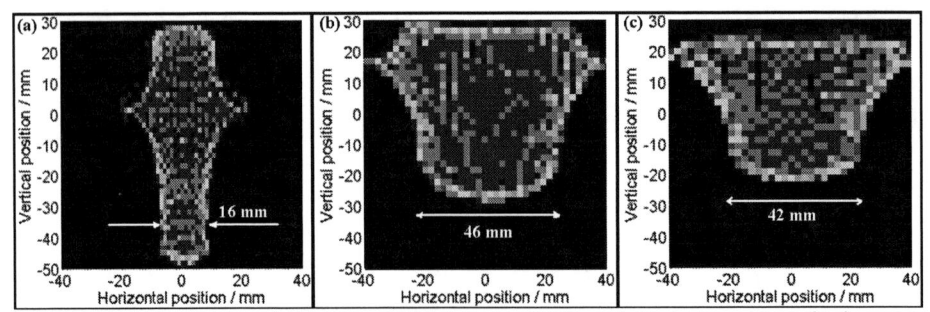

FIGURE 8. Predicted beam profiles for (a) Widened ISIS, (b) n = 1 and (c) n = 0.75 pole pieces. The $n = 1$ poles create the roundest beam. Note that the long tails above and below the beam core seen in Figure 6 are now absent due to the wider radial extent of the new pole pieces.

The reason in fact lies with the magnetic field index once more. The sector magnet can be thought of as a 90° portion of a cyclotron accelerator, where weak focussing forces are very important in the design. Particles in these accelerators undergo betatron oscillations around the equilibrium radius – both in the radial direction and the axial direction [7]. The frequencies of these oscillations, ω_R and ω_z respectively, with respect to the desired cyclotron orbit frequency, ω_c are

$$\omega_R = \omega_c \sqrt{1-n} \qquad (3)$$

$$\omega_z = \omega_c \sqrt{n} \qquad (4)$$

Radial orbital stability in a cyclotron only occurs for $0 < n < 1$, therefore the imaginary value obtained in Eq. 3 for the radially divergent beam from the $n = 1.35$ ISIS ion source is to be expected.

Using Eq. 4 it is clear that for $n = 1$ pole pieces in a cyclotron, a particle beam undergoes one axial oscillation for every orbit around the cyclotron. Therefore for the ion source sector magnet, which is one quarter of an orbit, the beam would spread to its maximum axial size. For the $n = 1.35$ pole pieces, the beam has gone past its maximum axial extent and started converging again, whereas for the $n = 0.75$ pole pieces the beam has not quite reached its maximum spread. These facts explain why the $n = 1$ beam profile in Fig. 8 has the largest horizontal (axial) width.

EXPERIMENTAL RESULTS USING THE NEW POLE PIECES

A comprehensive study of the beam profile, emittance and current for each set of pole pieces was performed. The results for a 17 keV extracted beam are shown in Figs. 9 and 10 for two of the three sets of sector magnet pole pieces. When compared to the old set of poles in Fig. 3, the vertical emittance is reduced due to the increased radial width, with a further reduction when $n = 1$.

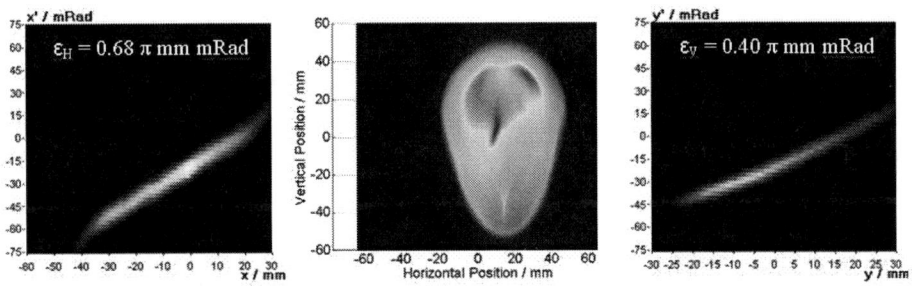

FIGURE 9. Beam emittance and profile from widened ISIS pole pieces with 17kV extraction energy.

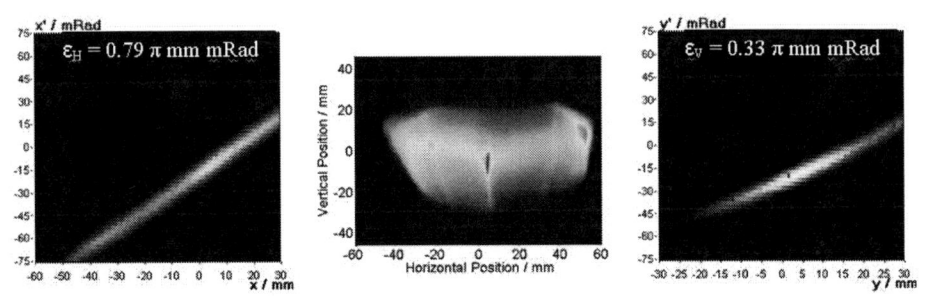

FIGURE 10. Beam emittance and profile from $n = 1$ pole pieces with 17kV extraction energy.

Unfortunately, power supply failures delayed testing of the $n = 0.75$ set of pole pieces. However the substantial agreement between the beam profiles measured in Figs. 9 and 10 to those calculated in Fig. 8 means the beam from the $n = 0.75$ poles is likely to have the shape predicted.

The horizontal emittance is largest for the $n = 1.0$ pole pieces, in agreement with Eq. 4. However, because the extraction electrodes also apply horizontal divergence to the beam, different extraction geometries were implemented in order to reduce the horizontal emittance. The electrode configurations used were: 0.6 x 10 mm slit aperture plasma electrode, with either open-ended or closed 2.1 mm separated extraction jaws (this is the standard ISIS extraction geometry); a 0.8 x 10 mm slit aperture, used to extract greater beam current; small circular apertures positioned along the slit; and extraction geometry with the Pierce angle [10].

As can be seen in Fig. 11, for the widened ISIS pole pieces the emittance continues to increase roughly linearly with extraction energy. However, the beam from the $n = 1$ pole pieces reaches a limit in its horizontal emittance. For both sets of pole pieces, the horizontal emittance is reduced by approximately 15% when using the Pierce extraction geometry. Therefore, the minimum normalised RMS emittance values currently achievable for a 17 keV extracted beam accelerated to a total energy of 35 keV are $\varepsilon_H = 0.59$ and $\varepsilon_V = 0.33$ π mm mRad. Emittance values and errors were calculated using the SCUBEEx algorithm [11].

The maximum beam current achieved at all extraction energies was when using the 0.8 mm wide plasma electrode aperture slit, up to a maximum of 76 mA at 20 keV. Nevertheless, over 60 mA was achievable with all extraction geometries.

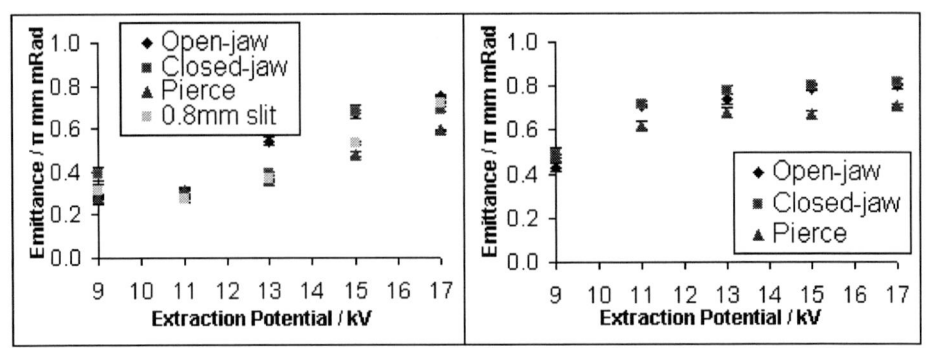

FIGURE 11. Horizontal emittance for widened ISIS (left) and $n = 1$ (right) pole pieces at different extraction energies and with various extraction electrode geometries used.

CONCLUSIONS

Modelling the ISIS ion source with space charge and drift space included in the simulation has successfully produced the 'cobra-head' beam profile which collimates at high extraction energies. Modifying the radial width, intersection angle and separation of the sector magnet pole pieces allows weak focusing forces to alter the shape of the beam and prevent collimation.

Experiments using the new pole pieces have shown the beam profile to be very similar to that predicted. By modifying the magnetic field index, the beam divergence has been reduced enough in the vertical direction to meet the FETS criterion for emittance. The horizontal emittance is still too large; however using the Pierce extraction electrode geometry, this has been somewhat reduced. The beam current requirements for FETS can be achieved using any extraction or sector magnet setup; but the 0.8 mm wide aperture slit produces the greatest beam current, which may become important when using longer pulse lengths of 2 ms.

The thorough investigation of the effects of both the extraction and sector magnet geometries has led to great progress in the understanding of beam transport through the ion source. This knowledge will help ensure that the H⁻ beam from the ISIS ion source has the correct properties for use in FETS and other future projects.

REFERENCES

1. R. Sidlow et. al., EPAC '96, THP084L.
2. D. C. Faircloth, "Commissioning of the Front End Test Stand High Performance H⁻ Ion Source at RAL", these proceedings.
3. J. W. G. Thomason et. al., EPAC '02, THPRI012.
4. D. C. Faircloth et. al., EPAC '08, MOPC142.
5. D. C. Faircloth et. al., Rev. Sci. Instrum. 79 (2008), 02B717.
6. D. C. Faircloth et. al., EPAC '08, MOPC143.
7. J. J. Livingood, "Principles of Cyclic Particle Accelerators", Van Nostrand, Princeton, NJ (1961).
8. CST Studio Suite 2008, www.cst.com
9. D. C. Faircloth et. al., Rev. Sci. Instrum, 75 (2004), 1735.
10. J. R. Pierce, "Theory and Design of Electron Beams", 2nd ed., Van Nostrand, Princeton, NJ (1954).
11. M. P. Stockli et. al., Rev. Sci. Instrum. 75 (2004), 1646.

ION SOURCES FOR FUSION

Plasma And Beam Homogeneity Of The RF-Driven Negative Hydrogen Ion Source For ITER NBI

U. Fantz, P. Franzen, W. Kraus, D. Wünderlich, R. Gutser, M. Berger, and the NNBI Team

Max-Planck-Institut für Plasmaphysik, EURATOM Association, D-85748 Garching, Germany

Abstract. The neutral beam injection (NBI) system of ITER is based on a large RF driven negative hydrogen ion source. For good beam transmission ITER requires a beam homogeneity of better than 10%. The plasma uniformity and the correlation with the beam homogeneity are being investigated at the prototype ion sources at IPP. Detailed studies are carried out at the long pulse test facility MANITU with a source of roughly 1/8 of the ITER source size. The plasma homogeneity close to plasma grid is measured by optical emission spectroscopy and by fixed Langmuir probes working in the ion saturation region. The beam homogeneity is measured with a spatially resolved H_α Doppler-shifted beam spectroscopy system. The plasma top-to-bottom symmetry improves with increasing RF power and increasing bias voltage which is applied to suppress the co-extracted electron current. The symmetry is better in deuterium than in hydrogen. The boundary layer near the plasma grid determines the plasma symmetry. At high ion currents with a low amount of co-extracted electrons the plasma is symmetrical and the beam homogeneity is typically 5-10% (RMS). The size scaling and the influence of the magnetic field strength of the filter field created by a plasma grid current is studied at the test facility RADI (roughly a ½ size ITER source) at ITER relevant RF power levels. In volume operation in deuterium (non-cesiated source), the plasma illumination of the grid is satisfying.

Keywords: Negative hydrogen ion, rf source, ITER, Neutral beam injection
PACS: 52.80.Pi, 52.27.Cm, 29.25.Ni, 28.52.Cx

INTRODUCTION

The international fusion experiment ITER requires for its heating and current drive neutral beam injection (NBI) systems based on negative hydrogen ion sources. One of the requirements is that the beam homogeneity has to be better than 10% for the source dimension of 1.9×0.9 m^2 with an extraction area of 0.2 m^2. In the ITER design review in 2007 the inductively RF-driven negative ion source was chosen as the ITER reference ion source based on the, in principle, maintenance-free operation and the success of the development of smaller prototypes at IPP [1]. Although the IPP RF source has made substantial progress towards ITER's requirements [2-4] there are still open issues to be addressed [4].

Key issues are the size scaling of the source and the long pulse stability (up to one hour). Size scaling concerns the question of plasma homogeneity, i.e. uniform plasma illumination, and thus electron density above the grid system. Since in fusion relevant sources negative ions are produced by conversion of hydrogen atoms and positive ions on cesiated surfaces uniform atomic and positive ion fluxes, as well as uniform and stable cesium coverage are essential for extracting a homogeneous beam.

In order to reduce the amount of co-extracted electrons, a magnetic filter field is used in negative ion sources. The field strength is typically 5 - 10 mT close to the grid system. In addition, a bias voltage is applied to the first grid of the three grid system with respect to the source body. Reduction of co-extracted electrons without affecting the extracted negative ion current is most effective for bias voltages between the undisturbed floating potential of the grid and plasma potential [4, 5].

The physics of the boundary layer which is formed near the grid and extends typically 3 cm is very complex. The magnetic field and the bias voltage can cause a vertical plasma drift and thus plasma non-uniformity [4]. In addition, negative ions produced at the plasma grid penetrate into the plasma and push the electrons away to keep the quasineutrality in the boundary layer [4]. In a well conditioned cesium seeded source, it has been measured that at 2 cm distance to the grid the negative ion density is comparable to the electron density and can even be the majority [4-7].

A further critical task is a controlled coverage of the plasma grid with cesium. Cesium is evaporated by a cesium oven mounted at the back plate of the source. The Cs amount in the source can be influenced by the oven temperature, the wall temperature which determines the cesium retention of the source walls, and the duty circle (plasma on/off time) [8]. Generally, the cesium dynamics is very complex in such sources and the interplay of the individual contributions and their control to establish optimum cesium coverage of the plasma grid is still an open issue.

Required beam homogeneity of better than 10% pre-requires surely a uniform negative ion production at the plasma grid surface which implies the above mentioned circumstances. The temporal stability of the ion beam, however, is, the operational source stability given, strongly dependent on the cesium dynamics, which affects also the temporal behavior of the co-extracted electron current [8]. The allowed amount of co-extracted electrons is limited by the acceptable power load on the second grid, the extraction grid. Therefore ITER requires that the co-extracted electron to ion ratio is kept below one. Poor cesium conditions imply low ion currents and an increase of the co-extracted electron current [8,9].

The paper is devoted to the measurement of plasma and beam homogeneity at two of the three test facilities at IPP: the long pulse test facility MANITU with a source of roughly 1/8 of the ITER source size and an extraction area of 206 cm^2 [8] and the test facility RADI which is equipped with roughly ½ the size of the ITER source but has no extraction system [10]. Special attention is given to the correlation of the plasma uniformity with the beam homogeneity investigated at MANITU. The basic understanding of the physical background has been, and is further, investigated at the third test facility BATMAN (1/8 ITER source, limited pulse length < 4s).

DIAGNOSTICS AND EXPERIMENTAL SETUP

The plasma emission and the plasma parameters, such as electron density and negative ion density, are measured by optical emission spectroscopy (OES) [11,12]. Spatial resolution is provided by using several lines-of-sight (LOS) arranged symmetrically to the grid centre at different distances to the grid (Figure 1). This allows the determination of a symmetry factor, top-to-bottom symmetry, at 2 cm (blue, and with eccentric optics at 1 cm) and 3 cm (red) distance to the grid. Their vertical distance is 14 cm and 30 cm, respectively. As standard, the emission of the Balmer line H_β is used, which depends basically on the atomic hydrogen density, electron density and temperature.

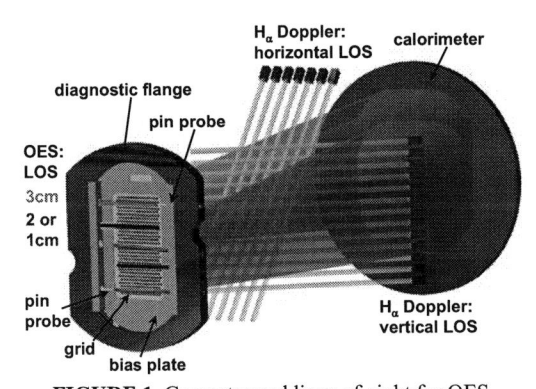

FIGURE 1. Geometry and lines-of-sight for OES (plasma) and H_α (beam) spectroscopy at MANITU.

The sources are equipped with fixed Langmuir probes (pin probes) operating in the ion saturation current (biased to -24V) and are placed in the outermost upper and lower diagnostic port at MANITU (Figure 1). At RADI an array of pin probes is embedded in the plasma grid (see below). The ion saturation current is a measure for the ion density; the absolute value, however, depends also on the square root of the electron temperature and the average ion mass of the hydrogen ions. The latter is enhanced by an increasing cesium amount.

The beam homogeneity is measured at MANITU with the H_α Doppler beam spectroscopy system using 13 LOS in vertical (40 mm spaced) and 7 LOS (46 mm spaced) in horizontal direction (Figure 1). The grids are vertically inclined by 0.9° which results in two distinguished beams. The viewing angles are roughly 50° at a distance of 1.2-1.5 m from the extraction system. The calorimeter is at a distance of 2.4 m. The temporal stability and spatial homogeneity of the beam can be determined by analyzing the width of the Doppler shifted H_α line which is proportional to the local divergence. The divergences obtained with the calorimeter agree rather well. MANITU is typically operating below or in the optimum perveance. Hence, the measured homogeneity reflects the homogeneity of the extracted negative ion current density which is a mean value averaged by the intersection of the respective LOS.

PLASMA AND BEAM HOMOGENEITY AT MANITU

The dependence of the source performance on the applied RF power is shown in Figure 2 for hydrogen and deuterium discharges at MANITU. As it is typical for the negative ion source the electron current is higher in deuterium than in hydrogen. This is correlated with the higher electron density measured in deuterium [4]. The plasma symmetry improves with RF power and is much better for deuterium. This is shown by the symmetry factor of the pin probes (I_{sat}) and H_β both at 3 cm distance to the grid. At a distance of 1 cm, however, the symmetry is unaffected by the RF power which demonstrates the dynamics of the spatial behavior of the boundary layer in vertical and axial direction. The data have been taken during a single 200 s pulse by varying the

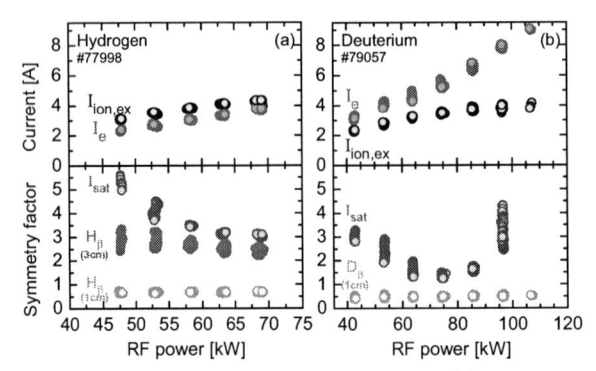

FIGURE 2. Extracted ion and electron current, and the symmetry factor for the ion saturation current of the pin probes and the H_β line measured at two distances to the grid for hydrogen (a) and deuterium (b) discharges at 0.4 Pa and 0.35 Pa, respectively.

F power. This is the reason for the slight drift of the signals (averaged over 1 s) for a given power step of several ten seconds.

The corresponding vertical and horizontal beam data are shown in Figure 3 for discharges with poor (#77970) and high (#77998) performance in hydrogen. The correlation of the beam homogeneity and plasma symmetry depends on the source performance. A deuterium discharge (good performance) is shown for comparison. The divergence decreases with increasing perveance, which is, in this case, caused by the increase of the RF power. For a discharge with good performance (#77998) the plasma and the beam homogeneity (defined by the RMS value) correlate. Both increase with increasing ion current. The contrary, namely a degradation of the homogeneity with improving plasma symmetry is observed for a discharge with poor performance (#77970, factor 1.5 lower ion currents). In this case the boundary layer is less affected by the negative ions which improve the plasma symmetry. For deuterium (#79057) a much better beam homogeneity is observed, correlating as well with the plasma symmetry, the divergence, however, is at 4.5° in this case.

The general trend that the plasma symmetry improves with increasing ion current accompanied by a decreasing electron current is demonstrated in Figure 4, which shows almost all hydrogen and deuterium discharges of the last experimental campaign with molybdenum coating of the inner surfaces [8].

Basically the plasma symmetry depends on the boundary layer which is determined also by the interplay between bias voltage at the plasma grid and the plasma potential. Figure 5 shows the dependency of the source performance and the homogeneity on the bias current. The data were taken before the molybdenum coating of the source; hence the ion current is low and the electron current is high (poor source condition-

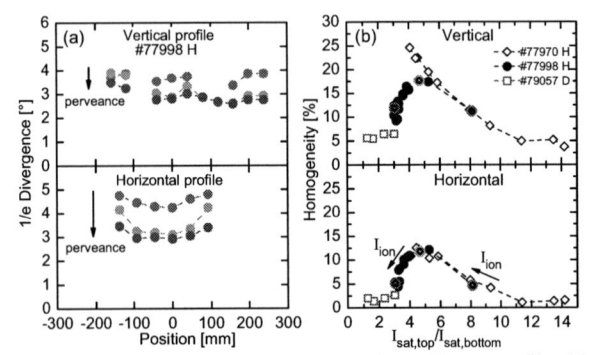

FIGURE 3. Examples of vertical and horizontal beam profiles (a), and homogeneity (RMS) for H and D discharges (b).

268

ing). The bias circuit was operated in current control mode; the bias voltage changed accordingly from 10 to 30 V.

The ion current does not change much with increasing bias current; the electron current drops by a factor of almost three. The pin probe signals show an opposite behavior: the ion saturation current of the upper probe increases with increasing bias current, whereas the ion saturation current of the lower probe decreases. This behavior is typical for a not well conditioned source as discussed for the #77970 discharge of Figure 3.

The Balmer lines have been observed at 2 cm distance; the line ratios H_β/H_γ and H_α/H_β have been used to determine the electron density and the density of negative ions, respectively [4]. The negative ion density is 30%-50% of the electron density.

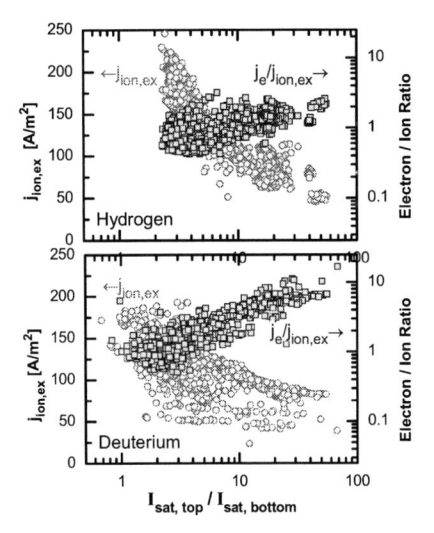

FIGURE 4. Extracted ion and electron current for H and D discharges at MANITU.

The electron density is larger in the upper half of the source than in the lower half and the electron density in the upper half clearly correlates with the co-extracted electron current. Since destruction of negative ions is correlated with electron density, the negative ion density is smaller in the upper half of the source.

With increasing bias current the plasma symmetry, in terms of electron density and negative ion density, improves. Thus, the plasma drift caused by the magnetic field is compensated for by shifting the potentials in the source, which in turn reduces the co-extracted electron current. The symmetry for the negative ions, which is much more important for beam extraction, is better than for the electrons and is less dependent on the bias current. This is reflected in the beam homogeneity (Figure 6) which is almost independent on the bias current.

In summary the plasma symmetry depends on the complex dynamics of the boundary layer near the plasma grid. The latter is influenced by the magnetic field, the bias voltage and the amount of negative ions. At high source performance, which

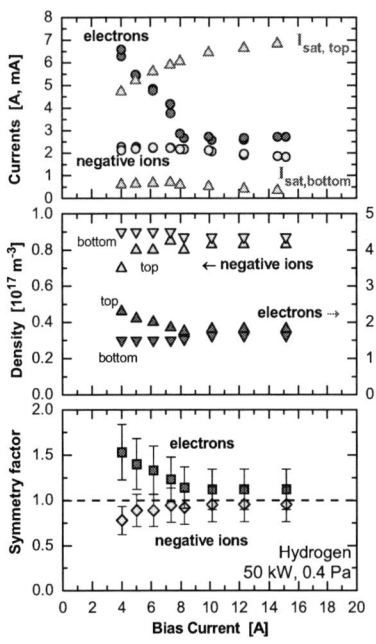

FIGURE 5. Dependence of the extracted currents, pin probe signals, and particle densities from OES at 2 cm distance, and their symmetry factor on the bias current.

269

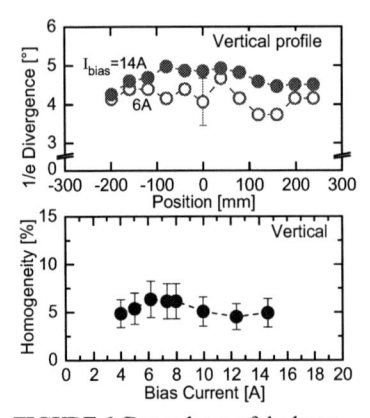

FIGURE 6. Dependence of the beam homogeneity (RMS value) on the bias current for the same parameters as Figure 5.

means high ion currents and low extracted ion currents, the plasma in this region is symmetric (electron density and much more important the negative ion density) and the beam homogeneity is typically 5-10%. This is within the ITER requirements.

SIZE SCALING AND PLASMA UNIFORMITY AT RADI

The test facility RADI is equipped with a source of roughly 1/2 the size of the ITER source, extraction however is not available [10]. The plasma uniformity is measured in front of the grid by OES using 3 LOS in vertical and 3 LOS in horizontal direction at a distance of 2 cm to the grid and by an array of 18 pin probes embedded in the grid measuring the ion saturation current. Figure 7 shows a sketch of the dummy plasma grid geometry, the position of the pin probes and the projection of the four drivers used for plasma generation. The pin probes are arranged to give a horizontal profile (in the projection of the upper pair of drivers) and a vertical profile (in the projection of the right pair of drivers) of the positive ion distribution. Two horizontal drivers, connected in series, are powered by one RF generator. This allows the usage of individual power levels for the upper ($P_{RF,top}$) and lower ($P_{RF,bottom}$) pair of drivers. Two movable Langmuir probe systems are installed near the grid (2 cm distance) in horizontal and vertical direction [4]. A Faraday cup for local extraction is installed at the position of the projection of upper right driver close to the centre and is presently commissioned.

RADI is operating in deuterium discharges for typically 6 s at pressures between 0.3 and 0.5 Pa without cesium evaporation up to now. This means the results obtained are measured under conditions for volume production of negative ions. After some initial problems with the RF generators and the RF coupling which limited the maximum achievable RF power to 40 kW, power levels of 100-120 kW from each generator, i.e. 50-60 kW per driver, are now used on a routine basis.

At RADI the magnetic filter field is created by a current (maximum of 5 kA) flowing through the plasma grid. By mounting permanent magnets outside the source body, a magnetic field similar to the ITER design (4 kA plasma grid current and permanent magnets) can be created. The adjustable current allows detailed studies of

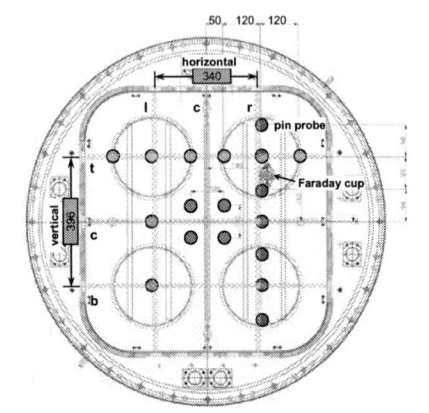

FIGURE 7. Sketch of the RADI geometry, position of the pin probes, and the vertical (t,c,b) and horizontal (l,c,r) LOS.

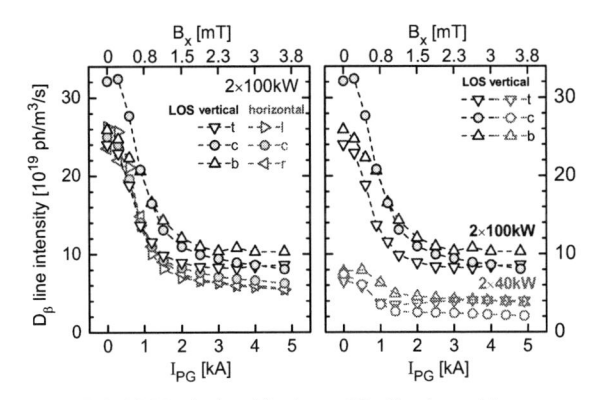

FIGURE 8. (a) Vertical and horizontal D_β line intensities at power levels of $P_{RF,top}=P_{RF,bottom}= 100$ kW at 0.4 Pa. (b) Vertical D_β line intensities for two power levels (0.55 Pa at lower power).

the effect of the magnetic field strength on the plasma uniformity.

The dependence of the vertical and horizontal D_β line emission on the PG current and thus the magnetic field is shown in Figure 8 for a total RF power of 200 kW. Without magnetic filter field almost the same D_β line emission is observed for all LOS indicating a uniform plasma illumination of the grid. An exception is the centre LOS in vertical direction. The increase is most probably caused by an overlap of the plasma from the vertical drivers although the distance between the vertical drivers is larger than the one for the horizontal drivers. This might indicate a different mutual coupling between drivers in vertical (individually powered) and horizontal (in series powered) drivers.

With increasing magnetic field strength the D_β intensity decreases, which is due to a reduction of the electron temperature from roughly 6 eV to 3 eV as measured with the movables Langmuir probes [4]. Obviously, the strongest effect takes place in the region between 0.5 mT and 2 mT; below 0.5 mT the electrons seem to be not magnetized. As expected, the horizontal distribution is almost not affected. The vertical distribution, however, shows a separation in the upper and the lower channels with a depletion of the signal of the centre channel. This behavior can be explained by a reduced plasma expansion out of the drivers in vertical direction with increasing filter field current. The reason is most probably the far reaching filter field which has a sufficiently large value in the drivers (almost 2/3 of the strength at the grid). The depletion is much more pronounced for lower power levels as the comparison of the vertical profiles shows in Figure 8 (b). The trends at low power are discussed in detail in [4]. For high power levels the overall plasma uniformity at maximum filter strength is much better than for low power and is in the same range as for the small sources in volume operation. Thus, the plasma illumination looks satisfying up to now.

The electron density decreases with increasing magnetic filter field from 4.5×10^{17} m^{-3} at 0 mT to 3.5×10^{17} m^{-3} at 5 mT for 2×100 kW (LOS at the bottom). For 2×40 kW electron densities of 1.8×10^{17} m^{-3} decreasing to 1.4×10^{17} m^{-3} are obtained from the OES as it has been observed with the movable Langmuir probes [4]. The electron density scales linearly with RF power both without and with magnetic field. The electron densities are enhanced at RADI compared to the densities in the respective volume in the small prototype sources at BATMAN and MANITU. In MANITU (where the magnetic field strength is about 8 mT) the electron density is 2×10^{17} m^{-3} for a power of 50 kW (one driver, Figure 5), whereas the corresponding value at RADI is 3.5×10^{17} m^{-3} at 3.8 mT for 2×100 kW for the four drivers. That means that less power

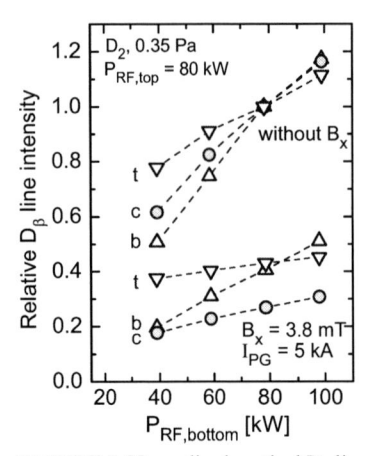

FIGURE 9. Normalised vertical D_β line intensities for a fixed $P_{RF,top}$ at different $P_{RF,bottom}$ with and without magnetic field.

is needed in the large RF sources for the generation of the same plasma. The reasons for that are the plasma overlap, the reduced losses and the influence of the different magnetic filter. The negative ion density however, is comparable to the one observed in the small sources operating in volume operation, being typically around 4×10^{16} m^{-3}. This indicates again, that lower power levels at the large RF source might be sufficient to achieve a similar source performance. In how far this argumentation will be transferable to plasmas with surface production of negative ions can only be studied after cesium seeding, which will be done in the near future.

In order to demonstrate the possibility to modify the plasma intensity in vertical direction the power of the bottom drivers ($P_{RF,bottom}$) has been varied at a constant power level for the top drivers ($P_{RF,top}$) with and without magnetic field (Figure 9). The reduction of the plasma overlap is again to be seen clearly. For example, the top channel increases without magnetic field by 40% and only 20% with magnetic field. The depletion in the centre channel still remains and can not be compensated for by using individual power levels but improves with higher power for both levels as shown in Figure 8 (b). The consequences on the negative ion production by surface processes will be investigated for cesium seeded discharges, as MANITU demonstrated that the boundary layer is strongly influenced by the negative ions.

In contrast to OES, which measures at 2 cm distance to the grid (diameter of the LOS approximately 10 mm averaged of the 80 cm length of the LOS) the pin probes measure the positive ion distribution at the grid. Figure 10 shows an example of the vertical distribution for powering solely the upper pair or the lower pair and powering both pairs with 80 kW each. The magnetic field has been switched off. As expected, a smooth decrease in vertical direction is observed showing the expansion from the cylindrical plasma into the expansion chamber. The contribution to one pair of drivers into the projection of the other pair is symmetrical; it is in the order of 15%. Powering all four drivers show a profile which is almost flat in the center; the same is measured in horizontal projection. The sum of the signals from powering the individual pairs is almost equal to the one measured for all four drivers powered. The absolute error of the ion saturation current is about

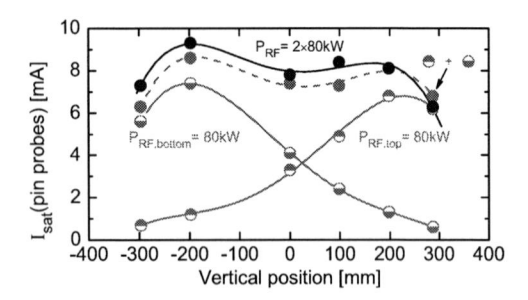

FIGURE 10. Vertical profiles at different power levels obtained with pin probes embedded in the grid.

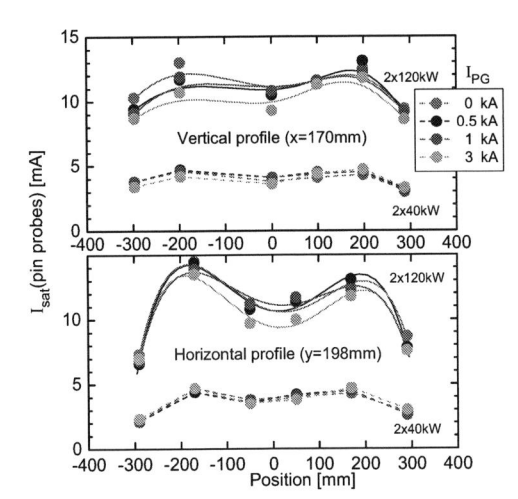

FIGURE 11. Vertical and horizontal profiles at two power levels with and without applying a magnetic field.

20% but the relative error is much lower.

The dependence of the vertical and horizontal profiles of the pin probe signals on the RF power and the magnetic field is shown in Figure 11. As for the OES signals, the ion saturation currents show a linear increase with RF power. However, only a slight dependence on the magnetic field strength is observed which can be explained by the dependence of the ion saturation current on the ion density and the square root of the electron temperature. The effective ion mass should be, in a first estimation, unaffected by the magnetic field. As shown by OES, the electron density decreases slightly and the strong decrease of the OES signal on the magnetic field is due to the electron temperature decrease, which has not much effect on the ion saturation current. Thus, the dependence of the positive ion density near the grid on the RF power and the magnetic field is the same as the one for the electron density in deuterium plasmas operating in volume formation of negative ions.

The vertical plasma drift observed at the small sources, equipped with one driver and permanent magnets, is counteracted at the large source by the reduction of the plasma expansion from the four drivers which is caused by the far reaching magnetic field created with the plasma grid current. To what extend this will influence the homogeneity of negative hydrogen ions produced at the plasma grid by the surface mechanism is part of the experimental program in future. Another factor is certainly the cesium coverage of the plasma grid, which should be as uniform as possible. How this can be achieved for the 80×80 cm² area will be tested at RADI using two cesium ovens mounted at the backplate of the source between the drivers. As a measure the cesium emission normalized to the D_β line emission will be used as it is regularly done at the small sources [4].

CONCLUSION

The plasma symmetry has been investigated in small prototypes of the RF source for ITER NBI and in a roughly ½ the size of the ITER source. The results measured with OES and pin probes show the same trends.

The symmetry improves with increasing RF power and is strongly dependent on the bias voltage applied to the plasma grid with respect to the source body. Near the plasma grid a boundary layer is formed which extends typically 3 cm into the source and is influenced by the bias voltage, the amount of negative ions with respect to the electrons, and the magnetic field strength and topology. At the small RF source prototypes the magnetic field created by permanent magnets causes a vertical plasma drift at high electron currents (poor source performance). In the large RF source, in which a weaker but far reaching magnetic field is created by the plasma grid current, the plasma expansion from the vertical pairs of drivers is reduced. However at ITER relevant power levels, the plasma illumination turns out to be satisfactory: no strong depletions and enhancements are observed in volume operation. At high power level the plasma drift is counteracted by the reduced plasma expansion magnetic field. A general plasma overlap of the plasma generated in four drivers is measured in the expansion region. At comparable power levels higher electron densities are measured in the large source than in the small sources, indicating that a lower power level might be sufficient for negative ion production in large sources.

The symmetry of the plasma depends on the source performance. For volume operation or poor source performance (low ion currents), the plasma symmetry is determined by the plasma flow out of the driver and the corresponding plasma drift. With increasing source conditioning (increasing ion currents) the symmetry improves due to the increasing relevance of the processes in the boundary layer.

The experiments at the small sources showed that in a well conditioned source, i.e. high negative ion currents and a low amount of co-extracted electrons, the plasma symmetry is given for both the electron density, and much more important, the negative ion density. The beam homogeneity correlates with the negative ion density and is better than 10% at high source performance.

ACKNOWLEDGMENTS

The work was (partly) supported by a grant from the European Union within the framework of EFDA (European Fusion Development Agreement). This continuous support is gratefully acknowledged. The authors are solely responsible for the content.

REFERENCES

1. R. Hemsworth *Rev. Sci. Instrum.* **79**, 02C109 (2008)
2. E. Speth et al., *Nucl. Fusion* **46**, S220 (2006)
3. P. Franzen et al., *Nucl. Fusion* **47**, 264 (2007)
4. U. Fantz et al., *Plasma Phys. Control. Fusion* **49**, B563 (2007)
5. S. Christ-Koch, "Laser photodetachment on a high power, low pressure rf-driven negative hydrogen ion source, submitted to *Plasma Sources Sci. Techn.*
6. U. Fantz et al., *Rev. Sci. Instrum.* **79**, 02A511 (2008)
7. M. Berger, "Cavity-Ringdown-Spectroscopy on a high-power RF-driven source for negative Hydrogen ions", submitted to *Plasma Sources Sci. Techn.*
8. W. Kraus et al., "Long pulse H- beam extraction with a RF driven ion source with low fraction of co-extracted electrons", this conference
9. W. Kraus et al., "Development of RF driven H-/D- sources for ITER", Proceedings of the 35th EPS conference 2008, Greece
10. P. Franzen et al. *Fusion. Eng. Des.* **82** 407 (2007)
11. U. Fantz et al., *Nucl. Fusion* **46**, S297-S306 (2006)
12. U. Fantz et al., *Rev. Sci. Instrum.* 77, 03A516 (2006)

Long Pulse H⁻ Beam Extraction With A RF Driven Ion Source With Low Fraction Of Co-Extracted Electrons

W. Kraus, M. Berger, U. Fantz, P. Franzen, M. Fröschle, B. Heinemann, R. Riedl, E. Speth, A. Stäbler, D. Wünderlich

Max-Planck-Institut für Plasmaphysik, Boltzmannstr. 2, D-85748, Garching, Germany

Abstract. IPP Garching is developing H⁻/D⁻ RF ion sources for the ITER neutral beam system. On the MANITU testbed the experiments are focussed on long pulse H⁻/D⁻ beam extraction with a 100 kW prototype source. The negative ion production is based on surface conversion of atoms and positive ions on Caesium layers. In long pulses with H⁻ beam extraction the ion currents were stable but with too high fraction of co-extracted electrons. The electron current could be lowered considerably by avoiding copper impurities from the Faraday screen in the plasma which was achieved by coating of the inner surfaces of the source with Molybdenum. A positive bias potential with respect to the source applied to the plasma grid, the bias plate or to a metal rod installed near the plasma grid enables regulation of the electron current during long pulses. In this way low values consistent with the ITER requirements can be achieved without significant loss of ion current.

Keywords: H⁻, Ion Source Development, Neutral Beam Heating, ITER, Radio Frequency Ion Source
PACS: 52.50.Dg, 52.50.-b, 52.59.-f, 52.70.-m, 52.80.Pi

INTRODUCTION

In 2007 the RF source was chosen for the reference design for the plasma generation in the ITER neutral beam system. The reasons for this decision were the in principle maintenance free operation, which is expected due to the filamentless discharge, the lower Cs consumption compared to arc sources because of the tungsten free plasma. Last, but not least, the results achieved with a smaller RF source prototype decided the issue, which demonstrated the suitability of an RF driven source to meet the ITER design requirements, concerning operation pressure, extracted ion current density and fraction of co-extracted electrons. But these experiments were carried out with the pulse length limited to less than four seconds and only an extraction area of ~70 cm² [1].

The long pulse "MANITU" test facility (multi ampere negative ion test unit) was built in order to demonstrate that the ITER requirements are also achievable in long pulses up to CW operation and with an enlarged extraction area.

It was found, that the negative ion current density at the same power did not change substantially, when the extraction area was enlarged by almost a factor three (to 206 cm²) [2]. After various technical modifications of the source and the testbed it was possible to perform pulses up to 3600s [3]. The ion current was remarkably constant, but the fraction of co-extracted electrons could not be kept on the low level known from short pulses. Within the first 100 s to 150 s the electron current increased, exceeded the ion current and reached a constant but too high level. The subsequent high power load on the extraction grid was one reason that prevented operating the source with high RF power; the other one was damage to the RF antenna from RF breakdowns.

This paper will concern itself mainly with reduction and stabilisation of the electron currents, with the source modifications aiming for higher reliability at high RF power and with experiments in Deuterium.

THE RF SOURCE

The RF power is inductively coupled into a circular volume of 25 cm diameter ("driver"), out of which the plasma is flowing into the main chamber (b x l x d = 30 x 60 x 25 cm^3) (Fig. 1). The ITER source will have eight of these "drivers".

The negative ions are produced by conversion of hydrogen ions and atoms on the cesiated surface of the plasma grid. The Cs is evaporated into the source by a Cs oven mounted onto the back plate. The level of the Cs content in the source is detected by measurement of the Cs852 line intensity (neutral Cs). A magnetic filter field of approx. 850 Gcm is used to reduce the electron temperature close to the plasma grid. The negative ions are extracted through 404 chamfered holes of 8 mm diameter in the plasma grid of a 3 grid extraction system. In this paper only extracted ion currents are quoted. The co-extracted electrons are deflected out of the beam by permanent magnets mounted in the second grid (extraction grid). More details of the set-up and of the modifications to the source design and the upgrade of the testbed for long pulse operation in Deuterium are described in [3].

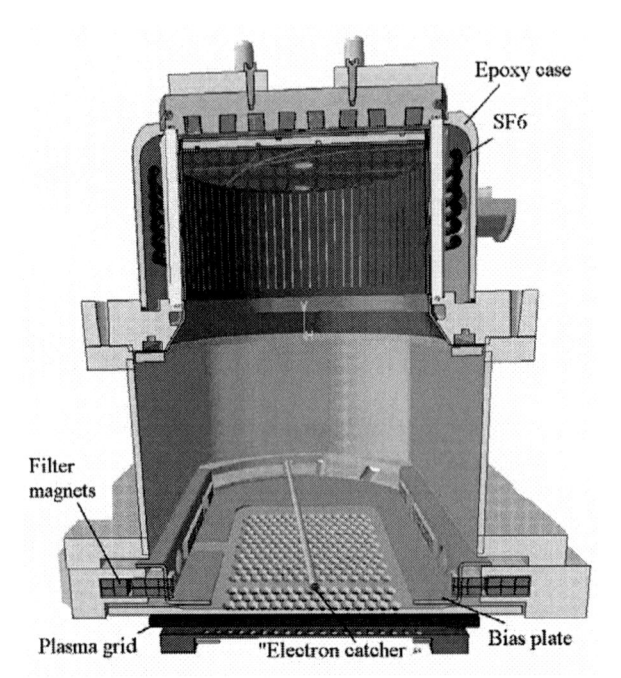

FIGURE 1. Cross section of the RF source

IMPROVEMENTS OF THE SOURCE DESIGN

Particularly in long pulse operation it was observed that after intense operational time all inner surfaces were covered by a thin layer of copper and copper lines were observed in the plasma light. The copper was apparently sputtered from the copper Faraday shield which protects the alumina insulator of the driver from plasma erosion. It was suspected that this would affect the work function and hence the e/H$^-$ ratio close to the plasma grid. To reduce the sputtering the Faraday shield and almost all inner surfaces have been coated with a 3 µm Molybdenum layer. This had great impact on the source performance:

- Almost no more Copper lines and no coatings were found in the source.
- The co-extracted electron current, which was an issue in previous long pulse experiments, now remains stable on a much lower level, in particular during long pulses. This is demonstrated in Fig. 2 by an example of two typical pulses, one before and one after the coating.
- The maximal achievable power efficiency (H$^-$ current density/RF power) simultaneously increased from 0.3 mA/cm^{2}kW to 0.4 mA/cm^{2}kW at 50 kW. At high power the efficiency decreases; however, a maximum extracted ion current of 27 mA/cm^2 has been reached (Fig. 3).

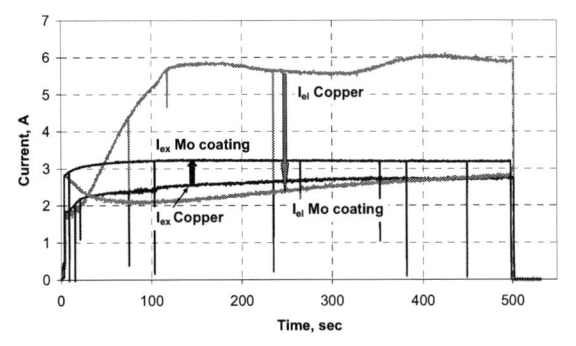

FIGURE 2. The arrows show the effect of the Mo coating of the inner source walls on the currents

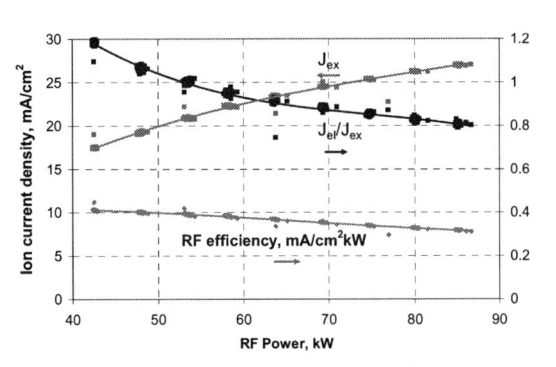

FIGURE. 3. Power scan performed during one pulse in Hydrogen at 0.4 Pa

In Fig. 4 the data of the last experimental campaign with a Mo coating are compared with previous results and show the progress in long pulse operation up to 1000 s, where for the first time an electron fraction below one is achieved, even at ion current densities of about 20 mA/cm^2.

Because of the changed surface conditions the role of the plasma grid temperature has been investigated again. A minimal temperature of 140° is required for the optimal source performance (Fig. 5); a further increase up to 220° has no significant effect on currents extracted from the source.

High power operation in particular in long pulse operation so far suffered greatly from RF breakdowns at the coil, which could happen above approx. 70 kW and can destroy the coil. This will not occur at the ITER source, because there the coil is in vacuum like has been the case for many years at the ASDEX-Upgrade NBI sources. At MANITU the problem was solved by a case made from epoxy resin filled with SF6, in which the coil is immersed (Fig. 1). This enabled to extend the operation range to >100 kW without breakdowns even with long pulse duration. Unfortunately this modification could so far not be tested with a well conditioned source, and so the maximal power was still limited by the electron current which in this case increases rapidly at high power.

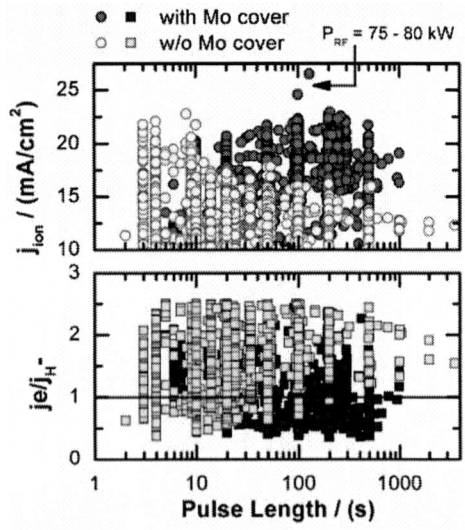

FIGURE 4. Comparison of the H⁻ current density and fraction of co-extracted electrons before and after the Mo coating of the inner source walls, at a RF power of 50 – 60 kW and 0.45 Pa, two pulses with 75 – 85 kW

CS HANDLING

To achieve high H⁻ currents by surface production it is necessary to generate a homogenous Caesium layer on the plasma grid surface. Many pulses are needed to distribute the Cs into the source. Parameters to control this "conditioning" are the Cs-oven temperature (evaporation rate), the wall temperature (Cs inventory on the walls) and the pulse length. Beam extraction can be useful to release the Cs which is trapped on the back plate by back streaming positive ions. This tedious procedure is finished when the electron current is lower than the ion current, and the electron fraction does not increase at high power (Fig. 3) and at low pressure. This indicates that the surface production is dominating the H⁻ yield.

In arc sources the high Cs

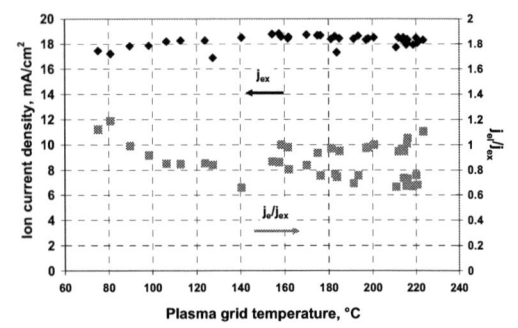

FIGURE 5. Dependence of extracted ion current and the fraction of co-extracted electrons current on the plasma grid temperature at 0.45 Pa and 55 kW

consumption is an issue because the tungsten evaporated from the filaments covers the Cs layers and makes it therefore necessary to evaporate Cs more frequently. The RF source does not have such a problem, apart from sputter products, now avoided by Mo coating, so the Cs consumption is expected to be much less. Unfortunately a common method for the estimation of the Cs consumption does not exist. It depends on the plasma on time, the total time, the evaporation rate and the wall temperatures and the avoidance of the copper sputtering will probably also have an effect. However, in an operation mode with pulse durations from 100 s to 500 s one g of Cs lasted for 12 experimental days within one month, corresponding to 20 hours plasma-on time or 14 µg/plasma-s Cs consumption. This is a pessimistic estimation, because the Cs oven used evaporates not continuously but in single bursts of very much Cs and it is not clear if all of the Cs was evaporated.

Experiments indicate that Cs evaporation during the pulses has no immediate impact on the source performance. In Fig. 6 the Cs is released from the source walls during a 500s pulse by raising the temperature of the cooling water and in this way of the wall temperature from 20°C to 40 °C. The currents did not change during this pulse, but in the next pulse.

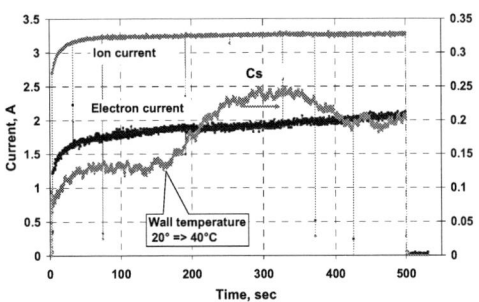

FIGURE. 6 Increasing Cs signal caused by a raised wall temperature

ELECTRON SUPPRESSION

Although the overall current of the co-extracted electrons could be reduced by the Mo coating to a remarkably low level, it can differ from pulse to pulse or even during one pulse due to changes of the surface conditions. Therefore it is still required to be able to control the j_e/j_H^- fraction. This can be done in different ways:

- The common method is to apply a positive bias voltage to the plasma grid with respect to the source
- Biasing the "bias plate" in the same way is a second method
- Shifting a Langmuir probe parallel to the plasma grid into the source reduced the electron current in previous short pulse experiments [4]. This was the motivation to

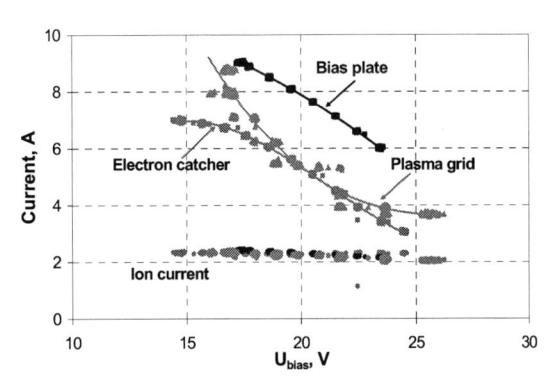

FIGURE 7. Reduction of the electron current by a positive bias voltage applied to the bias plate, the electron catcher or the plasma grid with respect to the source body; each of the scans was performed within one pulse

place a 6 mm metal rod ("electron catcher") in 2 cm distance from the plasma grid perpendicular to the filter field into the source at MANITU (s. Fig. 1). In this way the electrons which are trapped parallel to the filter field should be removed from the plasma more efficiently.

All these methods have been tested in various combinations. Fig. 7 shows an overview of the results in the case of a badly conditioned source, which is indicated by the high electron current. All methods seem to work and reduce the electron current without affecting the ion current. But no clear preference could be seen, because the differences can at least partly be caused by changes of the surface conditions, which can occur from pulse to pulse and lead to different electron currents.

OPERATION IN DEUTERIUM

Due to the legal restriction of the radiation dose the total beam-on time in Deuterium is limited at the MANITU testbed to 6 h/year [2]. In order to maximize the number of pulses the pulse length has been limited to less than 100 s. This is reasonable, because the currents are more or less stable after this time since the source has been coated with Mo. In addition the beam extraction was pulsed with typically 10s/20s beam/off time. The comparison of two 100 s-pulses in Fig. 8 shows no difference in source performance with and without interruption of the extraction and so for the conditioning the duration of the beam extraction can be reduced. This result is very important for the future Elise testbed [5], for which a pulsed operation mode is planned for technical reasons.

In previous experiments in Deuterium on MANITU and Batman the electron current was much higher as in Hydrogen. Therefore it was necessary to strengthen the filter field by

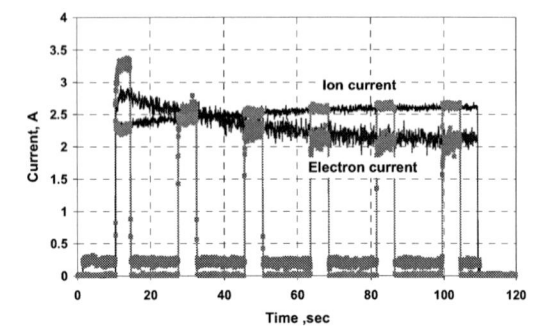

FIGURE 8. Two subsequent pulses in Deuterium with/without interruption of the beam extraction at 50 kW and 0.4 Pa

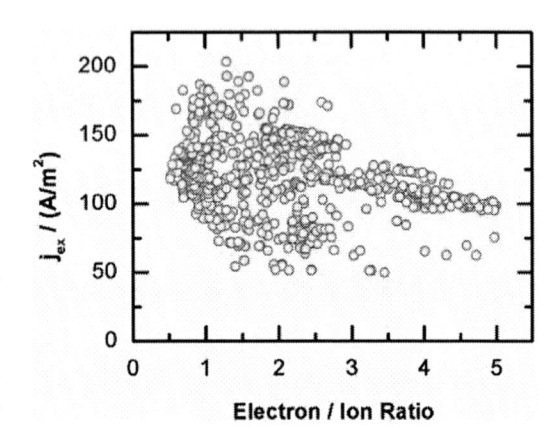

FIGURE 9. Electrically measured ion current density vs. electron fraction in Deuterium at 0.4 Pa and 50 - 60 kW

additional rows of permanent magnets at the sides of the source in order to suppress the co-extracted electrons [1]. After the Mo coating, the electron current is still higher in Deuterium, but it was possible to keep the electron fraction below 1 without changing the magnet configuration (Figs. 8 and 9). Unfortunately it was not enough time to finish the conditioning procedure in Deuterium, which is indicated by insufficient power efficiency and an increasing electron fraction in the beam at high power.

NEXT STEPS

The next step will be conditioning at high power, which is now possible with the improved antenna insulation, with the goal to reach the high ion current density according to the ITER requirements also in long pulses. Revising of the Cs oven design and investigating the Cs dynamics in the source are further important tasks in order to find reproducible conditioning and operation procedures.

ACKNOWLEDGMENTS

This work was supported by a grant (#TW6-THHN-RSFD4) from the European Union within the framework of EFDA (European Fusion Development Agreement). The authors are solely responsible for the content.

REFERENCES

1. E. Speth, et al., *Nucl. Fusion* **46** (2006) S220.
2. W. Kraus, et al., *Fusion Engineering and Design, Vol. 74, Issues 1-4, (2005)337-341*
3. W. Kraus, et al., *Rev. Sci. Instrum.* **79** (2008) 02C108
4. P. Franzen et al., *AIP Conf. Proc.,993, 51(2008)*
5. P. Franzen et al., this proceedings

Multi-antenna RF Ion Source at a High RF Power Level

Y. Oka[a], T. Shoji[b], O. Kaneko[a], Y. Takeiri[a], K.Tsumori[a], M. Osakabe[a], K. Ikeda[a], K. Nagaoka[a], E. Asano[a], M. Sato[a], T. Kondo[a], M. Shibuya[a], and S. Komada[a]

[a]*National Institute for Fusion Science / -NINS-, Oroshi, Toki 509-5292, Gifu, Japan*
[b]*Nagoya university, Chikusa, Nagoya 464-8603, Japan*

Abstract. A multi-antenna radio-frequency ion source with a Faraday shield is newly tested at a high RF power level in a large area negative ion source of 1/5th scale of the Large Helical Device-NNBI ion source. Inductively coupled dense hydrogen plasmas were generated uniformly over an area of 25 x 25cm^2 at an RF input power up to 300 kW for a 10ms pulse duration. A large negative plasma potential for the non-Faraday shielded antenna was remarkably reduced by introducing a Faraday shield. The positive ion saturation current density measured by Langmuir probe reached 148 mA/cm^2 at 174kW around the center of the plasma. The optimal hydrogen filling pressure ranged around 0.13 Pa- 0.4 Pa for the positive ions. Ion beam extraction with a single hole (ϕ 0.5cm) extractor has been studied systematically. A maximum H$^-$ ion beam current density of 1.6 mA/cm^2 was obtained preliminarily. It was confirmed that the plasma profile was controllable by both the number and configuration of the antennas.

Keywords: radio frequency ion source, NBI, multi-antenna, Faraday shield
PACS: 52.80.Pi, 52.50.Dg, 52.50.Gj

INTRODUCTION

A filament-less RF ion source has several advantages such as long operation time and less impurity (expectedly little poisoning of seeded Cs). In the National Institute for Fusion Science (NIFS), two concepts of a large area RF ion source have been proposed to generate uniform high dense plasmas: (1) Multiple small RF source where plasmas were expanded into a large area tandem-type magnetic bucket source, as was reported in [1-3] originally. This concept has been applied recently as a primary concept of an International Thermonuclear Experimental Reactor- RF ion source [4,5]; (2) A multi-antenna type tandem RF ion source (MATISE) - multi-antenna was installed in a large area magnetic bucket source [6-8], similar to the concept of a multi-filament arc discharge source [9]. Multi-antenna can reduce the antenna inductance and results in reducing the breakdown RF voltage [6] (due to parallel load connection) at the feedthrough. Therefore, high RF power can be input to the antennas to make dense uniform large plasmas. All metal inside antenna is considered to be preferable for the source under the radio-active circumstances without quartz or ceramic chamber as in the concept (1).

In the first MATISE source, a 10 linear antenna system was tested [6, 8] in a 1/5th scaled Large Helical Device-NNBI ion source to prove the concept and to characterize basic source operation. In the next experiment for high power, 4 long antennas were assembled, using antenna feedthroughs with a high voltage hold-off. Uniform profile of dense plasmas was checked and plasma parameters were studied with RF power up to ~ 50 kW [10, 11].

In this paper as second step of the MATISE source, a Faraday shield which covers the multi-antenna is devised to protect the antenna from plasma penetration and/or to work as an electrostatic shield. This leads to reduce the plasma potential [12], which is needed for good beam extraction optics without sputtering from antenna. Without Faraday shield, we observed the large negative potential dip near the plasma grid (PG) (due to the high energy electrons) and low beam extraction current. The reason for this is conjectured that the expansion of the sheath gap flattens the plasma boundary / meniscus without focusing near the extraction hole. To characterize the source performances, we study the operating parameters at high RF power levels for a 10 ms pulse duration, and the extraction of ions with a small extractor systematically. The profile of the ion saturation current is measured by a Langmuir probe. These results with and without a Faraday shield are discussed. The plasma performance parameters in detail are to be reported in another paper of Ref. [10] and [12].

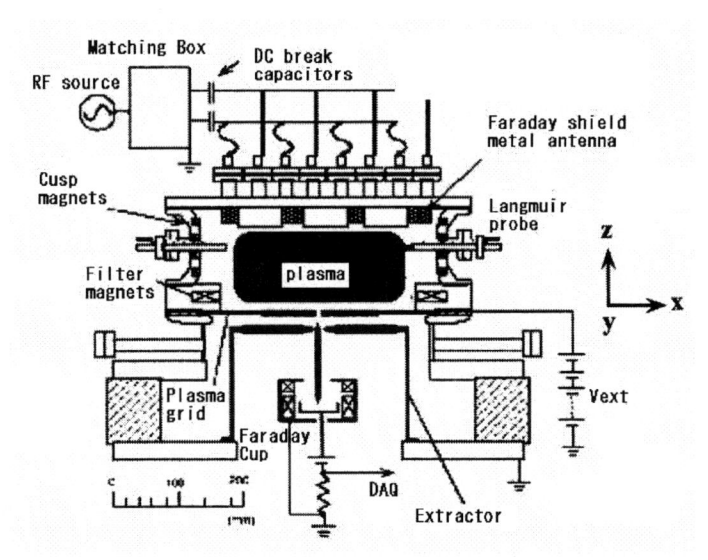

FIGURE 1. A schematic drawing of a multi-antenna type/tandem RF ion source (MATISE) and the major instruments.

ION SOURCE AND APPARATUS

A schematic diagram of the MATISE RF ion source and the major instruments are shown in Fig.1. A multi-antenna system was installed in the 1/5th LHD-NNBI ion source [13, 14] (of which size is 38 x 38 x 18.3 cm^3), equipped with external filter magnets (~800Gauss·cm) located at z=1.7 cm in front of the plasma grid (PG) and a superposable LMF filter magnet [15] (i.e., ~370Gauss·cm of localized virtual magnetic filter) embedded inside the PG (thickness of 0.4cm).

The RF antenna is made of a ϕ 0.6cm-copper tube of 20cm in length, passing through feedthrough ports on the back-plate of the plasma source chamber. Four antennas fit with a separation of 7 cm to the adjacent antenna. The antenna tube in direction is configured orthogonally across the field line of the weak field of the external magnetic filter.

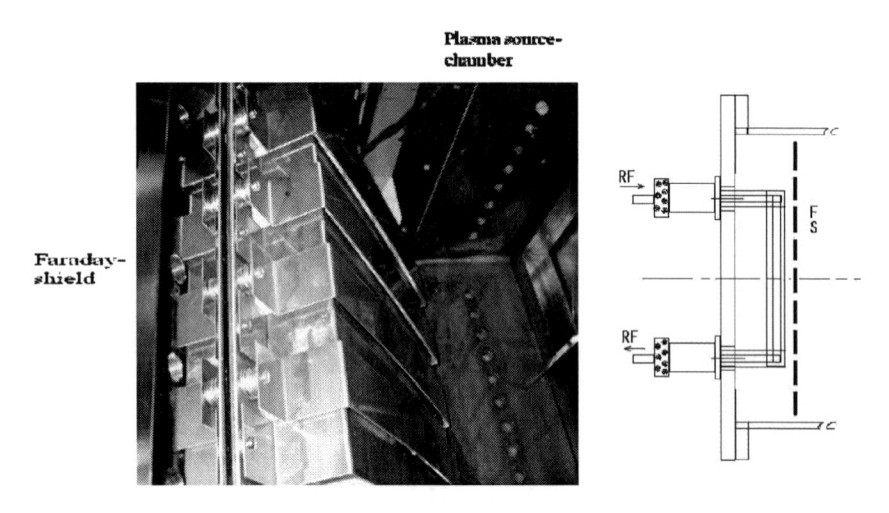

FIGURE 2. Photograph of a multi-antenna with a Faraday shield inside the plasma source chamber of the 1/5th scaled LHD-NNBI ion source.

Faraday shields (Fig.2) are composed of 9 plates of stainless steel with 4 cm in width, covering the RF antennas. Those are mounted inside on the back-plate. The RF power from the generator up to ~300 kW is coupled to the antennas for a 10 ms pulse duration every 50sec, at 9 MHz. A multi-antenna is connected in parallel to reduce the antenna inductance. RF power is coupled to the antennas via a matching circuit, and the antenna was isolated electrically by DC-break capacitors. A small extractor has a single ϕ 0.5cm-hole at the center of the PG. A grounded grid was equipped with pairs of magnets to separate co-extracted electrons. Extraction voltage is supplied to the PG at up to 15kV. The PG and the plasma source chamber are connected electrically. Plasma parameters and the profile are measured with a movable Langmuir probe (probe area of 0.061cm^2), having a 5.5 cm crank-shaped arm. The origin (x=0, y=0, z=0) of the coordinates of the probe position is taken at the center of the PG (i.e., at the extraction hole), as shown in Fig.1. The Faraday cup was located at ~10cm. The collector cup was immersed in a quasi-uniform [15] magnetic field of ~ 750 Gauss. When the cup was biased with -180 V to +180 V during the beam measurements, no changes of the ion beam current were detected. The cup collected only the ion beam component. The filling gas pressure on the plasma source was measured with a capacitance manometer. The pressure on the volume downstream from the extraction hole ranged from 1.3×10^{-5} to 1.3×10^{-3} Pa.

EXPERIMENTAL RESULTS

High-RF Power Operation

For the production of inductively coupled dense plasma, high RF power is transferred to the antennas through an impedance matching circuit. When RF power was pulsed for 10 ms at the timing before the end of fast gas puff for 2.2s, the plasma was ignited/generated for ~10 ms. The highest RF input power achieved up to 300kW in the discharge shots for probe measurement, and up to 200kW in the beam shots for ion extraction. The filling gas pressure in the plasma source chamber was in the range from several times 10^{-2} Pa to 1 Pa in H_2.

It can be seen from Fig.3 that the maximum ion saturation current density measured by a Langmuir probe at an RF power of 150-200kW reaches ~115mA/cm^2 at 0.13-0.4Pa at the axis-position (0, 0, 7.2 cm) near the center of the plasma source chamber. The maximum achieved ion current density was 148 mA/cm^2 with 174 kW at 0.4 Pa in the shots for RF power dependence. The current decreases gradually with the increasing gas pressure over a few times 10^{-1} Pa. The plasma potential was typically in the range of 5-10V [11, 12]. The floating potential with a Faraday shielded antenna became smaller by one order of magnitude compared to those for the non-Faraday shielded antenna.

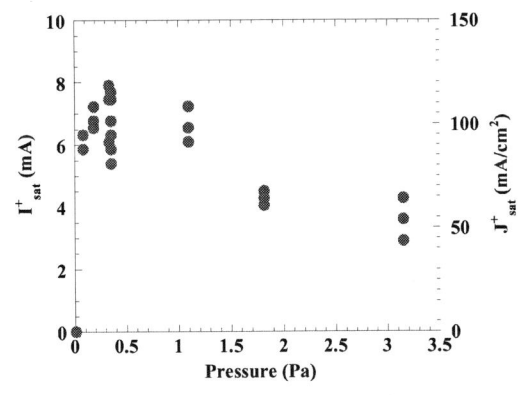

FIGURE 3. Ion saturation current I^+_{sat} and the current density, J^+_{sat} with a Langmuir probe vs. gas pressure in H_2. RF input power, P_{rf} = 150–200 kW.

Positive Ion Beam Current

For ion source characterization, positive ions were extracted at an extraction voltage of +15kV on the PG (Fig.4(a)), for configuration of a negative ion source with a magnetic filter. When the RF input power reaches about 50kW, the ion beam current begins to appear and increases rapidly with the RF power. This RF power could be a threshold power for inductively coupled plasma production. The ion beam current with the power of 140–200kW tends to be saturated.

The gas pressure dependence of the ion beam current shows a maximum value at about 0.27 Pa of a desirably low gas pressure (Fig.4(b)). The ion current decreases gradually with the increase of gas pressure. It corresponds to the dependence of the ion saturation current on the gas pressure in the source plasmas (see Fig.3). It is attributed that RF power absorption by collisional damping of electrons decreases with pressure in high pressure region in an inductively coupled plasma.

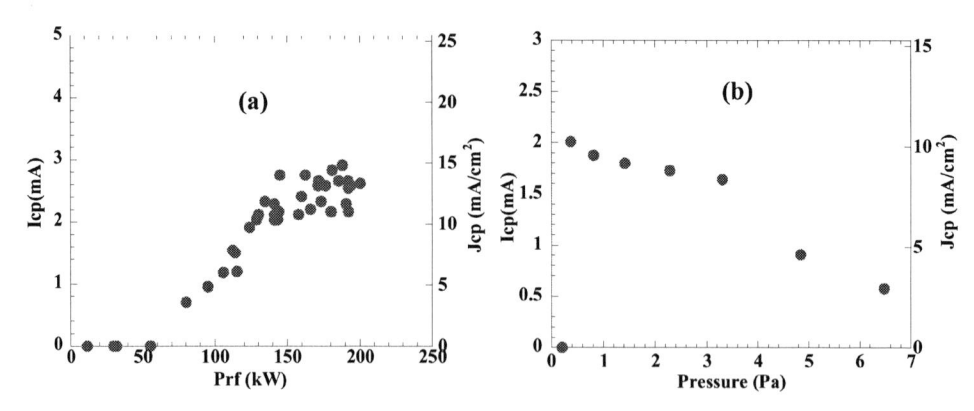

FIGURE 4. (a) Positive ion beam current, I_{cp} and the current density, J_{cp} vs. RF input power, Prf. At 0.97 Pa.; (b) Positive ion beam current and the current density vs. gas pressure. Prf= 150-200kW

In order to check the maximum achieved ion beam current under the same experimental condition used before, the thick PG (0.4cm) was changed back to previous thin one (0.2cm) without LMF filter magnets, while applying the external filter. The ion beam current density (Fig.5), which behaves with similar dependences to Fig. 4(a), reaches a maximum of 30 mA/cm^2. Correspondingly the ion beam current density of 105 mA/cm^2 is expected, if the external filter magnet for the tandem configuration is dismounted from the plasma source chamber because the ion saturation current density from the Langmuir probe near the center was larger by a factor of ~3.5 than the extracted ion beam current density [8]. This factor would come from a density distribution along the z-axis due to mainly the diffusion across the external magnetic filter field for the configuration of the negative ion source. Although the power efficiency of the source is low, it is anticipated that only a small fraction of the RF input power is absorbed in inductively coupled plasmas in the present source.

For the cases without and with a Faraday shield, the RF power dependence of the ion beam current is compared in Fig.5. Previous data are cited from Fig.2 of Ref. [11]. The ion current without a Faraday shield (RF current conductor is covered [10] by quartz tube) increases proportionally with the RF power in the low power region (0-50kW). The RF power cannot be increased above ~ 50 kW for antennas without a Faraday shield due to the RF breakdown between the antenna and the plasma. The Faraday screen shields the antenna voltage which makes plasma production by capacitive coupling difficult. Therefore, lack of the seed electrons for plasma production by inductive coupling leads to high power ignition threshold. So, the plasma ignition in present source starts from a relatively high power of 50 ~ 100 kW

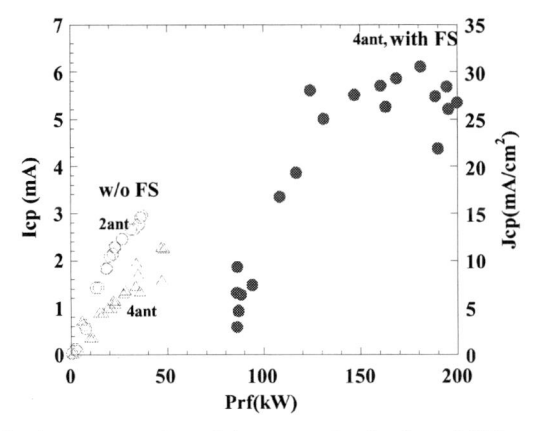

FIGURE 5. Positive ion beam current, I_cp and the current density, J_cp at 0.97 Pa vs. RF input power. Both data plotted only in this figure are taken for cases with the thin PG. Open symbols show the data without the Faraday shield at 0.83 Pa.

Negative Ion Beam Current

Negative ions were extracted at an extraction voltage of -15kV on the PG. In Fig.6(a), in addition to the high density inductively coupled plasma we add the low density mode due to the capacitively coupled plasma production. When the RF input power is raised over ~50kW, negative ion beam current starts to be extracted and increases with the RF power. When the RF power reaches 100-140kW, it appears as a rapid increase of the negative ion beam current (where two modes of the ion current seem to be in co-existence, i.e., high beam current and low beam current). This dependence of negative ion beam current should be similar in behavior to those of positive ion beam current as is seen from Fig.4(a) and Fig.5, although the extent of the data scattering is not always same. The highest negative ion beam current density can be seen to be 1.6 mA/cm^2 with about 140 kW at 0.33 Pa at a preliminary experiment.

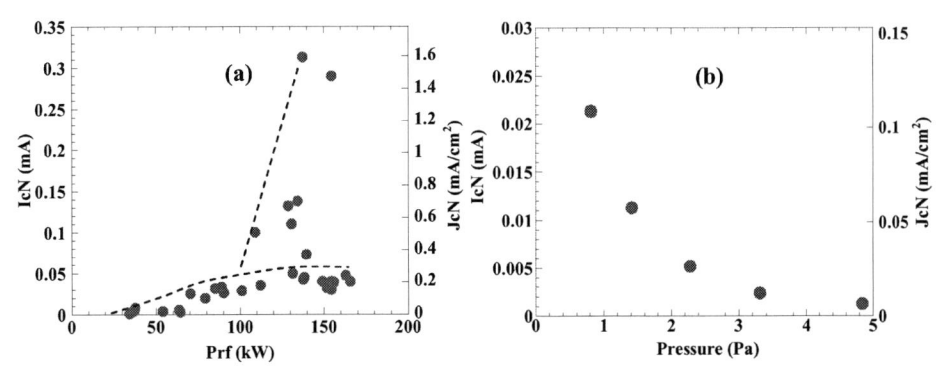

FIGURE 6. (a) Negative ion beam current, I_cn and the current density, J_cn vs. RF input power. 0.33 Pa.; (b). Negative ion beam current, I_cn and the current density, J_cn vs. gas pressure. 70-104kW.

When the gas pressure is increased (Fig.6(b)), the negative ion beam current decreases rapidly, where it looks inversely proportional to gas pressure. This means that this mainly comes from a stripping loss of H$^-$ ions colliding with residual gases. The extent of the variation in H$^-$ ion production and/or the destruction mechanism in the source plasmas with gas pressure seems to constitute only a small factor in the present RF driven source, operated on the basis of the volume production of H$^-$ ions [16].

As the maximum negative ion current density is still low, the production efficiency of the negative ion beam current in future studies should be improved by about one order more as well as being subjected to surface production with Cs seeding [17, 18]. In addition to the required improvement of the power efficiency of the source discussed above, plasma parameters, multi-antenna/Faraday shield design, and magnetic filter strength etc will be optimized. Electron components should be measured.

Spatial Profile Of Positive Ion Saturation Current

For a two antenna system, the profile of normalized (at x~0) positive ion saturation current from a Langmuir probe is shown in Fig.7(a). The plasma profile along the line parallel to the x-axis appears peaky near the two antennas in the plane of z=7.2cm next to the Faraday shield. In the plane of z=1.7cm near the PG, the profile seems to be diffusively uniform. Around the location of x~-11cm near the pole piece of the external filter magnets also in the plane of z=1.7cm, the plasma tends to be dense. The reason for this partly would come from the magnetic-filter flux concentration (resulting concentrated plasmas). Above the x=-12.5cm location very close to the pole piece, however the plasmas did not diffuse there (where the ion current value was nearly zero), due to the strong field.

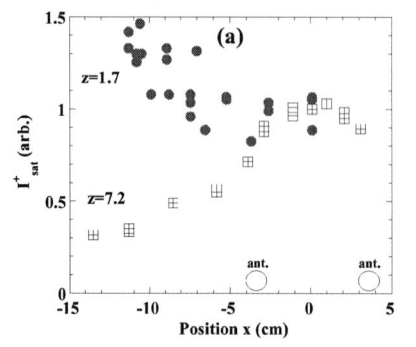

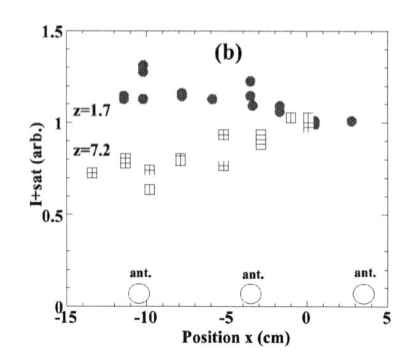

FIGURE 7. (a) Profile of normalized (at x~0) positive ion saturation current along the x-axis for two antenna case. Squares are data points along the x-line at y=5.5cm in the z=7.2cm plane next to the Faraday shield at 90-120kW and dots along x-axis at y=0 in z=1.7cm plane near the PG at 130-200kW; (b) Profile of normalized (at x~0) positive ion saturation current along the x-axis for 4-antenna case. Others are the same as those in Fig.7(a). At 160-200kW.

For a four antenna system, it can be seen from Fig.7(b) that the plasma profile becomes more uniform all over the source volume (25 x 25 cm^2), corresponding to one segment area of the LHD-NNBI accelerator. Therefore, it resulted that the plasma profile with the Faraday shielded multi-antenna system was controllable by antenna number/configuration, as was seen also in cases without a Faraday shield [10]. In this source, the necessary numbers of multi-antenna might be two to three antennas.

CONCLUSIONS

A new Faraday shielded multi-antenna RF ion source (MATISE RF ion source) was tested for characterizing the operating parameters. Inductively coupled dense plasma was produced especially at high RF power up to 300kW for a 10ms pulse duration. Negative and positive ions with small extractor were extracted from dense plasmas with a small plasma potential. Controllability of the plasma profile was proved. Further improvement will be needed along with extracted-negative ion beam current density, while positive ion beam current density in the present source appears to fulfill the level of the PNBI source, if the external magnetic filter is removed.

ACKNOWLEDGEMENTS

This work was supported by National Institute for Fusion Science collaborating research programs (NIFS06KLRR301 and NIFS05KCBB005).

REFERENCES

1. Y. Oka, *WS on Negative Ion Beams at JAERI*, Nov. 1988.
2. Y. Oka, T. Shoji, T. Kuroda, O. Kaneko, and A. Ando, Rev. Sci. Instrum. **61**, pp398-400 (1990).
3. Nikolai Gavrilov, "High Current Gaseous Ion Sources", in *The Physics and Technology of Ion Sources*, 2nd ed., edited by I. G. Brown, Weinheim, Wiley-VCH Verlag GmbH&Co, 2004, pp107-132.
4. D. Marcuzzi, P. Agostinetti, M. Dalla Palma, H. D. Falter, B. Heinemann, R. Diedl, *Fusion Eng. Design*, **82**, p798 (2007).
5. W. Kraus, H.D. Falter, U. Fantz, P. Franzen, B. Heinemann, P. McNeely, R. Riedl, and E. Speth, *Rev. Sci. Instrum.* **79**, 02C108 (2008).
6. Y. Oka and T. Shoji, *Fifth Joint JA-EU Workshop on NBI*, CIEMAT, Madrid, Sept. 2000 (umpublished).
7. Y. Setsuhara, Y. Takaki, S. Miyake, M. Kumagai, Y. Sakawa, and T. Shoji, Proc. of the 7th Int. Conf. on Plasma Surface Eng. Garmisch Partenkirchen, Germany, 2000 (unpublished), p17.
8. Y. Oka, T. Shoji, M. Hamabe, Y. Sakawa, C. Suzuki, K. Ikeda, O. Kaneko, K. Nagaoka, M. Osakabe, Y. Takeiri, K. Tsumori, E. Asano, T. Kawamoto, T. Kondo, and M. Sato, *Rev. Sci. Instrum.* **75**, pp1841-1843 (2004).
9. R. Limpaecher and K. R. Mackenzie, *Rev. Sci. Instrum.* **44**, pp'26 (1973).
10. T. Shoji, Y. Oka, and LHD NBI Group, *Rev. Sci. Instrum.* **77**, 03B513 (2006).
11. Y. Oka, T. Shoji, K. Ikeda, O. Kaneko, K. Nagaoka, M. Osakabe, Y. Takeiri, K. Tsumori, E. Asano, T. Kondo, M. Sato, and M. Shibuya, *Rev. Sci. Instrum.* **77**, 03B506 (2006).
12. T. Shoji and Y. Oka, in *Ann. Report of National Institute for Fusion Science*, Apr. 2007-Mar. 2008.
13. A. Ando, K. Tsumori, Y. Oka, O. Kaneko, Y. Takeiri, E. Asano, T. Kawamoto, R. Akiyama, and T. Kuroda, Phys. Plasmas, **1**, pp2813-2815 (1994).
14. Y. Takeiri, O. Kaneko, Y. Oka, K. Tsumori, E. Asano, R. Akiyama, T. Kawamoto, T. Kuroda, and A. Ando, Rev. Sci. Instrum. **66**, pp2541-2546(1995).
15. Y. Oka, A. Ando, O. Kaneko, Y. Takeiri, K. Tsumori, R. Akiyama, and T. Kawamoto, J. Vac. Sci. Technol. **A12**, pp3109-3114 (1994).

16 M. Bacal and G. W. Hamilton, Phys. Rev. Lett. 42, pp1538-1540(1979).
17 Yuri Belchenko, G. I. Dimov, and V. G. Dudnikov, Nucl. Fusion, **14,** p113 (1974).
18 Y. Okumura, M. Hanada, and T. Inoue, Proc. of the 16th Sym. on Fusion Technology, London UK., 1990 (umpublished) pp1026-1030.

Characteristics of rf H⁻ Ion Source by Using FET Power Source

A. Ando[a], C.H. Moon[a], J. Komuro[a], K. Tsumori[b] and Y. Takeiri[b]

[a] Graduate School of Engineering, Tohoku University, Aoba-yama, Sendai, 980-8579, Japan
[b] National Institute for Fusion Science, Oroshi-cho, Toki, 509-5292, Japan

Abstract. Characteristics of radio frequency(rf) plasma production are investigated using a FET inverter power supply as an rf generator. The matching circuit in the inverter system is simple compared to a conventional 50 Ohm matching system and only an imaginary part of the impedance of rf transmission should be matched by adjusting operating frequency or capacitance of the circuit. An electron density over $10^{18} m^{-3}$ is produced in argon plasma with 1kW rf power. Lower densities are obtained in helium and hydrogen plasmas compared to the argon plasma. Effect of axial magnetic field in driver region is examined. Electron density more than $10^{18} m^{-3}$ is obtained at the hydrogen gas pressure around 1Pa with the help of the axial magnetic field.

Keywords: ion source, rf plasma, FET inverter, neutral beam injector
PACS: 52.50.Dg, 52.25.Jm, 52.80.Pi

INTRODUCTION

Neutral beam injection (NBI) system is the most reliable and powerful heating methods for fusion devices. Neutral hydrogen isotopes are injected with the energy of more than 100keV for heating core plasmas in the devices. As neutralization efficiency of positive ions of hydrogen with these high energy is extremely low, less than 20%, negative ions should be accelerated and neutralized in order to obtain high energy neutrals.

The high power negative-ion-based NBI systems have been successfully developed in National institute for Fusion Science (NIFS) and Japan Atomic Energy Agency (JAEA) [1-4]. In NIFS, high current negative ion sources with 45A H- current per source and ion energy of 180keV are installed in the Large Helical Device (LHD)-NBI system, where total injected energy more than 16MW has been successfully achieved. In JAEA, 10A D- ion sources are operated with the beam energy of more than 300keV for plasma heating and current drive.

These high power negative ion sources developed in NIFS and JAEA are operated with arc discharge using tungsten filaments. Although it can be operated in several 10s seconds, damage by arcing and elimination of the filaments determine total operation duration. ITER requires ion source capable of delivering 40A D- with the current density of $200A/m^2$ with the energy of 1MV. It is expected to be operated in the gas pressure below 0.3Pa and long pulse duration up to 1 hour. In order to operate ion sources in such longer duration, radio frequency (rf) based discharge without any electrode for plasma production is one of the feasible methods for plasma production.

Researches of negative ion sources using high power rf have been performed in various groups. A large area rf source was developed using a vacuum-immersed rf coil and a reliable high H- ion current was obtained [5]. Although the extracted current was comparable to the same size source by arc discharge, the vacuum-immersed rf antenna coil was damaged by sputtering and breakdown in plasmas.

External rf antenna type sources has been successfully developed for the production of positive hydrogen ions at IPP (Max-Plank-Institut für Plasmaphysik), Garching. A rf plasma was produced in a cylindrical tube of insulator with a rf coil winded in several turns externally in order to avoid the direct interaction with produced plasmas. The source was modified and intensively developed for negative ion production with Cs additive. The negative ion rf source consists of driver, expansion and extraction regions. They developed a large area rf sources, which deliver long pulse H- and D- ion beam with an current density of more than $200A/m^2$ at a source filling pressure of 0.3Pa [6-9].

Conventionally, high power rf amplifier is composed of vacuum tubes and transmission line with matching impedance of 50 Ohm. It requires high voltage power supplies and the conversion efficiency from dc to ac (rf) is relatively low due to internal losses in the vacuum tubes. Serious rf breakdown problems are sometimes occurred in the 50 Ohm transmission lines or feed through to an rf antenna. Recently high frequency generators based on an inverter circuit are used in various industrial applications. High power switching device such as thyristors and insulated gate bipolar transistors (IGBT) are used in the frequency range lower than 100kHz due to their relatively long turn-on and off time. Recent development of metal-oxide-semiconductor field effect transistors (MOSFET) enables us to use an inverter circuit for plasma production and heating with a frequency around 1MHz. In this paper, we report initial results of plasma production using rf generator based on a FET inverter circuit [10,11]. Characteristics of hydrogen plasma production are presented. The source will be operated as an H- ion source after introducing cesium additive and optimization of operating parameters.

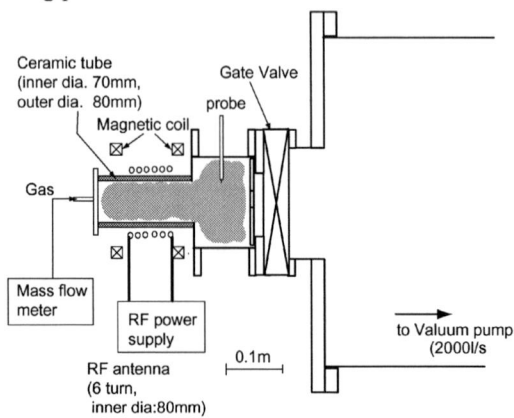

FIGURE 1. Schematic of the rf ion source.

EXPERIMENTAL SETUP

Experiments are performed using a small ion sources which consists of a driver and diffusion chamber. An rf coil is wound around ceramic (Al_2O_3) cylindrical tube (inner diameter: 70mm, outer diameter: 80mm, length: 170mm) in several turns. The driver is attached to a cylindrical chamber (diffusion chamber), where a cusp field is formed by permanent magnets. The source is installed at a test stand with a large cylindrical vacuum chamber (1m in diameter and 2m in length). The schematic of the experimental setup is shown in Fig.1. So far, plasma parameters are measured by a Langmuir probe set at the diffusion chamber. Characteristics of ion sources are investigated using the test stand.

An inverter circuit of the rf generator used for rf plasma production is shown in Fig.2(a). We utilize metal-oxide-semiconductor field effect transistors (MOSFET) as a switching device in the inverter circuit. An operation frequency can be adjusted by external signal generator in the range of 100-500kHz. A dc voltage up to 250V is fed to inverter units (PDM755HA), which compose a full bridge inverter circuit. The inverter circuit can switch on and off a maximum current of 225A The output of the FET units are combined and connected to a primary winding of an impedance transformer, which is employed a good impedance matching between the FET inverter and the low impedance antenna and transmission lines. The turn ratio of the transformer can be changed from 1:1 to 4:1 according to the matching conditions. The output voltage of transmission lines becomes lower than that of the primary voltage. Total rf power of more than 15kW can deliver to the antenna with a short duration of 5ms.

The secondary winding of the transformer is connected to a capacitor and rf antenna for plasma production. The rf antenna is a loop antenna, where the turn number is changed according to plasma conditions. A parallel plate balanced line made of copper is used as the transmission line to the rf antenna. The secondary circuit composed a LC matching circuit and capacitance of the capacitor is tuned to maximize the antenna current at LC matching condition. The rf frequency can be changed by adjusting an external signal generator and the capacitance. Fig.2(b) shows an example of the dependence of rf frequency on the capacitance.

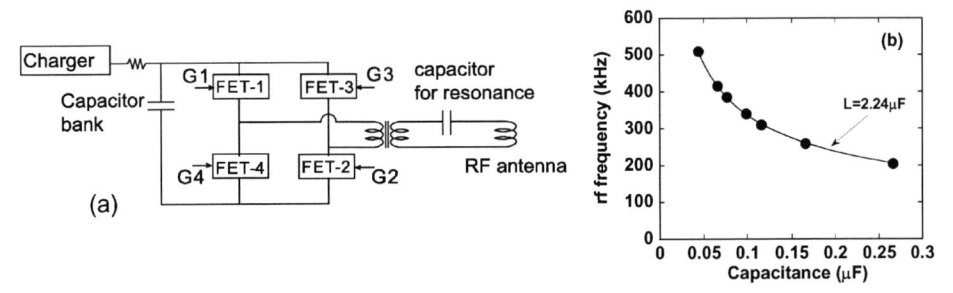

FIGURE 2. (a)Inverter circuit with an impedance matching transformer, a capacitor and the rf antenna. (b)Dependence of rf frequency on capacitance to maximum antenna current. Line is calculated resonance frequency assuming antenna inductance of L=2.24µF.

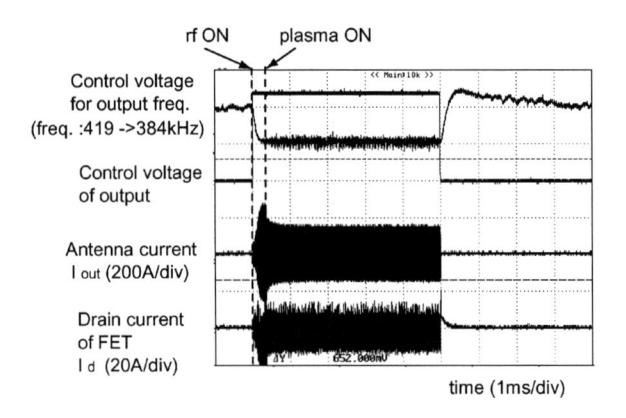

FIGURE 3. Temporal evolutions of rf inverter power supply with plasma production.

EXPERIMENTAL RESULTS

Figure 3 shows a typical waveform of FET inverter power supply with plasma production. The inverter switching frequency is automatically controlled with a phased-lock loop circuit in order to keep the matching condition. When the modulation signal is applied to the rf generator, rf frequency changes from preset frequency (419kHz) to matching frequency (384kHz). The antenna current increases as the frequency approaches to the matching frequency and a discharge occurs at the matching frequency. It is noted that the frequency does not change at the discharge occurrence, which indicates the imaginary part of antenna coupled with plasma does not change. At higher density plasma production, the frequency changes slightly.

Due to the increase of loading resistance by plasma production, the antenna current decreases as shown in Fig.3. The matching circuit based on the inverter switching system requires matching of an imaginary part of the impedance only. It is not necessary to change capacitance at the moment of plasma production in spite of the increase of the loading resistance.

Figure 4(a) shows produced plasma densities as a function of input rf power in

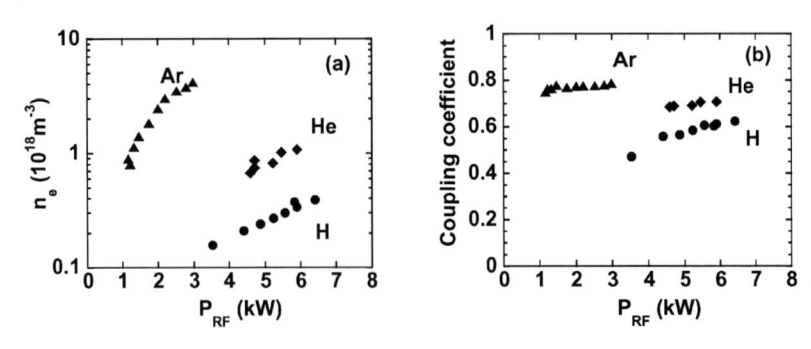

FIGURE 4. (a) Produced plasma density measured at diffusion chamber and (b) coupling coefficient as a function of rf input power. f=380kHz, 6 turn antenna, p=3Pa.

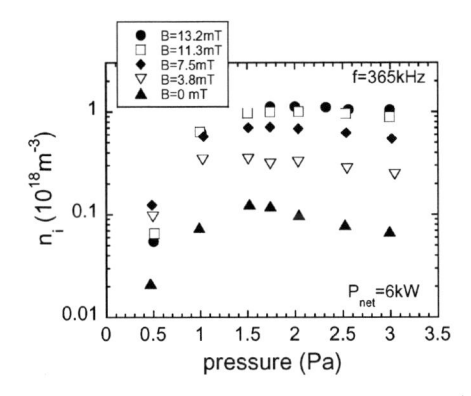

FIGURE 5. Produced plasma density measured at diffusion chamber as a function of pressure in various axial magnetic field strength. Hydrogen plasma, f=365 kHz, 20 turn antenna, P_{rf}=6kW.

argon, helium and hydrogen gases. The electron density is measured by a Langmuir probe with an rf filter located at the center of the diffusion chamber, where is 50mm apart from the exit of the driver. Although an electron density over $10^{18}m^{-3}$ is easily produced in argon gas, lower densities are obtained in helium and hydrogen gases. The coupling coefficient calculated by the following equation is plotted as a function of rf power in Fig. 4(b).

$$\eta = \frac{R_p}{R_{ant} + R_p}, \quad (1)$$

where R_{ant} and R_p are loading resistance of antenna and plasma, respectively. When the produced plasma density increases, the coupling efficiency increases.

In order to increase plasma density in hydrogen discharge, we have examined effects of axial magnetic field in the driver region. Magnetic field up to 15mT can be applied by external Helmholtz magnetic coil attached at the driver (Fig.1). As the magnetic field increases, higher density is obtained as shown in Fig.5. The density more than $10^{18}m^{-3}$ is obtained at the hydrogen gas pressure around 1Pa. The coupling efficiency decreases as the magnetic field increases. The effect of magnetic field should be pursued further with consideration of transport and rf coupling with plasmas.

So far, the duration of rf inverter power supply is limited up to several ms. This prevents low pressure operation of the source due to the difficulty of ignition. We are preparing a long pulse rf inverter, an extraction grid system, external magnetic field and cesium injection device in order to apply it as a negative hydrogen ion source. Optimization of the source operation with cesium additive will be performed in near future.

SUMMARY

We have performed initial experiments of a small rf ion source operated with a FET inverter power supply. Operating characteristics including a matching circuit in the inverter system are investigated. In the non 50 Ohm matching system, only an imaginary part of the impedance of rf transmission should be matched. Although an electron density over $10^{18} m^{-3}$ is easily produced in argon gas, lower densities are obtained in helium and hydrogen gases. Effect of axial magnetic field in driver region is examined and higher axial magnetic field results in higher plasma density in diffusion region. Electron density more than $10^{18} m^{-3}$ is obtained at the hydrogen plasma with the gas pressure around 1Pa.

ACKNOWLEDGMENTS

This work was supported in part by Grant-in-Aid for Scientific Researches from Japan Society for the Promotion of Science, and under the auspices of the NIFS Collaborative Research Program (NIFS07KOAB010).

REFERENCES

1. Y. Takeiri, *et al.*, *Nuclear Fusion* **47**, 1078–1085 (2007): *ibid.* **46**, S199-S210 (2006).
2. K. Tsumori, *et al.*, *Rev. Sci. Instrum.* **75**, 1847-1850 (2004).
3. Y. Ikeda, *et al.*, *Nuclear Fusion* **46**, S211-S219 (2006).
4. T. Inoue, *et al.*, *Nuclear Fusion* **46**, S379-S385 (2006).
5. T.Takanashi, *et al.*, *Jpn. J. Appl. Phys.* **35** 2356-2362. (1996).
6. E. Speth, *et al.*, *Nuclear Fusion* **46**, S220-S238 (2006).
7. P. Franzen, *et al.*, *Nuclear Fusion* **47**, 264-270 (2007).
8. W. Kraus, *et al.*, *Rev. Sci. Instrum.* **79**, 02C108 (2008).
9. U. Fantz, *et al.*, *Rev. Sci. Instrum.* **77**, 03A516 (2006).
10. S. Watanabe, *et al.*, *Rev. Sci. Instrum.* **69**, 3555-3557 (1998).
11. Y. Uesugi, *et al.*, *Vacuum.* **59**, 24-34 (2000).

Simulations for the generation and extraction of negative hydrogen ions in RF-driven ion sources

R. Gutser, D. Wünderlich, U. Fantz, P. Franzen, B. Heinemann, R. Nocentini and the NNBI Team

Max-Planck-Institut für Plasmaphysik, EURATOM Association, Boltzmannstr. 2, D-85748 Garching, Germany

Abstract. The injection of energetic neutral hydrogen atoms plays an important part for plasma heating in fusion experiments. In order to fulfill the requirements of the ITER neutral beam injection (NBI), a RF-driven ion source based on the generation of negative ions prior to neutralization has been successfully developed at IPP Garching. Negative hydrogen ions are generated on a cesiated converter surface (plasma grid) by neutral particles and positive ions and are then transported to the extraction apertures, where the ion beam formation process takes place. Numerical models are necessary to include the relevant physical aspects of these processes. The Monte Carlo transport code CSFLOW is used to describe the dynamical behavior of the cesium distribution on the source walls during vacuum operation. The negative ion transport process is simulated by means of the probabilistic ion transport code TRAJAN, focussing on the effects of aperture diameter variations in mono- and multiaperture extraction systems. A simulation of ion beam formation is carried out with the KOBRA3 ray tracing code, which allows a full 3d potential solution without any symmetry restriction. This is necessary to simulate beam steering effects by a non-axisymmetric magnetic field and electrode configuration for the design of the extraction systems for future negative ion experiments.

Keywords: Ion Optics, Steering, Particle Orbit and Trajectory, Cesium Flow, Negative Ion Sources
PACS: 29.25.Ni, 29.27.Ac, 52.65.Pp, 52.65.Cc, 51.10.+y

INTRODUCTION

Future fusion devices like ITER require high performance ion sources based on negative hydrogen ions to provide powerful heating and current drive capabilities. The ion source used for ITER should deliver an accelerated D^- beam current of 40 A with a current density of 200 A/m² at a source pressure of 0.3 Pa while maintaining an electron to ion ratio < 1 [1]. The RF-driven ion source was chosen to be the ITER reference source and fulfils the physical aspects of these requirements.
Detailed information regarding the individual components and operation parameters are available from [2]. Figure 1 gives a schematic overview of the source, which is divided into three parts: driver, expansion region and extraction system. A low pressure and low temperature plasma is generated by 1 MHz RF coils with a typical power of 100 kW. The plasma flow enters the expansion region where a magnetic filter field slows down the hot plasma electrons. A typical electron temperature of 2 eV and an electron density of $5 \cdot 10^{17}$ m^{-3} is obtained in the vicinity of the extraction system [3]. Negative ion production in the RF-driven ion source is dominated by the surface effect: positive and neutral hydrogen plasma particles are converted on a surface into negative ions by picking up one or more electrons. An enormous increase of the conversion rate can be

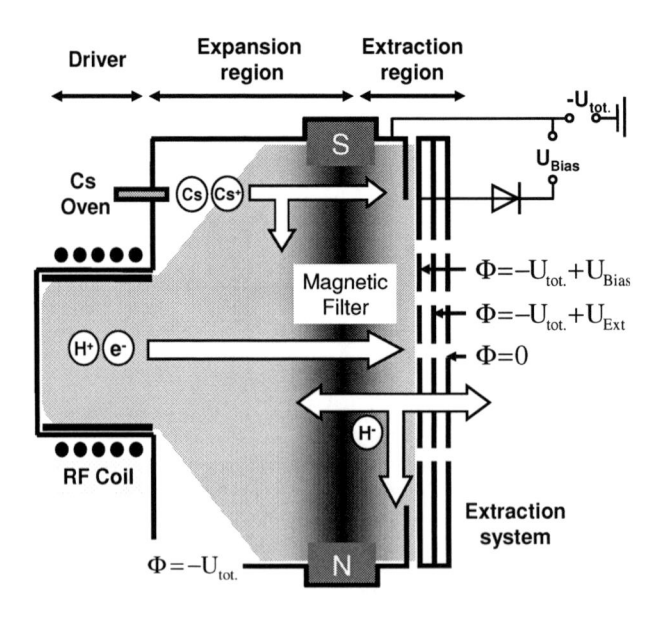

FIGURE 1. Schematic view of the RF-driven ion source. Typical operation parameter are: U_{Ext}=9-10 kV, $U_{tot.}$=25 kV and U_{Bias} ≈20 V.

achieved by lowering the work function of the converter material by a coverage with cesium [4], which is evaporated from an oven with a liquid Cs reservoir connected to the back flange of the source body.

Negative ions, which were generated at the surface of the plasma grid, are accelerated towards the plasma by the sheath potential, which is influenced by the bias voltage. This voltage is applied between the first grid of the extraction system and the source body. The negative ions heading away from the extraction system are either destroyed by collisions with plasma particles or redirected into the extraction apertures. These ions are immediately accelerated by the extraction voltage and the beam formation takes place.

Beam formation is accomplished in two stages by a three grid, multiaperture extraction system, which consists of the plasma grid, the extraction grid and the grounded grid. In the first step a voltage of 9-10 kV is applied between the plasma and the extraction grid in combination with a mass selective magnetic filter system to remove co-extracted electrons at reasonable energies. A second voltage of 20 – 30 kV is currently used for acceleration during the source development, while a stepwise 1 MV acceleration will be used for the ITER NBI.

An understanding of the complex physics of the generation and extraction of negative ions is essential for the further optimization of the source. The aim of this paper is to present three numerical models which can improve the insight in these effects.

CESIUM 3D TRANSPORT CODE

An efficient generation of negative ions on the plasma grid converter surface strongly depends on the condition of the Cs layer at the converter surface. The Cs layers on the inner walls of the ion source show a dynamical behavior and Cs redistribution inside the source takes place. As a result, the time scale of the ion source reaching high performance is strongly correlated to the time scale of Cs redistribution. An understanding of the time dependence and spatial distribution of the Cs, especially on the plasma grid, is therefore highly desirable.

The development of the Cs transport code CSFLOW is in progress. It calculates the Cs dynamics during vacuum and plasma operation of the negative ion source. Because of the low ionization energy of 3.89 eV, the Cs is ionized to a large amount during plasma operation. Therefore, neutral Cs particles have to be considered in the vacuum phase, whereas both Cs^+ ions and neutral Cs have to be treated during plasma operation. The current version of the code considers the Cs distribution for an evacuated source.

During the vacuum operation Cs redistribution is governed by thermal desorption processes from the source walls [5]. The atomic desorption rate of Cs is a sensitive parameter in the simulation. Results of the code using desorption rates of bulk Cs available in literature [6] and the temperature conditions inside the ion source indicate a very rapid loss of the Cs. The resulting depletion of the Cs from the source within the time scale of several hours is in total disagreement with experimental observation of Cs enhanced source operations.

As a result, an accompanied experimental investigation to determine the Cs desorption rate for the conditions of the ion source (background pressure and surface material) is in progress. Cs is evaporated into a small vacuum chamber and the time dependency of the Cs coverage on a coated quartz micro balance is measured for different temperatures under the relevant vacuum conditions of $p=10^{-5}$ mbar background pressure. The results of this studies indicate a desorption rate, which is reduced by a factor of 50 related to the literature value.

Simulation Model

The Cs flow from the oven into the evacuated ion source can be considered as a rarefied gas flow. While the oven and the plasma grid are heated to a typical temperature of 150 °C, the source walls are kept at 50 °C. The Knudsen number

$$Kn = \frac{\lambda_{Cs}}{L}, \qquad (1)$$

which is a characteristic quantity for the flow regime, is defined by the mean free path length λ_{Cs} and system dimension L. The low background pressure ($p=10^{-5}$ mbar) in the source volume results in a $Kn \gg 1$, which is a requirement to treat the Cs in the volume as a free molecular flow. Therefore, collisions of Cs atoms with other gas particles inside the ion source volume can be neglected. As a result, the trajectories of desorbed Cs atoms inside the source follow straight lines. Wall interactions are modeled by a cosine distribution of the Cs velocity vector with respect to the surface normal vector. A cartesian surface mesh is used to determine the amount of accumulated Cs on the source walls. Cesium atoms, which enter the aperture area of the extraction system are lost and therefore removed from the simulation.

Simulation Results: Cesium Vacuum Distribution

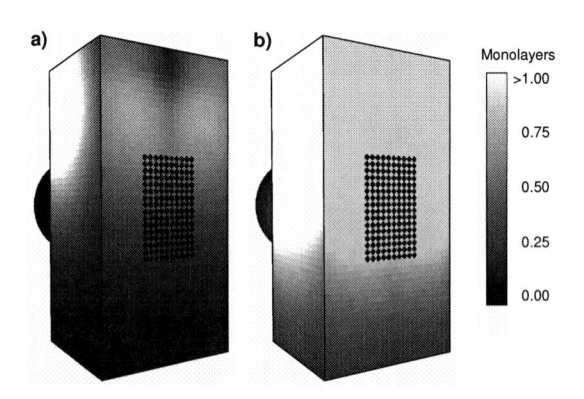

FIGURE 2. Contour plot of the Cs distribution on the source walls for 50°C wall temperature and 150°C plasma grid temperature. For a) 5 min Cs evaporation and for b) 20 min evaporation with focus on the Cs distribution on the plasma grid (< 1 ML). (Colored version online)

A flow calculation of the Cs within the full 3d geometry of the RF-driven ion source was done for the desorption rate from the experimental investigation and 10 mg/h Cs evaporation from the oven [2]. The unit monolayer ML is a convenient description of the Cs wall coverage, where 1 ML corresponds to a Cs layer of $4.5 \cdot 10^{14}$ Atoms/cm^2. Figure 2 a) shows the distribution of Cs on the side walls of the ion source after 5 minutes of Cs evaporation. Cesium accumulations of several ML are formed on the side walls of the ion source in the vicinity of the oven. Because of Cs desorption from the areas of high Cs accumulation, a Cs re-distribution between the inner walls takes place. Owing to the elevated temperature of the front wall, layer formation on the plasma grid is inhibited and the coverage is limited below 1 ML. The Cs coverage of the plasma grid is illustrated by figure 2 b), indicating a complete coverage of the aperture area after 20 minutes evaporation under vacuum conditions.

A comparison of the results with the Cs distribution actually observed after opening the ion source indicates, that a Cs distribution with high accumulations on the side walls in the vicinity of the oven, as observed in the simulation, was present in the experiment for (low) wall temperatures of 20 °C [2]. An elevation of the wall temperatures to 50 °C resolved this problem in the experiment and resulted in a more homogeneous Cs distribution.

While there is still a discrepancy regarding the temperature dependence of the desorption rate, the Cs transport code allows a realistic treatment of the Cs distribution inside the ion source. The validation of thick Cs layers by the computer simulation indicates a significant lower desorption rate of the Cs on the walls of the ion source than expected from literature values of bulk desorption. This effect can be explained by deviations of the layer from bulk Cs due to surface impurity compounds like CsOH and CsH resulting in an inhibited desorption.

A parameter variation of the desorption rate is in progress to reproduce the response of the ion source for altered wall-temperature conditions. The further development of the code will also consider effects related to the influence of the plasma on the Cs redistribution.

NEGATIVE ION TRANSPORT CODE

The distribution of Cs on the plasma grid enables an efficient conversion of positive ions and neutral particles to negative ions. A transport process from the converter surface to the extraction system has to take place in order to extract a negative ion beam. This involves a redirection of the ion velocity vector. The low binding energy (0.75 eV) of the addition electron, which is beneficial to neutralization, is a disadvantage for the transportation process. A theoretical description of the negative ion transport is therefore very important to optimize the ion current delivered by the source.

The geometry of the plasma grid converter, especially the diameter of the extraction apertures, is a question of particular importance. As a consequence, the probabilistic ion transport code TRAJAN was applied to investigate the effect of a variation of the aperture diameter on ion transport for mono- and multiaperture extraction systems.

Simulation Model

While a detailed description of the TRAJAN code is available from [7], a short summary of the code is given in the following paragraph. The transport code is based on the solution of the Lorentz equation for a statistically significant ensemble of independent test particles with different initial conditions, similar to the model described in [8]. The ions are started from the converter with an energy which represents the acceleration by the sheath voltage. A normal distribution of the spatial coordinates on the converter surface and a cosine velocity distribution with respect to the surface normal vector was chosen for the initial ion state [9]. The electrical fields of plasma sheath and the plasma-beam interface were considered in the simulation, while a zero electrical field was assumed in the plasma volume. Elastic and inelastic collisions of negative ions with plasma particles were simulated using Monte Carlo methods. A binary collision model [10] was used to treat elastic Coulomb collisions with H^+ particles. The probability of an inelastic collision was calculated by the path length estimator algorithm [11] including the relevant processes listed in table 1. The calculation of these probabilities requires knowledge of particle densities and temperatures, which were taken from experimental data [12]. The quasi-neutrality condition $n_{H^+} + n_{Cs^+} = n_e$ was used for the calculation. Negative ions are either destroyed by inelastic collisions with the plasma background or

TABLE 1. Atomic processes taken into account in the transport simulation.

Electron Stripping	$H^- + e \rightarrow H + 2e$	[13]
Mutual Neutralization	$H^- + H^+ \rightarrow 2H$	[14]
	$H^- + Cs^+ \rightarrow H + Cs$	[15]
Charge Exchange	$H^- + H \rightarrow H + H^-$	[16]

extracted by entering the plasma beam interface, while ions hitting the converter electrode are specularly reflected inside the plasma sheath.

An ion transport investigation of the converter geometry was done focussing on the effects of changes of the aperture diameter. For the sake of comparability an unidirectional, constant magnetic field of 8 mT was chosen.

Simulation Results: I. Monoaperture Extraction Systems

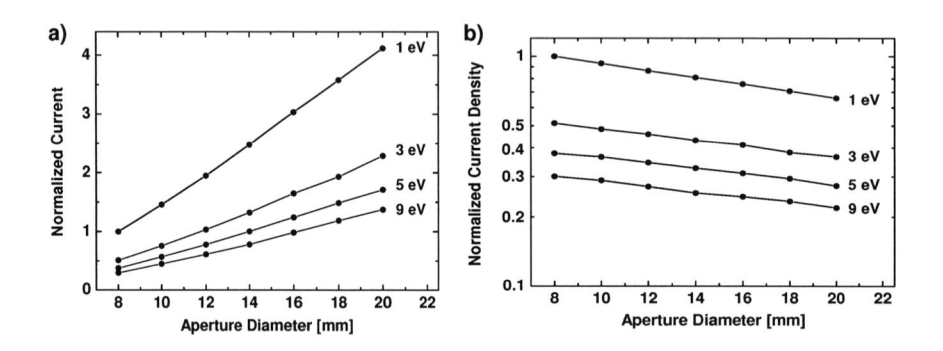

FIGURE 3. a) Normalized aperture current and b) current density versus the aperture diameter for different starting energies and a constant magnetic field of 8 mT.

Transport studies for 45° chamfered monoaperture systems with different aperture diameters (8 mm - 20 mm), but constant thickness of 2 mm were carried out. A computation domain of 5×5 cm^2 and a unidirectional filter field of 8 mT was used for the calculation. The number of extracted ions for different aperture diameters and starting energies was normalized with respect to the result for an extraction with 1 eV starting energy and 8 mm aperture diameter.

Figure 3 a) shows the dependency of the calculated extracted ion current on the aperture diameter for various starting energies. The amount of extracted ions is strongly influenced by the initial energies. Low starting velocities result in a higher retention time of the ions near the plasma grid, which enhances the probability of an extraction process.

The variation of the converter geometry manifests its influence in a direct proportionality of the extracted ion current to the aperture diameter. This effect can be explained by the linear correlation between the aperture diameter and the circumferential area encircling the aperture (effective production area). Ions, which are started on this inclined area, have a higher probability to be extracted.

Figure 3 b) shows the corresponding dependency of the current density, where the linear behavior of the ion current is exceeded by the quadratic rise of the aperture area. The current density depends therefore on the quotient of effective production and aperture area, which is indirectly proportional to the diameter.

Experimental investigations of aperture diameter variations in monoaperture systems show a similar dependence. [17]

Simulation Results: II. Multiaperture Extraction Systems

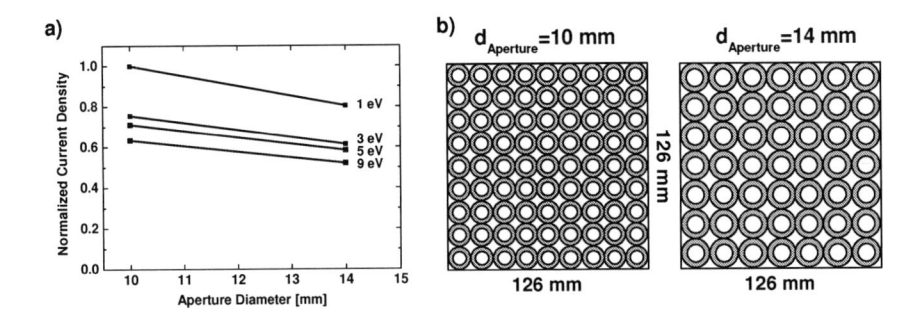

FIGURE 4. a) Normalized multiaperture current density versus the aperture diameter for different starting energies and a constant magnetic field of 8 mT. b) Illustration of extractions systems with 10 mm and 14 mm aperture diameter. The effective production area is marked in grey.

The effect of the aperture diameter in a multiaperture system is an important question. A converter geometry of a 126 mm x 126 mm array of 2 mm thickness and closely packed, chamfered apertures was investigated for two aperture sizes. Figure 4 b) illustrates these arrays of 49(81) apertures with 14(10) mm diameter. Similar to the monoaperture calculation, a higher current density is obtained in case of the smaller aperture diameter as illustrated by figure 4 a). An increase of the diameter within the constraints of the defined geometry results in a higher total aperture area, while the effective production area decreases.

Nevertheless, it is not possible to state a simple relation between the current density and the aperture diameter. Additional geometric constraints like the converter thickness and the aperture opening angle have significant influence on the current density by altering the effective production area. A comparison with experimental values requires an exact consideration of these parameters according to the individual geometries of the extraction systems.

ION BEAM SIMULATION

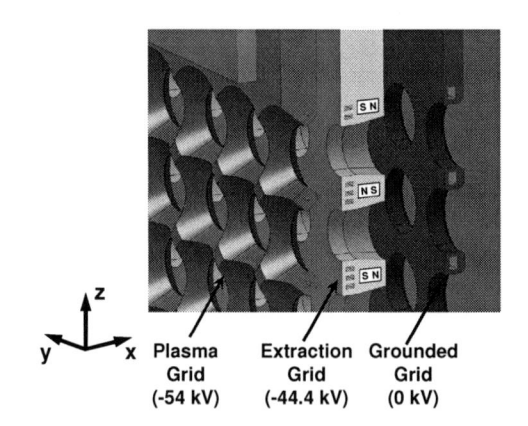

FIGURE 5. Design of the ELISE extraction system including the magnet rods the create the electron deflection field (orthogonal configuration).

A "half-size" ITER-like extraction system is being designed at IPP Garching for the ELISE (Extraction from a Large Ion Source Experiment) [18] test facility.
The ELISE extraction system, based on 14 mm diameter apertures is shown in figure 5 and will provide an ion acceleration of 60 kV. The x-axis is chosen to be the direction of the ion beam acceleration. A periodic arrangement of CoSm magnet rods with alternating magnetization is inserted into the extraction grid to generate the electron deflection field (EDF). This arrangement can be orientated either horizontally (y-direction, as seen in figure 5) or vertically (z-direction). The overall magnetic field results from a superposition with the filter field (FF), which points in the y-direction. This field is produced by an electric current flowing vertically in the PG and by additional permanent magnets positioned on the side of the source vessel. The orientation of the EDF is therefore termed according to the orientation to the FF: orthogonal and parallel.
In order to achieve optimum Cs conditions, the plasma grid will be operated at 150±50°C [19]. The expansion for the operating range of ± 50°C causes a maximum aperture offset for the outer most aperture of 0.56 mm in horizontal direction [18].
Both, the EDF and the plasma grid expansion cause steering effects on the extracted ion beam, which were analyzed by a Ray Tracing simulation.

Simulation Model

The non-axisymmetric configuration of the electrode geometry owing to the heat induced expansion of the plasma grid and the magnetic field configuration requires a fully 3d self consistent calculation of the ion beam formation process. The KOBRA3 code [20] satisfies this criterion. An ANSYS FEM 3d model was used to calculate the magnetic fields in the ELISE extraction system for the ion optics calculation [21].

Simulation Results: Beam Steering

A D$^-$ beam with a current density of 260 A/m^2 and an extraction voltage of 9.6 kV was simulated at the outermost aperture in horizontal direction for the orthogonal and parallel EDF configuration. The results of this investigation are given by figure 6. In the parallel case a), significant beam steering is caused solely by the expansion of the plasma grid in the horizontal direction and is found to be acceptable. The contribution of the FF to the steering is insignificant within the extraction system (domain of 6 cm). An intensified steering was observed in case b) of the orthogonal field configuration. In this case, the beam deflection owing to the EDF contributes to the steering by the plasma grid expansion. This results in an intensified steering and leads to minor collisions of the peripheral beam sections with the extraction grid electrode, which is found just acceptable in case of the outermost aperture.

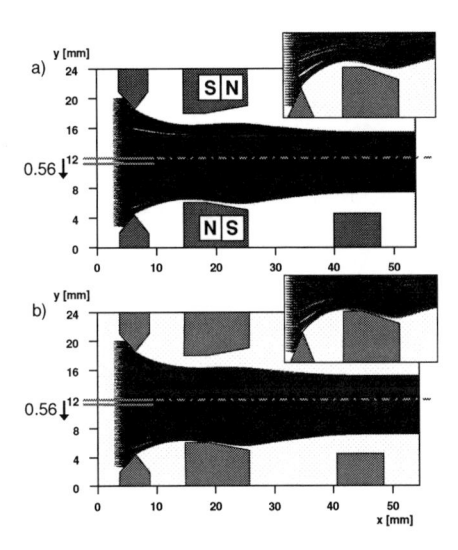

FIGURE 6. Beam steering for the ELISE extraction system for the outermost aperture with a D$^-$ current density of 260 A/m^2. a) Filter field and electron deflection field are parallel. b) Filter field and electron deflection field are orthogonal.

CONCLUSION AND OUTLOOK

Several physical aspects of the generation, transport and extraction of negative ions have been investigated by means of numerical simulations. A calculation of the dynamics of the Cs distribution for vacuum operation based on experimentally determined desorption rates showed, that large amounts of Cs accumulate on the inner walls of the ion source. The influence of Cs poisoning by formation of stable Cs compounds was clarified by comparison of experimental observation with results of the CSFLOW code. Studies of the influence of plasma operation on the Cs distribution are in progress.

Ion transport investigations with the TRAJAN code for monoaperture systems showed an indirect proportionality of current density and aperture diameter, that was also observed in the experiment. A relation between the current density and the quotient of effective production and aperture area was found within the scope of the transport simulation. This effective production area is affected by additional geometrical factors, like the converter thickness. Future comparisons with experimental results will therefore include detailed geometric features of the individual extraction systems.

Beam steering effects caused by a combination of the heat load induced plasma grid expansion and the electron deflection field were calculated self-consistently with the KOBRA3 code. An enhanced steering was found in case of the orthogonal magnetic field configuration due to the influence of the electron deflection field. Nevertheless, the steering for the outermost apertures (worst case) was considered just acceptable for both magnetic field configurations.

ACKNOWLEDGMENTS

This work was (partly) supported by a grant from the European Union within the framework of EFDA (European Fusion Development Agreement). The authors are solely responsible for the content.

REFERENCES

1. R. S. Hemsworth, A. Tanga and V. Antoni, *Rev. Sci. Inst.* 79(2008), 02C109
2. E. Speth, H. Falter, P. Franzen, U. Fantz et al, *Nucl. Fusion* **46** (2006), 220
3. U. Fantz et al., *Nucl. Fusion* 46 (2006), 297
4. W. G. Graham, "Properties of Alkali Metals Adsorbed onto Metal Surfaces" in Proc. 10th Symp. on Fus. Tech., Brookhaven National Laboratory, Upton, 1980
5. S. Krylov et al., *Nucl. Fusion* **46** (2006), 324
6. J. D. Levine et al., *Surf. Sci.* **1** (1964), 171
7. R. Gutser et al., Negative Hydrogen Ion Transport in RF-driven Ion Sources for ITER NBI, to be published
8. O. Fukumasa and R. Nishida, *Nucl. Fusion* **46**, (2006), 275
9. M. Seidl et al, *J. Appl. Phys.* **79** (1996), 6
10. S. Ma, R. Sydora and J. Dawson, *Comp. Phys. Comm.* **77** (1993), 190
11. C. Birdsall, *IEEE Trans. on Plasma Sci.* **19**, 1991, 65
12. U. Fantz and D. Wuenderlich, *New J. Phys.* **8** (2006), 301
13. R. Janev, "Elementary Processes in Hydrogen-Helium Plasmas", Springer, Berlin, 1987
14. M. Eeerden, M. van Sanden, D. Otorbaev and D. Schram, *Phys. Rev. A* **51** (1994), 3362
15. R. Janev and Z. Radulovic, *Phys. Rev. A* **17** (1978), 889
16. M. Huels, R. Champion, L. Doverspike and. Y. Wang, *Phys. Rev. A* **41** (1990), 4809
17. H.P.L. de Esch et al., these proceedings
18. B. Heinemann et al, Design of the "half-size" ITER Neutral Beam Source Test Facility ELISE, SOFT conference 2008, to be published in Fusion Eng. Des.
19. W. Kraus et al., these proceedings
20. P. Spädtke and S. Wipf, *GSI Report* **89-09** (1989)
21. R. Nocentini, R. Gutser and B. Heinemann, Plasma grid design for optimized filter field configuration for the the NBI test facility ELISE, SOFT conference 2008, to be published in Fusion Eng. Des.

BEAM FORMATION, ACCELERATION, NEUTRALIZATION AND TRANSPORT

Aperture Size Effect on Extracted Negative Ion Current Density

H.P.L. de Esch[a], L. Svensson[a] and D. Riz[b]

[a] *CEA Cadarache, IRFM, F-13108 St. Paul-lez-Durance, France.*
[b] *CEA, DIF, Bruyères-le-Châtel, 91297 Arpajon Cedex, France*

Abstract. This paper discusses experimental results obtained at the 1 MV testbed at CEA Cadarache that appear to show a higher extracted D⁻ current density from small apertures. Plasma grids with different shapes have been installed and tested. All grids had one single aperture. The tests were done in volume operation and in caesium operation. We tested four grids, two with ⌀14 mm, one with ⌀11 mm and one with ⌀8 mm apertures. No aperture size effect was observed in volume operation. In caesiated operation the extracted current density for the ⌀8 mm aperture appears to be significantly higher (~50%) than for the ⌀14 mm aperture. Simulations with a 3D Monte Carlo Trajectory Following Code have shown an aperture size effect of about 20%. Finally, as byproducts of the experiments, data on backstreaming positive ions and the temperature of the plasma grid have been obtained.

Keywords: neutral beams, negative ions, ion sources, Monte Carlo Code
PACS: 41.20Cv, 41.20Gz, 41.75Cn, 41.85Ar, 41.85Ja, 52.50Dg, 52.65Cc, 52.65Pp

INTRODUCTION

During operation of the 1MV testbed [1,2,3] at CEA-Cadarache, plasma grids with different shapes have been installed and tested. All grids had one single aperture. The tests were done in volume operation and in caesium operation. Considered was the arc efficiency (defined as the amount of accelerated D⁻ current density per unit of arc power). The accelerated D⁻ current was evaluated by measuring the thermal power on an inertial carbon target using an infrared camera. The current density is then defined as the measured current divided by the aperture area in the plasma grid.

We tested four grids, two with ⌀14 mm, one with ⌀11 mm and one with ⌀8 mm aperture and evaluated the current density.

Our simulations with a 3D Monte Carlo Trajectory Following Code have shown an aperture size effect of about 20%. Independantly, another group presented very similar calculation effects during this Conference.

Byproducts from these experiments include data on the effect of the temperature of the plasma grid and measurements of the backstreaming positive ions.

EXPERIMENTS

The four different plasma grids that have been tested on the 1 MV testbed at CEA Cadarache over a period of one year are shown in figure 1. The four grids are identical downstream of the knife edge. The aperture diameter of the grids are different and also the upstream parts of the grids (which see the source plasma) are not the same.

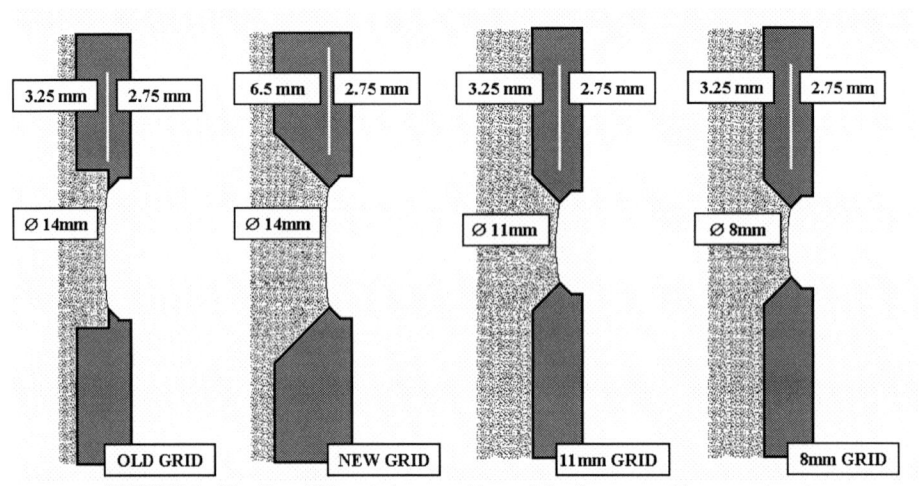

FIGURE 1. The four different plasma grids used for the comparitive experiments.

The D⁻ ions are provided by a revised [3] version of a small DRIFT source [4]. The original prototype source was equipped with proper water cooling. Permanent magnets incorporated in the side of this filamented source provide a magnetic filter of 300 Gauss.cm. The filter field at the plasma grid is 50 Gauss. Using a SINGAP accelerator [1,2,3] consisting of a 10 mm thick extraction grid (aperture diameter 13 mm widening to 15 mm) and a 10 mm thick pre-acceleration grid (⌀15 mm), D⁻ beams were accelerated to several hundreds of keV. The permanent magnets that are embedded in the extraction grid add another 100 Gauss.cm to the source filter and another 100 Gauss to the local magnetic field. Thus the magnetic fields are 400 Gauss.cm and 150 Gauss locally at the plasma grid.

After acceleration, the beam drifts 2.73 metres through a drift tube, after which it impinges on a carbon fibre calorimeter. The carbon fibres are oriented in the beam direction, which ensures a slow lateral heat conduction. The beam profile is then measured from behind with an infrared camera through a sapphire vacuum window.

The tests were done in volume operation and in caesium operation. Of particular interest is the arc efficiency. The accelerated D- current was evaluated by measuring the thermal power on the inertial carbon target using an infrared camera. The current density is then defined as the measured current divided by the aperture area in the plasma grid. For a correct evaluation of the arc efficiency, stripping losses have to be estimated.

In caesium operation we took care to reproduce the caesium conditions. For every new grid the source and accelerator were completely cleaned. During operation, the source walls were kept at 40°C all the time. Then 145 mg/hr Cs was injected until the Cs effect manifested itself. During the experiments 30 mg/hr was continuously injected.

Finally, we took care to operate the arc with an anode to cathode voltage V_{ac} between 66 and 81 Volts. Previous experimentation had shown that the arc efficiency is constant to within ±3% in this range of V_{ac}.

VOLUME OPERATION

The results for volume operation are given in figure 2. At constant arc power (17-18 kW) the gas filling pressure was increased. This is known to increase negative ion production [5]. At low filling pressure ($P_{sce} \leq 0.5$ Pa), no difference at all is visible between the apertures. At higher pressure the measured current density is higher for the smaller diameter apertures.

The reason for this effect are the stripping losses: the gas pressure in the accelerator is much higher for the big aperture due to the larger gas conductance. For 1.5 Pa, the stripping losses are calculated to be 34%, 46% and 53% for 8, 11 and 14 mm, repectively. These are entirely consistent with the measurements in fig. 2. The conclusion is that in volume operation no aperture size effect on D⁻ production is evident: the differences in measured power on the calorimeter are entirely explained by the stripping losses.

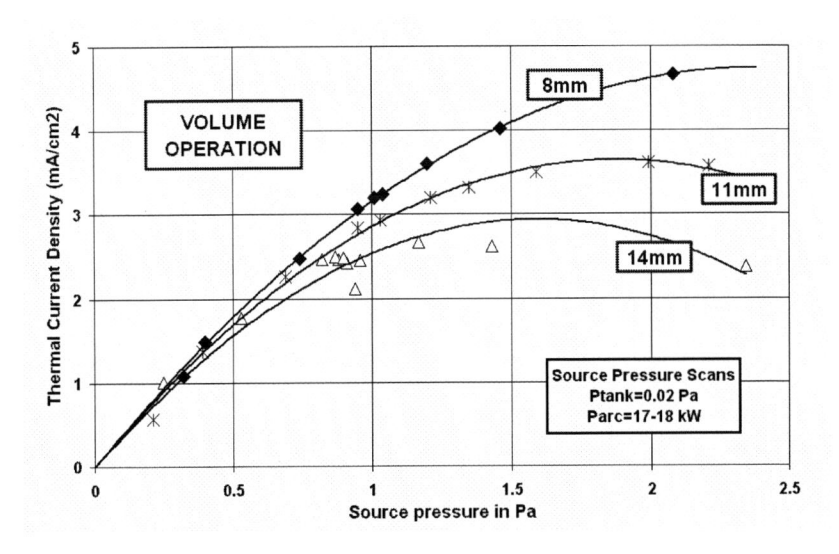

FIGURE 2. Thermally measured D⁻ current density on the calorimeter for different sizes of plasma grid aperture. Current density is plotted against source filling pressure.

The curves in fig. 2 suggest, but do not prove, a downward trend at high pressure P. The downward trend seems reasonable as production is proportional to P and stripping losses are proportional to $e^{-\alpha P}$, leading to a dependance $Pe^{-\alpha P} \approx P(1-\alpha P)$. For each aperture diameter, the coefficient α has a different value: according to the fit 0.84, 1.04, 1.17 for 8, 11, 14 mm, respectively.

Plotting the current density against arc power shows that the current density is constant above 5 kW in the arc although at high source pressure a slight increase with arc power is visible. At 0.5 Pa source pressure, the measured current density on the target is 1.7 mA/cm^2 for all aperture sizes and for all arc powers above 5 kW.

CAESIATED OPERATION

Reproducibility of the measurements under caesiated conditions is a problem as there can be too much caesium, too little caesium, it can be in the wrong place or it can be oxidised and/or nitrated. It was therefore decided to stick to one single procedure.

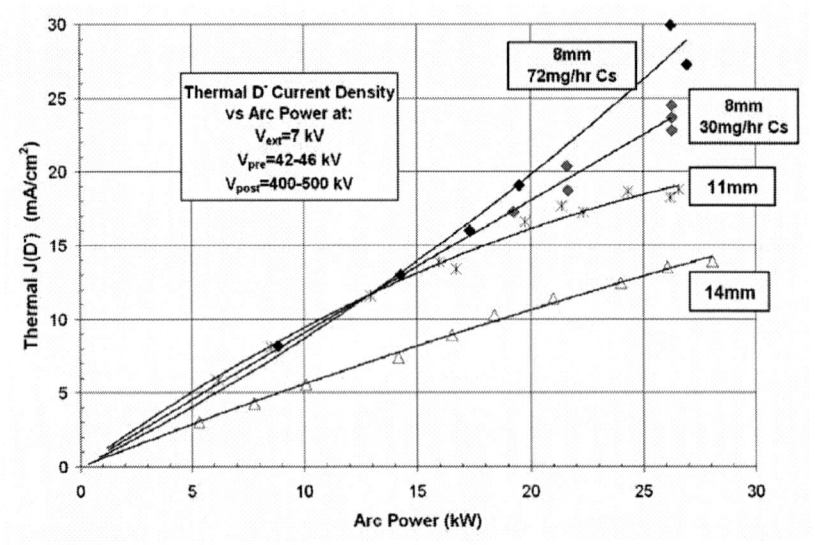

FIGURE 3. Thermally measured D$^-$ current density on the calorimeter for 7 kV extraction voltage and different sizes of plasma grid aperture. Source pressure was 0.3 - 0.4 Pa.

Each time when the vacuum tank has been opened the entire source and accelerator are thoroughly cleaned to make sure that no left-over caesium is present. After closure and pumping, volume operation is established and it is verified that previous volume operation is reliably reproduced. The source body water temperature is controlled and we operated always at 40 °C. This ensures that all the source walls are at 40 °C and Cs does not get trapped in cold areas. When establishing Cs injection, 145 mg/hour Cs is injected until the caesium effect clearly manifests itself. Once the Cs effect is there, the Cs flow is reduced to 30 mg/hour except for operations with 8 mm aperture where 72 mg/hr seeding was tried out as well. Continuous Cs seeding appears necessary to maintain constant performance.

In fig. 3 a number of arc efficiency curves is given for 8, 11 and 14 mm aperture diameter plasma grids. The extraction voltage was 7 kV in all cases and the source pressure was kept between 0.3 and 0.4 Pa. Stripping losses are 22%, 19% and 16% for the $\varnothing$14, $\varnothing$11 and $\varnothing$8 mm apertures, respectively. Thus differences in stripping losses account for no more than 8% in the measured thermal data.

There is a maximum current density that can pass through the extraction grid without beam interception. The calculation has been made with the 2D beam optics code SLACCAD [6]. A maximum of 20 mA/cm^2 D$^-$ current density (as measured on target, after stripping losses) from a $\varnothing$14 mm plasma grid aperture can pass through the extraction grid at 7 kV without suffering interception. If a safety margin of 1 mm between the beam and the extraction grid is kept, the maximum current density is 16 mA/cm^2, see figure 4. The 14 mm data in fig. 3 is thus OK.

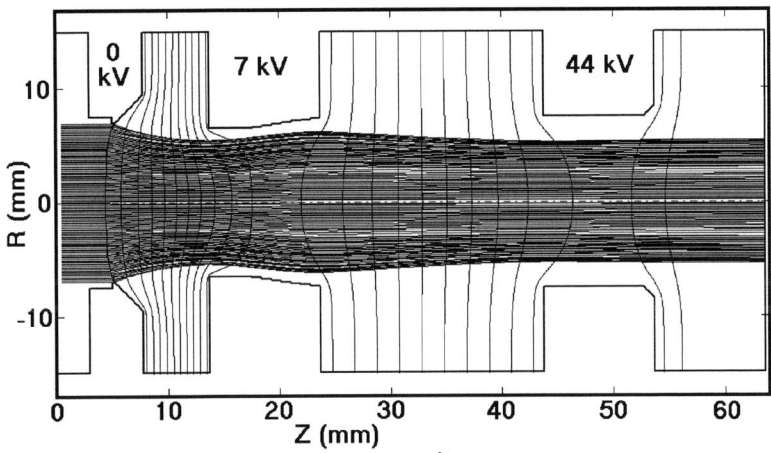

FIGURE 4. Simulation by SLACCAD for 20 mA/cm^2 D$^-$ extracted from a $\varnothing$14 mm aperture, being accelerated using 7 kV extraction voltage. After stripping losses, 16 mA/cm^2 D$^-$ is deposited on the calorimeter. There is 1 mm clearance to the extraction grid.

The same calculation was repeated for the $\varnothing$11 and $\varnothing$8 mm apertures. Because the plasma grid aperture is smaller, the beamlet can expand more before a danger of interception exists; a higher current density can pass through the extraction grid. For a clearance of 1 mm between the 7 kV extraction grid and beams the maximum current density on the calorimeter becomes 16, 25 and 35 mA/cm^2 for $\varnothing$14, $\varnothing$11 and $\varnothing$8 mm, respectively. All data in figure 3 is well below these limiting values.

The SLACCAD calculations have been repeated for an extraction voltage of 5 kV. Allowing a clearance of 1 mm between beam and extraction grid, the maximum current density for transmission through the accelerator is 9, 15 and 25 mA/cm^2 for $\varnothing$14, $\varnothing$11 and $\varnothing$8 mm, respectively. These numbers are as measured after transmission to the calorimeter.

Figure 5 shows the results for 5 kV extraction voltage. It can be seen that the limiting current for proper transmission is approached and that at higher arc power the current density as measured on the target goes down. Therefore, the 7 kV data in figure 3 is considered to be more meaningful and is used in the following discussions.

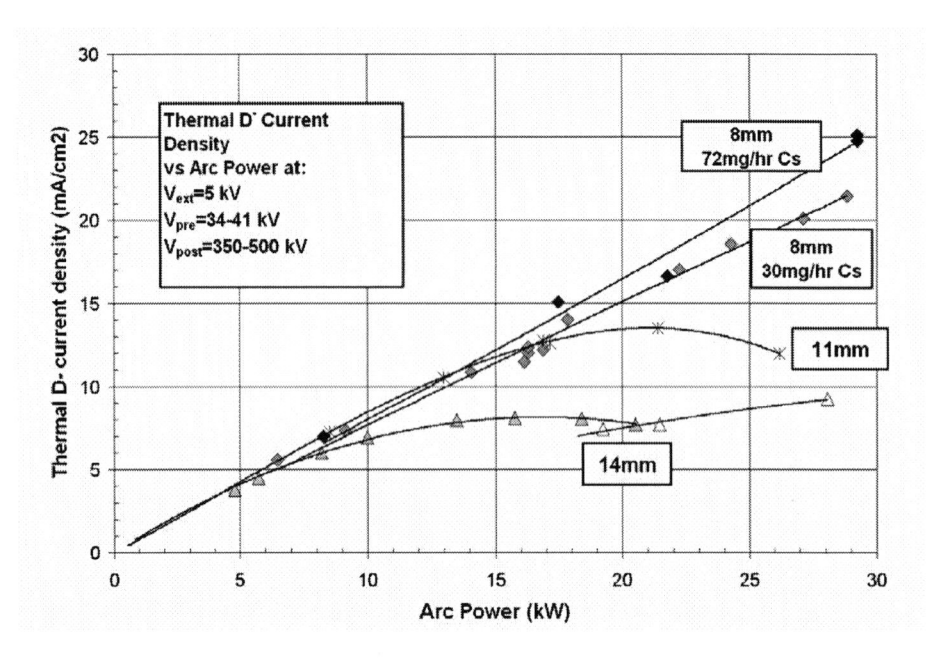

FIGURE 5: Thermally measured D⁻ current density on the calorimeter for 5 kV extraction voltage and different sizes of plasma grid aperture.

ADDITIONAL MEASUREMENTS

Plasma Grid Temperature Effect

As a byproduct of the experimental campaigns, measurements on the effect of the plasma grid (PG) temperature on the production of negative ions have been obtained. The PG is actively heated but there is no cooling, except by radiation and conduction. Data have been recorded when stable operation has been obtained at a PG temperature of at least 250 °C. Then the PG was allowed to cool down for a period of one hour, while Cs seeding was being maintained. After cooling, data was taken again while the temperature of the grid was being slowly increased. The data taken before cooling could be reproduced, showing that the Cs condition remained constant. The data shown in figure 6 confirms the PG temperature effect and suggests that there is a threshold between 180 °C and 260 °C. Increasing the temperature of the plasma grid to over 260 °C improves the arc efficiency by a factor of two.

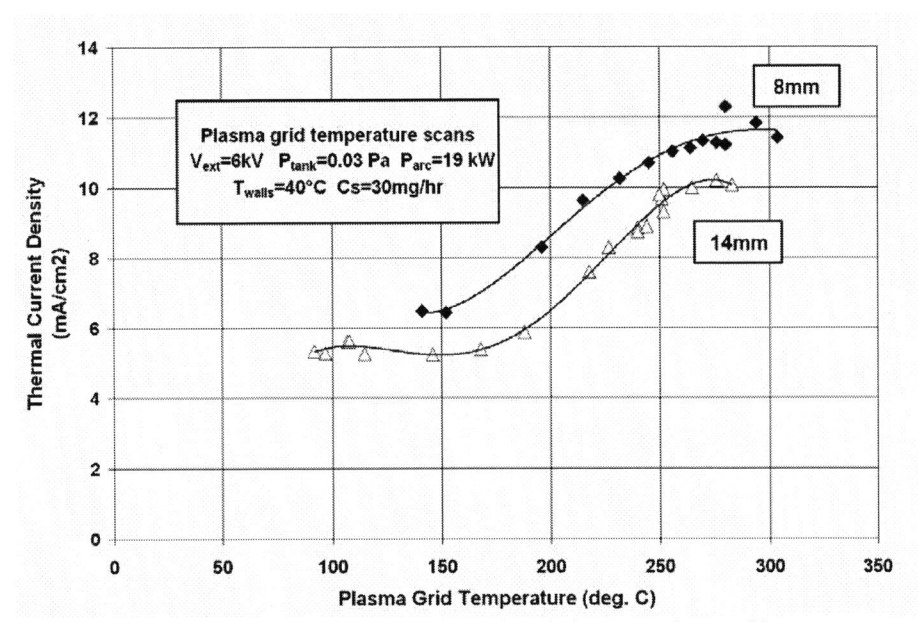

FIGURE 6: Thermally measured D⁻ current density on the calorimeter vs. plasma grid temperature under stable caesiated conditions.

The same experiments have been performed in volume operation (no Cs present). No effect of the PG temperature on the arc efficiency could be observed within ±10%.

Backstreaming Positive Ions

A small inertial copper calorimeter equipped with a thermocouple is installed in the source backplate, 10 cm behind its surface, directly opposite the single aperture in the plasma grid, on the beamline axis. It has been verified that operation of the arc has no influence on the thermocouple reading. Positive ions formed in the SINGAP accelerator (the single gap is 350 mm) are accelerated towards the source. If their energy is high enough (>100 keV), the magnetic fields present in source and accelerator will not bend their trajectories sufficiently for them to miss the calorimeter. Also positive ions formed in the drift space between accelerator and beam target can be measured if they happen to diffuse to near the aperture in the anode, which is 150x150 mm².

Experiments have been performed by filling the beam tank with helium gas. This choice of gas is most convenient as it is not pumped by the cryo pumps and a constant gas pressure is easily maintained. The measured data is in figure 7. Plotted is the power in the positive ions divided by the sum of the neutral and negative ion beam powers on the beam target. For ITER-like conditions (P_{tank}=0.03 Pa), 2.5% of the beam power comes back to the source in positive ions. This corresponds to 1.0 MW on ITER.

The measurements have not yet been repeated using D_2 filling gas. Data on electron production by Svensson et al[1] suggest that backstreaming ion production could be a factor of 2 higher in D_2 gas, whereas Fubiani et al [7] calculate 0.9 MW (2.3% of the beam power, 7.4% of the beam current) of backstreaming ions in MAMuG, a different type of accelerator considered for ITER.

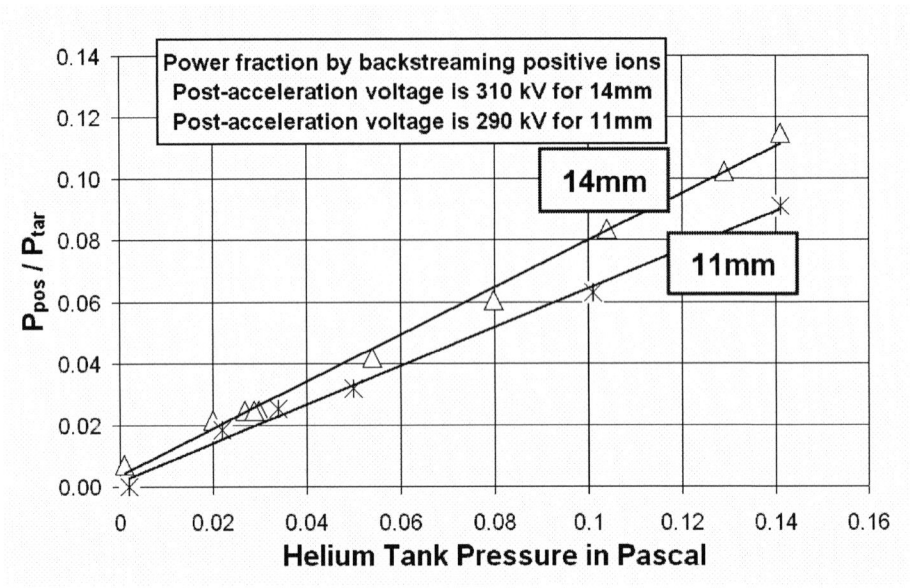

FIGURE 7: Power by backstreaming positives ions, divided by the beam power on the target.

MONTE CARLO CALCULATIONS

Calculations have been performed using a modified version of the 3D Monte-Carlo code DIANE [8,9]. This code is able to follow the ions until their destruction or extraction on an Euler meshgrid by solving the multigroup Boltzmann-Fokker-Planck equation. The code uses multigroup cross sections which are group-to-group scattering matrices calculated by another code, called ZADIG [10]. The processes taken into account are the destruction processes, the charge exchange with H^0/D^0, the Coulomb interaction with the background plasma and the magnetic field.

In the calculations presented here, negative ions are launched from the plasma grid (PG) surface with an initial energy (E_0). The calculations show that the extraction aperture is not uniformly illuminated (which would lead to beam halo) and that the smaller apertures are more uniformly illuminated than the larger ones. The calculations also show that the smaller apertures are more strongly illuminated than the larger apertures. Depending on E_0, this effect amounts to up to 20% as can be appreciated from fig. 8.

Monte Carlo calculations on the aperture effect have also been reported by Gutser et al [11] during this conference. Using the TRAJAN [12] code, their calculations suggest, depending on E_0, an effect of 15-25% for a single aperture. Similar results have been calculated by this group for multi-aperture grids [11].

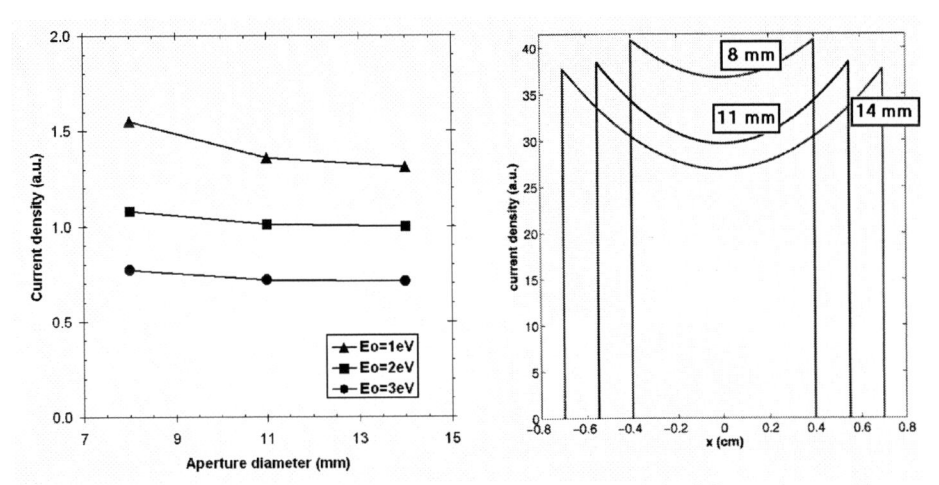

FIGURE 8: Relative current density (a.u.) calculated by DIANE/ZADIG for different plasma grid aperture sizes. Calculations have been performed for three different assumed H⁻ starting energies, 1 eV, 2eV and 3 eV. $n_e = n_{H+} = 10^{18}$ m^{-3}, $n_H = 10^{18}$ m^{-3}, $T_e = T_{H+} = T_H = 1$ eV, B=50 Gauss. The picture on the right gives the current density profile over the aperture for 1 eV starting energy.

CONCLUSIONS

An experimental campaign has been conducted to measure the effect of the plasma grid aperture size on the arc efficiency. Plasma grids with one single aperture were used throughout. In caesiated operation the smaller aperture (⌀8 mm) produces 50% more current density than the larger one (⌀14 mm), whereas in volume operation no effect at all was found. Monte Carlo simulations with DIANE/ZADIG also show an aperture size effect. The size of this effect is 8% to 20%, depending on which assumption was used for the initial energy of the D⁻ ion leaving the plasma grid. Independently, Gutser et al presented during this conference TRAJAN simulations showing an effect of 15% to 25%.

The plasma grid temperature effect was confirmed in caesiated operation (100% more current density above $T_{PG} > 260°C$) and significant backstreaming positive ions have been measured.

ACKNOWLEDGMENTS

This work, supported by the European Communities under the contract of Association between EURATOM and CEA, was carried out within the framework of the European Fusion Development Agreement. The views and opinions expressed herein do not necessarily reflect those of the European Commission.

REFERENCES

1. L. Svensson, D. Boilson, H.P.L. de Esch, R.S. Hemsworth and A. Krylov, *Nucl. Fusion* **46**, S369-S378 (2006).
2. L. Svensson, D. Boilson, H.P.L. de Esch, R.S. Hemsworth, A. Krylov and P. Massmann, *Fusion Engineering and Design* **66-68** 627-631 (2003).
3. P. Massmann, H.P.L. de Esch, R.S. Hemsworth and L. Svensson, *Fusion Engineering and Design* **74** 409-412 (2005).
4. A. Simonin, G. Delogu, C. Desgranges, M. Fumelli, *Rev. Sci. Instrum.* **70**, 4542 (1999).
5. T.S. Green, *Plasma Physics and controlled Fusion* **30**, 1505 (1988)
6. W.B. Hermannsfeld, Electron Trajectory Program, *SLAC report, Stanford Linear Accelerator Center*, SLAC-226 (1979).
7. G. Fubiani, H.P.L. de Esch, R.S. Hemsworth and A. Simonin, *Physical Review ST Accel. Beams*, **11**, 014202 (2008).
8. D.Riz and M.Chiche, « Simulation of energetic ions transport by a 3D Monte Carlo method », *Int. Conf. on Supercomputing in Nuclear Applications* (Paris) 2003, Am. Nucl. Soc. (CD-ROM).
9. D. Riz et al. , *Nucl. Fusion* **46**, 864-867 (2006)
10. D. Riz, *International Conference on the Physics of reactor*, Paris, PHYSOR 2000, *Am. Nucl. Soc.*, CD-ROM
11. R. Gutser, D. Wünderlich, U. Fantz, P. Franzen, B. Heinemann, R. Nocentini and the NNBI Team, *This Conference*.
12. R. Gutser *et al*, Negative Hydrogen Ion Transport in RF driven Ion Sources for ITER NBI, *to be published*.

How to Find Valid Parameters for the Modelling of H⁻ and D⁻ Ion Extraction with nIGUN[C]

Reinard Becker

Institut für Angewandte Physik der Goethe-Universität,
Max von Laue-Straße 1, D-60438 Frankfurt, Germany

Abstract. A dedicated interactive computer program nSHEATH has been written in order to facilitate the exploration of self-consistent solutions in the (at least) 9-D parameter space for the inverted plasma sheath. In the transition from the quasi-neutral plasma to the unneutralised acceleration space of ion sources 3 major failures can occur, if the parameters are not correctly chosen:
1) A virtual cathode situation may develop producing a potential maximum inside the sheath.
2) The density of the most abundant kind of thermal positive ions becomes negative.
3) The density of fast positive ions (protons) becomes negative.
The program displays the entry parameters, allows to change one of them at a time and performs a numerical integration through the sheath. The output is a graphical display of the densities through the sheath and a history list of parameter changes. The entry parameters can be selected for either a cesiated source or for a source without cesium. The maximum number of different thermal positive ions is arbitrary and set at present to be a limit of 10. The physics for the mathematical modelling has been fully described in the proceedings of PNNIB-2004 [1]. nSHEATH as well as the ion extraction simulation program nIGUN[C] [2] are based on this model and the final output of nSHEATH consistes of a file, which contains the important input part for running nIGUN with the found parameters.

Keywords: Inverted plasma sheath, H⁻ extraction, Simulations
PACS: 52.27,Cm, 07.77.Ka, 33.15.Fm

RECALL OF THE MATHEMATICAL FORMULATION FOR THE INVERTED SHEATH

The model for the transition of the ion source plasma to the acceleration space is shown in Fig. 1. Protons and other fast ions are born at the plasma potential and are accelerated to the region of interest, where again quasi-neutrality is obtained and from where the negative ions are extracted. The potential fall in the classical sheath on the left together with the potential increase for the extraction of negative ions on the right side forms an electrostatic trap for positive ions, which are born there and hence will obey a Boltzmann distribution. The negative ions as well as the electrons are assumed to be created in this trap and extracted towards the main plasma as well as towards the extraction side. The fast ions, born inside the source at plasma potential are reflected in the extraction region, where they are forming a virtual cathode.

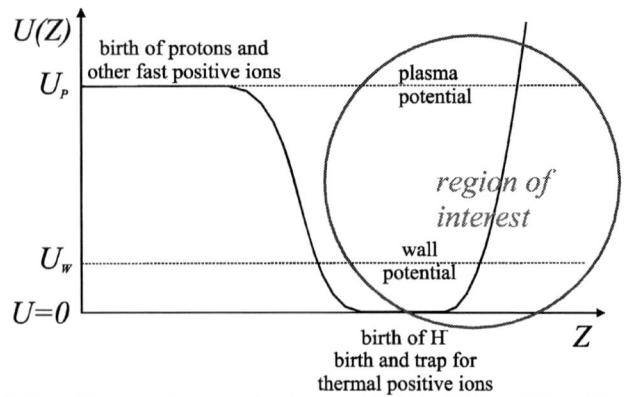

FIGURE 1. Electrostatic model for the numerical simulation of H⁻ and D⁻ extraction

The density of electrons with energy U_e will be reduced by acceleration in the same way as the H⁻ density:

$$n_e(U) = \frac{n_e(0)}{\sqrt{1+\frac{U}{U_e}}} \qquad n_-(U) = \frac{n_-(0)}{\sqrt{1+\frac{U}{U_-}}} \tag{1}$$

the density of fast protons (and other fast ions) is dying out by the virtual cathode process [3]:

$$n_p(U) = n_p(0)\left(1 - \frac{U}{U_p}\right) \tag{2}$$

Other positive ions are considered to be thermal and trapped between the extraction field and the plasma, hence will obey a Boltzmann distribution:

$$n_i(U) = n_i(0)\exp\left(1 - \frac{U}{U_i}\right) \tag{3}$$

This problem now has 2 variables for each charged particle, the density and the energy (either directed or thermal). While the energies must be known to obtain solutions, density relations can be expressed by the definition of the electron-to-H⁻ current Γ and by the condition of quasineutrality at $U = 0$

$$\frac{n_e}{n_-} = \Gamma \sqrt{\frac{m_e U_-}{M_- U_e}} \tag{4}$$

$$\frac{n_p}{n_-} = 1 + \frac{n_e}{n_-} - \frac{n_1}{n_-}\sum\frac{n_i}{n_1} \tag{5}$$

By the balance of charged particle currents to the plasma electrode with wall potential U_w the relation of the density of the first kind of positive thermal ions to the density of H⁻ ions is found in Eq. 6:

$$\frac{n_l}{n_-} = \frac{1+\Gamma-\left(1+\Gamma\sqrt{\frac{m_e U_-}{M_-U_e}}\right)\left(1-\frac{U_w}{U_p}\right)^{\frac{3}{2}}\sqrt{\frac{\pi M_-U_p}{M_p U_-}}}{\sum\frac{n_i}{n_l}\left\{\sqrt{\frac{M_-U_i}{M_iU_-}}\exp\left[-\frac{U_w}{U_p}\right]-\left(1-\frac{U_w}{U_p}\right)^{\frac{3}{2}}\sqrt{\frac{\pi M_-U_p}{M_p U_-}}\right\}} \qquad (6)$$

THE PROGRAM nSHEATH

All these formulae are essential for nIGUN[C]. The program nSHEATH also uses these formulae, but is restricted only to the numerical integration of the inverted sheath. Therefore a fast graphical answer for the potential function and the variation of all particle densities in the sheath is obtained. The program then offers interactive input of new parameter values or of a certain percentage of change for an input variable. This makes it very easy to quickly vary parameters and see the result. In addition to this a list for the history of changes is displayed, which allows easy walk back in case of sudden unexpected results.

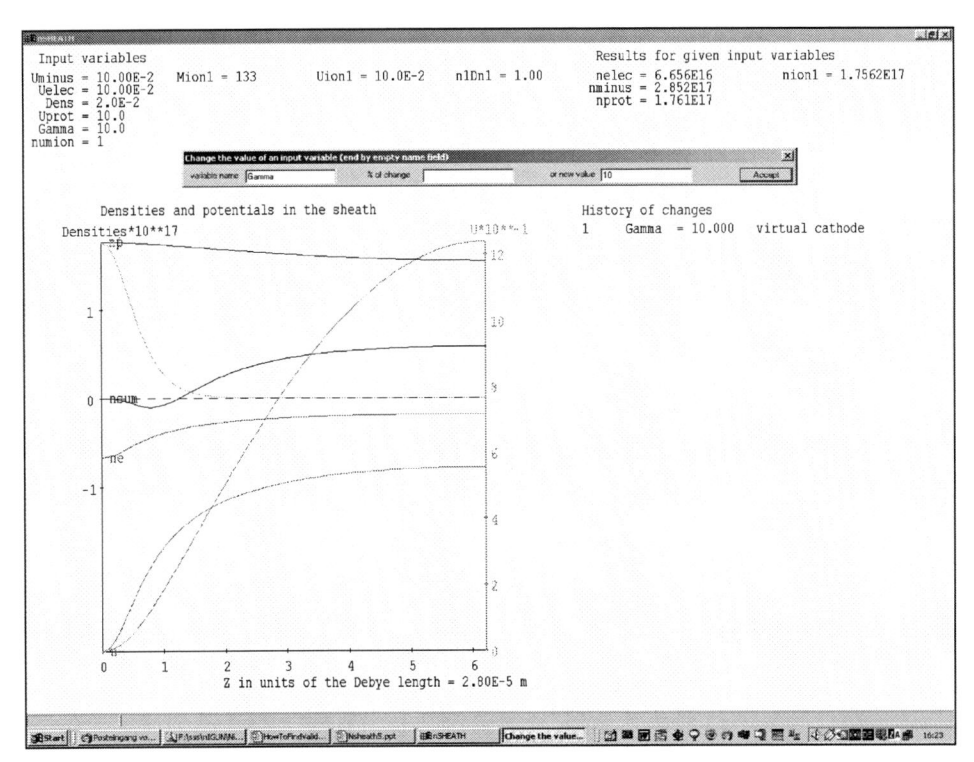

FIGURE 2. nSHEATH screen shot in the case of a virtual cathode

The integration for the potential function in the sheath may fail for 3 reasons, all depending on the choice of parameters:
1. development of a virtual cathode – oscillating potential function.
In Fig. 2 the output for a virtual cathode situation is shown. The characteristic of this is a potential function, which comes to a maximum in the extraction region instead of increasing further.

2. negative density for the first kind of positive thermal ions according to Eq. 6.
A negative density for the first kind of thermal positive ions (mostly protons) results in a break down of nSHEATH, because no integration may be performed in this case.

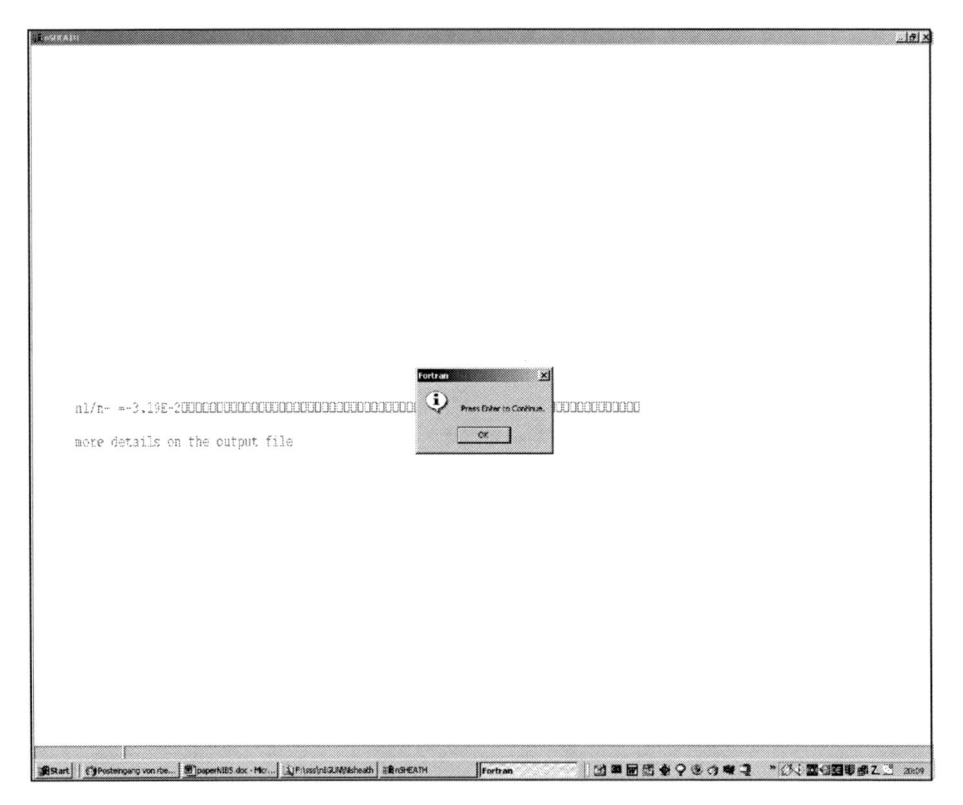

FIGURE 3. Screen shot of break down of nSHEATH in the case of a negative result in Eq. 6

3. negative density for fast protons.
As a result of the required quasi neutrality in the plasma by eq. 5 the density of fast positive protons may become negative. The integration is performed, however the output of nSHEATH in Fig. 4 shows a solution, which is not useful.

In all these cases nSHEATH issues a diagnostic and allows further change of parameters to return to the region of valid solutions. The final result of nSHEATH is a file, which contains the NAMELIST HMINUS, needed by nIGUN[(C)] to perform

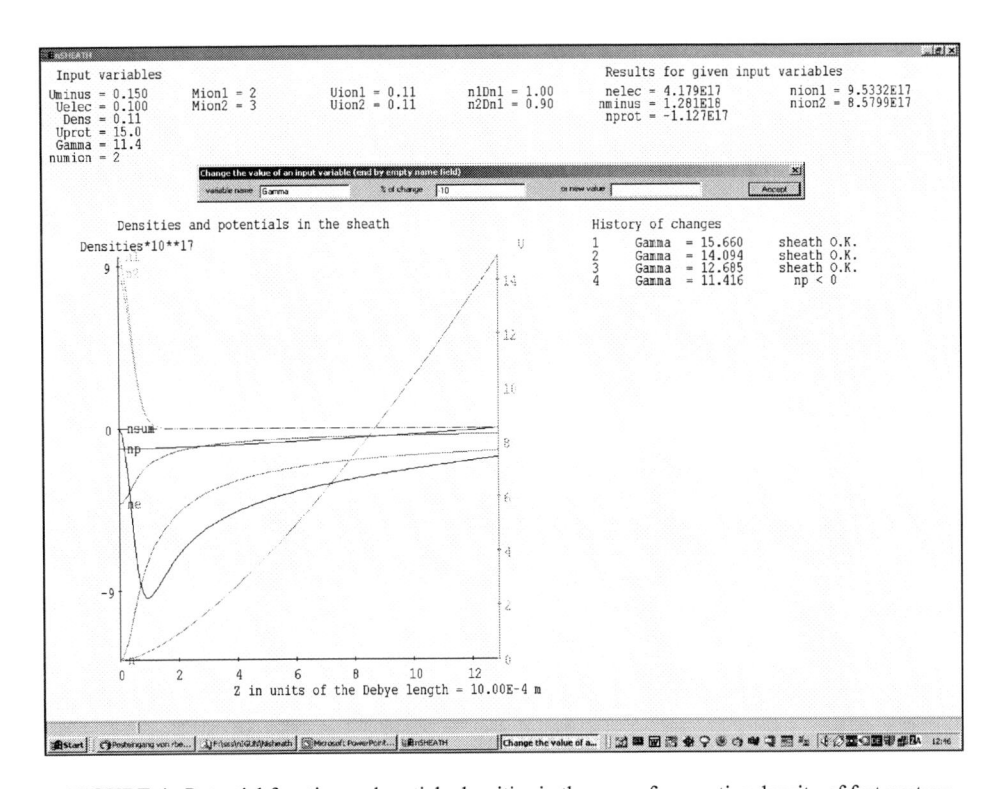

FIGURE 4. Potential function and particle densities in the case of a negative density of fast protons

simulations of negative ion extraction. With the aid of a text editor this can be copied and pasted into the nIGUN[(C)] input file to perform the self consistent calculations of the plasma sheath and the extraction region [4] – see Fig. 5.

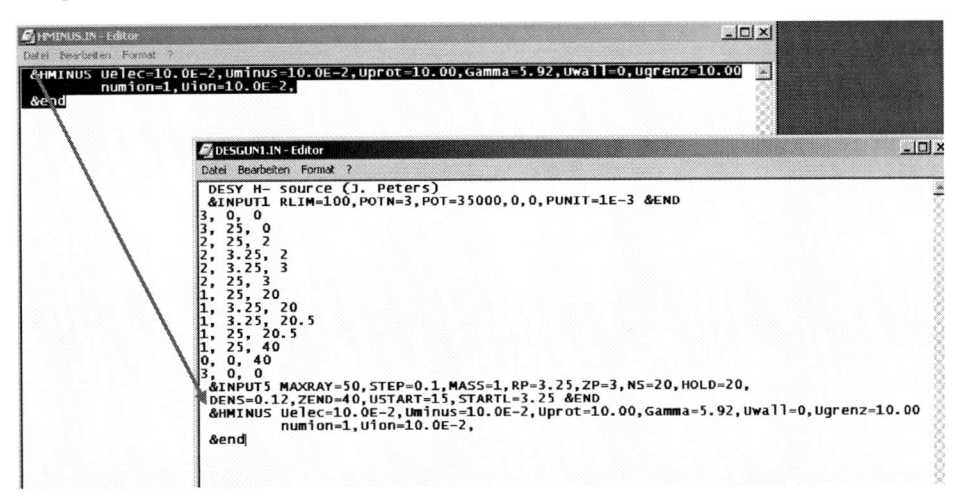

FIGURE 5. Copying and pasting the output of nSHEATH to the input file of nIGUN[(C)]

CONCLUSIONS

The simulation of H⁻ and D⁻ extraction with nIGUN$^{(C)}$ has at least 9 input variables with complex interplay for the electrostatic sheath calculation. In most cases no solutions exist. Therefore it is recommended to use the interactive program nSHEATH to explore the significance of a chosen parameter combination. This is greatly facilitated by the interactive graphics output of nSHEATH and the possibility to "see" at once the result of the change of one parameter. The output of nSHEATH consists of a suitably formatted text file for the NAMELIST HMINUS, which is needed in the input file of nIGUN$^{(C)}$ as well as of text and graphic files, which show the history and results of changing parameters.

REFERENCES

1. R. Becker, AIP-CP**763**, 194-200 (2005)
2. http://www.egun-igun.com
3. R. Becker, K.N. Leung, W. Kunkel, Rev. Sci. Instrum. **69**, 1107-1109 (1989)
4. R. Becker, Rev. Sci. Instrum. **77**, 03A505 (2006)

Design of a low voltage, high current extraction system for the ITER Ion Source

P. Agostinetti*, V. Antoni*, M. Cavenago†, H.P.L. de Esch**, G. Fubiani‡,
D. Marcuzzi*, S. Petrenko†, N. Pilan*, W. Rigato*, G. Serianni*, M. Singh§,
P. Sonato*, P. Veltri* and P. Zaccaria*

*Consorzio RFX, Associazione Euratom-ENEA sulla fusione, Padova, Italy
†INFN-LNL, Legnaro, Italy
**Association EURATOM-CEA Cadarache, IRFM/SCCP, France
‡LAPLACE/CNRS, University Paul Sabatier, Toulouse, France
§Institute for Plasma Research, Gandhinagar, India

Abstract. A Test Facility is planned to be built in Padova to assemble and test the Neutral Beam Injector for ITER. In the same Test Facility the Ion Source will be tested in a dedicated facility planned to operate in parallel to the main 1 MV facility. Purpose of the full size Ion Source is to optimize the Ion Source performance by maximizing the extracted negative ion current density and its spatial uniformity and by minimizing the ratio of co-extracted electrons. In this contribution the design of the extractor and accelerator grids for a 100 kV, 60 A system is presented. The trajectories of the negative ions, calculated with the SLACCAD code [1], have been benchmarked by a new 2D code (BYPO [2]) which solves in a self consistent way the electric fields in presence of electric charge and magnetic fields. The energy flux intercepted by the grids is estimated by using the Montecarlo code EAMCC [3] and the grids designed according to the constraints set by the permanent magnets and by the cooling channels. The interaction of backstreaming ions due to the ionization process with the grids and the Ion Source backplate is investigated and its impact on the project and performance discussed.

Keywords: ITER, ion, source, extraction
PACS: 41.75Cn, 52.59

INTRODUCTION

The ITER Neutral Beam Test Facility (NBTF) is planned to be built at Consorzio RFX (Padova, Italy) in the framework of an international collaboration, and has the purpose to test and operate the ITER Neutral Beam Injectors (NBIs) [4, 5, 6, 7]. A full size ion source is the first experimental device to be built and operated, aiming at testing the extraction of a negative ion beam (made of H^- and in a later stage D^- ions) from an ITER size ion source. The main requirements of this experiment are a current of 60 A H^- (and later 40 A D^-) and an energy of 100 keV. A good beam uniformity and optics are required to match the operating scenarios foreseen for the ITER NBIs.

DESIGN OVERVIEW

The extraction and accelerator system for the full size ion source, sketched in Fig. 1 is composed of three grids: the Plasma Grid (PG), the Extraction Grid (EG) and the

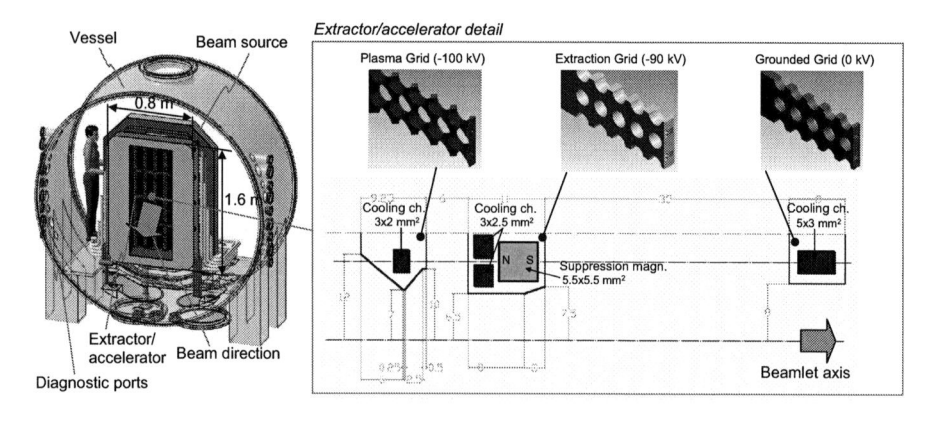

FIGURE 1. Design overview of the extractor/accelerator system

Grounded Grid (GG). Each grid features 1280 apertures, where the ion beamlets are extracted from the ion source and accelerated up to 100 kV.

All the grids are made by electrodeposition of pure copper onto a copper base plate. This technique permits to obtain a very complex geometric shape (with very small cooling channels that run inside the grid and embedded magnets) and to have good mechanical properties, due to the high purity and to the very small grain size.

The plasma grid is heated by the plasma inside the RF ion source, with a surface power density that is estimated to be about 20 kW m^{-2} (IPP experimental results [8]). This grid is required to operate at a temperature of about 150°C in order to enhance the caesium effect for negative ions surface generation. For the same reason, it is Molybdenum coated on the plasma side. The apertures are designed with conical chamfers on the upstream and downstream sides of the grid. A larger surface for ion production is obtained with this solution, and its efficacy has been demonstrated by experimental results on the Batman ion source at IPP Garching [9]. A 4 kA current flows in the vertical direction, to provide a horizontal magnetic field that reduces electron temperature and the number of co-extracted electrons.

The extraction grid has an electric potential that is about 10 kV higher than the PG, so that the negative charged ions (H$^-$ or D$^-$) can be properly extracted from the RF expansion chamber. Suppression magnets, embedded in the grid, have the function to deviate the trajectories of the co-extracted electrons, making them collide with the grid surface. The consequent power loads are quite high and concentrated, hence this grid is the most critical by the structural point of view, and is designed with a high performance cooling system.

The grounded grid has the function to accelerate the ion beamlets up to a potential of about 100 kV, and is also loaded by co-extracted and stripping electrons.

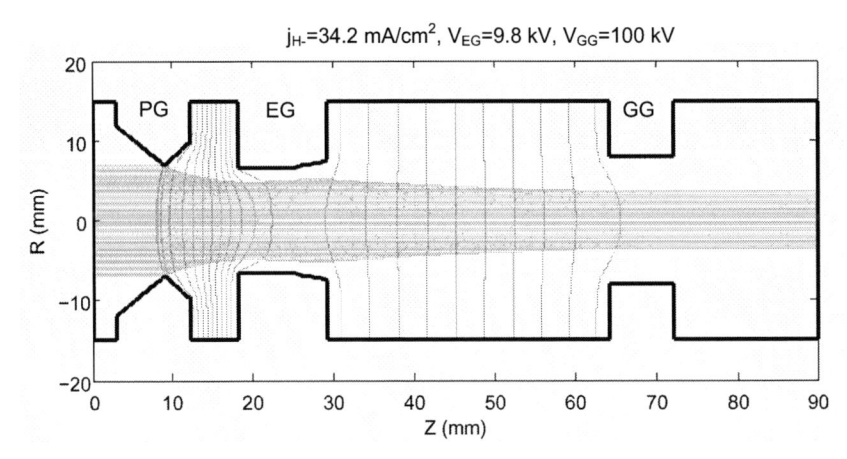

FIGURE 2. SLACCAD simulation of the beam optics: equipotential lines (blue) and particle trajectories (magenta) are estimated by integration of the Poisson's equation

PHYSICS OPTIMIZATION

The SLACCAD code was used to estimate the electric field inside the accelerator by integration of the Poisson's equation, with cylindrical geometry conditions [10]. This is a modified version of the SLAC Electron Trajectory Program [11], adapted to include ions, a free plasma boundary and a stripping loss module [12]. The geometry shown in Fig. 1, obtained after several optimizations, permits to obtain at the same time good optics and low stripping losses. Fig. 2 shows the equipotential lines and the corresponding beam trajectories obtained with SLACCAD using the geometry of Fig. 1. Ref. [1] gives a detailed description of these analyses.

Alternative approaches, ranging from fluid models [13] to ray tracing [14] and interpolation, to simulate a richer physical description of the plasma sheath-beam interface (traditionally called meniscus in source physics) were summarized elsewhere [15]. Several length scales coexist, from the Debye length λ_D to the size of the electrodes, for example the extraction diameter $2r_h$; this suggests to study the sheath formation in 2D models before generalizing to 3D models. Moreover, magnetic field effects on the space charge and on the selfconsistent electric field can be easily simulated in a planar geometry used in BYPO [15, 16]; here z is the beam axis and x is parallel to the long axes of the permanent magnet bars embedded into the EG: the model planar geometry is the $y = 0$ section of the extraction system. Magnetic field is assumed to be in the $\hat{y}$ direction at $y = 0$. Electron space charge is enhanced, since electron trajectories are lengthened, as apparent from the code BYPO, and computation time accordingly increases. Magnetic field is specified by suitable analytical approximations as

$$B_y + iB_z = \frac{ik_7 B_R}{\pi} \left[\arctan \frac{\sinh(a_z \pi / L_y)}{\cosh((z - z_m + i(y - a))\pi / L_y)} \right]_{a=-a_y}^{a=a_y} \quad (1)$$

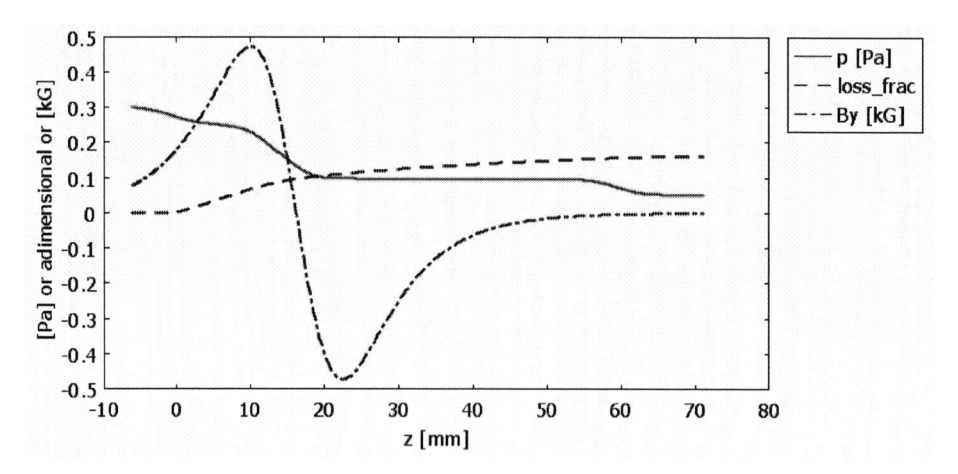

FIGURE 3. Profile of the gas pressure, of the stripping losses (loss_frac) and of the magnetic field vs z computed with BYPO

where $B_R = 0.96$ T is the residual flux density, $z = z_m$ is the middle plane of the magnet bars, $2a_z$ and $2a_y$ are the sides of a bar yz section, L_y is the distance between centers of bars and $k_7 \leq 1$ is a fitting parameter. Equation 1 with $k_7 = 1$ holds exactly for the case of a infinite array of infinitely long bars, with magnetization axis alternating between $\hat{z}$ and $-\hat{z}$. Another major input to BYPO is the ratio $R_j = j_e/j_H^-$ between current densities of electrons and ions at plasma border. Without magnetic field, we expect a ratio $R_c \equiv \rho_H^-/\rho_e \cong (m_H^-/m_e)^{1/2}/R_j$ between the negative ion and the electron space charge. We set $R_j = 2$ in the examples, so $R_c \cong 21$ without magnetic field.

BYPO represents plasma temperature by using trajectories (rays) with different starting angles α_s. It automatically refines ray spacing near the PG edges, to better resolve the ion beam envelope. It can be argued that plasma temperature is small compared to the rapid acceleration of the beam on axis, so that ion beam results approximately laminar; anyway beam acceleration is not uniform (near the extraction edges [15]), so justifying the computational effort of using several starting angles α_s. We here assume starting angles α_s of -0.1, 0 and +0.1 rad for the emitted ions (with current proportional to weights 0.2 , 0.6, 0.2 respectively), which corresponds to a 'transverse' ion temperature of 0.3 eV. The effects of extraction edges (affecting the beam halo) and of the meniscus curvature (affecting the whole beam) are then modelled by BYPO, with an extreme precision in the ray tracing Runge-Kutta integration (minimum step is about 2 micron). For comparison, meniscus curvature adds an energy in the order of 2 eV to the transverse motion of ions, in typical cases as the following simulations.

Fig 3 shows the gas pressure approximately computed by an additional module recently introduced in BYPO (assuming circular openings for the gas, as in the real 3D model). Ion stripping losses are then computed as for SLACCAD. Figure 4a shows ion rays computed from BYPO, with the Fig. 1 geometry and $k_7 = 0.5$ (see the resulting B_y in fig 3). Beam size is then similar to Fig. 2, even if some slightly different tuning of the voltages may be needed for an optimal beam. We also see that ion beam deviation

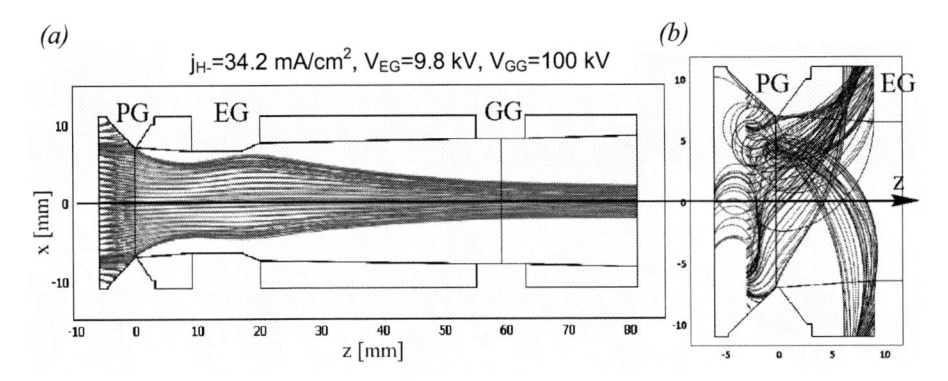

FIGURE 4. Beam simulations with BYPO: (a) Trajectories of H^- with $\alpha_s = -0.1$ rad (red), $\alpha_s = 0$ rad (brown), $\alpha_s = 0.1$ rad (green); (b) Trajectories of coextracted electrons (Editor's note: see online version)

is nearly critical at EG. Figure 4b shows coextracted electrons in the same simulation; note that these electrons are confined in the first gap. Periodic conditions at $x = \pm 11$ mm were used both for the selfconsistent potential ϕ and the electron trajectories; some electrons appear to return into the plasma. Electron load is confined (marginally) to the rear face of EG, that is good; moreover accumulation of electron space charge results tolerable in this simulation ($R_c = \rho_H^- / \rho_e \cong 10$). For the comparison with cylindrical electrode simulations, it should be taken into account that in the BYPO planar geometry the lens effects are twice as strong (for equal electrode angles).

ESTIMATIONS OF THE HEAT LOADS ON THE GRIDS

The interactions between particles inside the accelerator, like secondary particle production processes, were analysed with the code EAMCC [3]. This is a 3-dimensional (3D) relativistic particle tracking code where macroparticle trajectories, in prescribed electrostatic and magnetostatic fields, are calculated inside the accelerator. In the code, each macroparticle represents an ensemble of rays considering the time-independent physical characteristics of the system.

This code needs as inputs the electric and magnetic fields inside the accelerator. The former was calculated with SLACCAD, as explained above. The latter was calculated by summing the field given by the SmCo permanent magnets (calculated with the semianalytical code PERMAG [17]) and the field from the plasma grid filter current (calculated by assuming an infinitely thin electron sheath).

Also a numerical approach, using the code ANSYS, was considered in order to crosscheck the magnetic fields. The agreement between the two codes is quite good, provided that in ANSYS the mesh is sufficiently fine and the domain sufficiently large. So the ANSYS code can be considered as benchmarked against the PERMAG code regarding the magnetic field calculations inside the accelerator, as visible in Fig. 5. On the other hand, the PERMAG code has been benchmarked with success with many experimental measurements.

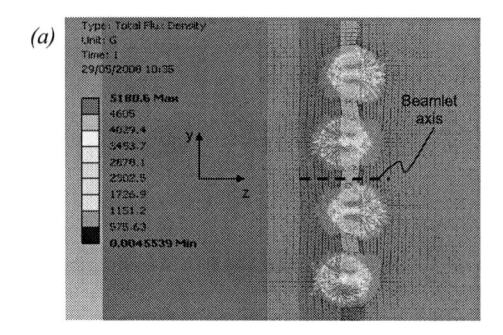

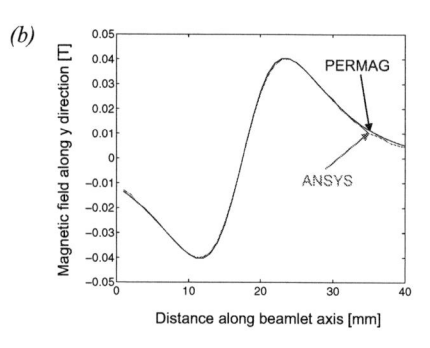

FIGURE 5. Estimations of the magnetic field inside the accelerator: (a) Vector plot (vertical section) of the magnetic field due to the suppression magnets inside the EG, calculated with ANSYS; the magnets are put in x direction (horizontal perpendicular to the beam axis) with polarities along z (beam axis direction) with alternated versus from row to row; (b) Comparison of the magnetic field B_y along the beamlet axis, calculated with PERMAG (semi-analytical approach) and ANSYS (FEM approach).

Collisions are described using a Monte-Carlo method. The various kinds of collisions considered in the code are: (i) electron and heavy ion/neutral collisions with grids, (ii) negative ion single and double stripping reactions and (iii) ionization of background gas [3].

Fig. 6 shows the main results of the EAMCC calculations. Two different simulations are performed to simulate the H^- ions (and related species generated by stripping) and the co-extracted electrons. The current density is assumed to be the same (34.2 mA cm^{-2}) in the two cases, considering an electron-to-ion ratio of 1. The heat loads on the grid surfaces are evaluated by summing the two contributions. While the EG is heated mostly by the co-extracted electrons, the heat on the GG is approximately half coming from the co-extracted electrons, and half from the secondary electrons due to stripping and surface reactions.

The transmitted beamlet power distribution features a ring that is hotter than the central part. These could be due to the chamfered shape of the PG apertures. In fact, this effect is reversed in case of a flat PG surface.

The backstreaming positive ions are quite concentrated in the center of the aperture area. The consequent heat power density is quite high, but covers only an area of some tens of square millimeters. These ions could give rise to sputtering phenomena on the plasma source back plate and on the driver Faraday shields, with a consequent decay of the plasma purity and problems of surface integrity. The sputtering yield due to the backstreaming deuterium ions is generally reduced by a factor of about 5 if the copper surface is coated with Molybdenum [18]. Hence, in order to minimize the detrimental effects consequent to sputtering, a layer of Molybdenum of some microns is foreseen to be applied on the plasma source back plate.

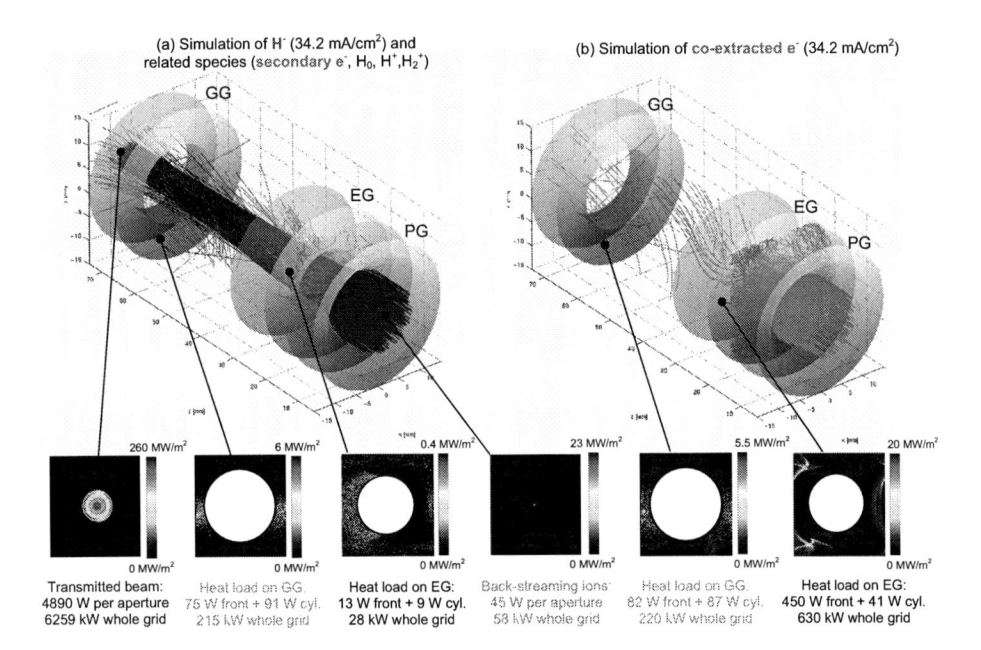

(a) Simulation of H⁻ (34.2 mA/cm²) and related species (secondary e⁻, H₀, H⁺, H₂⁺)

(b) Simulation of co-extracted e⁻ (34.2 mA/cm²)

Transmitted beam:	Heat load on GG:	Heat load on EG:	Back-streaming ions:	Heat load on GG:	Heat load on EG:
4890 W per aperture	75 W front + 91 W cyl.	13 W front + 9 W cyl.	45 W per aperture	82 W front + 87 W cyl.	450 W front + 41 W cyl.
6259 kW whole grid	215 kW whole grid	28 kW whole grid	58 kW whole grid	220 kW whole grid	630 kW whole grid

FIGURE 6. EAMCC simulation of the beam: the particle trajectories and stripping reaction are simulated with a Monte Carlo approach in a domain with electrical and magnetic fields. For the grids, the heat loads on the front surface (front) and on the cylindrical part (cyl.) are reported for a single aperture, as well as the load on the whole grid. For the transmitted and backstreaming beams, the power corresponding to a single aperture and the total power are reported.

THERMO-STRUCTURAL OPTIMIZATION

The grids must be designed in such a way that the corresponding apertures are well aligned during all the operating scenarios, in order to obtain good beam optics. For this reason and for manufacturing requirements they are vertically split in four segments, independently supported with a fixed pin at the left side and with a sliding pin at the right side, as shown in Fig. 7a [19].

The PG cooling system is designed with the goal to maintain the required temperature of 150°C ± 10°C on all the grid surface. The EG and GG have to withstand very high and localized peak power densities due to the co-extracted, secondary and stripping electrons, that are colliding in small areas located around the apertures. The heat loads calcluated with the EAMCC code are used to simulate the thermo-mechanical behaviour with the finite element code ANSYS. A limit value of 300°C, suggested by IPP experience, is considered for the copper peak temperature.

The maximum allowable misalignment between the corresponding apertures of the three grids is fixed to 0.4 mm, for optic reasons. Since 0.2 mm can be considered as a reasonable manufacturing tolerance, the remaining 0.2 mm is set as a maximum value for the misalignment due to thermal expansion [20].

Since the grids are working at different temperatures (150°C for PG and about 50°C)

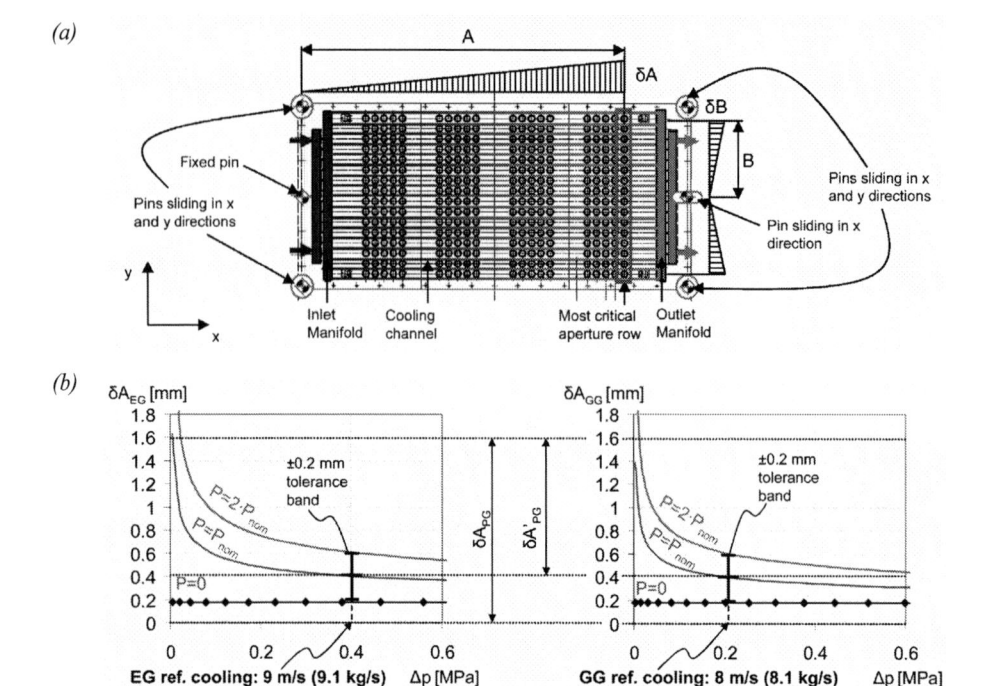

FIGURE 7. Alignment optimization for grids: (a) Cooling and fixing scheme; (b) In-plane deformation plots with identification of PG apertures pre-offset and of the minimum EG and GG cooling parameters. The nominal power is the sum of the power given by the secondary, stripping and co-extracted electrons, as calculated by the EAMCC code in the reference operating scenario (full power and 1:1 electrons-to-ions ratio)

for EG and GG) these alignment requirements can be met only by adopting a mechanical offset to the aperture positions. Imposing the alignment requirement under power loads ranging from zero to the double of nominal power (see Fig. 6 for nominal power), the horizontal pre-offset of the PG apertures is identified ($\delta A'_{PG} = 1.2\ mm$), as well as the minimum water velocities along the channels (9 m s^{-1} for EG and 8 m s^{-1} for GG) and corresponding pressure drops (see Fig. 7b).

The grids have to withstand two categories of stresses:

- Cyclic thermal stress due to the temperature gradients between hotter and colder zones. These stresses must be maintained low in order to satisfy the requirement on fatigue life.
- Static stress due to the water pressure. The local values of equivalent stress must be lower than the allowable values for electrodeposited copper (fixed at 100 MPa).

The position and dimensions of the cooling channels, as well as the water flow, were optimized in order to satisfy at the same time requirements on alignment as well as the

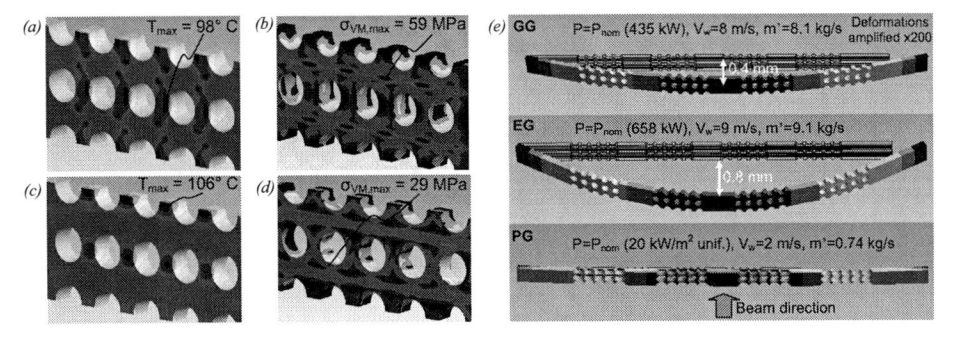

FIGURE 8. Main results of the thermo-structural analyses performed with the ANSYS code on the reference scenario: (a) and (b) Temperature and Von Mises equivalent stress on the EG; (c) and (d) Temperature and Von Mises equivalent stress on the GG; (e) Out of plane deformations of the three grids.

structural, ratcheting and fatigue verifications according to the ITER SDC-IC criteria and considering all the foreseen scenarios (conditioning, partial power, full power etc.). Hence, several analyses were performed to estimate the temperatures and stresses along the grids. Fig. 8 reports the main results for the the reference operating conditions. The thermo mechanical analyses are non linear elastoplastic, taking into account the kinematic hardening model for the material [21]. The copper properties are taken from the ITER Material Handbook [22].

As shown in Figure 8e the thermal stresses are causing also an out of plane deformation of the grids. These deformations give some deterioration of the beam optics, i.e. the beam divergence has an increase. For the case of the reference operating scenario, the average divergence (root mean square) increases from 3.3 to 5.1 mrad. This effect can be considered as acceptable for the full size ion source. Further analyses are advisable and foreseen on this aspect.

CONCLUSIONS

The design of the extraction and acceleration system for the full size ion source experimental device have been accomplished by taking into account at the same time physics and engineering requirements. Several aspects will be investigated during the experimental campaigns, like the magnetic configuration and the plasma source conditioning. For this reason, possible modifications and further optimization are foreseen during the experiments.

REFERENCES

1. M. Singh and H.P.L De Esch. *Accelerator system for DNB and ISTF*. Proceedings of the 1st International Conference on Negative Ions, Beams and Sources, 10-12 September 2008, Aix en Provence, France.

2. M. Cavenago et al. *Development of Small Multiaperture Negative ion Beam Sources and Related Simulation Tools*. Proceedings of the 1st International Conference on Negative Ions, Beams and Sources, 10-12 September 2008, Aix en Provence, France.
3. G. Fubiani et al. *Modeling of secondary emission processes in the negative ion based electrostatic accelerator of the International Thermonuclear Experimental Reactor*. Phys. Rev. Special Topics Accelerators and Beams 11, 014202 (2008)
4. P. Sonato et al. *The ITER full size plasma source device design*. 25th Symposium on Fusion Technology,15-19 September 2008, Rostock, Germany.
5. D. Marcuzzi et al. *Detail design of the RF Source for the 1 MV Neutral Beam Test Facility*. 25th Symposium on Fusion Technology,15-19 September 2008, Rostock, Germany.
6. T. Inoue et al. *Design of neutral beam system for ITER-FEAT*, Fus. Eng. and Design 56-57 (2001) 517-521
7. R. S. Hemsworth et al. *Neutral beams for ITER*. Rev. Sci. Instrum. 67, Issue 3, 1120 (1996).
8. P. Franzen et al. *Physical and experimental background of the design of the ELISE test facility*. Proceedings of the 1st International Conference on Negative Ions, Beams and Sources, 10-12 September 2008, Aix en Provence, France.
9. E. Speth et al. *Overview of the RF source development programme at IPP Garching*. Nucl. Fusion 46 (2006) S220-S238
10. J. Pamela. *A model for negative ion extraction and comparison of negative ion optics calculations to experimental results*. Rev. Sci. Inst. 62 (1991) 1163.
11. W.B. Hermannsfeld. *Electron Trajectory Program*, SLAC report, Stanford Linear Accelerator Center, SLAC-226 (1979).
12. H.P.L. de Esch, R.S. Hemsworth and P. Massmann. *Updated physics design ITER-SINGAP accelerator*. Fusion Eng. Des. 73 (2005) 329.
13. R. N. Franklin. *The plasma-sheath matching problem*. J. Phys. D, Appl. Phys., Vol 36, no. 22, pp. 309-320, 2003.
14. J. H. Whealton et al. *Computer modelling of negative ion beam formation* J. Appl. Phys., Vol 64, no. 11, pp. 6210-6226, 1998.
15. M. Cavenago. *Use of COMSOL Multiphysics in the Modeling of Ion Source Extraction*,8 pages, in *Comsol Conference 2006*, Milano, Nov. 2006 (Comsol, CDROM, ISBN 0-9766792-4-8).
16. M. Cavenago, P. Veltri, F. Sattin, G. Serianni and V. Antoni. *Negative Ion Extraction With Finite Element Solvers and Ray Maps*. IEEE Trans. on Plasma Science, vol. 36, pp 1581-1588 (2008).
17. D. Ciric, *Private communication*.
18. P. Agostinetti. *Methods for the Thermo-mechanical Analysis and Design of High Power Ion Beam Sources*. Ph.D. Thesis, University of Padua, 2008.
19. P. Agostinetti, S. Dal Bello, M. Dalla Palma and P. Zaccaria. *Thermomechanical design of the SINGAP accelerator grids for ITER NB Injectors*. Fusion Eng. Des. 82 (2007) 860-866.
20. P. Agostinetti, S. Dal Bello, M. Dalla Palma, D. Marcuzzi, P. Zaccaria, R. Nocentini, M. Fröschle, B. Heinemann and R. Riedl. *Thermomechanical design of the ITER Neutral Beam Injector grids for Radio Frequency Ion Source and SINGAP Accelerator*. 22nd IEEE/NPSS Symposium on Fusion Engineering, Albuquerque, New Mexico, 2007.
21. M. Dalla Palma. *Multiaxial Fatigue Design Criteria for High Heat Flux Components Actively Cooled in Nucleate Boiling Conditions*. Ph.D. Thesis, University of Padua, 2006.
22. Various Authors. *ITER MPH. Material Properties Handbook*. ITER Document No. G 74 MA 16.

Development of 1 MeV H⁻ Accelerator at JAEA for ITER NB

M.Taniguchi[a], H.P.L. de Esch[b], L.Svensson[b], N.Umeda[a], M.Kashiwagi[a], K.Watanabe[a], H.Tobari[a], M.Dairaku[a], K.Sakamoto[a], and T.Inoue[a]

[a] *Japan Atomic Energy Agency, Fusion research and development directorate*
801-1 Mukoyama, Naka, Ibaraki, 311-0193 Japan
[b] *CEA Cadarache, IRFM, F-13108 St. Paul-lez-Durance, France*

Abstract. This paper reports the recent activities at JAEA for the development of 1 MeV H⁻ ion accelerator toward the ITER NBI. For the development of MAMuG accelerator, a 320 mA H⁻ ion beam was successfully accelerated as the highest record in the world at the MeV class energy (796 keV). This was achieved by protecting the H⁻ ion source from the high heat load by the backstream positive ions, which was produced during the high current H⁻ ion acceleration. The SINGAP accelerator was tested at JAEA to compare the performance of the SINGAP and the MAMuG. This paper also reports the results of the SINGAP test and the comparison between the SINGAP and the MAMuG is discussed.

Keywords: ITER NBI, H⁻ ion accelerator, SINGAP, MAMuG
PACS: 41.75.Cn

INTRODUCTION

The neutral beam (NB) injection system for ITER is required to inject 16.5 MW of D^0 beams per one injector at the energy of 1 MeV [1]. To realize the ITER NB, high power H⁻ ion accelerator R&D is on going at Japan Atomic Energy Agency (JAEA). For the accelerator of the ITER NB, two concepts are proposed, one is the MAMuG (Multi-Aperture Multi-Grid) accelerator [2] and the other is the SINGAP (Single-Aperture Single-Gap) accelerator [3]. In JAEA, the MAMuG accelerator called "MeV accelerator" has been developed at the MeV test facility (MTF), whose power supply capability is -1 MV, 0.5 A for 60s [4]. The target of the MeV accelerator is to demonstrate the acceleration of 200 A/m² H⁻ ion beam at the energy of 1 MeV, which is required for the ITER accelerator. Since the construction of the MTF, the performance of the MeV accelerator has been improved step by step [5]. Recently, it was found that the countermeasure against the high heat load due to the backstream positive ions is essential for the stable beam acceleration at high power. This paper reports the successful acceleration of 140 A/m² (320 mA) at the energy of 796 keV.

In parallel to the development of the MAMuG accelerator at JAEA, the SINGAP accelerator has been developed at CEA Cadarache. The advantage of the SINGAP accelerator is its simplicity due to the absence of the intermediated grids. However, electrons produced in the accelerator are accelerated to full energy which causes decrease in acceleration efficiency as well as excess heat load on the beam line components. On the contrary, the MAMuG has a complicated structure compared with the SINGAP, but the electrons can be suppressed before acceleration to full energy by the intermediated grids.

In order to choose the accelerator type for the ITER NBI, a benchmark test was planed to test both accelerators at a same test facility, and the performance of both accelerators was compared utilizing the same diagnostics. For this purpose, the SINGAP accelerator was installed in the MTF at JAEA. This paper also reports the result of the SINGAP test at MTF.

RECENT PROGRESS IN MAMUG ACCELERATOR

Figure 1 shows a cross-sectional view of the MeV accelerator developed at JAEA. The whole accelerator was designed to be insulated by vacuum, where insulation gas is not utilized. To simulate this vacuum insulation by the experiment, the main structure of the MeV accelerator is installed in vacuum inside an FRP (fiber reinforced plastic) insulator stack. The H⁻ ions are generated in the KAMABOKO ion source mounted directly on the top of the accelerator. The extracted H⁻ ions are accelerated by four intermediate grids and a ground grid up to 1 MeV. At the beginning of the development of the MeV accelerator, performance was limited due to breakdowns at the inner surface of the FRP. To improve the voltage holding capability, a large stress ring was designed and installed to the MeV accelerator [6]. This countermeasure was effective to improve the voltage holding performance and the H⁻ ion beam of 836 keV, 146 A/m^2 (206 mA) was successfully accelerated in 2005 [7]. Since then, the R&D

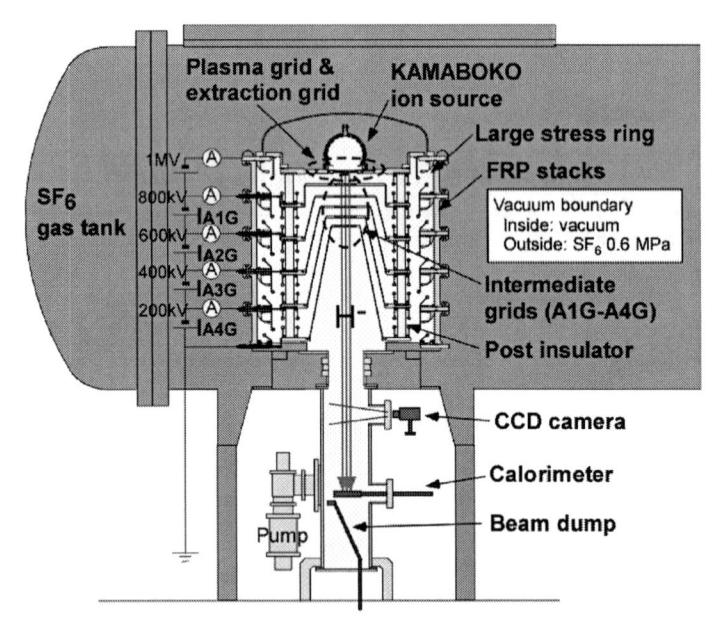

FIGURE1. A cross-sectional view of the MAMuG accelerator installed in the MTF.

was extended to increase the beam current maintaining the high current density. This was achieved by increasing the number of extraction apertures, from 3 x 3 (9apertures) to 3 x 5 (15 apertures). However, during the high current beam acceleration test, the MeV accelerator suffered from decrease in the negative ion production under Cs seeded condition. To clarify the reason, the interior of the MeV accelerator was carefully investigated, and it was found that this trouble was caused by an unexpected high heat load on the port plug of the KAMABOKO ion source due to backstream positive ions. The backstream positive ions are produced in the course of acceleration and downstream beam plasma by collision of negative ions with gas molecules. Then positive ions are extracted from the beam plasma at the grounded grid, and accelerated backward to high energy. Taking into account of the beam plasma production and loss of positive ions at sound velocity in a plasma column generated in the beam path, the heat load was estimated to be 3.6 MW/m^2. To solve this problem, water cooled backstream ion dump was installed and the un-cooled port plug was protected. As the result, operation of the MAMuG accelerator was improved even under high power operation and the H⁻ ion beam of 320 mA (140 A/m^2) was accelerated up to 796 keV. The accelerated drain current including co-accelerated electron (421 mA) reached close to the power supply limit of the MTF.

SINGAP TEST AT JAEA MeV TEST FACILITY

Figure 2 shows the comparison between the SINGAP and the MAMuG accelerator installed in the MTF. The negative ion source, the PG, and the EXG were the same as those of the MAMuG. Originally, pre-acceleration voltage of the SINGAP has been designed to 60 kV [3]. However, for the SINGAP test at JAEA, pre-acceleration voltage was changed to 200 keV and the pre-acceleration grid was connected to the -800 kV potential of the MTF. The SINGAP system was designed and implemented maintaining the existing 200 keV A1G from the MAMuG in place and incorporates it as the pre-accelerator for the SINGAP. The intermediate grids A2G (-600kV), A3G (-400kV) and A4G (-200kV) for the MAMuG were removed. The grounded grid has a one large aperture of 114 x 106 mm.

In the SINGAP accelerator, the negative ion beams are focused by a deep well structure at the pre-accelerator. For this purpose, the "Kerb" made of aluminum alloy is attached behind the pre-acceleration grid. The pre-accelerated H⁻ ion beams are accelerated up to the full energy by the post acceleration gap (the gap between the pre-acceleration grid and the grounded grid). The gap length between pre-acceleration grid and grounded grid was taken to be 307 mm from a result of a three dimensional beam analysis.

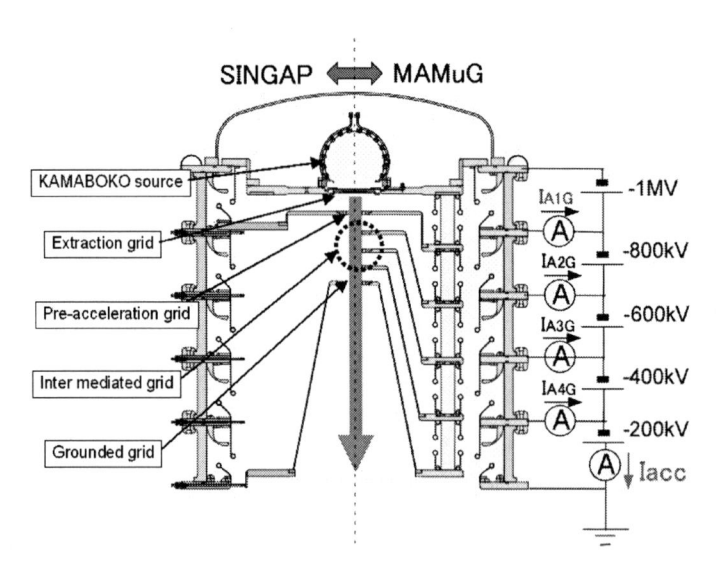

FIGURE2. A comparison of the SINGAP and the MAMuG accelerator with the power supply connection.

At the exit of the accelerator, a pair of small magnet was installed to deflect and separate co-accelerated electrons from the H⁻ ion beam for accurate measurement of the H⁻ ion current by a calorimeter. The deflected electrons are removed by two electron dumps installed downstream of the beamline.

High Voltage Conditioning of SINGAP Accelerator

After installation of the SINGAP accelerator and pumping down to vacuum, high voltage conditioning was started. The conditioning was performed in the same way as the MAMuG accelerator. Out gassing and subsequent pressure evolution, power supply drain current, intermediated grid current and X-ray emission were carefully monitored during the high voltage conditioning. Typically, when the applied voltage was increased, increases of these values were observed. Applied voltage was increased only when these values decreased to be negligible levels. Figure 3 shows the results of the voltage holding conditioning. The data of the MAMuG and the SIGMA (Single Gap Multi-Aperture) accelerator [8] is also plotted for comparison. The SIGMA accelerator has the similar geometry with the SINGAP. In the case of SIGMA, the H⁻ ions extracted from the multi-aperture are accelerated by a single gap as in the SINGAP. The gap length of the SIGMA was 520 mm.

The maximum voltage holding was 572 kV after 120 hours of conditioning and seems to be saturated at the level below 600 kV. With adding H_2 gas of 0.25 Pa, voltage holding was increased to 800 kV. The progress of voltage holding for the SINGAP was slow compared to the MAMuG. In the case of the MAMuG, 757 kV can be sustained after 60 hours of conditioning, and the rated voltage of -1 MV can be attained by adding H_2 gas of 0.20 Pa. For the SINGAP accelerator, breakdowns and over current started to emerge from around 150 kV, whereas in the MAMuG case, it

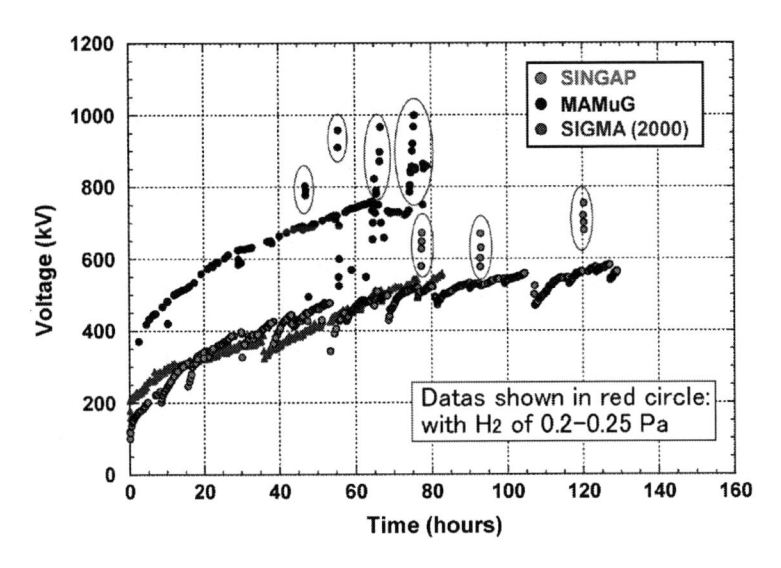

FIGURE 3. High voltage conditioning curve for the SINGAP accelerator tested at MTF.

appeared at around 350 kV. The interesting feature of the conditioning curve is that the difference in voltage holding at the initial stage for both accelerators was about 200 kV, and this difference is kept at all the time during the conditioning procedure.

The voltage holding was improved by adding H_2 gas to the accelerator. Figure 4 shows the gas pressure dependence of voltage holding for the SINGAP and the MAMuG accelerator. Pressure dependence showed similar tendency for both accelerators. At the pressure below 0.2 Pa, voltage holding was limited due to the flashover at the inner surface of FRP. By increasing the pressure, surface flashover was reduced and MAMuG could sustain 1 MV at the pressure range of 0.2 to 0.4 Pa. However for the SINGAP accelerator, maximum voltage was 800 kV with adding the gas to 0.25 Pa, and -1 MV could not be sustained at all the pressure range. The voltage holding of the SINGAP was 150 - 200 kV lower for all the pressure range. This is considered to be due to the discharge between the metal, as will be discussed in the next paragraph. At the pressure above 0.4 Pa, voltage holding rapidly decreased due to the Paschen discharge.

The voltage conditioning curve for the SINGAP was similar to that of the SIGMA accelerator (fig.3). This fact indicates that the poor voltage holding of the SINGAP and the SIGMA can be attributed to the long vacuum gap of both accelerators. Watanabe et.al. [9] investigated that the breakdown between

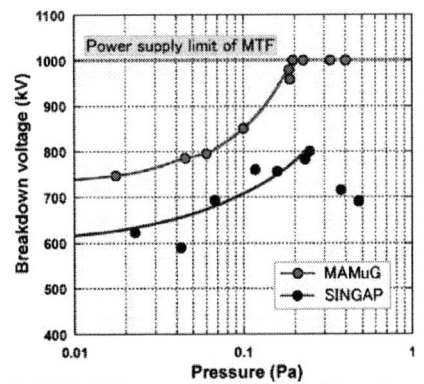

FIGURE 4. Pressure dependence of the voltage holding capability.

metals in vacuum obeys the clump theory and stable voltage holding (V) depends on the gap length (g) as follows;

$$V = V_0 \times g^z, \quad z = 0.5 - 0.6.$$

The gap length investigated in ref [9] was limited to 0 – 50 mm, but by extrapolating the data to the longer gap, the curve shown in fig.5 is obtained. From this curve, it is estimated that the voltage holding expected for the SINGAP with gap length of 307 mm is around 600 kV (z = 0.5 was assumed). This agrees well with the result of the high voltage conditioning (maximum voltage; 572 kV). The applicability of the clump theory to such a long vacuum gap is unknown; however, this fact might indicate that the SINGAP accelerator can not sustain 1 MV stably.

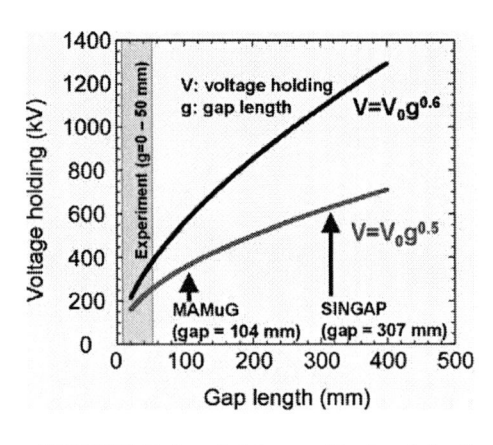

FIGURE5. Voltage holding v.s. Gap length for Cu electrode.

Beam Acceleration Test of SINGAP Accelerator

Following to the high voltage conditioning, beam acceleration test was performed. The KAMABOKO ion source was operated with Cs seeding. Figure 6 shows the accelerated H⁻ ion current as a function of beam energy. At the perveance matched condition, a 626 keV, 225 mA (97 A/m^2) beam was successfully accelerated. At under perveance, a 672 keV, 220 mA beam was obtained. However, higher energy above 700 keV could not be attained with a perveance matched beam due to the poor voltage holding. The maximum voltage with the H⁻ ion beam was 775 keV, but the current was limited to 75 mA at this voltage.

Figure 7 shows the amount of electrons accelerated during the H⁻ ion beam acceleration in the SINGAP. The electron current (I_e^-) was evaluated by the difference between the power supply drain current (I_{acc} in fig.2) and the H⁻ ion current (I_H^-) measured by the calorimeter ($I_e^- = I_{acc} - I_H^-$). Power supply drain current includes all the current of charged particles which exits from the accelerator. The data for the MAMuG was also plotted for comparison in fig.7. All the data in fig.7 were taken at the same gas pressure and at the perveance matched beams. The gas pressure in the beamline

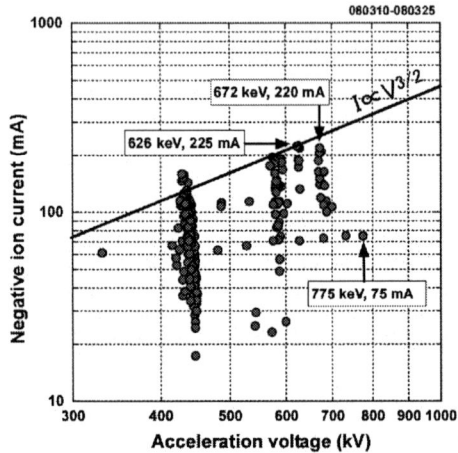

FIGURE6. Accelerated H⁻ ion current.

was 0.08 Pa (Source operation pressure: 0.16 Pa). The acceleration voltage for the SINGAP data was from 435 kV to 672 kV. The amount of electrons accelerated in both accelerators were proportional to the negative ion current. The ratio of the electron current to the H⁻ ion current (I_e^-/I_H^-) for the SINGAP was estimated to be 1.09. This is more than three times larger than that of the MAMuG ($I_e^-/I_H^- = 0.29$). Note that the data in fig.7 was taken at the relatively high pressure of 0.08 Pa compared with the ITER accelerator (0.03 Pa). However, it was found that the ratio of co-accelerated electrons for the SINGAP was much larger than that of the MAMuG.

Figure 8 shows the dependence of the pre-acceleration grid (A1G) current on the input arc power to the source. The positive grid current indicates the negative particles (H⁻, e⁻) impact on the A1G. On the contrary, the negative grid current indicates the emission of secondary electrons from the A1G grid. Thus the observed current is difference between the positive and the negative current. The same data was taken for several gas pressures in the beamline. At the normal operation condition, pressure in the source was 0.16 Pa and the pressure in the beamline was 0.08 Pa. The pressure in the beamline was able to be decreased to 0.013 Pa by a cryo-sorption pump. In the case of low beamline gas pressure ($P_{beamline}$: 0.013 Pa), A1G grid current was largely negative at the low arc power around 5 kW. At this arc power, the H⁻ ion density in the source was too low with respect to the extraction voltage to form a convergent beam. Subsequently, major part of the H⁻ ions hits the grid due to the poor convergence, which causes the secondary emission. With increasing the arc power, A1G grid current gets smaller, and at the optimum condition (around 20 kW), it becomes almost 0 mA. In this condition, it is expected that most of the H⁻ ions pass through the grid at the perveance matched conditions. However, with increasing the gas pressure in the beamline, A1G current tends to become negatively large even at a perveance matched condition. This fact indicates that the electron is emitted from the A1G grid at a high gas pressure in the beamline.

The production processes of co-accelerated electrons are considered as follows;

1) secondary electron emission due to the impact of the H⁻ ion on the pre-acceleration grid.
2) stripping of the H⁻ ion or

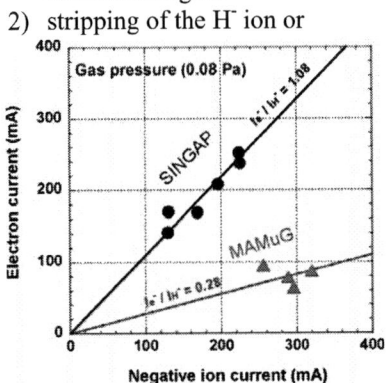

FIGURE7. Electron current accelerated by the SINGAP and the MAMuG.

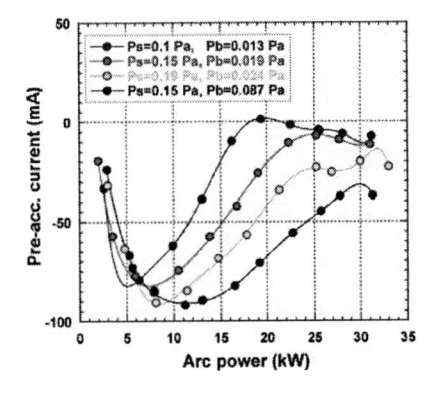

FIGURE8. A1G grid current of the SINGAP.

ionization of residual gas during the H⁻ ion acceleration.
3) secondary electron emission due to the impact of backstream positive ion on the pre-acceleration (A1G) grid.

As in the negative ion source, the identical extractor and the A1G grid were utilized both the SINGAP and the MAMuG, the amount of electrons produced by process 1) and 2) are considered to be the same level for both accelerators. Thus, the most of the co-accelerated electrons in the SINGAP might come from the process 3) secondary emission from the A1G due to the backstream positive ions. Most probably, the positive ions are extracted from beam plasma through the large opening of the SINGAP (grounded) grid. These positive ions are accelerated backwards with the energy of post acceleration voltage, to cause the large amount of secondary electron emission.

From the experimental results above, the accelerator type utilized for the ITER NB was assessed. Table 1 summarizes the comparison of the SINGAP and the MAMuG accelerator tested at the MTF. From the viewpoint of voltage holding, maximum beam current and electron acceleration, the MAMuG showed better performance. Based on these results, it was decided to adopt the MAMuG accelerator for the ITER NB on 28 May 2008 in SINGAP / MAMuG discussion meeting held at the ITER Organization.

Table 1 Comparison between MAMuG and SINGAP

	MAMuG	**SINGAP (at MTF)**
Voltage holding	1 MV with H$_2$ gas of 0.21 Pa	800 kV with H$_2$ gas of 0.25 Pa
Beam current	320 mA (140A/m^2) at 796 keV	220 mA (95 A/m^2) at 672 keV
Electron	$I_e/I_{H^-} = 0.28$	$I_e/I_{H^-} = 1.09$

SUMMARY

The development of the H⁻ ion accelerator for the ITER NBI at JAEA was summarized in this paper. The H⁻ ion current accelerated by the MAMuG accelerator is increasing year by year and reaches 320 mA (140 A/m^2) at 796 keV in 2007 by taking the countermeasures against the backstream positive ions. The power supply drain current including the electron current (421 mA) is close to the power supply limit of MTF.

The SINGAP accelerator has been tested at MTF for direct comparison of the two accelerator concepts (the SINGAP and the MAMuG) at the same test facility. The MAMuG showed better performance in voltage holding capability and electron acceleration. From the results of this test, it was decided to choose the MAMuG as the baseline accelerator for ITER NBI.

REFERENCES

1. ITER EDA Final Design Report, ITER technical basis, Plant Description Document (PDD), G A0 FDR 1 01-07-13 R1.0, IAEA EDA documentation No. 24 (2002).
2. T.Inoue, et al., Nucl. Fusion 46(6) (2006) S379-385.
3. H.P.L. de Esch et.al., Fusion Engin. Des., 73(2005)329.
4. K.Watanabe et al., Rev. Sci. Instrum. 73 (2) (2002) 1090-1092.
5. M.Taniguchi et.al., Nucl. Fusion 43 (2003) 665-669.
6. D. L. Davids, "Recovery Effects in Binary Aluminum Alloys", Ph.D. Thesis, Harvard University, (1998).
7. M.Taniguchi et.al., Rev.Sci.Instrum. 77 (2006) 03A514.
8 K.Watanabe et.al., JAEA-TECH 2005-002 (2005). (in Japanese)
9 K.Watanabe, et.al., J.Appl.Phys., 72 (9), (1992) 3949

Electrons in the negative-ion-based NBI on JT-60U

M. Kisaki[a], M. Hanada[b], M. Kamada[b], Y. Tanaka[b],
K. Kobayashi[b], and M. Sasao[a]

[a] *School of Engineering Tohoku University, Sendai 980-8579, Japan*
[b] *Japan Atomic Energy Agency, Naka 319-0913, Japan*

Abstract. The stripped electron trajectories in a large negative ion accelerator with multi-apertures and three acceleration stages, where non-uniform stray magnetic field is horizontally created, are calculated in the JT-60 negative ion source by a 3-D numerical code. The horizontal non-uniform stray field results in a significant power loss of the stripped electrons in the outmost acceleration channel on the grounded grid (GRG). The power loss in the outmost acceleration channel is more than twice higher than that in the central channel due to the weaker stray field although the total power loading on the GRG is by 25% larger than that by assuming a uniform stray field.

Keywords: NBI, Stripped electron, Electron trajectory, Grid power loading
PACS: 29.20.Ba, 29.25.Ni,

INTRODUCTION

Negative-ion-based neutral beam injection (N-NBI) system is one of the promising candidates for plasma heating and steady state operation of magnetic fusion machines. In JT-60SA, the required beam performance is 10 MW for 100s. To generate such a high power neutral beam, a negative ion source is required to generate a D⁻ beam of 22 A at 500 kV for 100s in JT-60SA [1].

One of the key issues for producing such a high power and long pulse beam is the power loadings of the acceleration grids and the beamline components. The power loadings of the acceleration grids and the beamline components are mainly dominated by electrons stripped from D- ions in the accelerator before full acceleration [2, 3].

In JT-60U negative ion source, a transverse magnetic field is produced by a plasma grid (PG) filter to suppress co-extracted electrons and/or to enhance negative ion production. The PG filter has been also designed in the ITER negative ion source [4]. The PG filter forms the long-range return magnetic field in the whole accelerator. By the stray magnetic field formed by the PG filter, the electrons in the accelerator are deflected, resulting in high power loading of the acceleration grids and the beamline components. Therefore, it is important to understand the electron trajectory in the accelerator and the beamline to design the acceleration grids in the JT-60 SA and ITER accelerators.

In the previous studies, the electron trajectory has been calculated in the single-aperture channel [3, 5]. However, the stray magnetic field in the accelerator on ITER and JT-60SA, formed by the PG filter, is foreseen to be non-uniform over the multi-aperture acceleration grids, so that electron trajectories are locally different.

To understand the electron trajectories through the multi-aperture grids, the calculation model with multiple apertures is established, by which the electron trajectories for different stray magnetic field are calculated and the grid power loading is estimated. In this paper, the electron trajectories for different stray magnetic field are reported.

NEGATIVE ION SOURCE ON JT-60U

A schematic diagram of the JT-60 negative ion source is shown in Fig 1. The ion source is composed of a plasma generator, an extractor and an accelerator. The arc chamber has the semi-cylindrical shape with dimensions of 0.68 m in inner diameter and 1.22 m in height. The plasma is generated by an arc discharge with 48 tungsten filaments and caesium is seeded to enhance a negative ion yield. The extractor consists of two grids, which are called as a plasma grid (PLG) and an extraction grid (EXG). To form the transverse magnetic field in the arc chamber, high current is applied to PLG in a longitudinal direction. In the EXG, permanent magnets (Sm-Co) are embedded to suppress the acceleration of the extracted electrons. The accelerator consists of 3-stage electrostatic acceleration grids, which are called as the first acceleration grid (A1G), the second acceleration grid (A2G), and the grounded grid (GRG). Each grid of the extractor and the accelerator is divided into 5 segments of dimensions of 515×230 mm^2 where 9 (vertical) $\times$ 24 (horizontal) apertures are drilled in the area of 480×180 mm^2. The diameter of the apertures is 16 mm. The pitches between the apertures in horizontal and vertical directions are 19 mm and 21 mm, respectively.

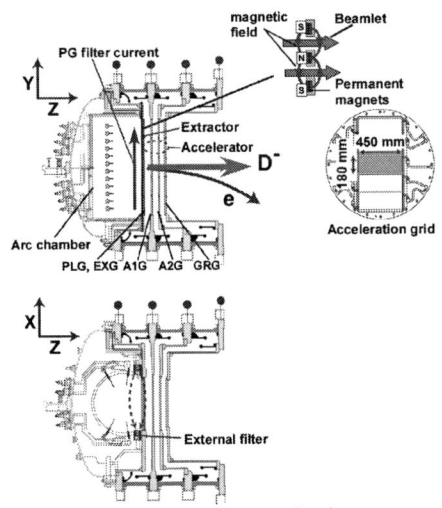

FIGURE 1. Schematic diagram of a negative ion source on JT-60U.

Figure 2(a) and (b) show the magnetic fields (B_y and B_x) of Y and X components as a function of the distance from the PLG in the accelerator at the operating PG filter current of 3.8 kA, respectively. The coordinate axis is indicated in Figure 1. In Fig. 2(a), a local strong magnetic field (B_y) at z = 20 mm is formed by the permanent magnets embedded in the EXG. In Fig. 2(b), a long-range transverse magnetic field (B_x) is formed by the PG filter. The B_x in the outmost channel is weaker than that in the central channel. The B_x monotonically decreases with Z. The B_x at PLG and GRG are ~50 Gauss and ~20 Gauss in the central channel, respectively. The vertical and horizontal profile of B_x in the first gap (at z = 50 mm), the second gap (at z =150 mm) and the third gap (at z = 225 mm) are shown in Fig 3. Although the vertical B_x profile is uniform within the acceleration grid, the horizontal B_x profile is uniform only at -130 mm <X<130 mm that is narrower than that ion extraction area. Since the magnetic field at X>130 mm and X< -130 mm is relatively small, the electrons stripped in the hatched region in Fig.3-(b) are less deflected vertically than that in the center region.

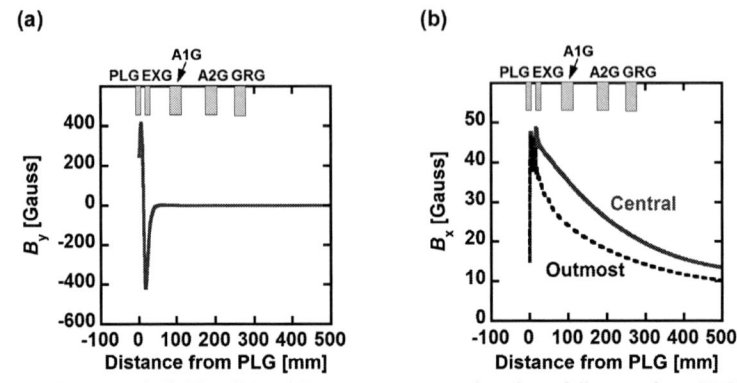

FIGURE 2. Magnetic fields of Y and X components as a function of distance from PLG in the accelerator at the operating PG filter current of 3.8 kA.

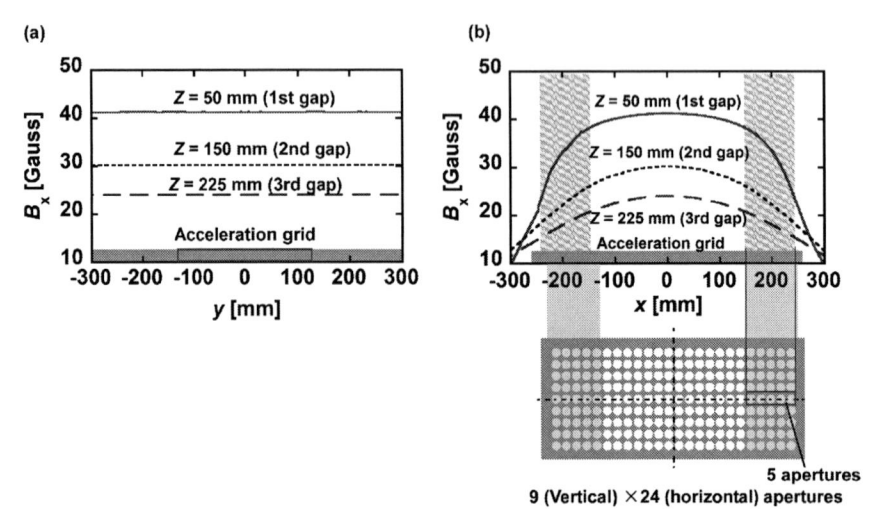

FIGURE 3. Vertical and horizontal profile of B_x at Z = 50 mm, 150 mm and 225 mm

CALCULATION

Stripping losses in the accelerator

For a 300 keV, 3.4 A D⁻ ion beam that was experimentally produced through the central segment at the operating pressure of 0.3 Pa, the stripping loss is calculated by a three dimensional Monte-Carlo code [6]. The collision considered in this study is negative ion single stripping reaction. Since no lateral pumping of the residual gas molecules through the gaps between the acceleration grids is expected in the JT-60U negative ion source, the residual gas molecular density is laterally assumed to be uniform in the accelerator. Figure 4 shows the total stripping losses and survival D⁻ ion current as a function of the distance from the PLG. The operating pressure in the arc chamber was set to be 0.3 Pa. The total stripping loss was 34%. The stripping losses in the PLG-A1G, A1G-A2G, A2G-GRG gaps were 24%, 8%, 2%, respectively. The D⁻ ion current at the PLG was estimated from the total stripping loss and the D⁻ ion current measured at the exit of the accelerator.

The D⁻ ion beam current at the PLG was estimated to be 5.2 A. The survived D⁻ ion currents at the EXG, A1G and A2G were 4.8 A, 3.9 A and 3.6 A, respectively. This leads to the stripped electron current of 0.9 A, 0.4 A and 0.2 A in the EXG-A1G, A1G-A2G, and A2G-GRG gaps, respectively. This shows that the stripped electrons are mainly generated in the EXG-A1G gap.

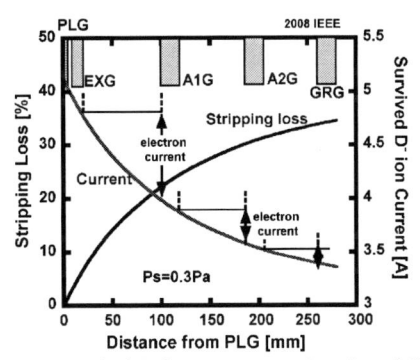

FIGURE 4. Stripping loss and the survival D⁻ ion current as a function of distance from PLG (Ref. 5).

Calculation model and electron trajectory

The trajectory of the stripped electron is calculated by the 3-D finite element method codes with three dimensions, ElecNet and MagNet [7]. The calculation model is shown in Fig. 5. The acceleration channel with the vertical 9 grid apertures is modeled. The dimensions of the apertures, the gap lengths between the acceleration grids, and the distance between the apertures in the model are the same as those of the JT-60 negative ion source. The stray field in the accelerator is widely simulated with the full-scale PLG composed of the vertical five segments and the permanent magnets embedded locally in the EXG. A pair of permanent magnets, served as "external filter" (see in Fig.1), is not taken in this model because the external filter field is locally created near the PLG, and hence relatively small in the acceleration gaps compared with the magnetic field by the PG filter.

Since the stray magnetic field is horizontally non-uniform, the electron trajectory is calculated by changing the horizontal positions of the acceleration channel. The horizontal positions of the acceleration channel are varied from X=9.5 mm to 218.5 mm where there is outmost aperture.

Along the Z axis of the central channel, the electrons are launched at 34 different positions, which are situated every 7 mm from Z = 28 mm. At each of the positions, 54 electrons are launched from an area of 8 mm in diameter with an initial energy of 1.0 eV. Namely, 1836 test particles are launched in one acceleration channel. The assumed launch area is evaluated from the calculated diameter of the D⁻ ion beam. One electron is allocated for an area of ~1 mm² except within the circle of 2 mm in diameter. Nine electrons are launched in the circle of 2 mm in diameter. Therefore, the equivalent current per one electron within the circle of 2 mm in diameter is assumed to be 0.33 of that for the other electrons.

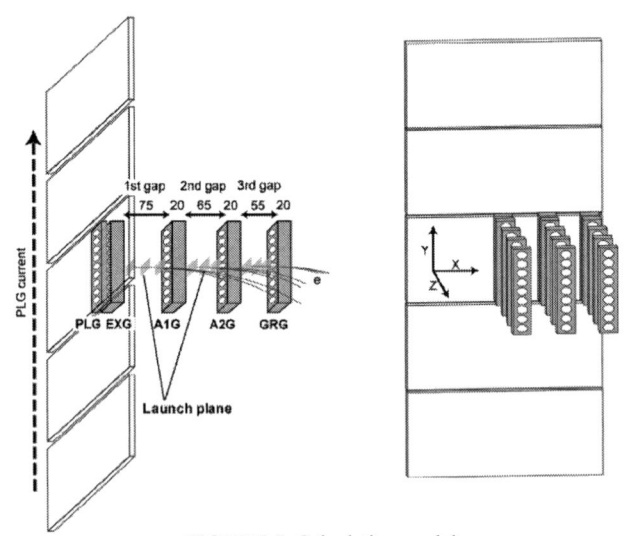

FIGURE 5. Calculation model.

Figure 6 shows the typical trajectories of the electrons in the central (X = 9.5 mm) and outmost channels (X = 218.5 mm) at the total acceleration energy of 300 keV and PG filter current of 3.8 kA. In Fig. 6(b) and (c), the electrons near the EXG are largely deflected by the strong magnetic field created by the permanent magnets. Due to the stray field from B_x, the electrons are vertically deflected. Some of the electrons born in the acceleration gaps are accelerated without interception by the intermediate acceleration grids. Since the stray field in the outmost channel is lower than that in the central channel, the vertical deflection of the electrons stripped in the outermost channel is less than in the central channel.

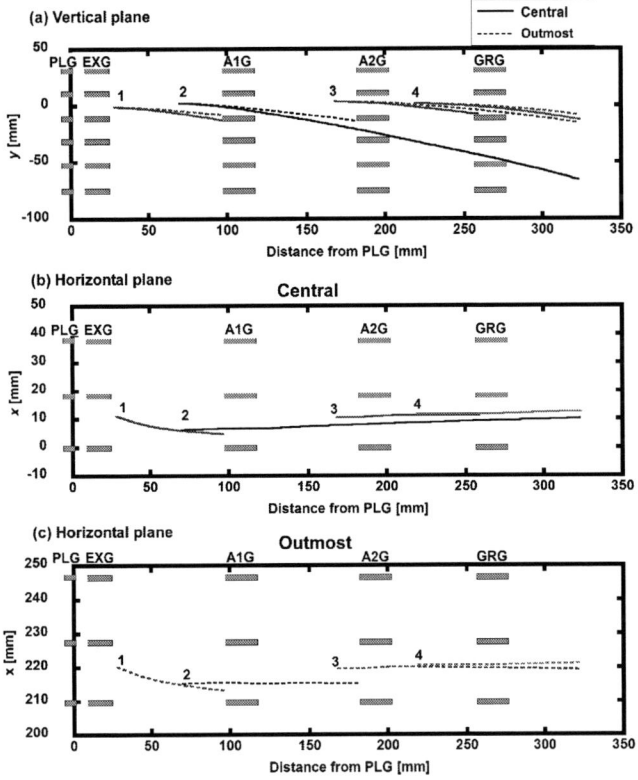

FIGURE 6. Typical trajectories of the electrons generated in the central channel and the outmost channel in the vertical plane and horizontal plane, respectively.

Figure 7 (a), (b), and (c) show the destinations of the electrons stripped in the first, the second, and the third acceleration gaps, respectively. The left and right graphs show the destinations of the electrons stripped in the central and outmost channel, respectively. As shown in Fig.7(a), in both of the central and outmost acceleration channel, most of the electrons are intercepted by the A1G in the first gap of Z=28 mm – 49 mm where the strong magnetic field is locally created by permanent magnets in the EXG. In Z = 56 mm to 112 mm, the electrons tend to escape to the downstream acceleration stages. In this region, the electrons stripped in the central channel are more largely deflected. Some of the electrons pass through the accelerator via the channels below the birth channel. In the same region, some of the electrons stripped in the outmost channel reach the A2G and GRG.

In the second gap (Fig. 7(b)), a half of the electrons stripped in central channel are intercepted by the A2G, and the other half reach the GRG and/or the beamline. In the outmost channel, electrons intercepted by the A2G and the GRG are less and more than those in the central channel, respectively because of the lower stray field.

In the third gap (Fig. 7(c)), most of the electrons in both of the central and outmost channels pass through the accelerator and reach the beamline.

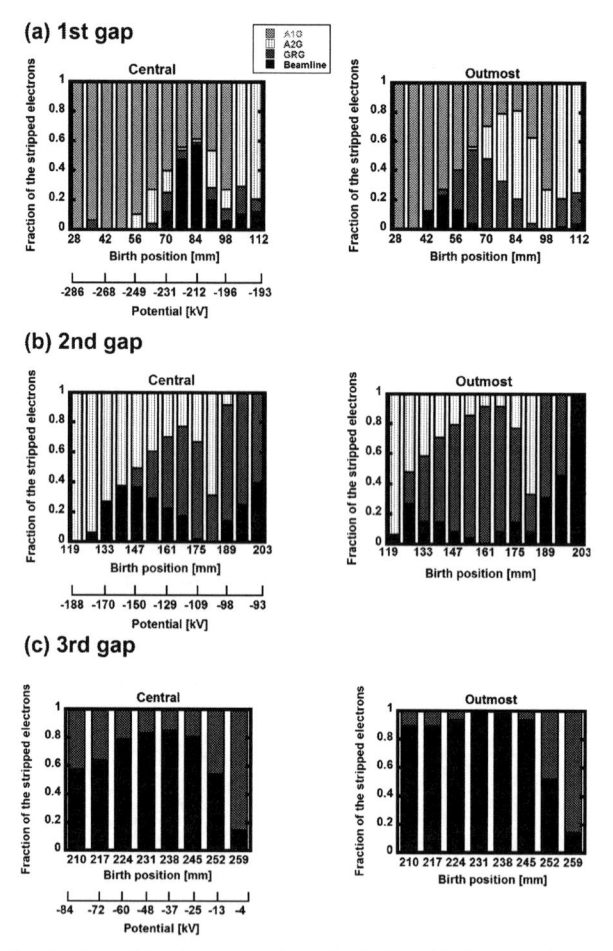

FIGURE 7. Destination of the electrons stripped in the first (a), the second (b), and the third acceleration gap (c)

Power losses of the stripped electrons on the acceleration grids

To estimate the power losses on the acceleration grids and beamline, the current and the energy of the stripped electron are evaluated. The electron current on the grids and the beamline is estimated from the stripped currents (Fig. 4) and fractions of the destinations (Fig. 7). The acceleration energy is evaluated from the potential difference between the birth and the destination positions (Fig. 7). Figure 8 (a), (b), (c), and (d) show the power loading caused by the stripped electrons on the A1G, the A2G, the GRG and the beamline, respectively. The horizontal axis indicates the location of the acceleration channels. On the A1G, the power loss slightly decreases by closing to the outmost aperture although the stray field in the center channel is 1.4 times higher than that in the outmost channel. This small difference is due that most of the electrons in the first gap are mainly deflected by the magnetic field created in the EXG and intercepted by the A1G.

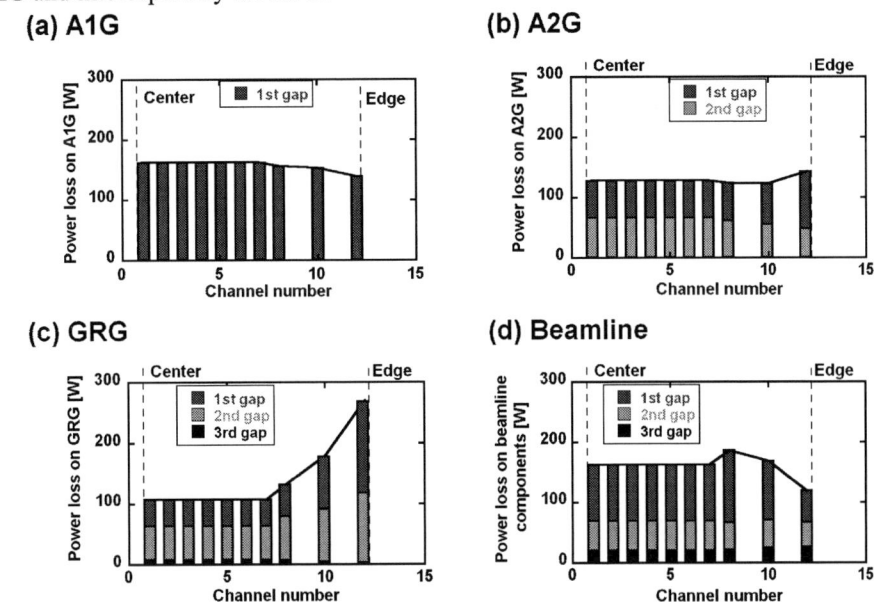

FIGURE 8. Power loss of the stripped electrons generated at each channel on the A1G (a), the A2G (b), the GRG (c), and the beamline components (d).

On the A2G, the power loss slightly increases by closing to the outmost channel. The increase of the power loss in the outmost channel is due that more electrons generated at the first gap are collected by the A2G without interception by the A1G. The increase of the power loss in the outmost channel is significant on the GRG. The power loss in the outmost channel is more than twice higher than that in the central channel. This is because that more electrons generated at the first and/or second gaps are collected by the GRG without interception by the intermediate grids due to weaker stray field. This significant power loss on the GRG in the outmost channels leads to less power loss in the beamline where the origin of the power loss is different from those on the acceleration grids. In the beamline, the stripped electrons come from all of the acceleration gaps. This shows that energy spectra of the electrons on the beamline is widely spread.

The total grid power loading for one segment is estimated from the summation of the power losses of 12 acceleration channels shown in Fig. 8. Namely, since total number of the apertures on one segment is 216, total grid power loading of one segment is simply estimated by multiplying the summation of grid power losses with 12 channels by 18 (=216/12). Figure 9 shows the total grid power on one segment. For the comparison, the grid power loading estimated only from the central channel is shown in the same graph. Solid circles show measured power loadings by the stripped electrons. Except for the GRG, the grid power loading for the non-uniform stray field is almost the same as that estimated from the central channel. The GRG power loading for the non-uniform stray field is higher by 25 % than that estimated from the central channel. The calculated power loading is higher than the measured value. This shows that the electrons are intercepted by the A2G and the GRG. Since in this calculation the blank space between segments is not taken into account, more electrons pass through the accelerator without interception by the A2G and/or the GRG.

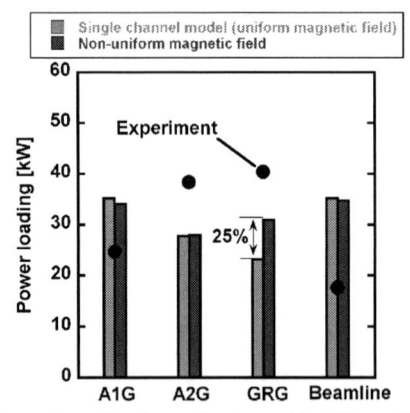

FIGURE 9. Power loss of each acceleration grid and beamline components

SUMMARY

The electron trajectories through the multi-aperture grids, where the stray magnetic field is horizontally non-uniform, are calculated by the 3-D numerical code. As the result, the followings are clarified.
1. The power loss of the stripped electrons on the GRG in the outmost channel is more than twice higher than that in the central channel.
2. The GRG power loading for the non-uniform stray field is higher by 25 % than that estimated from the central channel.
3. In the beamline, the stripped electrons come from all of the acceleration gaps.

ACKNOWLEDGMENTS

The authors would like to thank other members of JT-60U NBI Heating Group for their valuable discussion. They are also grateful to Dr. T. Tsunematsu and Dr. N. Hosogane for their support and encouragement.

REFERENCES

1. Y. Ikeda et al., Fusion Engineering and Design, 82, pp.791-797 (2007)
2. M. Hanada et al., Rev. Sci. Instruments, 69, pp.947-949 (1998)
3. G. Fubiani et al., Phy. Rev. STAB 11, 014202 (2008)
4. R S Hemsworth et al. Rev. Sci. Instruments, 67, 1120 (1996)
5. M. Hanada et al., IEEE Trans. Plasma sci. 36, pp.1530-1535 (2008)
6. M. Hanada et al., Fusion Eng. Design, 56-57, pp.505-509 (2001)
7. www. Infolytica. com, ElecNet and MagNet 3D, Version 6.25, 2007, Infolytica Corp, Canada

Results of the SINGAP Neutral Beam Accelerator Experiment at JAEA

H.P.L. de Esch[a] L. Svensson[a], T.Inoue[b], M. Taniguchi[b], N. Umeda[b], M. Kashiwagi[b] and G. Fubiani[c]

[a]*CEA Cadarache, IRFM, F-13108 St. Paul-lez-Durance, France.*
[b]*Fusion Research and Development Directorate, Japan Atomic Energy Agency, 801-1 Mukouyama, Naka 311-0193, Japan*
[c]*Laboratoire Plasma et Conversion d'Energie (LAPLACE) Université Paul Sabatier, Bt 3R2, 118 Route de Narbonne*

Abstract. IRFM (CEA Cadarache) and JAEA Naka have entered into a collaboration in order to test a SINGAP [1] accelerator at the JAEA Megavolt Test Facility (MTF) at Naka, Japan. Whereas at the CEA testbed the acceleration current was limited to 0.1 A, at JAEA 0.5 A is available. This allows the acceleration of 15 H- beamlets in SINGAP to be tested and a direct comparison between SINGAP and MAMuG [2] to be made. High-voltage conditioning in the SINGAP configuration has been quite slow, with 581 kV in vacuum achieved after 140 hours of conditioning. With 0.1 Pa of H2 gas present in the accelerator 787 kV could be achieved. The conditioning curve for MAMuG is 200 kV higher. SINGAP beam optics appears in agreement with calculation results. A beamlet divergence better than 5 mrad was obtained. SINGAP accelerates electrons to a higher energy than MAMuG. Measurements of the power intercepted on one of the electron dumps have been compared with EAMCC code [3] calculations. Based on the experiments described here, electron production by a SINGAP accelerator scaled up to ITER size was estimated to be too high for comfort

Keywords: Neutral beams, accelerator, negative ions, SINGAP, MAMuG.
PACS: 41.20Cv, 41.75Cn, 41.85Ar, 41.85Ja

INTRODUCTION

Two accelerator concepts were being considered for neutral beam injection (NBI) on ITER. These are the Multi Aperture, Multi Grid (MAMuG) [2] and the Single Gap, Single Aperture (SINGAP) [1] concepts. In MAMuG, 1280 D⁻ beamlets are accelerated in 5 intermediate steps to 1 MeV, whereas in SINGAP 1280 pre-accelerated beamlets are accelerated in one single step to 1 MeV. Total accelerated current is 40 A. The advantage of SINGAP over MAMuG is technical simplicity, which comes at the cost of more complicated physics. A particular disadvantage is that in SINGAP parasitic electrons are accelerated to high energy, whereas in MAMuG most are intercepted on the intermediate accelerator stages.

IRFM (CEA Cadarache) and JAEA Naka have entered into a collaboration in order to test a SINGAP [1] accelerator at the JAEA Megavolt Test Facility (MTF) at Naka, Japan. Whereas at the CEA testbed the acceleration current was limited to 0.1 A, at

JAEA 0.5 A is available. This allows the acceleration of 15 H⁻ beamlets in SINGAP to be tested, an increase by an order of magnitude. Normally the MTF testbed operates in the MAMuG [2] configuration and as the MAMuG accelerator had been extensively tested before on the same testbed, a direct comparison between SINGAP and MAMuG can be made.

Two experimental campaigns have been conducted in the SINGAP configuration: the first in August-October 2007, the second in February – March 2008. For the second campaign special electron dumps and magnets to deflect electrons on these have been installed.

THE MTF TEST FACILITY

A schematic of the beamline is in figure 1. Negative H⁻ ions are produced in a KAMABOKO type source [4]. Its filter strength ∫Bdl is 1000 Gauss.cm. Caesium is added to the source to enhance the production of negative ions. The SINGAP accelerator accelerates the pre-accelerated ions (typical pulse length 0.2 second) to high energy in a single step. After acceleration the ions drift towards a calorimeter located 2.69 metres downstream of the grounded grid electrode. If the calorimeter is not in use, the ions are intercepted on a beam dump.

During the second test campaign the beamline was equipped with electron deflection magnets. The electrons are deflected on two electron dumps. Electron dump I protects the gate valve and its edge is 8 cm away from the beamline centre. Electron dump II protects the downstream beamline vessel and was equipped with a thermocouple for calorimetric measurements. Its edge is 7 cm away from the beamline centre line.

The accelerator is shown in some more detail in figure 2. The left half of the figure shows the accelerator in SINGAP configuration, the right half shows it in MAMuG configuration. The pre-accelerator for SINGAP is identical to the extractor and first acceleration stage for MAMuG. In order to obtain the SINGAP configuration from the MAMuG accelerator, the following changes had been made.

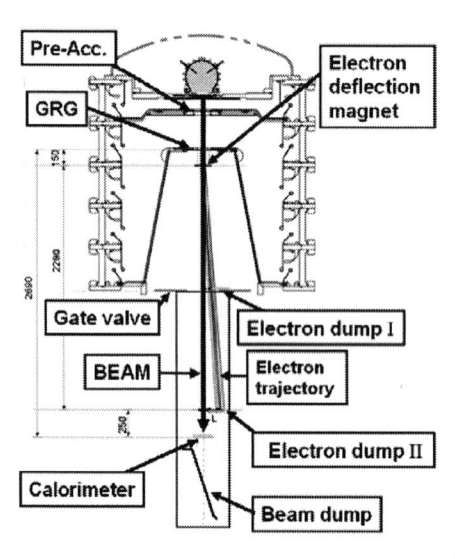

FIGURE 1. Schematic of the MTF beamline in SINGAP configuration at JAEA in Naka (Japan)

1. Remove the intermediate grids A2G, A3G and A4G,
2. Replace the 15 individual circular apertures in the grounded grid with one rectangular aperture, sized 106x114 mm²,
3. Attach a rectangular "well"[1], 132x150 mm² to the pre-acceleration grid A1G. The depth of this well was 33 mm in the first test campaign and 21 mm in the second

test campaign. Its purpose is to compensate for the beamlet-beamlet interaction by pre-focussing the beams.

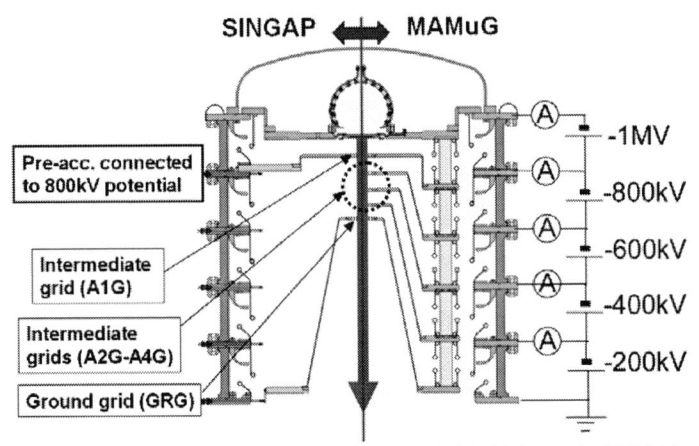

FIGURE 2. Accelerators in the MTF test bed. The left half shows the SINGAP configuration, the right half shows the MAMuG configuration.

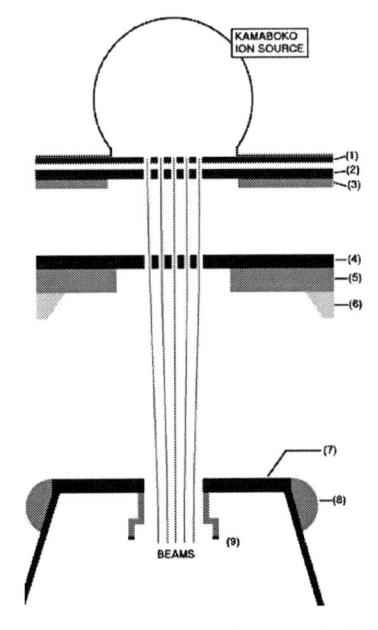

(1) The plasma grid contains 15 circular apertures (Ø14 mm) in a 3x5 pattern.
(2) The extraction grid contains 49 circular apertures (Ø11 mm) in a 7x7 pattern.
(3) The extraction grid (2) is recessed behind a circular structure (3) (Ø244 mm) that is 14 mm deep in the beam direction.
(4) The pre-acceleration grid A1G (20mm thick) contains 49 circular apertures (Ø16 mm) in a 7x7 pattern.
(5) The rectangular well structure (132x150 mm2) pre-focusses the beams. Its depth in the beam direction was 33 mm during the first test; 21 mm during the second test.
(6) A further circular structure, Ø1000 mm, is present. Its depth in the beam direction was 43 mm during the first test; 55 mm during the second test.
(7) The grounded grid contains one large rectangular aperture, 106x114 mm^2. The grid is 20 mm thick.
(8) A stress ring (Ø100 mm) is mounted on the grounded grid to alleviate the electric field stress.
(9) Electron deflection magnets (120x9.2x5.4 mm^3; SmCo, 0.96T) are mounted 160 mm apart and 150 mm downstream of the grounded grid. They were present only during the second test.

FIGURE 3: SINGAP accelerator at the JAEA MTF facility.

Gas pumping is provided by a 2000 litres/second turbo pump, which leads to a minimum pressure of 0.08 Pa in the beamline.

Figure 3 shows the salient features of the SINGAP accelerator as installed at JAEA. 15 H⁻ beamlets are accelerated in a 3x5 pattern to high energy. The apertures are 19 mm apart in the X-direction (3 apertures) and 21 mm apart in the Y-direction (5 apertures). The extraction gap is 4.7 mm, the pre-acceleration gap is 104 mm (90 mm metal to metal) and the post-acceleration gap between A1G (4) and grounded grid (7) is 328 mm. The geometry is aimed at providing low-divergence beamlets that are almost parallel at the same time.

HIGH VOLTAGE CONDITIONING

The high voltage conditioning in SINGAP configuration was done over a period of 4 weeks. Over 140 hours of HV was applied and the maximum voltage that could be held was 581 kV without adding gas in the accelerator. With added gas the voltage holding was increased to 787 kV. The SINGAP conditioning curves for the two test campaigns are in figure 4. They almost overlap.

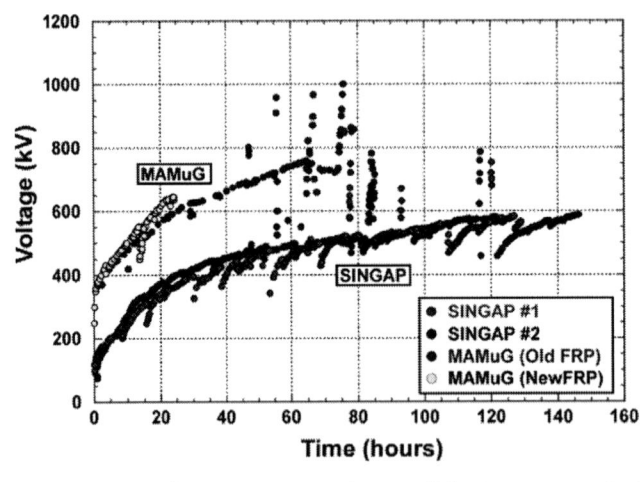

FIGURE 4. High voltage conditioning curves for the MTF testbed in SINGAP and MAMuG configuration.

For comparison purposes, the conditioning curves in MAMuG configuration are plotted as well. There are curves from earlier operation and also a curve from MAMuG operation that took place after SINGAP operation had ended. The required conditioning time in MAMuG configuration was shorter (60 - 80 hours) and the voltage that could be held is about 200 kV higher.

BEAM OPTICS

The beam profile was measured with a CEDIP infrared camera (wavelength 4 µm) through a saphire window. The beam pulse length was 0.2 seconds and the temperature profile was measured on a 14x14 cm² carbon fibre target. The fibres are orientated in the beam direction (1.0 cm long), which results in a poor heat conduction in the two perpendicular directions. This makes such targets quite suitable for beam profile measurements. The emissivity of the target was measured to be 0.89 and the transmission of the window appeared to be 0.42. The target has a carbon and a copper side. Calibration measurements between the infrared procedure on carbon and thermocouple measurements on copper gave a 15% discrepancy.

Due to an oversight, the feature (3) in figure 3 was not taken into account for the beam-optics calculations for the first test campaign. As a result, the kerb (5) was chosen too deep (33 mm) and all the beamlets were steered inwards. All the beamlet profiles became thus overlaid and not much could be concluded about the beam optics.

In the second test campaign the kerb (feature (5) in figure 3) was chosen 21 mm. This depth was chosen to make the beamlets slightly steered outwards. In this way the divergence of the individual beamlets could be studied.

Beamlet direction and beamlet divergence was calculated with the commercial code OPERA3D SCALA by Vector Fields [5]. The geometry shown in figures 2 and 3 was modelled in the code. The modelling started at the extraction grid (feature (2) in figure 3). To date there is no module in SCALA that calculates the extraction of beams in the aperture of the plasma grid (feature (1) in figure 3). The beam properties at the extraction grid had been calculated by the 2D code SLACCAD [6]. The result was then used as the starting condition for OPERA-3D. The 15 beamlets had been modelled simultaneously in OPERA-3D. This ensures that the interaction between the beamlets has been taken into account. The divergence of the individual beamlets is 3.5 - 4.0 mrad. The calculation is electrostatic only.

The magnetic deflection of the ion beams due to the filter field and the electron suppression magnets incorporated in the extraction grid has been calculated with an in-house developed particle trajectory code called TRACK [7]. It tracks the particles through the electromagnetic fields in the accelerator. The electric fields are imported from SLACCAD or SCALA, the magnetic fields from the permanent magnets are calculated by the code kindly supplied by D. Ciric [8]. The validity of this Ciric code is by now well established through direct measurements.

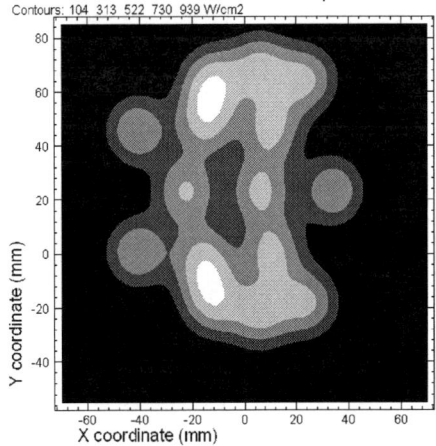

FIGURE 5: Calculated power density profile on the carbon target for 15 beamlets accelerated and transported in the calculated electric and magnetic fields. Individual beamlet divergence was 3.5 and 4.0 mrad in the X and Y direction, respectively.

The result of the beam-deflection calculations by TRACK is that all the fifteen 450 keV H⁻ beamlets are deflected in the filter field of the source by 8.9 mrad in the Y direction. In the X-direction deflection takes place by the magnets of the electron deflection system (the same for all 15 beamlets) and by the electron suppression magnets incorporated in the extraction grid. The latter provide an equal but opposite field in neighbouring rows of apertures. Some rows will see a magnetic deflection of -2.2 mrad, neighbouring rows see a magnetic deflection of +5.9 mrad. Providing all this information to a transmission code allows the expected beamlet pattern to be calculated (fig. 5). The divergence of the individual beamlets was assumed to be 3.5 and 4.0 mrad in the X and Y directions, respectively. This in accordance with the SCALA results.

Power density is determined experimentally from the target temperatures before and after the beam pulse. The beam target can be rotated in vacuum and temperature measurements from identical pulses have been obtained from the carbon side (by infrared measurement) and from the copper side (by thermocouple measurement). The infrared measurements indicate a 15% lower power than the thermocouple measurements. The thermocouple measurements are considered to be better, given the complications and the opaque window involved in the infrared measurement.

A good example of a perveance matched shot is pulse 372 done on 14 March 2008. A measured power density profile is shown in figure 6. It can be seen by comparing figures 5 and 6 that the calculated deflection by the magnets in the extraction grid is too small. This was also the case in the first experimental campaign. However, this deflection is most sensitive to the precise position of plasma meniscus and/or magnets inside the extraction grid. A power density profile where the deflection due to the suppression magnets is multiplied by a factor of 1.5 can be seen in figure 7. To obtain the stronger deflection, the position of the plasma meniscus or the magnets in the extraction grid have to be shifted by around 0.5 mm. Thus, within the error margins, the calculated and measured SINGAP beam profiles are in agreement.

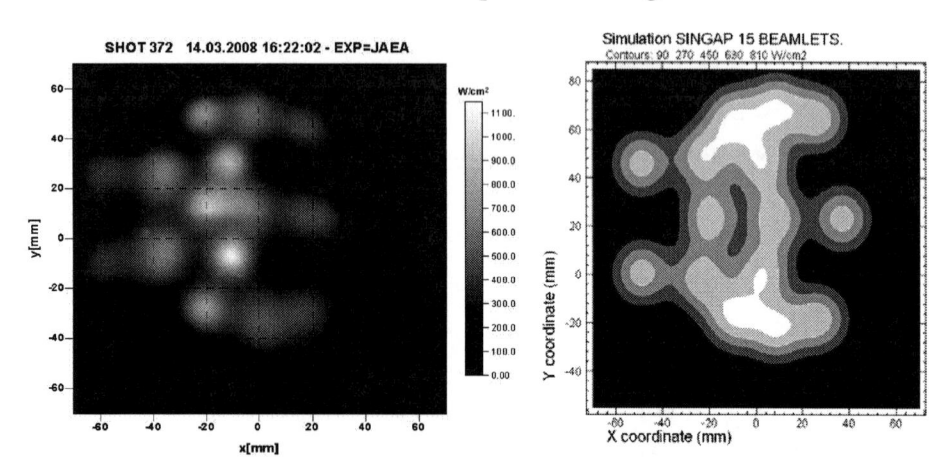

FIGURE 6 (left): Experimentally measured power density profile for shot 372 performed on 14 March 2008. The shot was at 450 keV and approximately perveance match (6 mA/cm^2 H$^-$).
FIGURE 7 (right): Calculated power density profile on the carbon target for 15 beamlets accelerated and transported in the calculated electric and magnetic fields. Beam deflection in the X-direction was increased by a factor of 1.5 to better match the data in figure 6. This fudge corresponds to a shift of 0.5 mm in the position of the magnets in the extraction grid or the plasma meniscus.

To assess the beamlet divergence from this, 1D data was taken at X=-20 mm and X=-12 mm. A 5-Gaussian fit was performed on these two slices of data, see figures 8 and 9. The data at X=-20 gives the beamlet divergence for the top, middle and bottom beamlets, whereas the data at X=-12 mm does the same for the beamlets at Y=-10 and +30 mm.

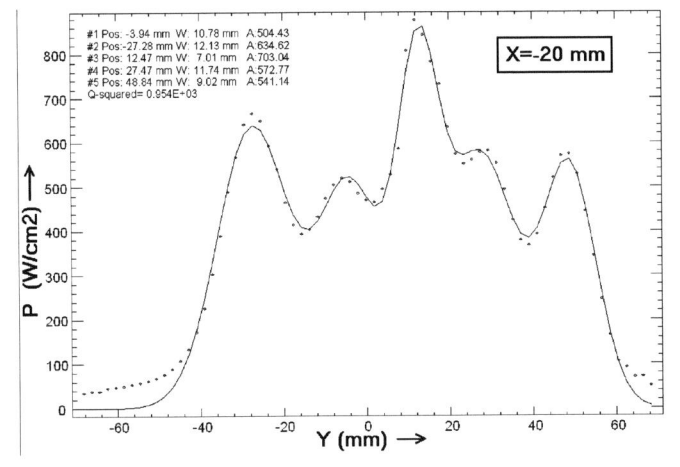

FIGURE 8: Experimental data and 5-Gaussian fit for shot 372 performed on 14 March 2008. The data is taken at X=-20 mm (see figure 8).

We see from figures 8 and 9 that going from top to bottom the beamlet widths are 9.0, 7.6, 7.0, 8.2 and 12.1 mm. As the target is 2.69 m from the accelerator, the corresponding beamlet divergences are: 3.3, 2.8, 2.6, 3.0 and 4.5 mrad, respectively. Because the accelerator pressure is high (0.08 Pa) substantial quantities of neutrals with partial energies are present, producing the halos seen at around X=-60 mm.

Similar beamlet divergences have been measured in the horizontal direction (not shown). In conclusion, SINGAP beam optics is as expected.

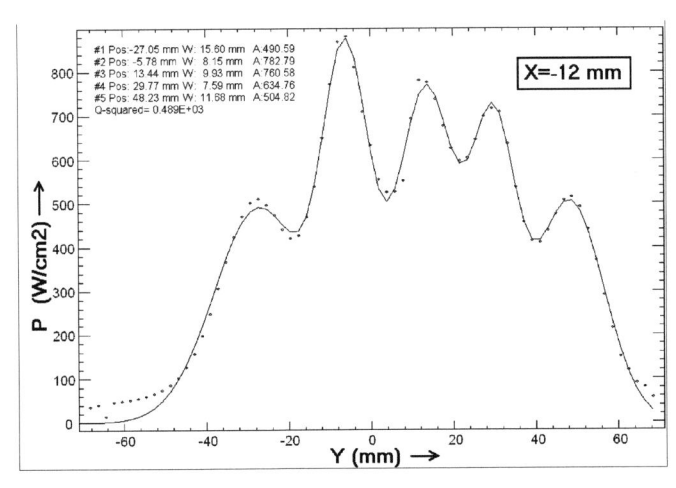

FIGURE 9: Experimental data and 5-Gaussian fit for shot 372 performed on 14 March 2008. The data is taken at X=-12 mm (see figure 8).

ELECTRONS

Whereas the amount of electrons produced by SINGAP is lower than that produced by MAMuG, the SINGAP ones are more troublesome because they are accelerated to almost full energy and leave the accelerator through the wide aperture in the grounded grid. Measurements have been made of the heat flux falling onto the downstream electron dump. By modelling this heat flux using different models of electron production, an idea can be formed about the electron production mechanisms in SINGAP and accelerators in general.

The downstream electron dump (electron dump II in figure 1) is a 2 mm thick piece of copper, 130x200 mm^2. The edge of it is 70 mm away from the beam centre (X>70 mm). A thermocouple is attached to it and the heat deposited during the 0.2 second beam pulse is measured with a thermocouple.

The power to electron dump II has been calculated using an electron trajectory following code (TRACK). The electron source is provided by stripping of the negative ion beams and by ionisation of background gas due to the beams. As the cross-section of ionisation by negative ions is poorly known we assumed that it is equal to the cross-section of ionisation by neutrals. There is justification of this from measurements done by Fogel et al [9]. In the model, electron production can take place inside the pre-accelerator and the post-accelerator. Some electrons produced in the pre-accelerator are intercepted on the pre-acceleration grid. The rest is accelerated to full energy. About 50% of the electrons produced inside the pre-accelerator leak through to the post-accelerator. Electron backscattering is taken into account, which reduces the calculated power to the dumps by around 25%.

Not included in the model described above are electrons from the ion source and extractor which are very well suppressed by the magnets in the extraction grid.

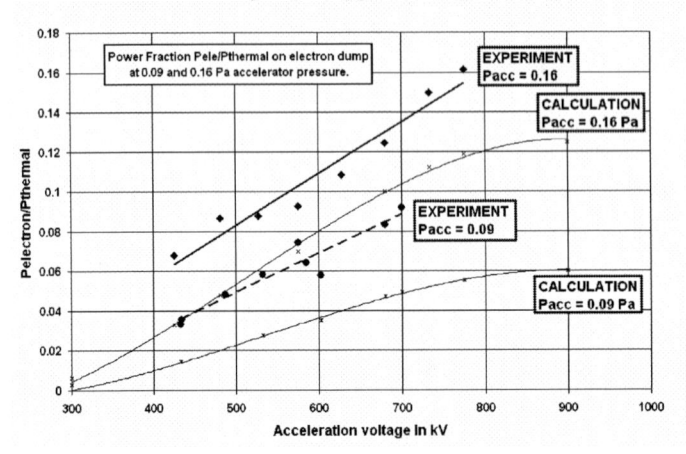

FIGURE 10: Power on the electron dump II. Experimental data and data by calculation of stripping losses and ionisation of background gas is shown for two different accelerator pressures.

Double stripping of the beams and especially ionisation of background gas cause positive ions to be generated in the accelerator. These are accelerated backwards into the source. In the SINGAP geometry they can hit the pre-acceleration grid on the downstream side. This causes emission of secondary electrons which are then accelerated to 80% of the full energy. Secondary electrons produced by backstreaming positive ions are not included in this model. They are, however, included in the EAMCC code [3], which has also been used to simulate this SINGAP experiment.

In figure 10 the measurements and the TRACK model calculations are shown. Measurements on the electron dump have been performed at accelerator pressures of 0.09 Pa and 0.16 Pa. The data taken at 0.16 Pa show significantly higher power than the data taken at 0.09 Pa, thus suggesting that stripping and ionisation is the cause of the measured power.

Not shown in figure 10 are the calculation results that take only the post-accelerator into account. Less than 20% of the power measured on electron dump II is calculated. It is necessary to include the pre-accelerator in the simulations.

There is a strong dependance on acceleration energy. The reason is that at low energy most electrons are intercepted on the uninstrumented electron dump I. Only at high energy, significant electrons make it past electron dump I and are subsequently intercepted on electron dump II.

The model results reproduce the strong energy dependance and the pressure dependance. However, by considering only stripping and ionisation, the calculated power on the dump is still too low. Hence, the other effects that are included in EAMCC are important.

In figure 11, EAMCC calculations for 0.09 Pa are shown. The EAMCC calculations reproduce the measurements better. The error bars shown for EAMCC correspond to two assumed positions for electron dump I, 7 cm and 8 cm (the electron dump I had been moving throughout the experimental campaign).

The physics included in EAMCC is:
- Stripping of negative ions (36%, 33% and 31% of the beam is stripped at 487, 603 and 700 keV, respectively).
- Ionisation of the 0.09 Pa background gas (electron production is 12%, 11% and 10% of the beam current at 487, 603 and 700 keV, respectively).
- electron production by secondary neutrals is significant at 0.09 Pa.
- electron production by backstreaming negative ions impinging on copper surfaces is considerable.
- direct heating of the calorimeter by negative ions and neutrals is included

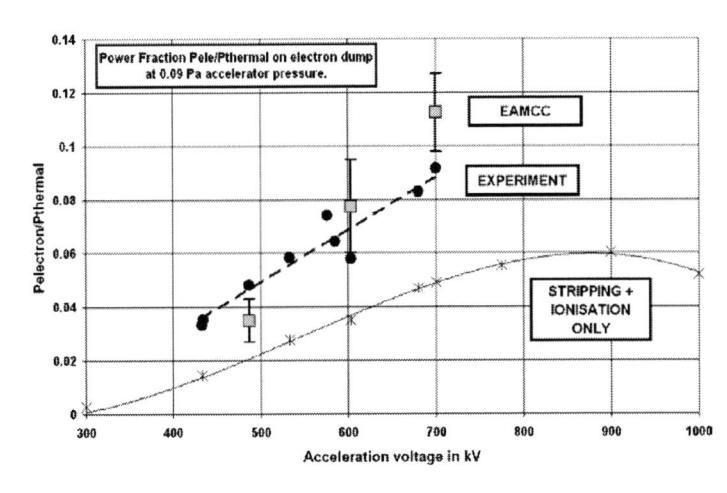

FIGURE 11: Power on the electron dump II for $P_{acc}=0.09$ Pa. Experimental data and calculations by EAMCC and the simpler stripping losses / ionisation code are shown.

The total electron power leaving the SINGAP accelerator (operated at 0.09 Pa) is 29%, 26% and 24% of the power on the calorimeter for 487, 603 and 700 keV, respectively. The power by secondary electrons released from copper surfaces due to positive ion impact is included in these figures. The power by these secondary electrons amounts to 7% of the beam power on the calorimeter, almost independent of beam energy.

ELECTRONS IN ITER

The accelerator pressure in the ITER beamline is not yet precisely known, but it will be much lower than the ~0.09 Pa used during these test campaigns. Asuming an accelerator pressure of 0.03 Pa in ITER, the electron power produced by a SINGAP accelerator can be estimated from the experimental results.

First, the electron dump II received only a small part of the total electron power: according to calculation 22% at 700 keV, 33% at 1 MeV. Secondly, the experimental measurements on the electron dump are around 50% higher than the calculation results. Thirdly, on ITER the accelerator pressure is lower by a factor of 3. Finally, ITER operates in deuterium, whereas present experiments had been performed in hydrogen.

We calculated electron production by stripping and ionisation of 0.03 Pa background gas through which hydrogen beams are accelerated to 1 MeV. We found that the total electron power in the system is 6.0% of the total beam power. In deuterium the number is 9.3%. If we assume, based on the present measurements, that an additional 50% of electrons is accelerated, then the electron power in the SINGAP accelerator is 14% of the accelerated D⁻ power (40 MW). Hence we expect 5.6 MW of electrons to be produced in ITER SINGAP, most of this power (90%) would be transmitted outside the accelerator and impinge on the neutraliser.

In the present experiment there were no magnets in the pre-acceleration grid to stop electrons passing from the pre-accelerator to the post-accelerator. If electrons passing from pre to post accelerator could be effectively stopped, the expected electron power produced by ITER SINGAP would reduce from 5.6 MW to 2.8 MW.

The electrons accelerated by ITER SINGAP have also been calculated using EAMCC. On ITER, SINGAP produces 8.0 MW of electrons, of which 7.3 MW are transmitted out of the accelerator [10]. Thus EAMCC predicts 7.3 MW of electron beams, whereas TRACK predicts 3.7 MW of electrons. In comparison, EAMCC predicts that the MAMuG produces 7.6 MW of electrons, but 7.0 MW of these are intercepted on the intermediate grids so that only 0.6 MW of electrons leave the accelerator[3].

Based on these results (electrons and voltage holding) a meeting held on 28 May 2008 at ITER decided to adopt the MAMuG accelerator for the ITER heating beams.

CONCLUSIONS

An experiment to test a SINGAP accelerator with 15 beamlets has been performed at the Megavolt Test Facility at JAEA in Japan with success. The following conclusions can be drawn.

- High voltage conditioning in the SINGAP configuration appears slower than in the MAMuG configuration.
- The beams were designed to be slightly divergent at perveance match as to enable an assessment of individual beamlet optics. Individual beamlet divergence is less than 5 mrad.
- Using a trajectory code (TRACK), 2/3 of the power on the downstream electron dump can be explained by stripping losses and ionisation of background gas. Ionisation by secondary fast neutrals and secondary electrons produced by ion impact (backstreaming positive ions and impact by beam particles) can account for much of the remainder (EAMCC).
- More than 50% of the electrons leak through the apertures in A1G and contribute out of proportion to the signal in the downstream electron dump.
- Extrapolation to ITER leads one to expect 3.7 MW (TRACK), 5.6 MW (experiment) or 7.3 MW (EAMCC) electrons.
- If electron leakage through the pre-acceleration grid could be stopped, the electron power would go down by half.

Based on the present results ITER adopted MAMuG on 28 May 2008.

ACKNOWLEDGMENTS

This work, supported by the European Communities under the contract of Association between EURATOM and CEA, was carried out within the framework of the European Fusion Development Agreement. The views and opinions expressed herein do not necessarily reflect those of the European Commission.

REFERENCES

1. H.P.L. de Esch, R.S. Hemsworth and P. Massmann, *Fusion Engineering and Design* **73**(2005)329.
2. M. Taniguchi, T. Inoue, N. Umeda, M. Kashiwagi, K. Watanabe, H. Tobari, M. Dairaku, and K. Sakamoto, *Rev. Sci. Instr.* **79**(2008)02C110.
3. G. Fubiani, H. P. L. de Esch, A. Simonin and R.S. Hemsworth, *Phys. Rev. ST Accel. Beams* **11**(2008)014202.
4. M. Taniguchi, M. Hanada, T. Iga, T. Inoue, M. Kashiwagi, T. Morisita, Y. Okumura, T. Shimizu, T. Takayanagi, K.Watanabe and T. Imai, *Nucl. Fusion* **43**(2003)664.
5. Vector Fields Ltd., 24 Bankside, Kidlington, Oxford OX5 1JE, UK. Tel.: +44 1865 854999. http://www.vectorfields.co.uk.
6. W.B. Hermannsfeld, SLAC report, Stanford Linear Accelerator Center, SLAC-226 (1979).
7. H.P.L. de Esch, Final report EFDA contract TW4-THHN-IITF2, Section 1 in CEA report SCCP/NTT-2006.003
8. D. Ciric (private communication and support).
9. Y. M. Fogel, A. G. Koval, Y. Z. Levchenko, and A. F. Khodyachikh, *Sov. Phys. JETP* **12**(1961)384.
10. G. Fubiani (to be published).

Lithium Jet Neutralizer to Improve Negative Ion Neutral Beam Performance

L. R. Grisham

Princeton Plasma Physics Laboratory, P. O. Box 451, Princeton, New Jersey 08543, USA

Abstract. Fusion neutral beam systems have conventionally used gas cells of the beam isotope as neutralizers. In the design of negative ion beam systems for ITER, and likely also for future reactors, the large gas efflux arising from these cells can lead to many problems. We discuss the possibility of decreasing the gas load by using a lithium jet neutralizer, which would also afford higher neutralization efficiency.

Keywords: Neutral beams, negative ions, neutralizer, supersonic lithium jet
PACS: 41.75.Cn

INTRODUCTION

Neutral hydrogen isotope beams have heated, fueled, and driven current in the plasmas of many fusion experiments over the course of the past half century. For decades, these beams were formed by electrostatically accelerating positive ions which were subsequently converted to neutral atoms in a closely coupled charge exchange cell fueled partially or entirely by the effluent gas from the ion source. As fusion experiments increased in size and density, the beam velocity required for adequate penetration increased well beyond the Bohr velocity, so that the neutralization efficiency of positive ions declined, leading to the necessity of negative ions as the precursors to fast neutral atom beams in the most recent generation of beam systems, and for future devices such as ITER[1]. Since negative hydrogen ions are easily stripped of their extra electron to become neutrals, and since the ultimate energies of negative ions are usually higher than was the case with positive hydrogen ions, necessitating longer accelerators for negative ions, the average gas pressure in negative ion accelerators is reduced by separating the neutralizer cell from the accelerator with a pumping region to greatly reduce the accelerator line density.

Accordingly, while positive-ion-based fusion neutral hydrogen beam systems were compelled to use a common gas for the neutralizer and ion source, since their closely-coupled neutralizers resulting in commingling of the gases, their was no such compelling reason for negative-ion-based neutral hydrogen beams to employ the same gas in the neutralizer cell as was used in the ion source. Nonetheless, the tradition of using the same gas for neutralization as for the source continued in the first generation of high power negative ion neutral beams for JT-60U[2] and LHD[3], and also for the present ITER neutral beam design.[4,5]

DIFFICULTIES WITH HYDROGEN ISOTOPE NEUTRALIZERS

Both the JT-60U and LHD negative ion neutral beam systems had sufficient space to allow adequate pumpout regions on each side of the neutralizer cell, as well as neutralizer cell lengths sufficiently long (10 m in the case of JT-60U) to reduce the gas efflux to a tolerable level, so their employment of the same gas for the neutralizer as for the ion source did not result in untoward consequences.

However, the spacious environment available for the JT-60U and LHD neutralizers and their adjacent pumping is unlikely to be the case on future fusion devices due to the necessity of minimizing the volume of space available for equipment near the machine within the radiation environment. Such is already the case for the megavolt neutral beams being developed for heating and current drive on the ITER tokamak, where the severe space constraints made it necessary to minimize the length of the neutralizer cell to only 3 m. However, the shorter the neutralizer cell, the greater the gas efflux for a given neutralizer line density. In order to reduce the neutralizer gas efflux into the beamline, the neutralizer duct is partitioned into 4 parallel ducts by 3 vertical vanes running the length of the neutralizer, which substantially reduce the overall conductance of the neutralizer cell. The vanes have the disadvantage that they will almost certainly intercept at least a few percent of the total beam power, and perhaps more if the alignment, divergence, and beamlet steering of the of the beam are worse than the design values. Moreover, even with the vanes, the neutralizer accounts for about 75% to perhaps as much as 80% of the total gas load in the beam system.[6]

This large amount of gas has several deleterious consequences, both for ITER, and for any future reactor beams. The efflux which enters the accelerator increases the fraction of the negative ions which are stripped to neutrals before they have fallen through the entire acceleration potential. These form a low energy tail on the beam energy distribution, but their principal hazard lies in the fact that they have not experienced the full set of electrostatic lenses in the accelerator, and thus have on average much greater divergence than the fully accelerated ions. As a result, many of these prematurely neutralized particles hit the accelerator, resulting in an increased heat load. The electrons which are stripped from ions in the accelerator then acquire energy as they fall through the remaining potential drop. Many of these also hit the accelerator grids, adding to the heat load, and the ones that don't hit grids escape downstream where they form a heat load on beamline components. Negative ions which lose two electrons through collisions in the accelerator become positive ions, which are accelerated backwards to strike the ion source backplate. The gas efflux downstream of the neutralizer reionizes part of the neutral beam, which in the tokamak stray fields then deflects onto the beam duct wall. The downstream neutralizer efflux also increases the pressure within the residual ion dump region. Since the present ITER design uses an electrostatic dump with high voltage between the vanes, it is desirable to keep the pressure low in order to avoid plasma formation, which could partially shield the region from the applied electric field and result in the residual ions being misdirected.

Because the neutralizer accounts for the major portion of the beamline gas load in the case of ITER and any similar future negative ion neutral beam designs, it is the dominant factor which determines the operating cycle of the beamline. The primary vacuum pumping for neutral beam systems has, for the past thirty-odd years, been cryogenic: initially cryocondensation pumps, which use liquid helium to condense hydrogen, or more recently, cryosorption pumps, which usually use chilled helium gas, but which will use liquid helium for ITER. In either case, the cryopumps must be warmed to drive off the accumulated hydrogen whenever enough has accumulated that, if it were released in an up-to-air incident, the hydrogen concentration within the beam system would be high enough to explode (for safety reasons, the permissible accumulation of hydrogen is taken to be at most half of the explosive limit). Thus, the beam operating time between required cryopump regenerations is determined by the volume of the beam system, which is usually limited by space constraints, and by the volume of hydrogen released into the system per second of beam time.

The need to regenerate the cryopanels when they reach half the explosive limit (typically after a beam operating period of several thousand seconds) is an additional operating constraint for ITER, but it could prove a serious limitation for the possibility of using neutral beams on fusion reactors, where continuous operation of the beams is likely to be required. In present day systems, the beam cannot operate while the cryopanels are being regenerated (typically requiring tens of minutes to several hours, depending upon the system), because the pressure in the beamline rises much too high for operation, and the cryopumps are of course not pumping when they are warm.

A way to get around the regeneration quandary, and make continuous beam operation possible, would be to use continuously regenerable cryopanel arrays in which, at any given time, a portion of the cryopumps are being regenerated while the remainder continue to pump. This would require that the cryopanel array be comprised of a number of sectors, each of which could be independently isolated from the beamline vacuum system while the deuterium released during regeneration is pumped away. This might, for instance, be accomplished either by having each cryopump sector in its own volume adjoining the beamline with a large valve for isolation during regeneration, or by making the liquid-nitrogen-cooled chevrons, which protect the helium-cooled cryopanels from the ambient gas and radiation thermal environment, be movable, so they can seal the region between the cryopanels and the nitrogen chevrons while the regenerated gas is pumped away. In either scheme, it is desirable to keep the volume into which the regenerated gas is released small, so that it can be pumped away at high pressure in viscous flow during most of the regeneration cycle, which would minimize the required regeneration time.

The principal impediment to implementing continuous regeneration of cryopump arrays on a neutral beamline is that the beamline must have excess pumping capacity; it requires sufficient capacity that the pressure throughout the beamline stays at acceptable levels even while part of the pumping array is not pumping because it is being regenerated. Providing adequate excess pumping capacity in all portions of the beamline is difficult due to space constraints. These constraints are such that, for instance, the ITER beam design has barely enough pumping to meet its pressure requirements, so reducing the pumping capacity during regeneration would not be an option if it were desired in the future. Because reactor beamlines will likely face similar space constraints, they will encounter similar difficulties in implementing continuous regeneration. Even with unlimited space, it would not necessarily be easy to add effective pumping. Because the cryopumps, which have pumping speeds of megaliters per second, need very high conductance paths to the beam gas in order to use their speed effectively, they require clear sightlines to the beam path.

Techniques to Reduce Neutralizer Gas Efflux

It is apparent that significant advantages would accrue to not only the ITER beam system, but also to any future negative ion neutral beam systems, if the gas load from the neutralizer cell could be greatly reduced. In a recent publicaton,[7] we examined a number of possible techniques to reduce the gas load arising from the neutralizer in the ITER or similar designs. Although we considered cryogenic cooling of the gas cell to reduce gas mobility, heavy gases with lower mobility, foils, plasmas, photodetachment, and transverse supersonic jets of deuterium or heavier, more easily condensed, gases, they all appeared either infeasible, unlikely to result in enough gas reduction to matter, fragile in the ITER or fusion reactor environment, or inappropriate for the range of beam energies intended for ITER or similar future reactors. An approach which appears to offer major gas reduction, excellent beam conversion from negative ions to atoms, and to probably be technically feasible within the near term is a transverse supersonic metal vapor jet, of which lithium appears to be the best choice.[7]

SUPERSONIC LITHIUM VAPOR JET NEUTRALIZER

A supersonic metal vapor jet oriented transverse to the direction of beam propagation would be easily frozen by surfaces at non-cryogenic temperatures, and the relatively good thermal conductivity of metals would permit the accumulation of a thick layer without the vapor pressure of the top of the layer rising much. The question then arises as to what would be the best metal to use.

Why Lithium is Probably the Optimum Material for a Vapor Jet

Since only a small fraction of the elements in the periodic table, and an even smaller fraction of all volatile compounds, have been the subject of measurements to determine F_0^{max} for H⁻, it is not absolutely certain what would be the best possible material to use in a vapor jet. However, it is quite probable that lithium is the best choice for two reasons.[7]

The first reason is that F_0^{max} for H⁻ or D⁻ should be largest for materials with which the ratio of the cross section for stripping one electron from H⁻ to form H⁰ relative to the cross section for forming H⁺ from H⁻, either directly in a single collision or by two collisions going from H⁻ to H⁰ to H⁺, is largest. The most optimum atomic electronic configuration to accomplish this would likely be a weakly bound valence electron outside a small low-Z core. Lithium's valence electron is bound by only 5.39 eV, the first ionization energy, and it has an inner shell of just two electrons, with an ionic radius of 76×10^{-12} m. The heavier alkali metals have slightly lower first ionization energies, ranging from 5.14 eV for sodium to 3.89 eV for cesium, but their Z is much larger than lithium's 3, ranging from 11 for sodium to 55 for cesium, and their first-ionic radii are much larger, varying from 102 x 10^{-12} m for sodium to 174 x 10^{-12} m for cesium. Only hydrogen and helium have lower Z than lithium, but they both have much larger first ionization potentials: 13.6 eV for hydrogen, and 24.59 eV for helium. Outside the alkali metals, only a few elements have first ionization potentials that are lower than lithium's, and these are only very slightly lower, but these elements are all much higher Z than lithium, and also would have much larger first-ionic radii. The elements with slightly lower first ionization potentials than lithium's are barium (5.21 eV), radium (5.28 eV), and actinium (5.28 eV), the latter two of which would also be unusable for practical reasons. Thus, it appears likely that lithium offers the highest neutralization efficiency of any material, even though measurements have been done for only a few.

The second reason that lithium is probably the best neutralizer material has to do with scattering of the deuterium beam. Because it is one of the lightest elements with a Z of only 3, the contribution of scattering to the divergence of the resulting neutral beam will be less than would be the case for higher Z elements. As would be expected, D'yachkov[8,9] found negligible growth in the divergence of a 10^{-3} radian 100 keV H⁻ beam converted to H^0 in a lithium jet. In the course of doing a literature search for publications on supersonic lithium jets, we discovered that in 1966 D'yachkov proposed a supersonic lithium vapor jet as a neutralizer for H⁻, but, due to the fact that high current negative hydrogen beams still lay far in the future, his idea seems to have been forgotten. Fortunately, before it was apparently forgotten, D'yachkov made many of the fundamental measurements necessary for a neutralizer concept.[8,9]

Neutralization Efficiency

The efficiency of converting an H⁻ or D⁻ beam to neutral atoms has been measured for a number of metal vapors, as well as a variety of other vapors.[8,10] In each case the authors varied the line density of the neutralizer to find F_0^{max}. Anderson et al.[10] found that for a variety of organic vapors, as well as several fluorocarbons, oxygen, water and the halogens iodine and bromine, F_0^{max} varied across a fairly narrow range of 44 – 52%, but in all these cases F_0^{max} was significantly less than the 58% they found for a deuterium neutralizer. These values were for an H⁻ beam at 100 keV (equivalent to a 200 keV D⁻ beam), but should be reasonable for comparison purposes since for most neutralizer materials F_0^{max} changes little at higher H⁻ energies. Anderson et al. also measured F_0^{max} for all the alkali metals except Francium. At the highest energy they measured, 200 keV H⁻, equivalent to 400 keV D⁻, F_0^{max} was about 52.5% for potassium and rubidium, 57% for sodium, 58% for cesium, and 62.5% for lithium. Anderson et al estimated that their accuracy on the measurements of F_0^{max} was about +/- 1%.

Thus, it appears that lithium would be a very good candidate for a supersonic metal vapor jet neutralizer. Not only would it fulfill its principal mission of eliminating the gas load from the neutralizer; as a bonus, it would offer somewhat better neutralization efficiency than deuterium. For magnesium, zinc, and lithium, D'yachkov[8] found that F_0^{max} was nearly independent of beam energy for H⁺ over the range he measured, which was 100 – 400 keV (equivalent to 200 – 800 keV D⁻). With 400 keV H⁻ (equivalent to 800 keV D⁻), D'yachkov measured F_0^{max} to be 58% for zinc, 60% for magnesium, and 65% for lithium, which is slightly higher than the 62.5% found by Anderson et al. with 200 keV H⁻, but is similar within the likely experimental accuracy.

Lithium Jet Neutralizer Architecture

Based upon the optimum line density measurements of D'yachkov[8] and of Anderson et al.,[10] and taking into account that optimum neutralizer line density with lithium was observed to vary linearly with beam velocity, the lithium line density needed for 1 MeV D⁻ (which has the same velocity as 500 keV H⁻) should be 6 x 10¹⁵ atoms/cm². This is similar to the line densities which were produced in the alkali metal vapor jets of the experiments of the 1960's and 1970's. D'yachkov's[8] supersonic lithium vapor jet produced 5 x 10¹⁵ cm⁻², which was the highest line density he needed for 400 keV H⁻. The supersonic sodium vapor jet used by Semashko et al. for conversion of H⁺ to H⁻ [11] operated continuously at line densities up 5 x 10¹⁵ cm⁻², and could have been easily increased, according to the authors, if they had required higher density. Hooper's sodium jet for H⁺ to H⁻ conversion operated at about 1.6 to 2 x 10¹⁵ cm⁻² [12], while supersonic cesium vapor jets used for the same purpose by Hooper et al.[13] and by Bacal et al.[14] both operated at 2.5 x 10¹⁵ cm⁻².

With a well-directed supersonic jet, it should be practical to keep the overwhelming majority of the lithium confined within a modest-sized collection space. This is because the vapor pressure of molten lithium at 180 C is only 10⁻¹⁰ torr, and it is about 10⁻²⁰ torr at 20 C. Thus, the basic components of a lithium vapor jet would include a lithium boiler connected to a transfer line with a valve that feeds a long slot nozzle spraying the lithium vapor jet as a curtain across the ion beam. The jet would deposit in a collection basin, from which the recovered lithium would drain into a pump that would transport it back to the boiler. The lithium circuit would need to be constructed of materials which are compatible with high temperature alkali metals. These include tungsten, which could be expensive to fabricate, inconel, and iron, which is much easier to fabricate into a variety of shapes. Using iron for the lithium circuit should not cause appreciable perturbations to ITER's magnetic field, since the iron of the lithium circuit will be inside the magnetic shielding of the neutral beamline. This would also be the case for any future reactor on which a lithium-neutralized neutral beam might be used. This is because the neutralizer always has to be well-shielded from the stray field of the tokamak to prevent deflection of the beam accompanied by a large increase in the divergence of the resulting neutral beam.

The lithium circuit would operate at high temperature (D'yachkov's boiler operated at 1100 – 1200 K, and in alkali vapor circuits the temperature of the vapor transport after the boiler is generally a little hotter than the boiler to guard against metal deposition in any cooler spots). The boundary sections of the lithium neutralizer module would be at a much lower temperature (300 K or less) to trap most of any lithium which escaped the collection region. The lithium vapor line running from the boiler to the nozzle would need to contain a hot lithium-compatible valve so that the jet could be started just before a beam pulse and terminated just afterwards. The neutralizer line density could be controlled by either varying the temperature of the lithium in the boiler, or by changing the diameter of a hot lithium-compatible orifice. Only the lithium circuit would need to be made of iron. The overall housing for the neutralizer could be stainless steel or any other vacuum-compatible material.[7]

Lithium Recovery Approaches and Dimensions

There are two possible lithium recovery scenarios for a lithium jet neutralizer: batch recovery or continuous recovery. The batch recovery mode would spray the lithium jet into a water-cooled basin for later recovery. So long as the surface temperature did not rise much above room temperature, where the vapor pressure is 10^{-20} torr, this strategy would result in the smallest amount of lithium loss from the neutralizer. At intervals, the collection basin would be covered by a movable plate or diaphragm, and the lithium would be heated to a high enough temperature (over 180 C) to melt it, and it would flow down to a pumping reservoir, which would send it back to the boiler for reuse. This batch recovery scenario was employed in, for example, the sodium vapor jet of Hooper and Poulsen[12,13,15] which they used in 1979 to convert H^+ to H^-. The continuous recovery mode differs from the batch mode in that the lithium collector surface is kept at a temperature (about 180 C) just sufficient to keep the lithium liquid, but with the vapor pressure as low as possible. The liquid lithium flows continuously to a collection reservoir, from which it is pumped back to the lithium boiler. Some examples of systems which used the continuous recovery mode were the cesium vapor jet of Bacal et al.[14,16] and the sodium vapor jet of Semashko et. al.[11,17,18] Both these groups used the jets to convert H^+ to H^-, and both pumped the alkali metal from the collection reservoir to the boiler with an electromagnetic pump. For continuous beam operation, or for the long pulses (1000 seconds or more) that will be used with the ITER beams, the continuous recovery mode will probably be the most practical.[7]

The dimensions transverse to the ion beam's trajectory spanned by these early supersonic jets were somewhat smaller than would be required for an ITER neutralizer. Semashko et al's sodium vapor jet[17,18] converted ions from an H^+ source with accelerator dimensions of 22 x 48 cm, while the sodium vapor jet of Hooper et al.[13,15] spanned the beam from an accelerator with dimensions of 7 x 35 cm. For comparison, the ITER gas neutralizer cell is 42 cm wide at the entrance and 38 cm at the exit, with a height of 1.7 m.[6] The total length parallel to the beam trajectory of the units housing the sodium jets, their collectors, and plumbing was about a meter for both the experiments of Semashko et al. and Hooper et. al., but the thickness of the jets themselves was less. In Hooper's experiment, the distance along the beam propagation direction to the point where the sodium flux had fallen by a factor of 10^3 was 20 cm (for a total width of 40 cm), and the Semashko jet seems to have occupied a similar or smaller span along the beam propagation direction. D'yachkov's lithium jet[8], Fogel's mercury jet[19], Bacal's cesium jet[14] were much narrower along the beam axis (a few cm), but they spanned smaller beams, with less distance for jet expansion.

Lithium Jet Neutralizer Development Issues

Developing a lithium jet neutralizer for ITER, or for future fusion beams, will require finding satisfactory engineering solutions to a number of issues. Although a new measurement of the optimum neutralization efficiency would be reassuring, since it has only been measured twice long ago, it is probable, for the physics reasons discovered earlier, that the neutralization efficiency is higher than that for any other non-ionized medium, and any ionization of the jet is only likely to increase this efficiency. Aside from the choice of materials, lithium recovery scenario, and lithium jet design, the major issues which a development program will probably need to address are lithium compounds, lithium containment, and lithium jet ionization.

Lithium, like all alkali metals, is highly reactive, and forms oxides, nitrides, and hydrides (with H, not H_2), which all have higher melting points than does metallic lithium. Depending upon how rapidly these compounds build up in the recirculating lithium inventory, it might be necessary to have a purification mode in which the lithium is heated to a high enough temperature for the contaminants to be driven off and collected. Alternatively, if the buildup of contaminants is sufficiently slow, then they might be removed when the neutralizer is serviced at maintenance intervals.

When metal vapor jets were used to convert positive ions to negative ions, very little vapor migration along the beam path could be tolerated, since the jets faced high voltage accelerators on both sides, and ground planes on neither. Thus, to some extent, the containment requirement is simpler for a neutralizer jet. If necessary, components such as the residual ion dump vanes at elevated voltages could also have elevated temperatures to preclude their voltage holding being compromised by any lithium that did exit the neutralizer.[7]

However, substantial lithium loss from a neutralizer unit would still be a problem for a beam intended for ITER or a reactor, since such systems must be capable of operation for long intervals with infrequent maintenance. This implies that the lithium collector basin should be sufficiently wide to catch the outlying portions of the jet. How wide this would need to be depends upon the details of the jet design, but, given that the earlier sodium jets were about 40 cm thick, something like 80 cm might make a good starting point for the design of the collection basin. The best lithium containment could probably be achieved through a hybrid operating mode, in which the cesium collection basin was warm enough (180 C) so that the condensed lithium flowed continuously to the collection reservoir, but the surrounding surfaces of the neutralizer unit were sufficiently cold (20 C) that any cesium vapor which touched them would be trapped. At intervals the entrance and exit to the neutralizer could be blocked, and these cold surfaces heated so the stray lithium could be recovered.

Most of the design issues of a lithium jet neutralizer can be addressed on a modest-sized model (full height, but reduced width transverse to the beam trajectory) without exposing it to an ion beam. The one which can not is assessing the effect upon lithium migration due to impact ionization of the lithium jet. Stripping of electrons from the D^- beam will provide space charge neutralization of the Li^+ plasma, reducing its expansion. However, the Li^+ could decelerate the jet through resonant charge exchange, dropping the Mach number, and thus increasing the deposition width of the lithium jet.

When supersonic alkali jets were developed to convert H^+ to H^- in the 1970's there was some concern that jet deceleration as a result of impact ionization might appreciably spread the alkali jet deposition, degrading voltage holding of adjacent surfaces at high voltages. Hooper et al.[13] and Semashko et al.[18] both mentioned this as a possible concern in their sodium jet experiments. Hooper et al. did not find it to be a serious problem, but did not make detailed measurements of the phenomenon. Semashko et al. said they observed some increased sodium migration due to ionization of their sodium jet. According to a model of Krylov et al.,[20] the intensity of interaction between the ion beam and a supersonic jet is proportional to the product of the beam current per unit width across the neutralizer times the ionization cross section, divided by the jet velocity. Applying this model, we[7] estimated that the intensity of interaction between the beam and the jet should be a factor of 100 or more less severe for a lithium jet neutralizer for an ITER or reactor-relevant beam than was the case for the H^+ to H^- conversion experiments using sodium jets Nonetheless, it would be prudent to eventually test a full size version of an ITER lithium jet neutralizer on the ITER test bed, and to make the lithium collection basin somewhat wider than would be deemed necessary from tests without a beam.

CONCLUSION

Supplanting the conventional hydrogen isotope duct neutralizer by a lithium vapor jet neutralizer has the potential to make negative ion based neutral beam systems for heating and current drive more reactor-relevant, as well as easier to operate on ITER and future fusion experiments. In reducing the beamline gas load by a large factor (about 75% for ITER or any similar design)[6], the beam power lost through stripping on gas in the accelerator is reduced, thus decreasing the power loads this diverts to the ion source backplate, accelerator grids, and downstream beamline components due to divergent energetic neutrals, detached electrons, and backstreaming positive ions. Beam power lost through reionization of beam neutrals through collisions with gas molecules is also decreased. The greatly reduced gas load opens the possibility of using sector-regenerable cryopumps or other innovative pumping techniques to permit continuous beam operation on reactors. Because lithium affords greater neutralization efficiency that does hydrogen, the power and electrical efficiency of negative ion neutral beam systems will be enhanced. Estimates by Hemsworth[6] for an ITER beamline indicate that eliminating the neutralizer gas load with a lithium jet neutralizer would reduce the heat load on the accelerator grids by more than 1 MW, the heat load on the ion source backplate by more than 200 kW, and the reionized beam loss downstream of the neutralizer by about 0.4 MW. The beam power to ITER could increase by about 3 MW (of which about 1.4 MW would be from enhanced neutralization efficiency, and the rest from reduced stripping and reionization) if the source were run at full nominal current, or allow full specified power at lower voltage and current, enhancing performance robustness. Since a lithium jet neutralizer does not require the conductance-limiting vanes of the ITER deuterium neutralizer, the delivered power could be increased by an additional 1.7 MW if the electrostatic residual ion dump with its vanes were replaced by a magnetic deflection system, eliminating beam losses on the vanes.[6] Eliminating the vanes would also widen the operating window of transmittable beam optics and grid alignment. Since beamlines for reactors will likely face much the same space and design constraints as the ITER beamlines, it appears probable that, if a reliably robust lithium jet neutralizer can be developed, it could enhance the relevance of neutral beams to fusion power plants.

ACKNOWLEDGMENTS

It is a pleasure to acknowledge helpful conversations with Dr. R. S. Hemsworth, Dr. R. Majeski, Dr. J. Timberlake, and A. von Halle. This work was supported by USDOE contract no. AC02-CHO3073.

REFERENCES

1. R. Aymar, et al, *Nucl. Fus.* **41**, 1301-1311 (2001).
2. M. Kuriyama et al, *Fus. Sci. & Tech.* **42**, 410-419 (2002).
3. O. Kaneko et al, *Nucl. Fus.* **43**, 692-696 (2003).
4. R. S. Hemsworth, *Nucl. Fus.* **43**, 851-859 (2003).
5. T. Inoue, M. Hanada, et al, *Fus. Eng. & Des.* **66-68**, 597-607 (2003).
6. R. S. Hemsworth (private communication).
7. L. R. Grisham, *Phys. of Plasmas* **14**, 102509-1 – 102509-8 (2007).
8. B. A. D'yachkov, *Zh. Tekh. Fiz.* **38**, 1259-1267 (1968).
9. B. A. D'yachkov, *Sov. Phys. Tech. Phys.* **22**, 245-248 (1977).
10. C. J. Anderson et al, *Phys. Rev. A* **22**, 822-832 (1980).
11. N. N. Semashko, V. V. Kusnetsov, A. I. Krylov, *Proc. Symp. onProd. & Neut. of Neg. Ions and Beams*, Brookhaven Nat. Laboratory, 1977, edited by K. Prelec (IEEE, NY 1977) pp. 170-176.
12. E. B. Hooper et al, *J. Appl. Phys.* **52**, 7027-7034 (1981).
13. E. B. Hooper et al, *Proc. Symp. onProd. & Neut. of Neg. Ions and Beams,* Brookhaven Nat. Laboratory, 1977, edited by K. Prelec (IEEE, NY 1977) pp. 163-168.
14. M. Bacal et al, *Rev. Sci. Instrum.* **53**, 159-166 (1982).
15. P. Poulsen & E. B. Hooper, in *Proc. 8th Symp. On Engineering Prob. of Fus. Res.*, San Franciso, 1979, (IEEE, NY, 1979) pp. 676-682.
16. R. Geller et al, *Nucl. Inst. & Meth.* **175**, 261-268 (1980).
17. N. N. Semashko, V. V. Kuznetsom, A. I Krylov, in *Proc. 8th Symp. On Engineering Prob. of Fus. Res.*, San Franciso, 1979, (IEEE, NY, 1979) pp. 853-860.
18. N. N. Semashko, V. V. Kuznetsom, A. I Krylov, *Proc. Symp. onProd. & Neut. of Neg. Ions and Beams,* Brookhaven Nat. Laboratory, 1977, edited by K. Prelec (IEEE, NY 1977) pp. 334-340.
19. Ya. M. Fogel' et al, *Zh. Tech. Fiz.* **26**, 1208-1212 (1956).
20. A. I. Krylov & V. V. Kuznetsov, *Fiz. Plazmy* **11**, 1508-1514 (1985).

Kinetic study of the secondary plasma created in the ITER neutraliser

F. Dure*, A. Lifschitz*, J. Bretagne*, G. Maynard*, K. Katsonis*, A. Simonin[†] and T. Minea*[1]

*LPGP, CNRS-Université Paris Sud, Orsay, France
[†]DRFC, CEA Cadarache, 13108 Saintt-Paul lez Durance, France

Abstract. The properties of the secondary plasma created inside the ITER Neutral Beam Injector (NBI) neutraliser, through the interaction of the high energetic hydrogen beam with the molecular hydrogen gas, have been analysed. Starting from the results of our OBI-2 PIC Monte-Carlo numerical code, detailed kinetic of the hydrogen plasma has been studied using a Collisional-Radiative model. In this model, the electron distribution function is determined by solving a Boltzmann equation, whereas main plasma species are derived from balance equations.This paper presents preliminary results obtained in a 0D geometry, boundary conditions being introduced through effective rates for gain and loss of particles at the neutraliser walls. It has been found that the main ion specie is H_2^+, essentially coming from the ionisation of the target gas. The electron energy distribution function is not maxwellian and its mean energy is about 5 eV. The plasma-wall interactions yield a strong contribution, in particular regarding the density of molecular ion H_3^+. Assuming several independent slices of plasma along the negative ions beam axis, the axial profile of the secondary plasma has been analysed. It has been found that the density and mean energy profiles of the plasma electrons are directly related to the plasma potential profile, which in turn closely follows the gas density one.

Keywords: ITER Neutral Beam Injection, hydrogen plasma, Collisional-Radiative model, Boltzmann equation
PACS: 52.25.Dg,52.40.Mj,52.50.Gj

INTRODUCTION

The heating system for ITER includes, in addition to microwave antennas, high energetic Neutral Beam Injectors (NBI, figure 1). The negative ions (H^-) extracted from the source are accelerated by a multigrid system up to 500 keV/amu. Then, these are neutralised by collisions with a molecular target gas (H_2) in a structure called neutraliser. After passing between the plates of the neutraliser and before the injection inside the ITER plasma, the beam still has energetic charged particles (stripping electrons, H^-, H^+). In order to avoid damages in the confinement chamber, these particles are deflected before arriving to the plasma torus. Electrons are deflected from the beam by a magnetic field after the exit plan of the neutralized, whereas ions are deflected by the Residual Ion Dump (RID).

The ITER neutraliser is composed of four equivalent sections divided by parallel metal plates, separated by 10 cm. The collisions of negative ions (H^-) with the molecules

---

[1] Correspondig author email: tiberiu.minea@u-psud.fr

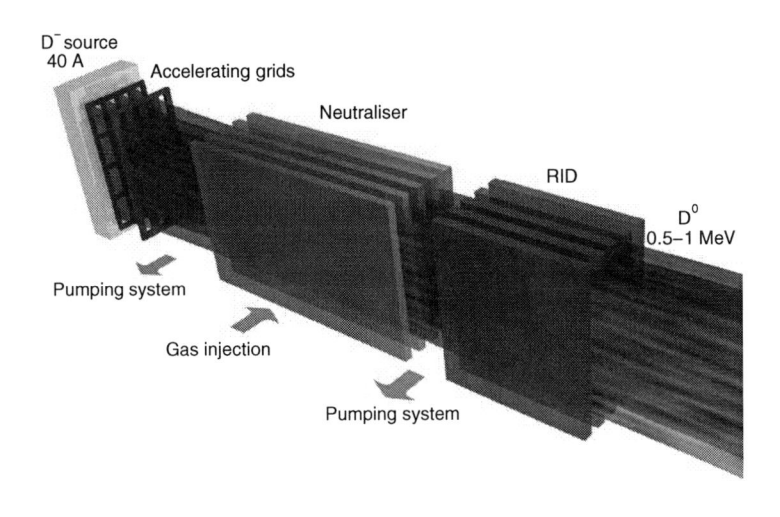

FIGURE 1. Sketch of the Neutral Beam Injector for ITER

(H_2) are mainly single stripping giving high energy neutrals, which in turn can be further ionised by subsequent collisions. Therefore the highest efficiency of the neutraliser is obtained for a given value of the linear gas density, which is close to 10^{20} m^{-2} with a corresponding maximum density at the middle of the neutraliser of 8×10^{19} m^{-3}. For this value of density, the composition of the beam is 60% of H^0, and 20% of H^- and of H^+ at the exit of the neutraliser.

The pioneer model of the ITER neutraliser [1] has shown that a secondary plasma builds up during the beam neutralisation. This result was confirmed by Particle-in-Cell Monte Carlo Collision (PIC-MCC) simulation using the code OBI-2 (Orsay Beam Injector - 2 dimensional) [2]. This secondary plasma is weakly ionised but still has a maximum of density one order of magnitude higher than the beam one. Therefore it can have non-negligible influence on the NBI efficiency. In particular, it has been shown, using the OBI-2 code, that the plasma fills all the volume inside the neutraliser, forming sheaths in front of the grounded metal plates. Moreover, it extends upstream and downstream the neutraliser. These consequences can be very important. Indeed, the plasma outflow downstream can affect the regular working of the RID and upstream the positive ions can damage the accelerator and even the negative ion source. A detailed analysis of the secondary plasma is therefore required.

In general, hydrogen (or deuterium) non-equilibrium plasmas have a quite complex kinetic connecting several species (H, H_2, H^+, H^-, H_2^+ and H_3^+) with rates that can heavily depend on the ro-vibrational excitation of the H_2 molecule. Moreover, in our case, collisions with the walls play an important role. It is not possible to introduce all these collisions processes into full 2D and 3D PIC calculations. Therefore we have developed a separate code, to treat the kinetic, but dealing with the transport in a very simplified way and neglecting any gradient effect. The main objectives of this kinetic code is to perform parametric analysis of the plasma in order to derive the rate of the most important processes that should be included within the PIC code through a Monte-Carlo Collision method. The kinetic model is presented in the next section, the results are discussed in the third section and the last section presents the concluding remarks.

MODELLING BACKGROUND

In our 0D model, we consider a plasma of uniform density, boundary conditions being introduced through effective gain and loss rates. For this plasma, collisions with energetic ions of the beam, with an energy of 500 keV/amu, introduce an external source term for plasma electrons and ions. The variables to be determined from our calculations are the electron energy distribution function (eedf) and the densities of each species. They are obtained by determining the steady state of the plasma.

Boltzmann equation

For a uniform plasma in a steady state, the electron Boltzmann equation reduces to the following form:

$$\frac{\partial f}{\partial t}(\varepsilon) = \left[\frac{\partial f}{\partial t}(\varepsilon)\right]_{collisions} = 0 \quad (1)$$

$$\left[\frac{\partial f}{\partial t}(\varepsilon)\right]_{collisions} = G_b(\varepsilon) + G_{pl}(\varepsilon) + C_{ee}(\varepsilon) + C_{ex}(\varepsilon) - L_{rec}(\varepsilon) - L_w - L_{out}(\varepsilon) \quad (2)$$

where ε is the electron energy and f is the eedf. The collision terms detail as following: G_b denotes the gain of electrons due to ionisation of the plasma molecules and atoms by collision with the beam particles H^0, H^+, H^- and fast stripping electrons e_s (see below); G_{pl} denotes the gain of electrons due to the ionisation of plasma molecules and atoms by collision with energetic electrons of the plasma itself; L_{rec} is the rate for electron-ion recombination inside the plasma; C_{ee} is related to the rate for elastic collision between electrons and C_{ex} comes from excitation of plasma atoms and molecules by electron collisions. Note that both C_{ee} and C_{ex} can be either positive or negative. Two additional rate terms have been introduced in the Boltzmann equation to take into account boundary conditions: L_w that represents the electron escaping rate up to the neutraliser walls (plates) and L_{out} for the electrons leaving the simulated plasma in a direction parallel to the plates.

We can separate the electrons into two groups: the plasma electrons coming from the ionisation of plasma particles with a maximum energy well below 100 eV, on one side, and the beam electrons generated by stripping of beam ions through collision with the buffer gas (H_2), on the other side. The latter group are fast electrons with an initial energy close to 272 eV. In fact, this value is obtained by the total transfer of momentum from a beam particle at 500 keV/amu (or 1 MeV for deuterium particles) to its stripped electron. The mass ratio between atomic hydrogen and electron is about 1836. These fast electrons are not taken into account in the eedf. As their collision probability is smaller than 1, we can assume as an roughly approximation that they cross the neutralizer without losing energy nor being angularly deflected. They are considered in our calculations through their contribution to the ionisation term G_b and to the dissociation of molecules.

The ionisation source term G_b is determined from the sum of the rates of the following reactions:

$$H_2 + B \rightarrow H_2^+ + e + B, \qquad (3)$$

with

$$B = H^0, H^+, H^- \text{ or } e_s \qquad (4)$$

Each of the four beam particles has an ionisation cross section that is close to 10^{-20} m^2. As the linear density of the neutraliser is 10^{20} m^{-2}, each beam particle creates roughly one plasma electron when flowing through the neutraliser. A beam particle loses only a small amount of its initial energy inside the neutraliser, therefore the cross sections for collisions with a beam particle can be kept constant inside the whole volume of the neutraliser. Each beam specie yields a contribution to the source term that is proportional to the differential ionisation cross section, which has been put in the simplified form:

$$S(\varepsilon_1, \varepsilon_2) = \frac{\sigma_{iz}(\varepsilon_1) B(\varepsilon_1)}{\left(\arctan(\frac{\varepsilon_1 - E_{iz}}{2B(\varepsilon_1)})\right) \left(\varepsilon_2^2 + B(\varepsilon_1)^2\right)} \qquad (5)$$

where ε_1 is the energy of the incident particle, ε_2 is the energy of the electron created, the constant $B(\varepsilon_1)$ depends on the nature of the target particle, E_{iz} is the ionisation energy of the target particle. From equation (5) we obtain an average energy of 50 eV for collisions with heavy beam particles and of 30 eV for collision with e_s.

For a hydrogen plasma at low temperature, radiative and dielectronic recombination have very low rates, therefore the (L_{rec}) term in equation (2) refers to the dissociative recombination that provides also the main channel for the atomic hydrogen formation in volume.

The electron losses to the wall (L_w) are possible only for energetic electrons, which are able to overcome the electrostatic trap imposed by the plasma potential V_p. There are no consideration of angle so that an electron with an energy a little higher than $e \times V_p$ will go to the wall. This case is not necessarily true in a multi-dimension geometry, only electron with a sufficient velocity into the wall direction will cross the barrier of potential. On the direction perpendicular to the neutraliser plates, fast electrons created by collision with the beam particles are quickly lost after making very few collisions (mostly zero or one), while low energy electrons remains trap for a much longer time so that they can be thermalized through elastic Coulomb collisions (C_{ee} term in equation 1) and eventually are lost by thermal diffusion with a rate depending on the volume on surface ratio. For the considered plasma parameters, the total rates for elastic (C_{ee}) electron-electron collisions and inelastic (C_{ex} and G_{pl}) electron-molecule collision have similar amplitude. However, their maximum influence concerns different energy domains. Elastic collisions are dominant at low energy, while inelastic collisions and loss to wall are related to energies mainly above 10 eV. Therefore one can expect that the eedf is close to be Maxwellian at low energies up to a given threshold, above which it becomes strongly reduced.

TABLE 1. Volume interaction processes

$H + e$	$\rightarrow$	$H^+ + 2e$	$H_2^+ + e$	$\rightarrow$	$H + H$
$H_2 + e$	$\rightarrow$	$H_2^+ + 2e$	$H_3^+ + e$	$\rightarrow$	$H^+ + 2H + e$
$H_2 + e$	$\rightarrow$	$H + H + e$	$H_3^+ + e$	$\rightarrow$	$H_2 + H$
$H_2^+ + e$	$\rightarrow$	$H^+ + H + e$	$H_2^+ + H_2$	$\rightarrow$	$H_3^+ + H$

Kinetic equations

Volume interactions

For the density regime of the neutraliser, three body collision processes can be safely neglected, therefore all particle interactions inside the plasma are binary collisions; the corresponding reactions included in our model are summarised in Table 1. In the case of H_2, 14 vibrational states have also been taken into account. Adding a correction term, these states modify the ionisation rate of H_2. Details on these reactions and on the corresponding rates can be found in [3]. In our density regime, we can also safely neglect electronic excited states of atoms and molecules, because radiative-decay rates are much higher than collisional ones, so that we can suppose that once an atom/molecule has been excited through an electron collision (C_{ex} term in equation 1) it returns instantaneously to its ground state.

Let us note that the code is flexible and it can be easily enriched with new reactions if they are shown relevant for experimental validation of the results.

Surface interactions

From the electrical point of view, the potential of the neutraliser copper plates is constant and equal to zero (grounded plates), but the plasma potential changes along the beam axis because its value is closely relate to the gas density. The values of the plasma potential along the beam axis have been extracted from our OBI-2 calculations; the obtained results are reported in figure 2.

Low energy electrons are mainly reflected by the sheath so that they remain confined inside the plasma, while electrons with higher energy than the plasma potential (V_p) can hit the copper plate at which they can either be absorbed, or reflected or also generate secondary electrons [5]. These secondary electrons will re-enter the plasma with an energy of at least $e \times V_p$. The particle-surface collision processes introduced in our calculations are reported in Table 2.

TABLE 2. Surface interaction processes [6]

e + wall	$\rightarrow$	e	H_2^+ + wall	$\rightarrow$	H_2
$2H$ + wall	$\rightarrow$	H_2	$2H_3^+$ + wall	$\rightarrow$	$3H_2$
H^+ + wall	$\rightarrow$	H	H_3^+ + wall	$\rightarrow$	$3H$

The wall interactions regenerate the target molecular gas, here H_2, as shown in Table 2. Two reactions give atomic hydrogen, and their efficiency is weak because it depends on the fraction of H^+ and H_3^+ ions in the plasma. Note that positive ions hit the wall

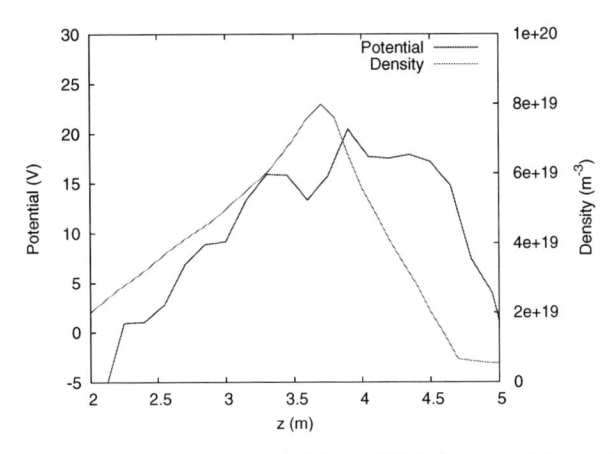

FIGURE 2. Axial profiles of the plasma potential (from OBI-2 simulation[2]) together with the target gas density from reference [4].

with an energy of $e \times V_p$. In our calculations, we have assumed that there is no direct interaction between the beam particles and the copper plates.

NUMERICAL APPROACH

The Collisional-Radiative code has been constructed by adapting, to the ITER neutralizer conditions, the numerical code of reference [3], initially written to simulate hydrogen discharges. The time dependent Boltzmann equation in addition to the balance equations related to reactions of Tables (1-2) are solved assuming the local plasma homogeneity (0D model). At the initial time of the calculation, a very weakly ionised H_2 plasma is assumed (10^{-5} lower electron density than in the final plasma state) with a Maxwellian eedf having an average energy of 1 eV. It has been checked that the obtained final state is independent of these initial values. The input parameters of one kinetic calculation are: the beam current density (60 A m^{-2}), the plasma potential deduced from OBI-2 calculation (Fig. 2), the gas density, the neutral gas temperature, the ion temperature, and the absorption coefficient on the wall. The calculations have been performed for different slices of plasma each one corresponding to a given z value on figure 2. The neutral gas temperature affects the ion density via ro-vibrational excited species. These species contribute also to ionization process, but due to the low gas pressure, their effect for the neutraliser is estimated at $\sim 1\%$ level. The ion temperature is assumed homogeneous with a value of 2 eV. The modification of this parameter changes the fraction of the different positive ion species through the modification of the diffusion rate to the walls by an amount of only few percents.

The balance equations of heavy species are coupled to the Boltzamnn equation. Each specie is created or converted into another specie mainly by electron impact and also it can diffuse up to the walls and in the axial direction, depending on the surface on volume ratios. Two types of diffusion are considered. First, the ion flux to the wall is given by the density of the corresponding species multiplying the corresponding Bohm velocity. Here is a factor 0.6 coming from the Boltzmann distribution of charged particles in the presheath. Second, the thermal diffusion acts for all neutral species and for ions in the axial direction (parallel to the walls). Boltzmann and balance equations are time integrated up to reach a steady state. We have checked that in the obtained steady state, the condition of quasi-neutrality of the plasma is satisfied.

RESULTS AND DISCUSSION

Some of the parameters used in this model were compiled, so it is interesting to check the sensitivity of these parameters on the final simulation results. From OBI-2 results, it comes out that the plasma potential and the target gas density are strongly corelated (figure 2). To determine the influence of each of the two parameters (gas density and plasma potential), a parametric study was done changing these two parameters independently. In Figure 3 we have reported the influence of the gas pressure, which is similar to the one of the gas density if the temperature is assumed constant, on the electron energy distribution function for a fixed value of $V_p = 18$ V. One can observe on figure 3 that the eedf does not follow a Maxwell law due to the scarce elastic collision processes inside the plasma. The elastic Coulomb collisions smooth the eedf only in the low energy part, whereas a sharp drop is recorded at energies above $e \times V_p$. In fact, with an energy above the plasma potential, the energetic electrons make very few collisions inside the plasma, so that they can escape quickly from the electrostatic trap created by the plasma potential and reach the walls. Note that only few of them (secondary electrons) come back into the plasma.

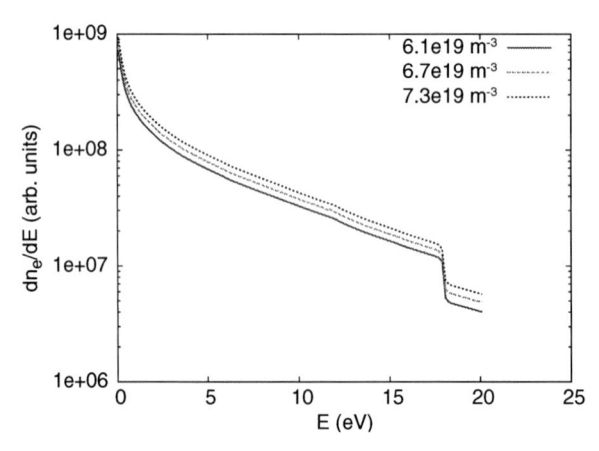

FIGURE 3. Energy Electron distribution function of the plasma inside the neutralizer for three values of the pressure and a plasma potential value of 18 V.

For the considered values of the gas density, the mean energy of the plasma electron is almost independent of the pressure, being its value $3/2E = 5.49$ eV. The main positive species are H_2^+ ions (1.19×10^{15} m^{-3}), representing an ionization degree of $\sim 10^{-4}$.

The second positive species are H_3^+ ions (3.1×10^{14} m^{-3}), representing $\sim 30\%$ of the H_2^+ ions. The atomic ion density is much smaller than the H_2^+ one (2.7×10^{12} m^{-3}).

In figure 4, we have reported the eedf keeping constant the pressure and changing the plasma potential. We can observe on this figure that only the energy domain around V_p is affected by the value of the plasma potential. On the low energy side, for energy below or close to the average energy, the density of low energy electrons is little modified by changing the cut-off at high energies, because the number of energetic electrons affected by this cut-off is much lower than the number of low energy electrons. On the high energy side, as soon as the electrons has an energy higher than V_p, then its probability to escape is one, which is independent of V_p. In conclusion to the results reported in figures 3-4 we can state that, staying in the range of realistic values for the plasma close to the centre of the neutraliser, the pressure and the plasma potential have only a relatively small influence on the plasma parameters.

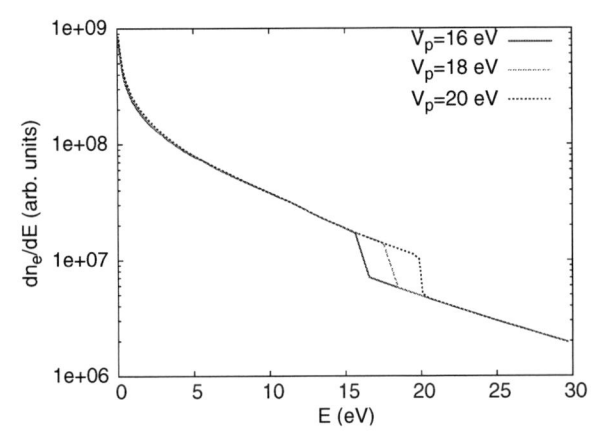

FIGURE 4. Energy Electron distribution function of the plasma inside the neutralizer for three values of the plasma potential and gas density of 6.7×10^{19} m^{-3}.

The variation of the plasma parameters along the z-axis has been obtained by determining, with the kinetic code, the steady state of the plasma for the values of density and plasma potential given in figure 2. The obtained results are reported on figure 5 for the electron density and on figure 6 for the densities of heavy species. The beam composition (H^+, H^0, H^-) changes along the neutraliser as given by the OBI-2 simulation.

From figure 5 we observe that, as expected, the electron density follows closely the evolution of the gas density, the additional factor induced by the variation of the plasma potential being relatively small. At the entrance and at the exit planes of the neutraliser, both the density and the plasma potential have small values; in this case, the eedf is directly related to the source term in equation 2. Closer to the center of the neutraliser, the eedf shape is similar to the one of figure 4. Variation with the plasma potential is little , excepted for energy around $E \times V_p$.

As can be seen in figure 6, the density of plasma heavy species is at least three order of magnitude smaller than the gas density (H_2). Therefore the gas profile is not affected by the plasma formation. This low ionisation degree implies that beam and plasma particles

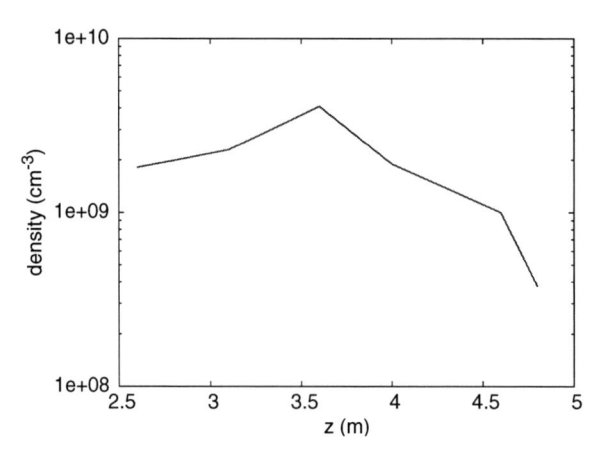

FIGURE 5. Electron density along the z-axis using the values reported on figure 2 for the gas density and for the plasma potential.

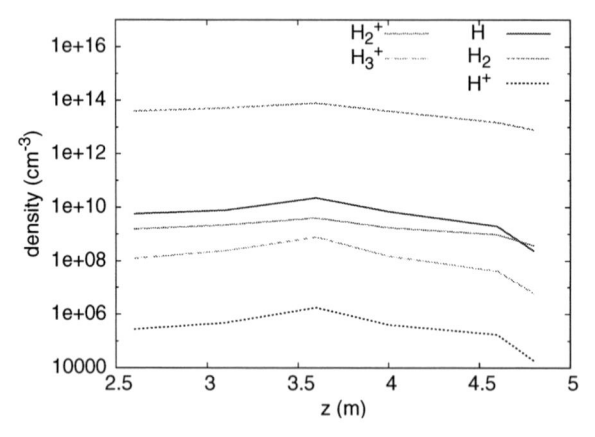

FIGURE 6. Heavy particle densities along the z-axis using the values reported on figure 2 for the gas density and for the plasma potential.

interact mainly with gas molecules. Comparing the H_2 profile with the others in figure 6, one gets a direct measure of the influence of the plasma potential. We can observe that the H_2^+ profile follows closely the one of H_2. It is due to the fact that the main source term for H_2^+ is direct ionisation of H_2 by the beam particles, therefore the density of H_2^+ is proportional to the one of H_2. The other three species, H, H^+ and H_3^+ have a profile, which differs significantly from the H_2 one, indicating a larger influence of the plasma potential than for H_2^+. In fact, we can see from Table 1, that the source terms for these particles are reactions with rates that are nearly proportional to the square of the plasma density, as it is the case for most reactions in volume. Due to the quasi-neutrality of the plasma, the density of electrons can be put equal to the sum of densities of H_2^+ and of

H_3^+ (H^+ density being negligible). As H_3^+ density is one third of the H_2^+ one, the electron density profile should be intermediate between the H_3^+ and H_2^+ ones, in accordance with figure 5.

In our 0D calculation, the diffusion up to the boundaries is treated in a simplified way. To analyse the influence of this diffusion, we have performed a second calculation in which the loss terms at the boundaries have been multiplied by a factor of 2. The results for H_2^+ and for H_3^+ densities are reported on figure 7. We can see on this figure that H_3^+ is much more affected by the boundary condition than H_2^+. It is in accordance with the relative dependence of the densities of the two molecular ions with the plasma potential. H_2^+ is little affected by the plasma potential and also by the boundary conditions while H_3^+, which is more dependent on the plasma properties, is sensitive both to the plasma potential and to the boundary conditions.

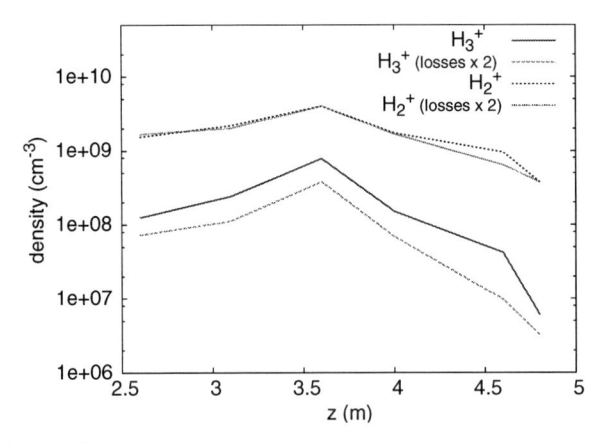

FIGURE 7. H_2^+ and H_3^+ densities along the z-axis using the values reported on figure 2 for the gas density and for the plasma potential. For each specie a second result is shown considering the wall loss twice higher.

CONCLUDING REMARKS

We performed a detailed analysis of the kinetics in the plasma created in the neutraliser via a 0D-Boltzmann code. Information about the plasma potential was obtained from PIC-Monte Carlo simulations made with the code OBI-2. From the kinetic model it is possible to quantify the relative role played by the various interaction processes and the main plasma species, allowing then to identify a minimal set of kinetic processes to be included in OBI-2. Our computational results show that a low-temperature plasma is created inside the neutralizer (ionization degree $\sim 10^{-4}$, electron mean energy ~ 5 eV). This plasma is dense enough to screen efficiently the beam spatial charge, allowing its correct focalization. The plasma positive ions are mainly molecular ($H_2^+ + H_3^+ > 99\%$) and these can escape from the neutralizer, reaching the accelerator after some μs. Simulations show that the induced current inside the accelerator depends strongly on the configuration of the electric field close to the accelerator. Our kinetic calculations show that the densities of some signicant particles such as H_3^+ strongly depends on the boundary conditions, therefore our 0D results have to be confirmed using more realistic 1D or 2D calculations.

ACKNOWLEDGMENTS

Part of this work has been support by the European Union under contract EFDA 06-1501 and by the "Fédération de Recherche Fusion Magnétique", France.

REFERENCES

1. E. Surrey, Nucl. Fusion **46** S360-S368 (2006)
2. T. Minea, A. Lifschitz, G. Maynard, K. Katsonis, J. Bretagne, and A. Simonin, J. Optoelectron. Adv. Mater. **10** 1899-1903 (2008)
3. D. Jacquin, J. Bretagne, and R. Ferdinand, Plasma Chem. Plasma Proc. **9** 165-188 (1989)
4. M. Dremel, EFDA - CCNB meeting, Culham, 22-24 May 2007
5. M.A.Furman and M.T.F. Pivi, Phys. Rev. ST Accel. Beams **5** 124404-17 (2002)
6. M. Bacal and D.A. Skinner, Comments At. Molec. Phys **23** 283-299 (1990)

Photo-neutralization of Negative Ion Beam for Future Fusion Reactor

W. Chaibi[*], C. Blondel[†], L. Cabaret[†], C. Delsart[†], C. Drag[†] and A. Simonin [*]

[*]*CEA, IRFM, F-13108 St Paul lez Durance FRANCE.*
[†]*Laboratoire Aimé Cotton, CNRS-Université Paris XI, Bât 505 Campus d'Orsay 91405 Orsay FRANCE.*

Abstract. An exploratory study of negative ion beam photo-neutralization for future fusion reactors is explained. A refolded Fabry-Perot cavity system is proposed, with which a 60% neutralisation efficiency could be reached with low electric power consumption. The system would make use of sophisticated optical-cavity locking systems, which have been developed recently for gravitational-wave optical detection. The ITER Neutral beam Injector is taken as an example.

Keywords: ITER, Neutral beam injection, Laser, Photodetachement, Fabry-Perot cavity
PACS: 28.52Cx, 32.80Gc, 41.75Cn, 42.60-v

INTRODUCTION

Neutral Beam Injection (NBI) involves injecting a high-energy beam of neutral atoms, typically deuterium, into the core of the fusion plasma. The neutral beam penetrates the tokamak plasma insensitive of the presence of the magnetic field, which makes it a very efficient tool for pla

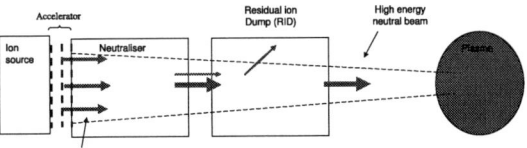

FIGURE 1. Neutral beam injection system

The components of a typical injector are (cf. figure 1): an ion source which generates a cold hydrogen (or deuterium) plasma ($T_e \simeq 2$ to 20 eV, $n_e \simeq 10^{18} - 10^{19}$ m^{-3}). The ions extracted from the source are accelerated to the required high energy (around 1 MeV for a fusion reactor) by an electrostatic accelerator. The ion beam passes through a gas cell (neutraliser) where inelastic collisions with the background gas (D$_2$) occur, giving rise to a partial neutralisation of the beam. The remaining ions (non neutralized D$^+$ and D$^-$) are magnetically (or electrostatically) deflected onto a water-cooled copper surfaces (Residual Ion dump), leaving the neutral beam to continue to the torus through the NB duct.

At energies higher than 100 keV, only negative ions (D$^-$) can be efficiently neutralised (about 60% of neutralization rate on a gas target at 1 MeV). But the extra electron captured in the negative ion has a very low binding energy (affinity 0.754 eV), and can be easily detached by a collision with a molecule (stripping). The drawback of using accelerated negative ions is thus a relatively high destruction rate along the beam line, especially within the accelerator channel. About 30% of the negative ions are lost by stripping reactions during acceleration (for instance in the ITER injector). These losses are mainly due to gas injection within the ion source and the neutralizer, and cannot be reduced in the usual NBI systems.

In this paper we study the feasibility of neutralizing the ion beam by laser photodetachment, which doesn't require any gas injection, and, consequently, should significantly reduce the negative ion losses due to stripping reactions. Moreover, photodetachment should not suffer from any fundamental limit to reach a nearly 100% efficiency, which would be an interesting advantage compared to gas neutralizing systems. An optical system is proposed and its electric efficiency is estimated.

ESTIMATION OF THE LASER POWER FOR A FULL PHOTO-NEUTRALIZATION OF THE ION BEAM

Photodetachment cross section

Photodetachment is the process of detaching the extra electron from a negative ion with light [1]. It is a threshold process, which means that it only occurs when the photon has a sufficient amount of energy, greater than a minimum value that is the electron affinity of the neutral species.

The electron affinity ^eA of H has been found experimentally to be 0.754 eV [2], with an isotope shift from H to D of ca. 0.4 meV [2]. A study of the total photodetachment cross section behaviour for either H^- and D^- is conducted in [3] and shows a similar dependence on the photon energy for the two species.

For maximum neutralization efficiency, the laser energy should be on the top of the photodetachment cross section curve. An absolute measurement of the photodetachment cross section was conducted by Smith et al. [4](see also [5]). It shows that the maximum is practically flat for a 1.2 eV to 2 eV photon energy, i.e. a 1100 nm to 650 nm laser wavelength, which broadens the choice of lasers to be used. Afterwards, the D^- photodetachment cross section will be taken equal to $\sigma = 4\,10^{-21}$ m^2.

Laser power

The following calculation will be based on the figure 2. For an estimation of the laser power, we assume that both the negative ion beam and the laser beam are rectangular and fully overlap.

In the deuterium case and for 1 MeV acceleration energy, the ions velocity is $\|\vec{v}\| \simeq 10^7$ ms$^{-1} \ll c$, where $c = 3\,10^8$ ms^{-1} refers to light speed. Therefore, negative ions will be assumed stationary while being photodetached. Also, the laser power P should be assumed unchanged after interaction with the negative ion beam, which will be justified

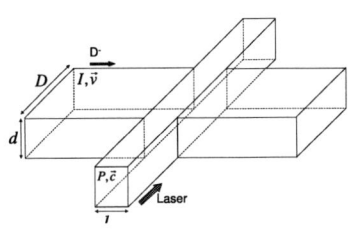

FIGURE 2. The ion beam is rectangular. The ion velocity is $\vec{v}$ and the integrated intensity is I. The laser beam is also rectangular and its power is P.

afterwards. Therefore, the laser power required should be independent of the ion beam intensity and the calculation can be done for a single ion. With n the quantum population of the negative ion state, the cross-section definition gives $\frac{dn}{dt} = -\frac{\sigma}{S}\frac{P}{h\nu} \times n$. This results in an exponential decrease of n, so after the time Δt needed to cross the laser beam, the negative ion population is reduced to $exp(-\frac{\sigma}{S}\frac{P}{h\nu}\Delta t)$. Correspondingly, photodetachment will enter its saturation regime when the laser power becomes greater than :

$$P_s = \frac{hc\|\vec{v}\|d}{\sigma\lambda} \quad (1)$$

with lambda the laser wavelength. This power would produce a $1 - 1/e \simeq 63\%$ photodetachment rate, and twice this power would be enough to exceed a 86% photodetachement rate, which shows that P_s provides the proper order of magnitude for a nearly complete photo-neutralization of the ion beam. P_s will thus be called the saturation laser power.

From the equation (1), we can point out the following ideas:

- the important velocity of the ion beam is responsible of the high laser power needed for a full neutralization.
- In order to decrease the laser power, we shall choose the largest possible laser wavelength and the smallest ion beam width.

In the following, we apply our calculation on the ITER geometry which is displayed in figure 3. The I =40 A 1 MeV D$^-$ beam at the exit of the accelerator is subdivided in 4 independent columns (segments) ; each segment section is 7 cm width and 1.5 m height[1]. The laser power required to reach the saturation regime of photoneutralization on one beam segment is $P = 30$ MW for $\lambda = 1064$ nm which is the NdYAG laser wavelength.

The most powerful D.C laser in this wavelength range available on the laser market is the "YLR-HP Series: 1-50kW Ytterbium Fibre Lasers" model proposed by IPG Photonics [2]. It is a doped ytterbium fibre laser with several fibre amplification stages. It provides a 50 kW laser at the wavelength $\lambda = 1060$ nm, which is far away from

---

[1] ITER geometry consists of 4 beam columns to accomodate three buffle plates in the neutralizer in order to lower the gas flow rate. In the case of a laser neutralizer, the buffle plate would not be necessary and so the column geometry. We choose to keep it for calculation anyway
[2] see the website http://www.ipgphotonics.com.

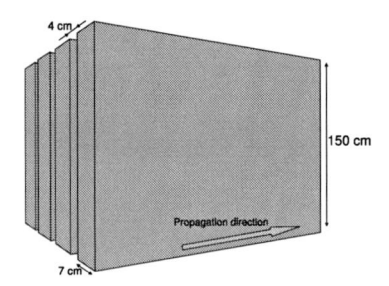

FIGURE 3. ITER negative ion beam

the needed laser power. Nevertheless, the fraction ξ of photons which interact with the negative ions is $\xi = \frac{I\sigma}{e\|\vec{v}\|d} = 1,5\,10^{-6}$ for $P \leq P_s$ and $\xi = \frac{I h c}{\lambda P e}$ for $P > P_s$. Thus, only 50 W of the laser power would be consumed for a full beam neutralization, which demonstrates the hypothesis that the laser power remains unchanged after crossing the ion beam. It would be then interesting to reflect the laser back and forth so it can cross the ion beam several times. In the next section we study the feasibility of such a configuration.

Multireflection configuration

This idea was first explored by J. H. Fink [6] in the eighties. He imagined a laser layer retroreflected in the longitudinal direction by a system of two mirrors, so it can cross the negative ion beam several thousands times. The two mirrors have to be 1 meter big and not spherical in order to compensate the natural divergence of the laser beam. These characteristics, among others, still not feasible twenty years later.

Systems based on open cavities are used in molecular spectroscopy when the photon-molecule interaction cross section is limited which is similar to our case. Some of these systems are outlined in [7]. It is a two mirrors optical system capable of reflecting the laser beam in all controlling its divergence.

For our purposes, it may be possible to use this kind of system, since it uses spherical mirrors. In fact, it is possible to manufacture 40 cm mirror diameter [8] with high quality dielectric coating, which are used in gravitational antenna[3]. The Gaussian shaped laser shall be reflected in the transversal direction in such a way that each portion sees a smaller width of the ion beam. In the following, we demonstrate that in the state of the art of mirrors coating, these systems can not be used in our case.

Mirrors coating have a light intensity damage threshold I_{th} close to 10 MW/cm^2 for $\lambda = 1,06$ μm coatings. For a Gaussian beam, $P_{th} \simeq I_{th} \times d_{th}^2$. The equation (1) gives:

---

[3] see the websites http://www.ligo.caltech.edu and http://www.virgo.infn.it

$$d_{th} = \frac{hc\,\|\vec{v}\|}{\sigma \lambda I_{th}} \tag{2}$$

In our case, the minimal width of the laser beam is $d_{th} = 4.5$ mm. Therefore, for a full neutralization efficiency, it comes $P_{th} > 2$ MW, much larger than the 50 kW available, which make an open cavity not feasible.

Moreover, we may be faced with the issue of using optical components close to the ion beam. Indeed, with the high electric intensity of the ion beam (40 A) and its high energy (1 MeV), sputtering the vacuum vessel wall may occur. The metallic gas produced may cover the optics.

In the following, we propose an optical system which should respond to all the issues we presented above.

THE REFOLDED FABRY-PEROT CAVITY

Let us first remind the issues the system should be able to respond to.

- optics should not be close to the ion beam
- optical components should be flat or spherical which guarantee their availability.
- the minimal width of the laser beam is 4.5 mm and it must carry 2 MW at least.

The system geometry

To protect the optical components from the background gas in the ion beam chamber, it would be advantageous to set the whole optical system in independent chambers where the pressure can be much lower. Laser beams could pass through several pipes (particle traps) separating the ion beam chamber from the two optical chambers (see figure 4) where we should obtain a differential vacuum [9]. We think that with $l \simeq 1$ m and $d \simeq$ few centimetres, the optical chamber vacuum should be low enough for the system to work properly.

The optical system

The optical system shall be able to reflect the laser beam several times in order to decrease the required power. Since the Gaussian beam width is largest than 4.5 mm, it should carry 2 MW light power. We propose to use a Fabry-Perot cavity where we can concentrate such a large laser power. The general system schema is illustrated on the figure (5).

To cover a whole ion beam segment width, the laser beam should have about d_0/d refoldings inside the cavity. Since each beam is at least 4.5 mm beam width, the Rayleigh length [10] is $z_R = 15$ m. Each beam width is then practically constant for 4 m propagation, but the retroreflection mirrors shall be spherical to compensate the beam divergence.

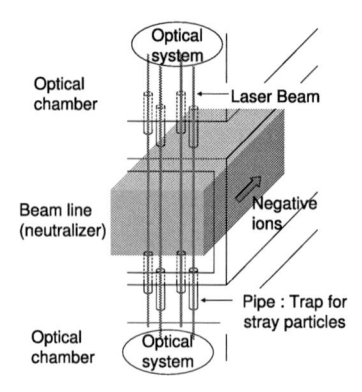

FIGURE 4. Vacuum system: a differential vacuum is created in the two optical cells

The only problem is to find a compromise between the beam width d and the intracavity power according to available technology, so we can decrease the number of reflexions and then simplify the whole system.

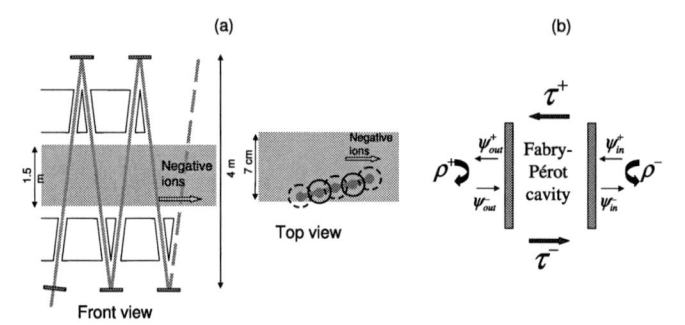

FIGURE 5. (a)The optical system: Neutralization beams belong to the same Fabry-Perot cavity. Retroreflexion mirrors are spherical in order to compensate the laser beam divergence.(b)The general Fabry-Perot cavity

The Fabry-Perot cavity

We assume our laser in a single space Gaussian mode and a single frequency mode.
Let consider a Fabry-Perot cavity made up of two mirrors separated by a propagation space presented in the figure 5.
It comes:

$$\psi_{out}^+ = \tau^+ \psi_{in}^+ + \rho^+ \psi_{out}^-$$
$$\psi_{in}^- = \rho^- \psi_{in}^+ + \tau^- \psi_{out}^- \qquad (3)$$

Fabry-Perot cavity includes two highly reflective mirrors bordering a free propagation space. Let $r_{1,2}$ the amplitude reflectivity of the two mirrors, $t_{1,2}$ their amplitude transmission, $q_{1,2} = 1 - r_{1,2}^2 - t_{1,2}^2$ the mirror losses, α the phase propagation and $1 - v^2$ the propagation loss in the intra cavity space. We then obtain [11]:

$$\tau = \tau^- = \tau^+ = -\frac{t_1 t_2 v e^{i\alpha}}{r_1 r_2 v^2 - e^{2i\alpha}}$$

$$\rho^- = \frac{r_2 e^{2i\alpha} - (1-q_2) r_1 v^2}{r_1 r_2 v^2 - e^{2i\alpha}}$$

$$\rho^+ = \frac{r_1 e^{2i\alpha} + (1-q_1) r_2 v^2}{r_1 r_2 v^2 - e^{2i\alpha}} \qquad (4)$$

Our cavity is working in transmission $r_1 \simeq r_2 \simeq r$ with a single entry $\psi_{in}^+ \neq 0$, $\psi_{out}^- = 0$. We first assume that the cavity is ideal, i. e. $q_{1,2} = 0$, $v = 1$, $r_1 = r_2 = r$ and and $t_1 = t_2 = t$. The transmission has a resonance behaviour for $\alpha \equiv 0[\pi]$. Its half width is $\delta\alpha = \eta \simeq \frac{t^2}{2}$. The finesse cavity is the given by $F = \frac{\eta}{2\pi} \simeq \frac{\pi}{t^2}$. For real cavity, i.e. mirrors slightly different and with losses, the half width increases and then the finesse decreases. For a real cavity let $\delta R = (r_1 - r_2)/2 << 1$, $r = \sqrt{R} = (r_1 + r_2)/2$, $w = 1 - v << 1$. At the first order ($q_{1,2} << 1$), we have $q_1 = q_2 = q$. The mirrors losses and the propagation losses have the same effect. We shall then define the total cavity loss $p = q + w$. The intra-cavity wave is stationary, i.e. it is a sequence of power maxima and minima. If the ion beam is perfectly perpendicular to the laser beam, too few negative ions will fell the maximum of the intracavity power. But thanks the 5 mrad negative ion beam divergence, the negative ion beam passes through more than 20 maxima. At the first order, the maximum intracavity power is :

$$P_{int} = \frac{4}{t^2}\left(1 - \frac{2p}{t^2}\right) P_0 \qquad (5)$$

where $P_0 \propto |\psi_{in}^+|^2$ is the laser power. The difference between the two mirrors has no effect on intra cavity power. For given total loss factor p, we should choose the energetic transmission of the mirrors as $t^2 = 4p$ to have the maximum intra cavity power $P_{int}^{max} = P_0/2p$. Usually, the loss factor can only be approximately estimated.

Losses sources and laser specifications

In order to determine the beam width, we have to evaluate the losses which gives the maximum power we can reach. The most important losses are :

- intra cavity losses due to the photodetachment process $p_{int} \simeq \frac{d_0}{d} \times \xi$.
- light scattering by local defect of the mirrors coating $p_{loc} \simeq \alpha_{loc}\left(\frac{d_0}{d} + 1\right) d^2$ with $\alpha_{loc} \simeq 6\,10^3$ ppm/m^2 [12].

- light scattering by polishment defect of the mirrors substrate $p_{pol} \simeq \alpha_{pol} \left(\frac{d_0}{d}+1\right) d^2$ with $\alpha_{pol} \simeq 5.2\,10^4$ ppm/m² [12].
- mirrors absorption and retroreflection mirrors transmission $p_{abs} \simeq \frac{d_0}{d}\beta$ ppm with $\beta \simeq 0.5$ ppm.
- mirrors borders losses $p_{diaph} \simeq \left(\frac{d_0}{d}+1\right)\gamma$ with $\gamma \simeq 1.5\,10^{-2}$ ppm for mirror radius $r = 1.5d$.

The total loss p shall be the sum of the terms we just explained. Using the equation (1) where we identify P_s to $P_{int}^{max} = P_0/2p$, we infer a relationship between the laser power available P_0 and the beam width d which is displayed on the figure 6. For a given laser power P_0, d refers to the maximum beam width compatible with the saturation of the photoneutralization process.

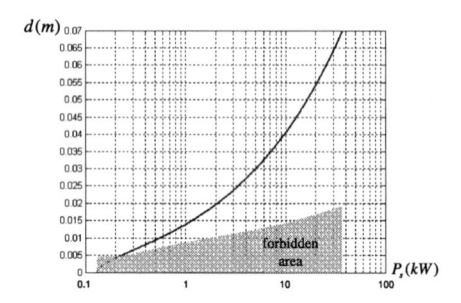

FIGURE 6. Variation of the laser beam width versus the laser power. The forbidden area corresponds to the mirror coating damage threshold.

As we shall see, the laser we use has to be highly stabilized. The most powerful stabilized lasers are used on gravitational antenna and provide 10 to 20 W. Several 200 W stabilized laser are being developed for the next generation of these antennas [13]. $P_0 = 200$ W corresponds to $d = 4.1$ mm < 4.5 mm (see figure6). This means that with this laser, saturation of the photodetachment process can not be reached. But using a laser beam width $d = 4.5$ mm, we reach we reach 90% of the saturation power with 16 reflexions.

The spectral width of the laser δv_L has to be smaller than the spectral width of the Fabry-Perot cavity δv given by $\delta v = \frac{c\eta}{\pi L_c}$ where L_c is the optical cavity length. For the maximum intra cavity power $\eta = p/2$. The cavity length is $L_c \simeq 16 \times 4 = 64$ m. For instance, we can assume $\delta v_L \simeq \delta v/10$ and we finally have $\delta v_L \simeq 7$ Hz and for $\lambda = 1.064$ μm $\delta v_L/v_L \simeq 2\,10^{-14}$, which means that the laser has to be highly stabilized.

Control system of the cavity

In order to keep the resonance condition, the cavity optical length shall be controlled. The mirror undergo three kinds of perturbation with various frequency bandwidths:

- thermal expansion: $v < 0.1$ Hz.

- seismic perturbation: $v \simeq 1$ Hz.
- vibration perturbation: 10 Hz$< v <$ few tens of kHz (pumping system...)

Moreover, intracavity refraction index perturbations can occur due to fluctuations in the density of the negative ion beam. In fact, negative ions in a ICP were the discharge is created by a 1 MHz microwave [14]. Thus, the optical length of cavity can undergo 1 MHz fluctuations.

Therefore, no noise exists in the GHz range, to which the photodetachement process is sensitive since the negative ion passes through the laser beam in 0.5 ns. Two control systems could be added to the optics. On the one hand, the tilt of retroreflection mirrors shall be controlled by PZT ceramics in order to insure the geometrical stability of the whole system. Nevertheless, these PZT ceramics limit the control bandwidth to a few kHz. Therefore, it can not eliminate the whole noise frequency range. The optical system shall thus be isolated mechanically so the high frequency of the vibration noise could be vanishing. Several isolation system exist with cut off frequencies in the (0.1-10) Hz range [15, 16]. On the other hand, the central frequency of the laser shall be controlled in the 1 MHz bandwidth to guarantee the resonance condition.

ELECTRICAL POWER

To evaluate the whole reactor efficiency, the electric efficiency of the laser system should be discussed. In order to reach high light power, a laser system is made of a master laser which provide few watt, followed by several amplification stages. These subsytems use diode-laser pumps with a 40% efficiency with respect to the electrical power. The master laser could be a solid state laser (Nd-YAG crystal for instance) with a 10% efficiency with respect to the pump power, or a doped fiber laser with 70% efficiency. The last one is also used in the amplification stages. Therefore, a laser system delivering more than few tens of watts has an efficiency higher than 0.25% with respect to the electric power. We demonstrated that the laser power required may not exceed few kilowatts, which obviously corresponds to an insignificant electric power compared to the power required by the NBI system (few tens of Megawatts).

CONCLUSION

In this paper, we propose an exploratory study of a photodetachement neutraliser for future NBI system, based on the present available technology and the know-how of high power optics. It consists of a refolded Fabry-Perot cavity able to concentrate high laser power. The main issue comes from mirrors losses which limit the neutralizing efficiency to 60%. Moreover, this system is highly complex especially with all the control systems it has to include. It has to be noted that only orders of magnitude has been calculated since we have used simplified theoretical models.

Nevertheless, the system would be more interesting if the laser power is increased with a factor of ten. For a 2 kW laser, the laser beam become 2 cm width and only 4 refoldings are required (see figure 6). This kind of laser doesn't exist yet, but it may be possible to be developed. It is also possible to replace retroreflexion mirrors with cube corners which are insensitive to the tilt noise and doesn't need any control systems. But cube corners can not compensate the beam divergence and thus would be only interesting for larger laser beams. In all cases, experimental studies should be conducted to confirm the ideas we presented above. In fact, problems as thermal effects may appear on the mirror surfaces.

On top of increasing the laser power, it is theoretically possible to increase the photodetachement cross section using magnetic resonances corresponding to Landau levels in the free photoelectron continuum[17]. These resonances have never been demonstrated experimentally, but it would be an interesting way to bring fundamental research to solve NBI problems.

ACKNOWLEDGMENTS

We express our gratitude to A. Brillet for stimulating conversations. This work, supported by the European Communities under the contract of Association between EURATOM and CEA, was carried out within the framework of the European Fusion Development Agreement. The views and opinions expressed herein do not necessarily reflect those of the European Commission.

REFERENCES

1. C. Blondel, *Physica Scripta* **T58**, 31 (1995).
2. K. R. Lykke, K. K. Murray, and W. C. Lineberger, *Phys. Rev. A* **43**, 6104 (1991).
3. L. M. Branscomb, and S. J. Smith, *Phys. Rev.* **98**, 1028 (1955).
4. S. J. Smith, and D. S. Burch, *Phys. Rev. Lett.* **2**, 165 (1959).
5. T. Ohmura, and H. Ohmura, *Phys. Rev.* **118**, 154 (1960).
6. J. H. Fink, *Photodetachment now*, 12th Symposium on fusion engineering, Monterey, CA, 12-16 October, 1987.
7. C. Robert, *Appl. Opt.* **46**, 5408 (2007).
8. F. Beauville, et al., *Class. Quant. Grav.* **21**, S935 (2004).
9. A. Öttl, S. Ritter, M. Köhl, and T. Esslinger, *Rev. Sci. Instrum.* **77**, 063118 (2006).
10. A. E. Siegman, *Lasers*, University Science Books, USA, 1986.
11. N. Hodgson, and H. Weber, *Laser Resonators and beam propagation : Fundamentals, Advanced Concepts and Applications*, Springer Series in OPTICAL SCIENCES, USA, 2005.
12. Z. Yan, L. Ju, C. Zhao, S. Gras, D. G. Blair, M. Tokunari, K. Kuroda, J.-M. Mackowski, and A. Remilieux, *Appl. Opt.* **45**, 2631 (2006).
13. B. Willke, K. Danzmann, C. Fallnich, M. Frede, M. Heurs, P. King, D. Kracht, P. Kwee, R. Savage, F. Seifert, and R. Wilhem, *J. Phys. : Conf. Series* **32**, 270 (2006).
14. M. A. Lieberman, and A. J. Lichtenberg, *"Principles of plasma discharges an materials processing"'*, JOHN WILEY & SONS, INC, New York, 1994.
15. M. Barton, T. Uchiyama, K. Kuroda, and N. Kanda, *Rev. Sci. Instrum.* **70**, 2150 (1999).
16. G. Ballardin, et al., *Rev. Sci. Instrum.* **72**, 9 (2001).
17. W. A. M. Blumberg, W. M. Itano, and D. J. Larson, *Phys. Rev.* **19**, 139 (1979).

Model of a SNS Electrostatic LEBT with a Near-Ground Beam Chopper

B. X. Han and M. P. Stockli

Spallation Neutron Source, Oak Ridge National Laboratory, Oak Ridge, TN 37831, USA

Abstract. The low energy beam transport (LEBT) of the Spallation Neutron Source (SNS) accelerator consists of two electrostatic lenses, of which the second is split into four electrically-isolated segments. Adding fast pulsed voltages to the lens high voltage creates the transverse fields required for beam chopping. Electric sparks, however, create transients that enter the fast high-voltage switches, which are occasionally damaged and cause machine downtime. This work models a new configuration of the electrostatic LEBT, which chops the beam with four shielded, near-ground electrodes between the two lenses. The model shows that the new configuration can match the RFQ injection requirements and sufficiently deflect the beam in the phase-space using the same chopping voltages as in the baseline LEBT.

Keywords: H⁻ Ion Source, Electrostatic LEBT, Ion Optics, Beam Chopper
PACS: 41.75.Cn, 41.85.Ja

INTRODUCTION

The low energy beam transport (LEBT) of the Spallation Neutron Source (SNS) accelerator focuses, steers and chops the 65-kV H⁻ beam before it is injected into the RFQ. The present baseline LEBT [1] consists of two electrostatic lenses. Its second lens (lens-2) is split into four electrically-isolated segments. Four 3-kV supplies are mounted on the 60-kV platform that has the same potential as the second lens center electrode, producing the necessary voltages for steering the beam. Fast high-voltage (HV) switches located on ground and capacitively coupled to the four lens-2 segments add ±2.5 kV, which chops the beam by steering it outside of the RFQ acceptance ellipse. Lens-2 sparks generate large transients that pass through the coupling capacitors into the fast HV switches. Some of the transients damage the fast HV switches, which causes downtime.

This problem could be mitigated by placing a shielded, near-ground chopper between the 2 electrostatic lenses, as previously proposed [2-3]. The previously proposed model achieved with ±2 kV chopping voltage a beam steering angle of 5.5 mrad and a transverse displacement of 1.5 mm [3].

With the SNS baseline LEBT chopper, ±2 kV is almost sufficient for the desired extinction rate, but the ±2 kV cause a beam steering angle of 105 mrad and a transverse displacement of 3.6 mm at the upstream face of the RFQ entrance aperture.

This paper describes the model of a newly configured electrostatic LEBT, which 1) meets the SNS RFQ beam injection requirements, 2) steers and chops the beam with

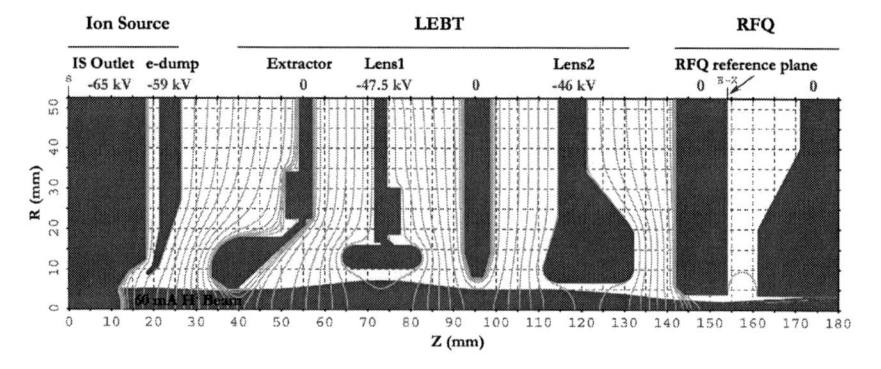

FIGURE 1. PBGUNS simulation of the ion source and the present LEBT for beam injection.

near-ground potentials, and 3) sufficiently deflects the beam in the phase-space with similar chopping voltages as in the present configuration, and therefore can be implemented without large changes to the existing SNS front-end infrastructure.

THE BEAM DYNAMICS OF THE PRESENT SNS LEBT

The present SNS LEBT transports and matches a 50-60 mA H⁻ beam into the RFQ. In addition, every ~1 μs it chops ~300 ns of beam when the fast-pulsed chopping voltages with a nominal amplitude of ±2.5 kV are applied to the opposing pairs of the lens-2 segments. The beam dynamics and chopping performance of the baseline LEBT satisfy the SNS baseline operational requirements, but failures of the LEBT chopper HV switches continue to accumulate downtime [4].

Figure 1 shows a PBGUNS [5] simulation of the SNS ion source and the baseline LEBT for beam extraction, transport and injection. The PBGUNS code supports negative ion extraction from plasma, and incorporates iteratively calculated space-charge force in the beam transport simulation. It has been very successful in reproducing experimental results in our previous work [6]. Figure 2 is the x-x'

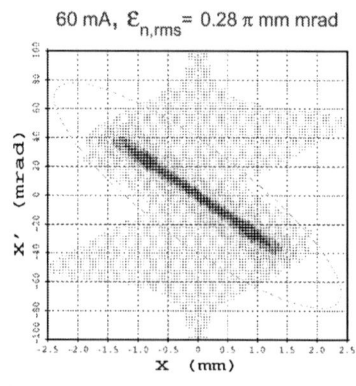

FIGURE 2. x-x' emittance plot at the RFQ reference plane generated from the PBGUNS simulation shown in figure 1

projected emittance plot from the PBGUNS simulation at the RFQ reference plane, demonstrating the matched beam size and convergence at the downstream end of the RFQ entrance aperture.

In the SNS H⁻ ion source, co-extracted electrons are swept away from the H⁻ beam by a dipole field in the extraction region. A slight tilting of the ion source against the LEBT axis corrects the dipole field effect imposed on the H⁻ beam [7]. The co-extraction and removal of electrons has been modeled previously [6] and is not included in this model because the study focuses on the downstream beam transport and chopping. Accordingly the H⁻ beam was initially assumed to be aligned with the LEBT axis.

Figure 3 shows the beam chopping in the present LEBT simulated with 3-D code SIMION [8]. In the SIMION model, the ion source and lens voltages were set to the same values as used in the PBGUNS simulation, and ±2.5 kV chopping voltages were applied to the opposing pairs of the lens-2 segments as illustrated in Fig. 4a, e.g. +2.5 kV on A and D, −2.5 kV on B and C. The Coulomb repulsion (a method used in SIMION to approximate the space-charge effect) feature was enabled in SIMION to resemble the PBGUNS beam profile. Figure 4b shows emittance plots of the injected and the chopped beam at the upstream face of the RFQ entrance aperture. As seen in Fig. 3 and Fig. 4b, the chopper voltages deflect the beam by ~120 mrad resulting in a ~4 mm dislocation at the entrance to the RFQ. All trajectories are intercepted by the cooled RFQ entrance flange, yielding 100% controlled chopping.

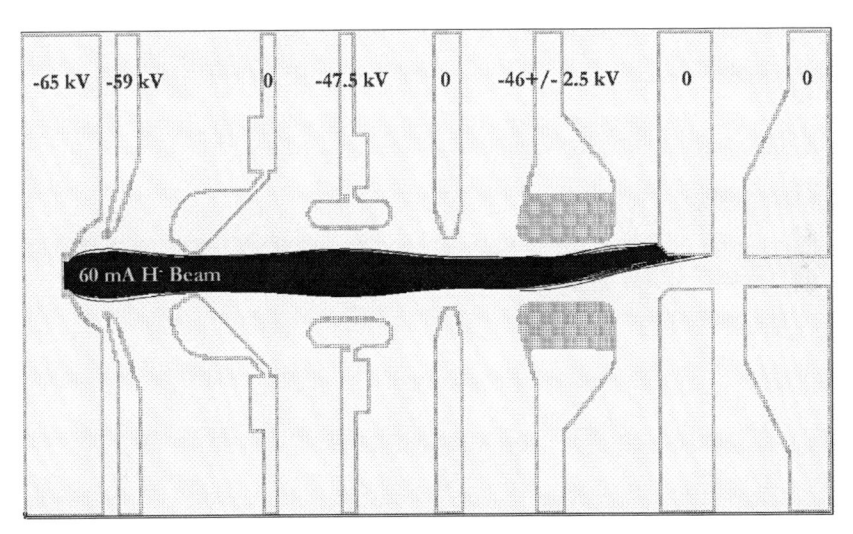

FIGURE 3. SIMION simulation of the ion source and the present LEBT for beam chopping.

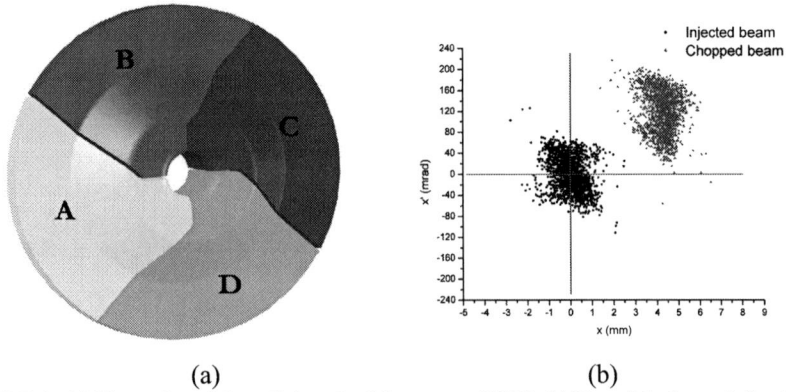

FIGURE 4. (a) Illustration of the split lens-2 of the present LEBT, (b) Plot of the beam deflection in phase-space at the RFQ entrance generated from the SIMION simulation shown in figure 3.

A NEW SNS LEBT WITH A NEAR-GROUND BEAM CHOPPER

In establishing a new electrostatic LEBT with a near-ground beam chopper, the design criteria are achieving a similar beam dynamics performance as the present LEBT configuration with minimal changes in the existing infrastructure of the SNS front end. In the proposed new configuration, the ground electrode between the two lenses is extended and a tube peninsulated inside it is split into four electrically-isolated quadrants to serve as electrodes for beam steering and chopping, as shown in Fig. 5. In this configuration, large transients from HV sparks are unlikely to be found on the chopper electrodes because they operate only with the few kV needed for chopping and steering, and because high voltage sparks from the adjacent lenses are intercepted by the grounded shield.

When the chopper is placed between the lenses, the second lens drastically reduces the beam steering angle established by the chopper, which is the primary reason that the original LEBT proposal [3] does not work for SNS. The reduction of the beam steering angle can be mitigated by a sufficient transverse displacement before the beam enters lens-2. This can be accomplished with a longer chopper and shorter lens-2

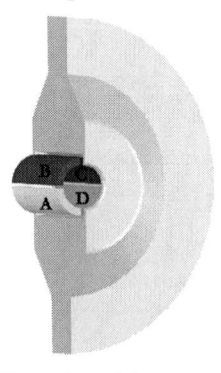

FIGURE 5. Illustration of the proposed new chopper.

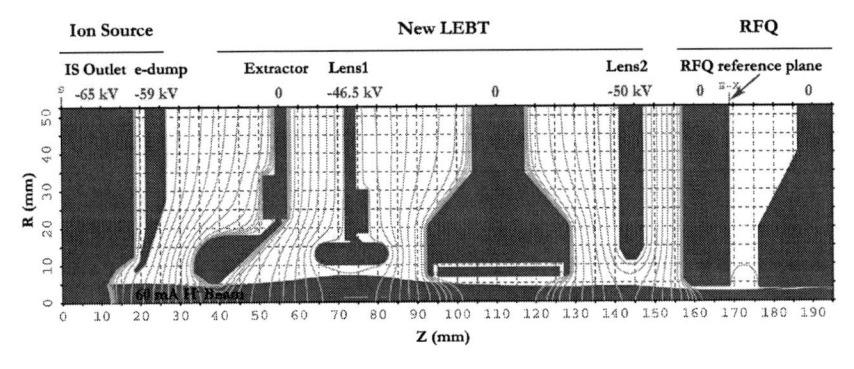

FIGURE 6. PBGUNS simulation of the ion source and the new LEBT for beam injection.

featuring a larger inner diameter. It is important that no trajectories are intercepted by the uncooled chopper or lens-2 electrodes. Several iterations led to the LEBT shown in Fig. 6, which is 15 mm longer than the baseline LEBT. The increased length of the LEBT can easily be accommodated by shifting the ion source backward with a 15 mm-thick spacer flange.

The same simulations as described for the baseline LEBT were conducted for the new LEBT to determine its dimensions and to assure identical injection into the RFQ. Accordingly, Fig. 6 shows the PBGUNS beam transport simulation through the newly configured LEBT with a longer, near-ground chopper and thin second lens featuring a larger aperture. As seen in Fig. 7, it matches the RFQ injection requirements at the downstream face of the RFQ entrance aperture.

Figure 8 shows the SIMION simulation of beam transport while the beam is being chopped. None of the particle trajectories end on the uncooled chopper or the uncooled second lens. Almost all the trajectories are intercepted by the water-cooled RFQ entrance flange. A very few trajectories enter the RFQ, where they hit the cavity walls between the vanes, because in reality the chopper is rotated by 45° with respect to the RFQ vanes.

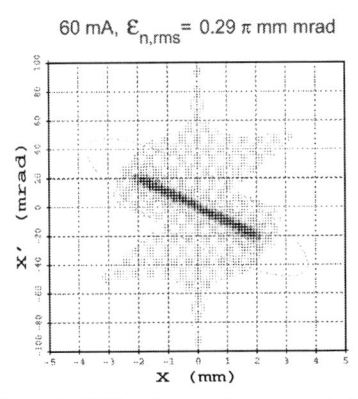

FIGURE 7. x-x′ emittance plot at the RFQ reference plane generated from the PBGUNS simulation shown in figure 6.

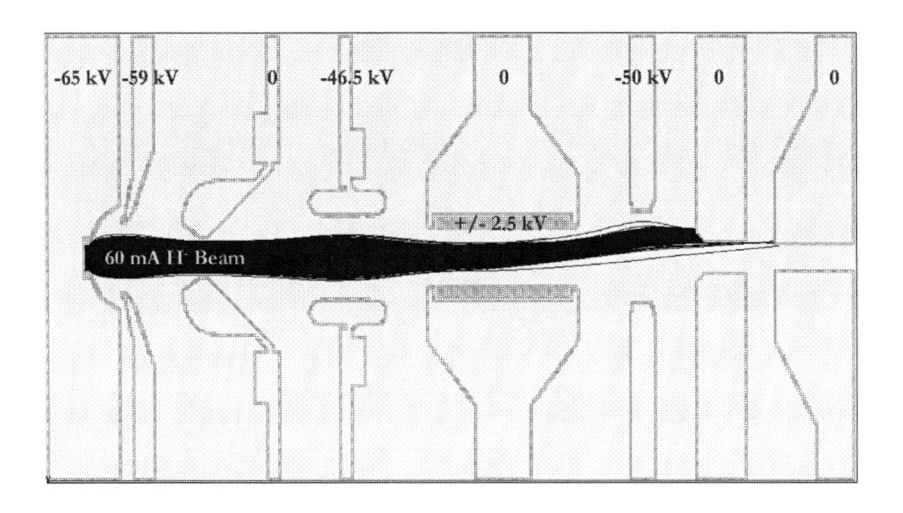

FIGURE 8. SIMION simulation of the ion source and the new LEBT for beam chopping.

Figure 9 shows the phase-space of the transmitted and the chopped beam at the upstream face of the RFQ entrance aperture. The second lens nearly cancels the steering angle applied to the beam by the ±2.5 kV chopper voltages, but the transverse offset has already reached more than 5 mm at this location, thus preventing the beam from entering the RFQ.

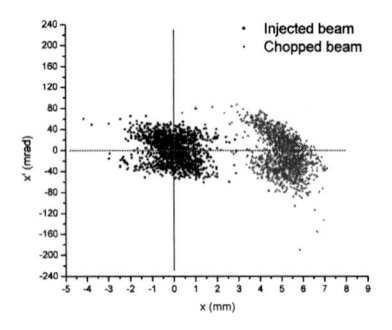

FIGURE 9: Plot of the beam deflection in phase-space at the RFQ entrance generated from the SIMION simulation shown in figure 8.

ACKNOWLEDGMENTS

This work was supported by SNS through UT-Battelle, LLC, under contract DE-AC05-00OR22725 for the United States Department of Energy.

REFERENCES

1. J. Reijonen, R. Thomae, and R. Keller, *Proc. Linac 2000 Conf.*, pp. 253-255
2. R. Keller and S. K. Hahto, *Proc. 34th ICFA Advanced Beam Dynamics Workshop on High Power Superconducting Ion, Proton, and Multi-Species Linacs,* 2005.
3. S. K. Hahto, D. G. Bilbrough and R. Keller, *AIP Conf. Proc.* 925, 2007, pp. 318-323.
4. M. P. Stockli, B. X. Han, S. N. Murray, D. Newland, T. Pennisi, M. Santana, R. Welton, these proceedings.
5. PBGUNS 5.04, available through Thunderbird Simulations, 4704 Downey Street NE, Albuquerque, NM 87109, USA.
6. B. X. Han, R. F. Welton, M. P. Stockli, N. P. Luciano, and J. R. Carmichael, *Rev. Sci. Instrum.* 79, (2008) 02B904
7. R. Keller, R. Thomae, M. Stockli, and R. Welton, AIP Conf. Proc. 639, 2002, pp. 47-60.
8. SIMION 8.04, available through Scientific Instruments Services Inc., 1027 Old York Road, Ringoes, NJ 08551, USA.

Beam Induced Effects in the ITER Electrostatic Residual Ion Dump

E Surrey[a], AJT Holmes[b] and TTC Jones[a]

[a] *EURATOM/UKAEA Fusion Association, Culham Science Centre, Abingdon, Oxfordshire, OX14 3DB, U.K.*
[b] *Marcham Scientific, Sarum House, 10 Salisbury Road, Hungerford, Berkshire, RG17 0LG, U.K.*

Abstract: The effect of beam generated plasma in the channels of the ITER electrostatic residual ion dump (ERID) is investigated. The space charge due to the plasma and the separating positive and negative ion beams could cause the ERID to fail if the plasma density is sufficiently high. The gas density at which this occurs is derived for both the heating and diagnostic beam systems. For the heating beam system this critical density is a factor of 75 higher than the normal operational gas density, so no problems are anticipated. For the diagnostic beam, the critical density is only a factor of 3 higher and under certain circumstances could be exceeded.

Keywords: neutral beam, beam induced plasma, space charge
PACS 29.27Eg, 52.20Hv, 52.40Kh, 52.80Tn

INTRODUCTION

The ITER neutral beam injection system is comprised of two 1MeV injectors for heating and current drive (HNB) and a single 100keV injector for charge exchange recombination spectroscopy (DNB). Each HNB injector generates 16.7MW of neutral deuterium beam and the DNB generates 2.5MW of neutral hydrogen beam from negative ion precursor beams via the stripping reaction:

$$D^- + D_2 \rightarrow D^0 + e + D_2$$

which occurs in the gas filled neutraliser. An additional reaction:

$$D^0 + D_2 \rightarrow D^+ + e + D_2$$

also occurs in the neutraliser, so that the beam is a three component system. The optimum gas density in the neutraliser gives rise to 58% neutralisation and equal quantities of residual negative and positive ions, which must be removed in a controlled manner in the residual ion dump. In the ITER design [1], it is proposed that the residual ions be removed by an electrostatic field transverse to the beam axis and collected on suitably designed panels. This electrostatic residual ion dump (ERID) is formed from five panels, giving four beam channels in accordance with the subdivision of the neutralizer as shown in Fig1.

The beam, comprised of neutral, positive and negative charged particles, enters from below. The electric field across the channel deflects the positive and negative ions in opposite directions. Secondary electrons will be suppressed by the electric field at the grounded panels but accelerated by the field at the negative panel. The dimensions of the ERID for both the HNB and DNB are given in Table 1 together with other parameters of relevance to the model. Gas from the source and the neutraliser contribute to the background pressure in the ERID and a fraction of this gas will be ionised by the beam, creating a plasma between the ERID panels, which will also be affected by the applied potential. A plasma sheath will form at the cathodic side of the RID, across which most of the applied voltage will fall. The width of this sheath depends upon the plasma density, the degree of separation of the positive and negative residual ion beams and the space charge due to ion formation in the sheath. If the sheath width is equal to or greater than the ERID channel width then the deflecting potential will be dropped across the whole of the channel. However, if this condition is not fulfilled, then the potential will be dropped across a fraction of the channel and part of the beam will see no deflecting electric field.

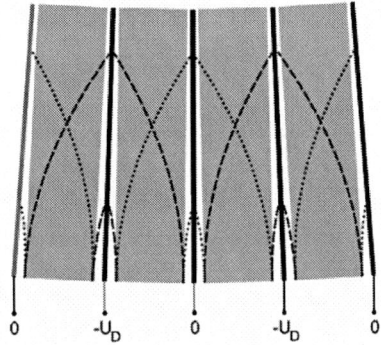

FIGURE 1 Schematic of the ERID. All four channels are shown with the panel biasing arrangement. The grey blocks represent the neutral beam; the dashed lines indicate the extreme trajectories of the positive ions and the dotted lines of the negative ions

TABLE 1 Dimensions of the HNB and DNB ERID and Model Parameters

	HNB	DNB
Length (mm)	1800	1000
Channel Height (mm)	1745	1500
Channel width entrance/exit (mm)	103.8/91.5	104/96
Deflection voltage (kV)	20	4
Gas density (m^{-3})	4x10^{18}	2x10^{18}
Current per channel (A)	10	15
Beam energy (keV)	1000	100
Species (%): D$^+$:D^0:D$^-$/ H$^+$:H^0:H$^-$	D: 20:60:20	H: 20:60:20

This study investigates the conditions for which the ERID sheath will be sufficiently large to allow full deflection of the residual ion beams and predicts conditions for which the ERID may fail.

THE ERID MODEL

Modelling of the ERID is complicated by the possibility that there may be no plasma present at all, i.e. the applied potential is sufficient to drain the beam generated plasma from the beam channel. Under these conditions the plasma sheath is equal to or wider than the ERID channel and free fall of all ions to the RID panel occurs. The model therefore calculates the plasma sheath width and compares this to the channel width in order to establish the presence of plasma. Three sources of space charge in the sheath are considered: beam ions deflected by the electrostatic field, slow ions created in the sheath and slow ions emerging from the neutral plasma. It is assumed that the space charge of each of these components can be considered separately and the potentials simply summed.

The Formation of Ions by the Beam

The plasma ions are formed by collisions between gas molecules and the fast D^+, D^- and D^0 particles in the beam. We assume that when the ion beams deflect in the RID, they deflect collectively so that the current density of each beam ion species is preserved. The beam currents (except for D^0) reduce as the beam ions are lost on the RID. The volumetric rate of charge production by ionization in the RID, Ψ, is:

$$\Psi = N(J_+\sigma_+ + J_-\sigma_- + J_0\sigma + J_e\sigma_e) \qquad (1)$$

where the subscripts "+", "0", "-" and "e" denote the positive ion, neutral, negative ion and stripped electron components of the beam, J is the current density of a given beam component and σ is the cross section for ionisation of the background gas by that beam component. The gas density, N, is averaged over the axial region of interest because the high axial mobility [2] of plasma particles along the beam axis means that this production process should not be considered a purely local effect. Typical axial fields of the order of a few volts per metre have been observed in beams at UKAEA and a gas density of $4 \times 10^{18} m^{-3}$ leads to an ionic drift velocity of 5500 m/s per volt/m. This is sufficiently large to average out plasma production over a significant part of the beam axis and may even result in coupling of the ERID and neutralizer plasma systems. Values of the average gas density for the ERID and neutralizer-RID combination are given in Table 2 for both HNB and DNB systems.

The Potential of Ions Created in the Sheath

Ions of mass M created at a distance ξ from the fully neutralised beam region are accelerated towards the cathodic panel by the potential $V(\xi)-V(y)$ where y is the local distance from the fully neutralised beam region. Using Poisson's equation to obtain the electric potential as a function of distance gives:

$$-\frac{d^2V}{dy^2} = \frac{\psi}{\varepsilon_0 (2e/M)^{1/2}} \times \int_0^y \frac{d\xi}{(V(\xi)-V(y))^{1/2}} \qquad (2)$$

where $\psi = \Psi - NJ_e\sigma_e$ as the stripped electrons are deflected in the same direction as the negative ion beam component but by a larger angle and are removed from the RID in the first few centimetres so do not contribute to the ionisation in the sheath. The integral allows for the fact that the plasma ions do not have the same energy at the same value of y because of their different points of creation. Langmuir [3] gives the solution of equation (2):

$$U_1 = \left[\frac{\pi\psi}{4\varepsilon_0(2e/M)^{1/2}}\right]^{2/3} D^2 \qquad (3)$$

where U_1 is the potential arising from this form of space charge across the sheath of thickness, D.

The Potential Created by Ion Flux from the Neutral Beam Plasma

The current density of plasma ions in the background gas inside the RID is:

$$J = (J_+\sigma_+ + J_-\sigma_- + J_0\sigma + J_e\sigma_e) \times \frac{Hw}{2HL}\int_0^L Ndl = \Psi \times 0.5w \qquad (4)$$

where H is the RID channel height, w the width of the plasma region and L its axial length. Here there is a contribution from the fast stripped electrons in addition to the ionization created by the beam itself. Note that the factor of 2 arises from the fact that the slow ion current diffuses towards both RID plates equally. For simplicity, w is assumed to be unchanged along the RID axis, although in practice, it will reduce as the beam is cleared from the RID. The slight enhancement of plasma production due to negative ions deflected from the sheath volume into the plasma volume has also been neglected. The potential, U_2, due to the planar sheath is given by the Child-Langmuir equation:

$$U_2 = \left[\frac{4.5\Psi w}{4\varepsilon_0(2e/M)^{1/2}}\right]^{2/3} \times D^{4/3} \qquad (5)$$

The Potential Formed by the Separated Beam Ions

Initially the positive and negative ion beams overlap entirely, so that no potential is generated. However, as the RID is traversed, the positive ion beam is collected on the cathodic panel and the negative ion beam moves out of the sheath into the neutral plasma (which remains neutral by definition). If the beam movement is s (D^+ into the RID plate and D^- into the neutral plasma (and eventually into the grounded plate), then a charged zone of thickness, s, next to the negative RID plate, will exist where positive beam ions exist alone. The analogous zone for negative ions does really not exist as the anode sheath is very narrow and only has a potential of a few fold the electron temperature across it.

A schematic for the model is shown in Fig 2 for the condition where plasma and the plasma sheath are sustained in the beam channel. The beam is taken to have a uniform core with a half-gaussian profile, of gaussian radius, ρ, over a distance s_0 next to the RID plate. The grey solid line represents the boundary of the plasma which is formed to the right of this plane between it and the grounded panel. The sheath lies to the left of this plane and is of width D (so that the plasma occupies a width (W-D) where W is the RID channel width). The dotted curve indicates the trajectory of the edge of the core negative ion component deflected by the field across the sheath. There is no curve corresponding to the other core edge as this falls within the plasma where the electric field, E, is zero. The dashed curves indicate the trajectories of the two extremes of the positive ion component that experience the deflecting field. Part of the positive ion current lies in the plasma region and is not deflected.

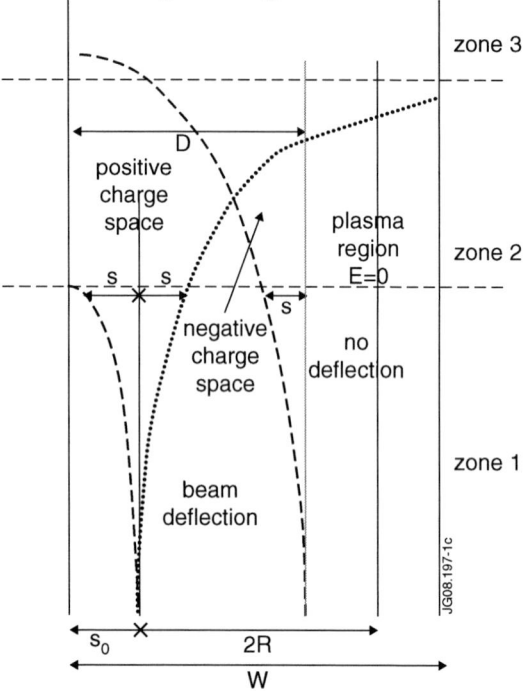

FIGURE 2 Schematic of the ERID model showing plasma and sheath. The solid black lines — represent limits of the uniform beam core of width 2R, the grey solid line — represents the edge of the plasma sheath, the dashed line the extreme trajectories of positive ions and the dotted line the extreme trajectory of the negative ions.

There are two parts to the calculation; firstly the beam displacement is derived assuming a uniform electric field across the sheath and the collected beam current is also derived at the same time. Secondly, from this displacement the net excess width of positive beam space charge is found as the difference between the exposed positive and negative charges. A new sheath width is calculated from Poisson's equation and is used in the next step of the beam displacement. The positive beam deflection is derived in a step-wise way from the expressions:

$$E = U/D \qquad (6)$$
$$\Delta v_\perp = eE\Delta z/mv_b \qquad (7)$$
$$\Delta s = v_\perp \Delta z/v_b \qquad (8)$$

where $v_\perp$ is the ion velocity perpendicular to the RID axis, v_b is the beam velocity and Δz is the incremental axial step. This deflection applies to all beam ions outside the beam plasma, whether positive or negatively charged.

In each zone the charge per unit length outside the plasma region is calculated (allowing for current collected on the RID panels) for both the positively and negatively charged beam components. The total net charge is simply the difference of these two quantities and Gauss's Law gives the electric field due to this charge. By integrating over the total distance that this field applies, the resulting potential, U_3, is found to have the form

$$U_3 = \frac{J}{\varepsilon_0 V_b} F(s, s_0, \rho) \qquad (9)$$

where the function $F(s,s_0,\rho)$ varies along the RID length.

The full potential U is given by the sum of the potentials due to the three components of space charge:

$$U = \left[\frac{\pi \psi}{4\varepsilon_0 (2e/M)^{1/2}}\right]^{2/3} D^2 + \left[\frac{4.5 \Psi w}{4\varepsilon_0 (2e/M)^{1/2}}\right]^{2/3} \times D^{4/3} + U_3 \qquad (10)$$

Equation (10) is solved in a stepwise manner moving along the RID to give the sheath width when U is equal to the applied potential. At each step the new value of s is evaluated from equations (6-8) and this allows a new value of D to be calculated.

Effect of Electrons

Electrons Created by Beam Ionization

The form of Poisson's equation given in equation (2) ignores the effect of the electrons produced by ionization of the background gas by the beam. These electrons will obviously be accelerated in the opposite direction to the ions. Including the flux of electrons results in an equation that is intractable analytically. However, Langmuir [3] has considered just this case and concludes that unless the linear production rate exceeds that defined by the limiting current the effect of the electron space charge is minimal. In the presence of a source of ions distributed between the plates of a diode, the Child-Langmuir equation is modified [3] so that the limiting current is given by:

$$I_C = \psi_0 A d = \frac{4\varepsilon_0}{\pi} \sqrt{\frac{2e}{M} \frac{A}{d^2}} V_0^{3/2} \qquad (11)$$

where V_0 is the potential between diode plates of area A and separation d. If the actual volumetric production rate, ψ, given by equation (1) is less than ψ_0 then the maximum in the parabolic potential described by equation (3) occurs at the cathode and the electrons flow to the anode under free fall conditions. For equal ionic and electronic currents it follows that the respective densities are in the ratio

$$N_e/N_i = \sqrt{m/M} \qquad (12)$$

and hence the contribution of the electrons to the space charge is tens of times smaller than that of the ions.

Secondary Electrons Created at the RID Panels

It has been noted that secondary electrons would be generated at the ERID panels by the residual positive and negative ion beams. At the grounded panels, the electric field suppresses secondary emission but at the cathodic panels the field would accelerate the secondary electrons to energies at which they would contribute to the ionisation. The secondary emission coefficient at off-normal incidence for most metal surfaces is greater than one, so that the current density of secondary electrons would be several times that of the collected ions providing a further enhancement. Estimates of the residual magnetic field at the ERID [1] indicate a value of approximately 1mT in the vertical direction and calculation of the trajectory of an electron emitted from the cathodic panel shows a displacement of only a few millimeters. Thus there will be a current flowing to the grounded grid equivalent to several times that of the residual positive ion current.

The cross section for ionisation of the background deuterium (or hydrogen) gas by electrons has a maximum value of $\sim 1 \times 10^{-20} m^2$ for electron energy between 40eV and 100eV, above which it falls approximately linearly and is an order of magnitude lower at 2keV[4]. This 2keV energy is attained by the secondary electrons after a distance of 18mm in the HNB RID and 59mm in the DNB case. The average ionization cross section over this distance can be obtained from [4] and used to estimate an average probability of ionisation of the order 10^{-4} for both the HNB and DNB cases. This is approximately one order of magnitude smaller than the equivalent probability of ionisation by the beam ions and so, unless the secondary emission coefficient is of the order of ten, secondary electron ionisation can be ignored. For clean, metallic surfaces the secondary emission coefficient for protons of energy in the 100-500keV range is approximately three [5] but this can vary depending on the condition of the surface. It should be noted that [1] estimates the secondary emission coefficient to be a factor of 6 or 7, in which case ionisation by secondary electrons may be significant.

The effect of secondary electrons on the space charge is similar to that of electrons created by beam ionization, i.e. the space charge is reduced by a factor proportional to their current density. This will be proportional to the positive beam ion current density incident on the panel multiplied by the secondary emission coefficient. Thus it will act to increase the sheath width at a given deflecting potential and so aid the operation of the ERID.

Results of the Modelling

Sheath Width and Sheath Potential

The general form of the potentials U_1, U_2 and U_3 along the length of the RID are shown in Fig 3(a) for the HNB with a gas density of $2 \times 10^{21} m^{-3}$, sufficient to cause the RID to fail with a sheath width of 44mm. The contributions to the sheath space charge arising from ions extracted from the neutral beam plasma and those created in the sheath are approximately constant, with the former dominating by a factor of approximately two. The contribution to the space charge potential of the separated positive ion beam is negligible, only a few volts. Fig 3(b) shows the sheath width, D, beam separation, s, and the collected positive ion current, I_c, for the same beam parameters. Note that when s = D, there is no further collection of ion current. In this case only 0.88A from a possible 2A of residual positive ion current is collected by the cathodic panel, so the RID is only 44% efficient.

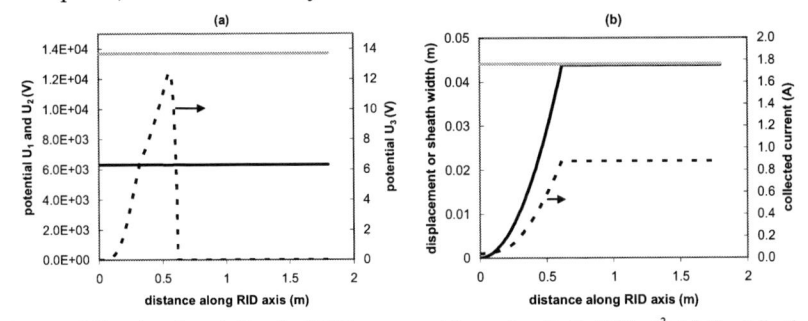

FIGURE 3 Results of modeling for HNB system with gas density $2 \times 10^{21} \text{m}^{-3}$. (a) Contribution of the three sources of space charge to the sheath potential: ionization in the sheath ——— U1, ions from the plasma ∼∼∼∼ U2 and separation of the positive and negative beam ions ········ U3. (b) ∼∼∼∼ sheath width, ——— beam displacement and ········ collected positive ion current.

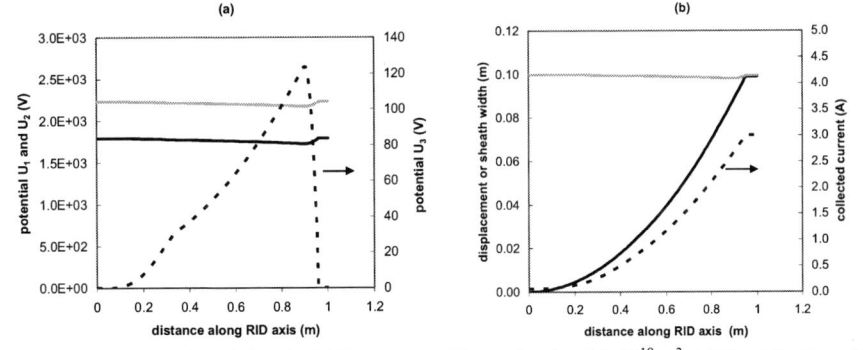

FIGURE 4 Results of modeling for DNB system with gas density $5.6 \times 10^{18} \text{m}^{-3}$. (a) Contribution of the three sources of space charge to the sheath potential: ionization in the sheath ——— U1, ions from the plasma ∼∼∼∼ U2 and separation of the positive and negative beam ions ········ U3. (b) ∼∼∼∼ sheath width, ——— beam displacement and ········ collected positive ion current.

Figures 3(a) and (b) represent a case of high gas density; the opposite extreme is shown in figures 4(a) and (b) in this case for the DNB with a gas density of $5.6 \times 10^{18} \text{m}^{-3}$, on the limit of operation of the RID. Now the full residual positive ion current of 3A is collected on the cathodic panel when both the sheath width and the beam separation are equal to the width of the RID channel. Note that in this case the space charge contribution due to ionization in the sheath is a much larger fraction of the sheath potential, reflecting the increase in the source term due to the entire positive ion beam contributing to ionization. This also increases the contribution of the space charge due to separation of the positive and negative ion beams and as a consequence, the potential derived from the space charge contribution due to ionization and ions flowing from the plasma decreases slightly.

Critical Density

The code was used to investigate the evolution of sheath width as a function of the background gas density for the HNB with parameters given in Table 1. The results are shown in Fig 5, where it can be seen that the plasma sheath is equivalent to the RID channel width for a gas density of 3×10^{20}m^{-3}, well in excess of the density of 4×10^{18}m^{-3} anticipated in the ITER DDD [1]. Considering the combined system of ERID and neutraliser, which may be relevant in view of the ionic mobility, the average density is raised to 1.5×10^{20}m^{-3} and the deflection voltage would still be sufficient to clear the ERID channel of plasma, although the margin for error is small.

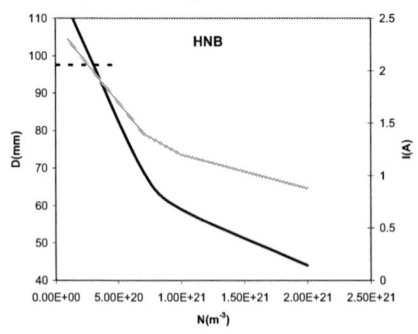

Figure 5 Sheath width ——— and positive ion current collected on the RID panel ········ for the HNB with 20kV applied potential as a function of gas density. The dashed line indicates the average width of the ERID channel.

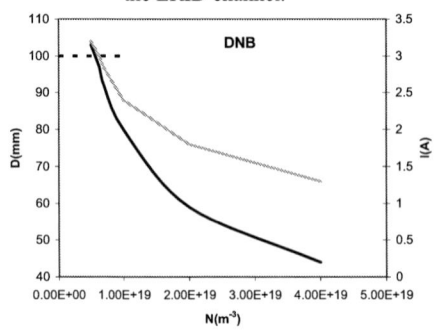

FIGURE 6 Sheath width ——— and positive ion current collected on the RID panel ········ for the DNB with 4kV applied potential as a function of gas density. The dashed line indicates the average width of the ERID channel

The equivalent plot for the DNB with parameters given in Table 1 is shown in Fig 6. Again the limiting gas density of 5.6×10^{18}m^{-3} is above the value of 2×10^{18}m^{-3} anticipated in the ITER DDD [1], although a factor of 2.8 may not be considered sufficient margin to ensure correct operation under all circumstances. For the combined ERID and neutralizer system of the DNB the average gas density is 9.5×10^{18}m^{-3} and is somewhat larger than the threshold value, so there may be a problem for the DNB system if the RID and neutralizer plasmas are fully coupled.

CONCLUSIONS

The effect of ionisation in the ERID channel leading to the formation of a plasma sheath narrower than the RID channel has been assessed. Under these circumstances, the deflecting potential applied across the channel would be mainly dropped across the sheath and may result in some of the residual ions not being deflected. Despite the dynamics of the system being complex it has been shown that in the case of the HNB, there is considerable margin in gas density before the onset of sustained plasma. For the DNB the situation is not so unequivocal; the threshold density is only a factor of three above the expected density. In addition, the mobility of the plasma ions created in the RID and neutralizer channels could result in these two components being coupled and this, in effect, raises the average gas density. Estimates for the gas density in these two cases are given in Table 2, together with the critical density, N_c, at which a sustainable plasma is formed. The "safety factor", S, also given in Table 2 is defined as $S=N_c/N$ if $N_c/N \geq 1$ and $S=0$ if $N_c/N<1$.

Thus for the HNB the critical gas density of $3 \times 10^{20} m^{-3}$ is not exceeded even for the worst case of RID and neutralizer coupling. However the safety factor is only 2 and this might be considered marginal. For the DNB coupling of neutraliser and RID panels, the model estimates a voltage of 7kV will be required to clear the channel of plasma. This is almost twice the specified voltage, so it would seem prudent to increase the specification of the DNB supply.

TABLE 2 Summary of Operating Conditions of HNB and DNB Residual Ion Dumps

Condition		HNB	DNB
Critical gas density N_c		$3.0 \times 10^{20} m^{-3}$	$5.6 \times 10^{18} m^{-3}$
ERID alone	density, N	$4.0 \times 10^{18} m^{-3}$	$2.0 \times 10^{18} m^{-3}$
	safety factor, S	75	2.8
ERID + neutralizer	density, N	$1.5 \times 10^{20} m^{-3}$	$9.5 \times 10^{18} m^{-3}$
	safety factor, S	2	0

ACKNOWLEDGEMENT

This work was funded jointly by the United Kingdom Engineering and Physical Sciences Research Council and by the European Communities under the contract of Association between EURATOM and UKAEA. The views and opinions expressed herein do not necessarily reflect those of the European Commission.

REFERENCES

1. ITER Design Description Document 5.3, N 53 DDD 29 01-07-03 R0.1
2. GWC Kaye & TH Laby, *Tables of Physical and Chemical Constants 14th Edn*, Longman (1973), p 259
3. I Langmuir, Phys. Rev., **33**, 954 (1929)
4. RL Freeman & EM Jones, *Atomic Collision Processes in Plasma Physics Experiments*, CLM-R 137, UKAEA (1974)
5. EJ Sternglass, Phys. Rev., **108**, 1 (1957)

Steering of Multiple Beamlets in the JT-60U Negative Ion Source

Masaki Kamada[a], Masaya Hanada[a], Yoshitaka Ikeda[a]
and Larry R. Grisham[b]

[a]*Japan Atomic Energy Agency, 801-1 Mukoyama, Naka, Ibaraki 311-0193, Japan*
[b]*Princeton Plasma Physics laboratory, P.O. Box451, Princeton, New Jersey 08543, USA*

Abstract. The direct interception of D⁻ ion beam by the acceleration grid was reduced by modifying field shaping plates (FSPs) in JT-60 negative ion sources to inject powerful neutral beams for long pulse duration of > 10 s. The modified FSPs are designed by a 3D simulation code to properly steer the outermost beamlets that are deflected outward by space charge of the inward beamlets. The measured steering angles were -1.2 mrad and -6.4/-4.5 mrad in the vertical and horizontal direction, which were in agreement with the designed value calculated by the 3D simulation code. The proper steering of the outermost beamlets allowed a significant reduction of the power loading of the grounded grid. The power loading was successfully reduced to an allowable level of 5 % with respect to the acceleration beam power.

Keywords: Negative ion source, multi-aperture grid, beam steering, field-shaping plate
PACS: 41.75.Cn

INTRODUCTION

Negative ion based neutral beam injector (N-NBI) is essential for steady-state operation of magnetic fusion machine. Negative ion beams are produced in a negative ion source, and then neutralized in a gas cell so that the resulting neutral beams can cross the confinement magnetic fields. In the JT-60U N-NBI, D^0 beams of 10 MW are required to inject into plasmas for heating and current drive. To produce such powerful D^0 beams, the JT-60U N-NBI has two negative ion sources, each of which is designed to generate 500 keV, 22 A, D⁻ beams for 10 s [1, 2]. Recently, the JT-60U N-NBI is required to inject D^0 beams for 30s. Furthermore, the JT-60SA N-NBI is required to inject D^0 beams for 100 s [3, 4].

One of key issues to produce such powerful beams for long pulse length is a reduction of power loadings of multi-aperture grids in the negative ion source by direct interception of D⁻ ions. In the previous studies, it is found that the direct interception of D⁻ ions is mainly caused by deflection of outer beamlets due to repulsion from space charge of inner beamlets [5]. To suppress deflection of outermost beamlets, thin plates for modifying the acceleration electric field (FSPs) were installed in the JT-60 negative ion sources [2].

However, the highest grid power loading was measured to be 7-10 % of the accelerated beam power at optimum perveance, and the half by the direct interception of D⁻ ions and the other half by stripped electrons [2]. This high direct interception was found from the experiment and simulation results to be caused by the overfocus of the outermost beamlets due to non-optimized geometry and mounted location of the FSPs [2, 6].

To reduce the grid power loading to 5 % as an allowable level for long pulse operation of the negative ion source for JT-60SA, FSPs were newly designed by using 3D simulation code and tested in the negative ion source for JT-60U. Deflection angle of outermost beamlet and grid power loading were measured. In this paper, the effect of the FSP on the grid power loading is reported.

JT-60U NEGATIVE ION SOURCE

Figure 1 shows a schematic diagram of the negative ion source for JT-60U. The negative ion source is composed of an arc chamber and a three-stage electrostatic accelerator. The arc chamber has a semi-cylindrical shape with dimensions of 0.68 m in diameter and 1.22 m in height. Intense arc plasmas are driven by tungsten filaments The D⁻ ions are produced via volume and surface production process in arc plasmas with cesium seeding. The D_2 gas is fed into the arc chamber to maintain the arc discharge. The arc chamber is typically operated at 0.3 Pa in the arc chamber.

The accelerator is composed of five grids, named as the plasma grid (PLG), the extraction grid (EXG), the first acceleration grid (A1G), the second acceleration grid (A2G) and the grounded grid (GRG). These grids are divided into five segments, each of which has 9 (in row) x 24 (in column) apertures in the area of 180 mm (in height) x 450 mm (in width). In two ion sources named as U and L ion sources, all five segments and central three segments are utilized for the ion extraction, respectively. The smaller ion extraction area of the L ion source is due to reduce the power loading of non-water cooled beam limiters that are located at 20 m downstream from the ion sources. In this study, the D⁻ ions were extracted at several keV and accelerated up to 280-350 keV, where the D⁻ ion beam current density was optimized to be 55-110 A/m² by tuning the arc discharge power to minimize the direct interception of D- ions with the acceleration grids.

To reduce the co-extracted electrons from the arc chamber into the accelerator, transverse magnetic field was locally formed near the PLG in the arc chamber by flowing high current to the PLG itself. In the EXG, the permanent magnets were also embedded to dump the co-extracted electrons onto it. The stray magnetic field calculated with a 3D finite element method code of Magnun [7] is shown in Fig.2. At 6 kA of the operating PLG current, the horizontal stray field (B_X) is ~ 75 Gauss near the PLG, and monotonically decreased to 30 Gauss with the distance from the PLG. The horizontal stray field (B_Y) locally becomes large near the EXG.

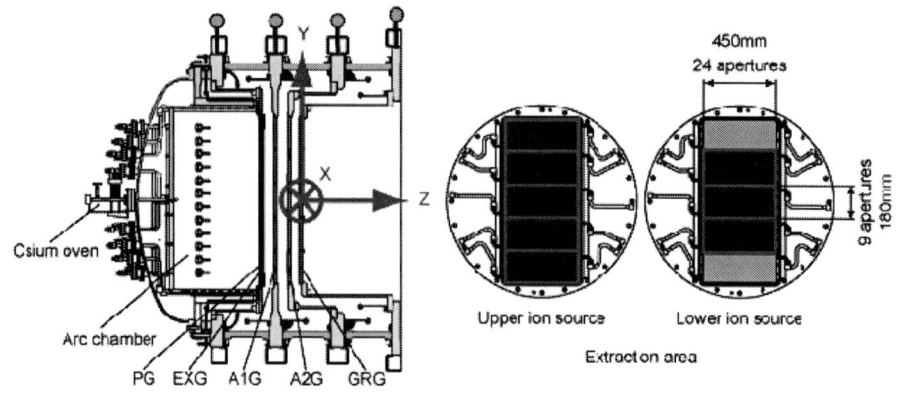

FIGURE 1. JT-60U negative ion source.

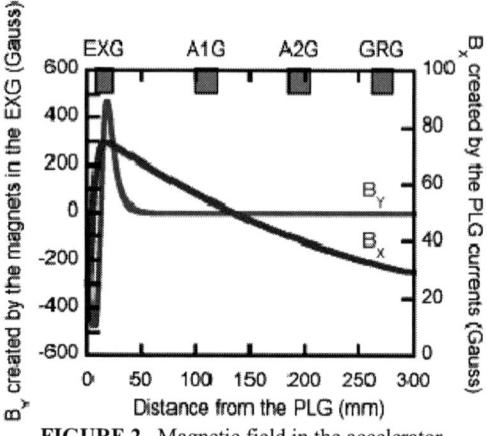

FIGURE 2. Magnetic field in the accelerator.

CALCULATION MODEL AND DESIGN OF FSP

In the negative ion source for JT-60U, FSPs were attached on the EXG to suppress outward deflection of the outmost beamles due to repulsion from space charge of inner beamlets [2]. The previous FSPs were too thick to form a strong focusing electric field. This resulted in the overfocusing of the outermost beamlets and interception by the GRG [2, 6]. To properly steer the outmost beamlets, the FSPs were newly designed by using 3D simulation code (AMaze series, Field Precision LLC) [8]. This code solves the electromagnetic field with test beam particles by finite element method and iterates this procedure to find the beam trajectory.

Calculation Model

Figure 3 shows a schematic diagram of 3D simulation model where space charge of multi-beamlets, magnetic fields and aperture displacements are taken into account [6].

To design the new FSPs, the acceleration voltage of 350 kV and extraction voltage of 6 kV were applied. The D⁻ beam current density was assumed to be 107 A/m² at the PLG. Stripping loss was not taken into account. Two kinds of magnetic fields shown in Fig.2 were also applied. Dipole magnetic field created by permanent magnets was fully solved in 3D space. PLG filter magnetic field created by PLG filter currents was modeled as simple transverse magnetic field. An example of calculated trajectories of D⁻ ions is shown in Fig.3. Deflection angle of each beamlet was defined as an averaged value of deflection angles of all ions in each beamlet. The FSP mounted on the side of the segment was designed by using the model in Fig.3. By changing the aperture arrangement from horizontal direction to the vertical one, the FSP mounted on the top and bottom of the segment was designed.

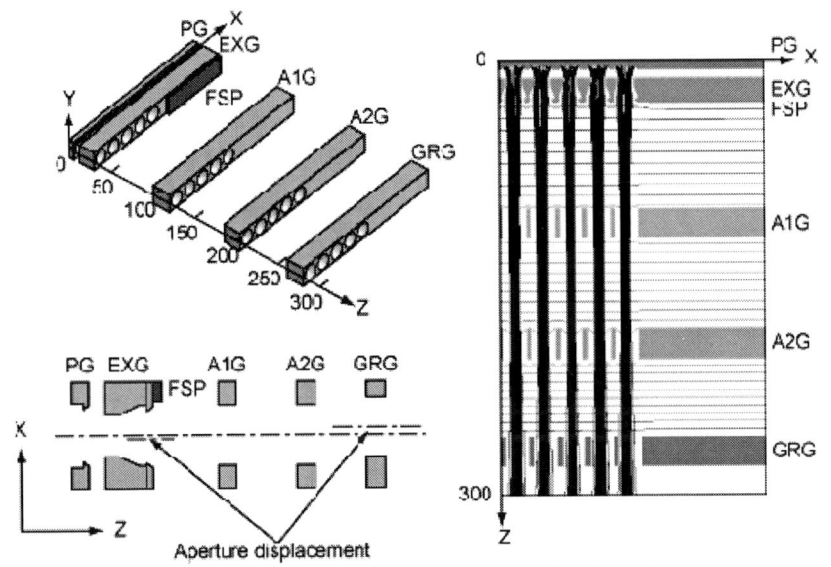

FIGURE 3. 3D simulation model for multiple beamlets in the JT-60 negative ion source and an example of calculated trajectories of D- ions.

Design of FSP

Because of the geometrical limitation, the distance between the outermost apertures and the FSP on the top and bottom of the segment is restricted to be 13 mm. Under this limitation, thickness of the FSP is determined from the calculation results. Figure 4 (a) shows the vertical deflection angle of the outermost beamlet as a function of the thickness of FSP. The sign of minus indicates inward deflection. The vertical deflection angle decreases linearly with thickness of the FSP. Allowable deflection angle is ranged from -2 mrad to 0 mrad to avoid the direct interception of D⁻ ions by the GRG and beam-line components. However, the thickness of the previous FSP was 1.5 mm where the deflection angle was -3 mrad and out of the allowable range. In this design, 1 mm is chosen as the thickness of the FSP on the top and bottom of the segment in order to steer the outermost beamlets with -1 mrad.

The thickness of the FSPs on the side of the segment is limited to be ~1.5 mm by the length of bolts. Under this limitation, distance between the outermost apertures and the FSP is determined from the calculation results. Figure 4 (b) shows deflection angle of the outermost beamlet as a function of the distance between outermost aperture and FSP. Two lines in this figure correspond to the beamlets deflected by dipole magnetic field (see upper part of Fig.4(c)). Both of the horizontal deflection angles decrease linearly with distance between the outermost apertures and the FSP. Allowable deflection angle is ranged from -7 mrad to -4 mrad to avoid the direct interception of D^- ions by the GRG and beam-line components. However, the previous FSP was 2 mm thickness and located at 11 mm where the deflection angles were -7 or -10 mrad and out of allowable range. For the outermost beamlets, the influence of dipole magnetic field can be suppressed by adjusting the distance between outermost aperture and FSP (see lower part of Fig.4(c)). In this design, 11 mm and 14 mm are chosen as the distances of the FSP and the outmost apertures in order to steer the outmost beamlets with -5 mrad.

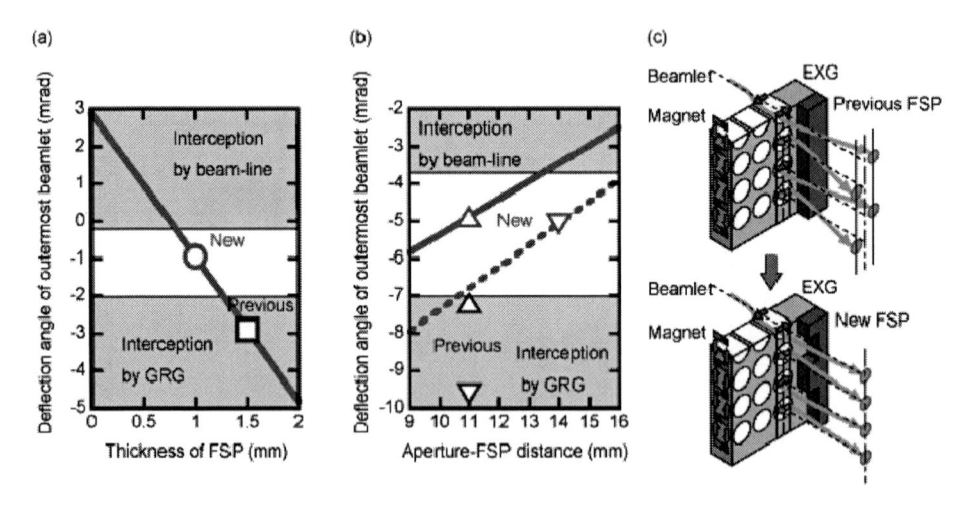

FIGURE 4. (a) Deflection angle of outermost beamlet as a function of thickness of FSP in vertical direction. (b) Deflection angle of outermost beamlet as a function of distance between outermost aperture and FSP in horizontal direction. (c) Schematic diagram of modification of FSP in horizontal direction.

Figure 5 shows a photograph of the FSPs mounted on a segment of the EXG and a detailed schematic of corner of the segment. On the top and bottom of the segment, FSPs with 1mm in thickness were mounted. The distance of outermost aperture and the FSP was set to be 13 mm. On the side of the segment, FSPs with 1.5 mm in thickness were mounted at right and left side of the EXG. The distances between outermost apertures and FSP were set to be 11 mm and 14 mm, respectively.

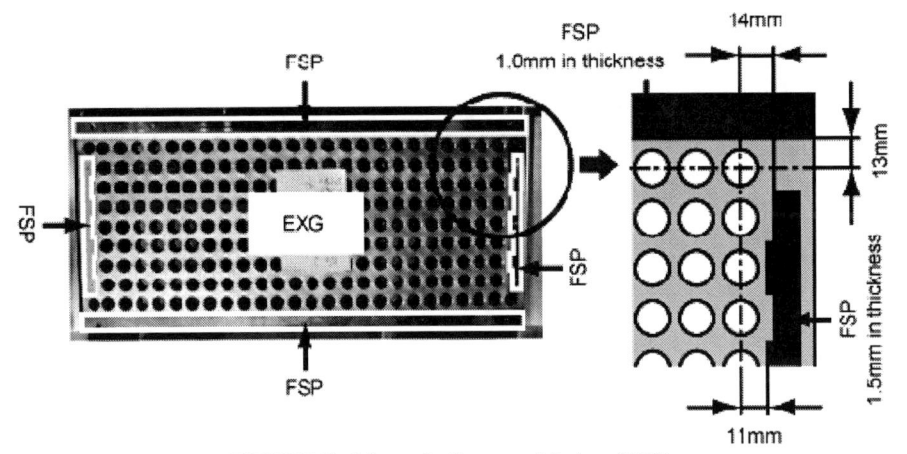

FIGURE 5. Schematic diagram of designed FSPs.

TEST IN JT-60 NEGATIVE ION SOURCE

The newly designed FSPs were installed to the JT-60 negative ion sources, and then the deflection angles and the grid power loading were measured. Early results of these measurements are shown further below.

Steering Angle Measurement

The deflection angle was evaluated from center position of the outmost beamlets on a beam target that was placed at 3.5 m downstream from the GRG. The temperature rise on the beam target was measured using an infrared camera (TS9100 model, NEC Avio Infrared Technologies Co., Ltd.) with 240 x 320 pixels [8]. The angle resolution of the infrared camera is 0.6 mrad. Since the infrared camera was set at 3 m away from the target, the spatial resolution was 1.8 mm on the target.

While the 280 keV, ~9.5 A D⁻ ion beam was produced under the optimum perveance, the deflection angle of outermost beamlet were measured and compared with 3D simulation. Figure 6 (a) and (b) show the vertical and horizontal deflection angles of the outmost beamlet as a function of acceleration voltage, respectively. The ratio between acceleration voltage and extraction voltage was fixed to be the same as that in the simulation. The extraction voltage was 4.9 kV for the acceleration voltage of 280 kV. The measured vertical deflection angle was -1.2 mrad and nearly in agreement with the calculation results. The measured horizontal deflection angles were -6.4 mrad and -4.5 mrad and also in agreement with the calculation results. These results show that new FSPs properly steer the outmost beamlets as predicted by the simulation.

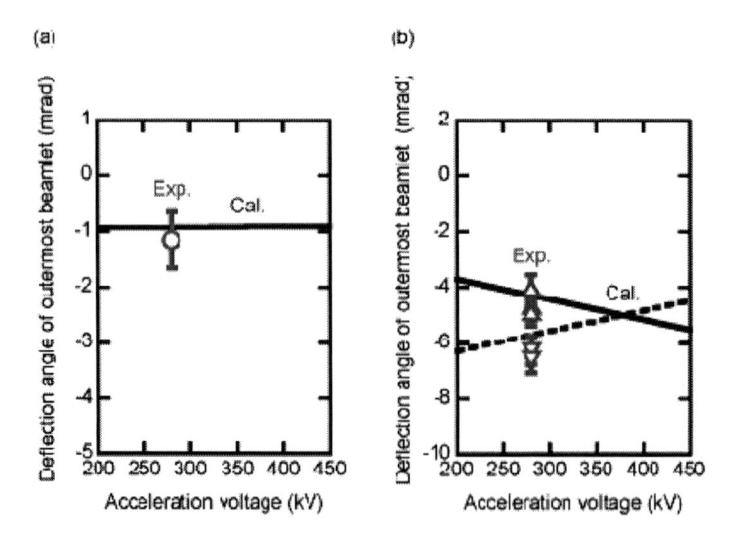

FIGURE 6. (a) Deflection angle of outermost beamlet as a function of acceleration voltage in vertical direction. (b) Deflection angle of outermost beamlet as a function of acceleration voltage in horizontal direction.

Grid Power Loading Measurement

While 300 keV, ~10.5 A D- ion beam was produced under the optimum perveance, the grid power loading of the GRG was measured and compared that before the modification of the FSP. The highest grid power loading was on the GRG, which was the same as that in the previous FSP. The power loadings of the GRG before and after the modification of the FSP are shown in Fig. 7 (a) and (b), which are the dependences of the power loadings of the GRG in U and L ion sources as a function of the operating pressure in the arc chamber, respectively.

In both of the ion sources, the grid power loading increased linearly with the pressure in the arc chamber. This is due that the stripped electrons increased linearly with the pressure in the present range of the pressure. By using the modified FSP, the GRG power loading decreased from 10 % to 7 % of the accelerated beam power at a typical operation pressure of 0.3 Pa in the U ion source (see Fig.7 (a)). Also in the L ion source, the use of the modified FSP reduced the GRG power loading from 7 % to 5 % at 0.3 Pa of an allowable level for long pulse injection during 100 s in JT-60 SA.

The difference of the grid power loadings in the L and U ion sources is due to the difference of the ion extraction area. The D- ions in the L ion source are extracted through the central three segments that are illuminated by spatially uniform D- ions, however, the profile of the D- ion density in the U ion source is non-uniform along full five segment. This non-uniformity of the D- ion density causes the significant direct interception of D- ions by the acceleration grids due to the local mis-match of the perveance of the multiple beamlets. In JT-60SA, the magnetic field configuration of the arc chamber will be changed to "tent-shaped filter configuration" where the beam uniformity has been successfully improved in the 10 A ion source [9]. By using FSP with the tent-shaped filter configuration in JT-60SA, the grid power loading could be reduced to the acceptable level.

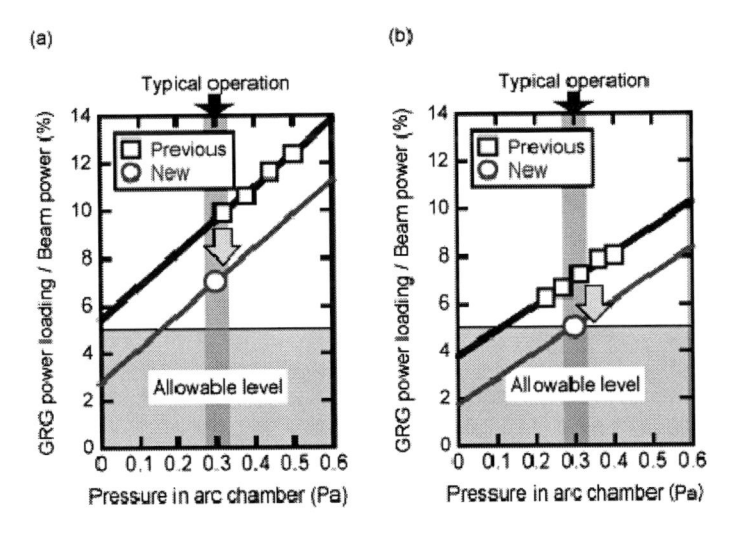

FIGURE 7. GRG power loading of GRG as a function of operating pressure of (a) U ion source and (b) L ion source.

SUMMARY

To reduce the grid power loading, FSP was newly designed and tested in the JT-60U negative ion sources. As the result, the followings are presented.

1. It was confirmed that new FSPs properly steered the outermost beamlet as predicted by the calculation result.

2. When negative ion beams with high spatial uniformity were extracted from central three segments, grid power loading was successfully reduced to an allowable level in JT-60SA.

These results show that long pulse injection can be achieved by new FSP in JT-60SA.

ACKNOWLEDGMENTS

The authors would like to thank other members of NBI Heating Group of JAEA for their valuable discussion. They are also grateful to Dr. T. Tsunematsu and Dr. N. Hosogane for their support and encouragement.

REFERENCES

1. M. Kuriyama, N. Akino, T. Aoyagi, N. Ebisawa, N. Isozaki, A. Honda, T. Inoue, T. Itoh, M. Kawai, M. Kazawa, J. Koizumi, K. Mogaki, Y. Ohara, T. Ohga, Y. Okumura, H. Oohara, K. Ohshima, F. Satoh, T. Takenouchi, Y. Toyokawa, K. Usui, K. Watanabe, M. Yamamoto, T. Yamazaki and C. Zhou, *Fusion Engineering and Design*, **39-40**, 115-121 (1998).
2. Y. Ikeda, N.Umeda N.Akino, N.Ebisawa, L.R. Grisham, M.Hanada, A.Honda, T.Inoue, M.Kawai, M.Kazawa, K.Kikuchi, M.Komata, K.Mogaki, K.Noto, F.Okano, T.Ohga, K.Oshima, T.Takenouchi, Y.Tanai. K.Usui, H.Yamazaki and T.Yamamoto, *Nucl.Fusion*, **46**, S211-219 (2006).
3. Y.Ikeda, N.Akino, N.Ebisawa, M.Hanada, T.Inoue, A.Honda, M.Kamada, M.Kawai, M.Kazawa, K.Kikuchi, M.Kikuchi, M.Komata, M.Matsukawa, K.Mogaki, K.Noto, F.Okano, T.Ohga, K.Oshima, T.Takenouchi, H.Tamai, Y.Tanai, N.Umeda, K.Usui, K.Watanabe and H.Yamazaki, *Fusion Engineering and Design*, **82**, 791-797 (2007).
4. M. Hanada, M. Kamada, N.Akino, N.Ebisawa, L.R.Grisham, A.Honda,M. Kawai, M.Kazawa, K. Kikuchi, M. Komata, K. Mogaki, K.Noto, K. Ohshima, T. Takenouchi, Y. Tanai, K. Usui, H. Yamazaki and Y. Ikeda, *Rev. Sci. Instrum.*, **79**, 02A519 (2008).
5. Y. Fujiwara, M. Hanada, Y. Okumura and K. Watanabe, *Rev. Sci. Instrum.*, **46**, S211 (2000).
6. M. Kamada, M. Hanada, Y. Ikeda and L. R. Grisham, *Rev. Sci. Instrum.*, **79**, 02C114 (2008).
7. Field Precision LLC, (Online) available from <http://www.fieldp.com/index.html> (accessed 2008/08/28).
8. NEC Avio Infrared Technologies Co., Ltd., (Online) available from < http://www.nec-avio.co.jp/en/index.html> (accessed 2008/08/28).
9. H. Tobari, M. Hanada, K. Kashiwagi, M. Taniguchi, N. Umeda, K. Watanabe, T. Inoue and K. Sakamoto, *Rev. Sci. Instrum.*, **79**, 02C111 (2008).

Compensation of beamlet repulsion in a large negative ion source with a multi aperture accelerator

M. Kashiwagi[a], T. Inoue[a], L. R. Grisham[b], M. Hanada[a], M. Kamada[a], M. Taniguchi[a], N. Umeda[a] and K. Watanabe[a]

[a]*Japan Atomic Energy Agency, 801-1, Nukoyama, Naka,311-0193, Japan*
[b]*Princeton University, Plasma Physics Laboratory, Princeton, USA*

Abstract. Excess heat loads to accelerator grids limit extension of pulse length in operation of the large negative ion sources with multi aperture accelerator. Part of the heat loads is caused by interception of deflected beamlets due to their space charge repulsion. In this paper, a beamlet steering technique using aperture offset was examined for compensation of the beamlet deflections utilizing a three dimensional beam analysis simulating the D⁻ negative ion source of JT-60U. The beamlet deflection was analyzed in detail using fifty beamlets, which were extracted from apertures arranged in a lattice pattern of 10 x 5. The simulation showed successful compensation of the beamlet deflection by aperture offsets defined according to the thin lens theory. Even if the beam energy was changed, the necessary aperture offset would not be changed maintaining the perveance and a ratio of extraction and acceleration voltage. In JT-60U, it was shown that the aperture offset of less than 1.0 mm would be enough to compensate the repulsion of all beamlets. When the magnetic field was applied for suppression of co-extracted electrons, necessary aperture offset was estimated to be ± 0.5 mm for 500 keV D⁻ ion beam in JT-60U, in addition to the offset for the space charge repulsion. This result showed good agreements with the previous experimental results and design study of the JT-60U N-NBI.

Keywords: NBI, negative ion source, aperture offset
PACS: 41.75Cn

INTRODUCTION

In a large negative ion source in a neutral beam (NB) system of JT-60U, excess heat loads on the accelerator grids and downstream components are one of issues in extension of beam pulse length. Part of heat loads is generated by direct interception of deflected beamlets due to their own space charge repulsion [1, 2]. In the JT-60U negative ion source, metal bars were attached around the aperture area at the exit of the extractor, namely, the electron suppression grid [3]. Distorted electric field formed around the metal bars steers the beamlets extracted from the outermost apertures to counteract the beamlet deflection. However, this beamlet steering is not enough since the field distortion propagates only to peripheral apertures in the extraction area.

As a useful beamlet steering technique, aperture offset steering [4, 5] has been utilized widely in positive ion sources in many fusion devices to focus multi beamlets

[6, 7]. The present paper proposes application of the aperture offset steering to the negative ion accelerators of JT-60U. Analytical results obtained with a three dimensional beam calculation code, OPERA-3d, are discussed to steer each beamlet precisely straight forward. At first, the beamlet deflections by the space charge repulsion are presented, and then their compensation by the aperture offset are discussed. The discussion follows to the beamlet deflection by magnetic field for electron suppression, in addition to the space charge deflection and offset aperture steering.

NUMERICAL MODEL AND APERTURE OFFSET

Figure 1 shows an illustration of a numerical model of the three stage <u>M</u>ulti <u>A</u>perture and <u>M</u>ulti <u>G</u>rid (MAMuG) accelerator simulating the JT-60U three stage D⁻ ion beam accelerator [8]. The aperture is arranged in a lattice pattern of 10 x 5 every 19 mm in X direction and 21 mm in Y direction as explained in next section, through the real aperture arrangement is 24 x 9. The negative ions are extracted by voltage difference (extraction voltage, Vext) applied between a plasma grid (PG) and an extraction grid (EXG). Permanent magnets for suppression of co-extracted electron are embedded in the EXG. This dipole magnetic field as shown in Fig.1 was 450 Gauss at the maximum on aperture axis and deflects the beamlets in ±X directions alternatively in each aperture line. An electron suppression grid (ESG) is attached at the backside of the EXG and connected to the EXG electrically. Filter field, which is generated by a plasma grid current to suppress the electron extraction and is perpendicular to the dipole magnetic field, decreases to a few tens gauss in the extractor. It was relatively smaller than the dipole magnetic field. Then, the filter field was not taken into account in this calculation. The accelerator consists of three grids, namely, first and second acceleration grids (A1G, A2G) and a grounded grid (GRG). One-thirds of the acceleration voltage (Vacc) is applied to each grid. The maximum rated Vacc is 500 keV. This accelerator was designed so that the optimum beam optics was obtained at

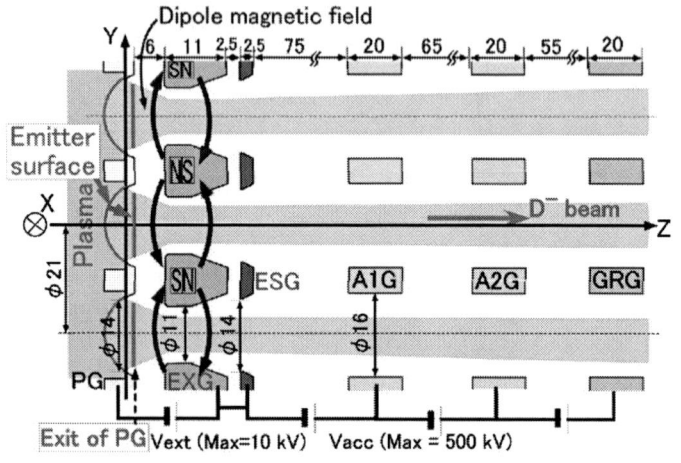

FIGURE 1. A numerical model of the three stage MAMuG accelerator. In long acceleration gaps, scales of length are made shorten in the figure.

the D⁻ ion current density (J_D⁻) of 200 A/m², at Vext = 8 kV and Vacc = 500 kV [8]. At a typical operation condition of Vacc = 340 kV, maintaining the optimum perveance (P = J_D⁻/Vacc^1.5), these parameters are J_D⁻ = 110 A/m² and Vext = 5.4 kV.

To compensate the beamlet deflection, the aperture offset in the ESG is examined in this paper. The steering angle by the aperture offset in the three stage negative ion accelerator was estimated according to the thin lens theory [4]. The focal points can be expressed as,

$$F = \frac{4V_{BE}}{E_2 - E_1}, \tag{1}$$

where V_{BE}, E_2 and E_1 are beam energy (keV) and electric field strength after and before the gird with aperture offset. In the aperture offset in the ESG, it was shown experimentally that E_1 could be negligible because the EXG is thick enough to be neglected E_1 as lens [9, 10]. Then, steering angle θ_0 at the ESG is expressed as,

$$\theta_0 = \frac{\delta_0}{F} = \frac{E_2}{4V_{BE}}\delta_0 = C_0\delta_0. \tag{2}$$

δ_0 is distance of aperture offset in the ESG. This deflected beamlet at the ESG is displaced from the axis of apertures in the accelerator even without aperture offset. Considering the beamlet steering by the substantial aperture offsets at A1G, A2G and GRG, the beamlet steering angle at the exit of accelerator, θ, is shown as follows.

$$\theta = C(\alpha)\delta_0 = 8.73\delta_0, \tag{3}$$

where, $C(\alpha)$ is a constant value as a function of $\alpha = \frac{Vext}{Vacc}$. Thus the steering angle by aperture offset in the ESG is independent of the beam energy when the ratio of extraction and acceleration voltage is kept constant.

THREE DIMENSIONAL BEAM ANALYSIS

Multi beamlets analysis was carried out utilizing the three dimensional beam calculation code, OPERA-3d on PC with the operating system of Windows [11]. The beam trajectories start from a fixed beam emitter position. An iterative calculation of beam trajectory and electric potential generated by space charge of beams are continued till obtaining the convergence.

In order to define the beam emitter surface of negative ions, a two dimensional beam calculation code, BEAMORBT [12], which was developed for positive ion beam calculation, was used. Then the emitter of the position ions, which was decided by the

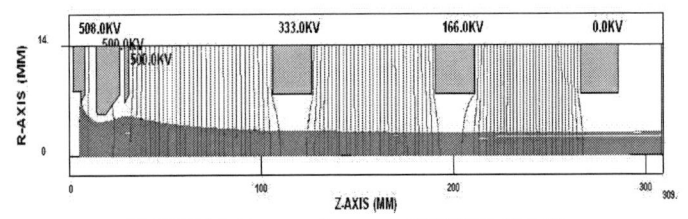

FIGURE 2. Beam trajectory in 2D BEAMORBT.

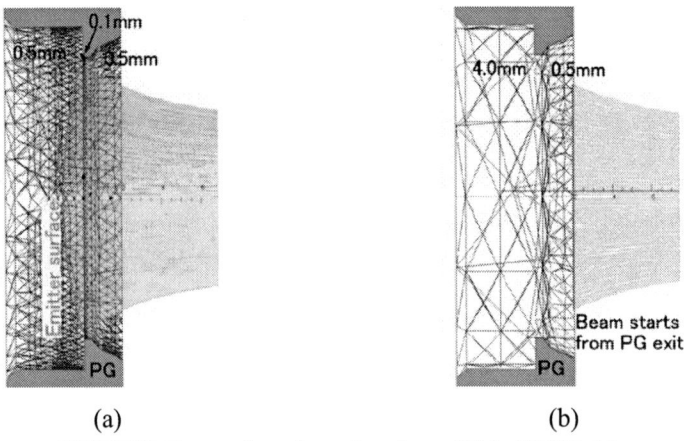

FIGURE 3. Beam trajectories and mesh near PG in OPERA-3d.
Beam particles start from (a) emitter surface and (b) plasma grid exit,
Where initial particle position and velocity were transferred from Fig. 2.

Child-Langmuir law, was utilized for the negative ion extraction, simply assuming that the negative ions are extracted from the emitter of the positive ions with same shape and position. Accordingly, the beam halo, which was formed by low dens ions extracted near aperture edge, could not be taken into account. The beam optics obtained in BEAMORBT could represent that of the beam core and showed good agreements with experimentally measured ones [13]. Then, BEAMORBIT has been used in the design study of negative ion accelerators in MTF [14], JT-60U [8] and ITER [15]. Figure 2 shows the beam trajectory in BEAMORBT at Vacc = 500 kV, 200 A/m^2 D$^+$ beam.

Figure 3 shows beam trajectory and mesh around the beam emitters in OPERA-3d. When the emitter surface in BEAMORBT was directly transferred to OPERA-3d as shown in Fig.3 (a), the emitter surface was close to the PG surface. Then, the mesh size around PG had to be subdivided into 0.1 mm to calculate accurately the electric field near the grid. Consequently, huge amount of memory was required. The number of beamlets was limited to only five even thought the memory limitation for one

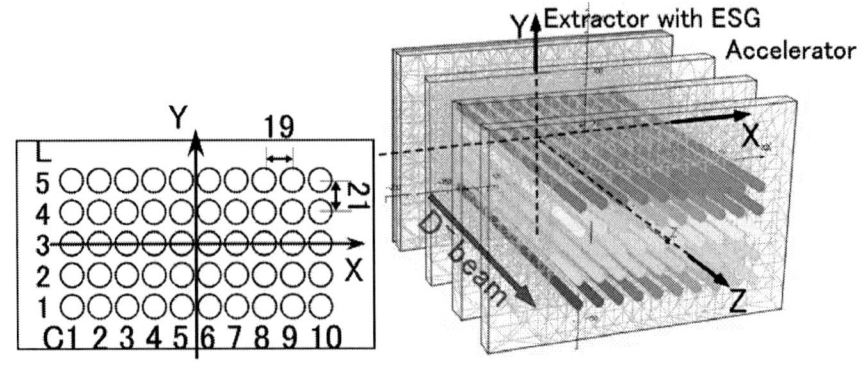

FIGURE 4. Fifty beamlets in the 3D beam calculation.
Beamlets were extracted from apertures in a lattice pattern of 10 x 5.

application program in Windows OS was 3 GB. As a solution to increase the number of beamlets, the beam emitter was changed to PG exit maintaining the particle position in BEAMORBT as shown in Fig.3 (b). At the PG exit, the beamlets were far from the grid and the mesh size was made relatively rough, 0.5 mm, maintaining the accuracy. Finally, the number of beamlets has been increased from five to fifty beamlets consisting of 400 particles in each.

Figure 4 shows fifty beamlets of 340 keV, 110 A/m^2 D$^-$ ion. In this study, apertures to calculate fifty beamlets were arranged in a lattice pattern of 10 x 5. The aperture positions are numbered by column, C1 – C10 and line, L1 – L5.

RESULTS

Beamlet Deflection by Space Charge Repulsion

Beamlet deflections and beam footprints were obtained from the beam calculations as shown in Fig.4. Figure 5 shows the beam footprints of 340 keV, 110 A/m^2 D$^-$ ion beam before the aperture offset at 3.5 m downstream from the GRG, which corresponds to the position of the beam footprint measurement in the JT-60U experiment. As you can see in the figure, the beamlet center (red points) were displaced from the aperture position (white circles), indicating that the beamlet deflections were different in each aperture position. The beamlets from the middle of peripheral aperture lines/columns were most deflected outward, 6 mrad. The deflection angle of beamlets on the corner was 4 mrad. The beamlets inside the lattice pattern of apertures were deflected with the angle of 2 – 3 mrad.

Figure 6 shows the change of beamlet deflection angle from peripheral apertures of the aperture area, L8, L9 and L10 on column 10. The beamlet deflection was not significant in extraction gap and increases gradually in the first acceleration gap

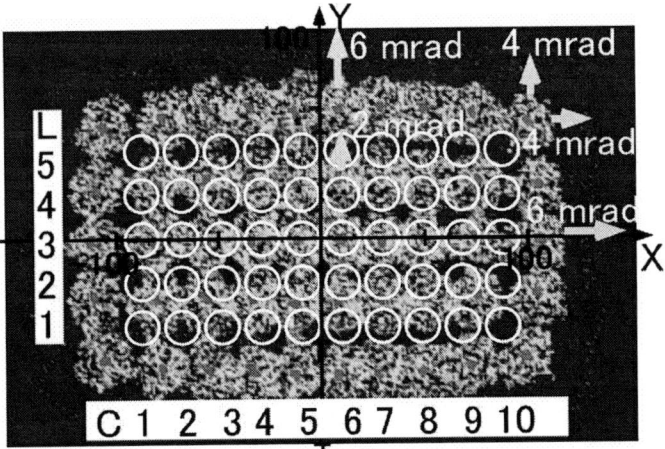

FIGURE 5. Beam footprint at 3.5 m downstream from GRG before setting aperture offset. Red points indicate the center of each beamlet at the downstream, whilst the white circles represent the original position of the apertures. The beamlets are deflected outward.

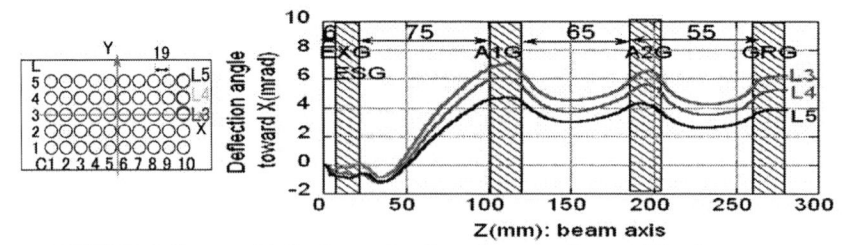

FIGURE 6. Change of deflection angle of beamlet center from peripheral apertures.

between the ESG and the A1G. It is considered that the aperture offset according to the thin lens theory should be applied to the ESG because the centers of beamlets are still close to the aperture axis.

Figure 7 shows the aperture offset required to compensate the beamlet deflection for each aperture. Proper aperture offset was calculated from Eq. 3. To compensate the beamlet deflection of 6 mrad, the necessary aperture offset was 0.7 mm.

Figure 8 shows the beam footprint after applying the aperture offset shown in Fig.7. The beamlets flied straight from the GRG to downstream and still within the projected

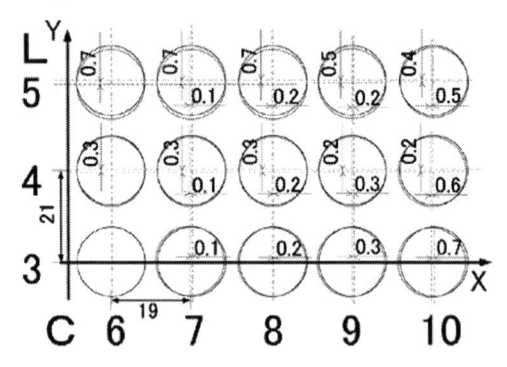

FIGURE 7. Aperture offset to compensate the beamlet deflection shown in Fig.5.

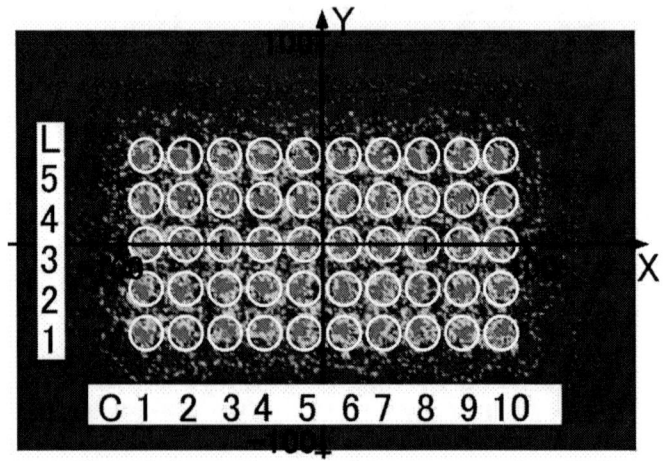

FIGURE 8. Beamlet footprint at 3.5 m downstream from GRG. This is after setting proper aperture offset as shown in Fig.7.

position of each aperture even at 3.5 m downstream from the GRG. Thus 3D simulation at present showed that the aperture offset in the ESG obtained by the thin lens theory would be effective to compensate the deflection of beamlets by space charge repulsions for all the apertures in the wide grid area.

Figure 9 shows the beamlet deflection angles of 340 keV, 110 A/m² D⁻ ion beam shown in Fig. 5 and of the 500 keV, 200 A/m² D⁻ ion beam. The perveance and the ratio of extraction and acceleration voltage in both beam conditions were the same. It is clear in the figure that the beamlet deflection angles were almost the same in each aperture position. Thus the beamlet deflection angle by the space charge repulsion did not vary with the beam energy if the beam optics (perveance and voltage ratio of extraction and acceleration) was maintained [16].

Figure 10 shows the beamlet deflection angles shown in Fig. 9 as a function of electric field formed by all beamlets' space charge as shown in Ref. [16]. Here the distance between beamlets was assumed to be the pitch of apertures. At a beamlet position, the electric fields formed by all beamlets' space charge, E_X in X and E_Y in Y, were shown in an arbitrary unit, which was normalized by the electric field formed by

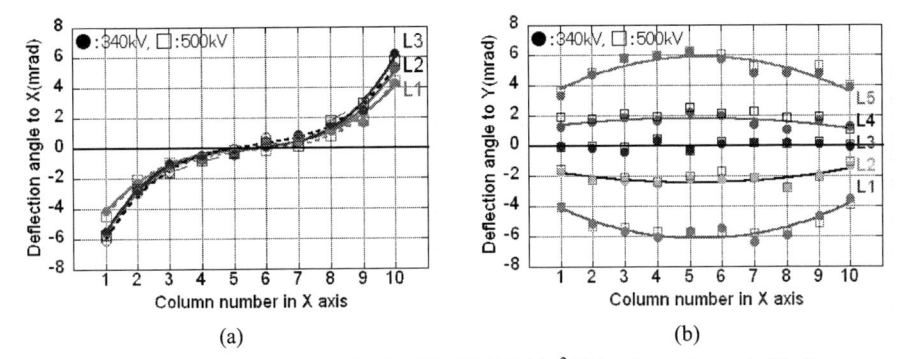

FIGURE 9. Deflection angle of 340 keV, 110 A/m² D⁻ ion beam shown in Fig.7 and 500 keV, 200 A/m² D⁻ ion beam.
The perveance is maintained in these two beam conditions.

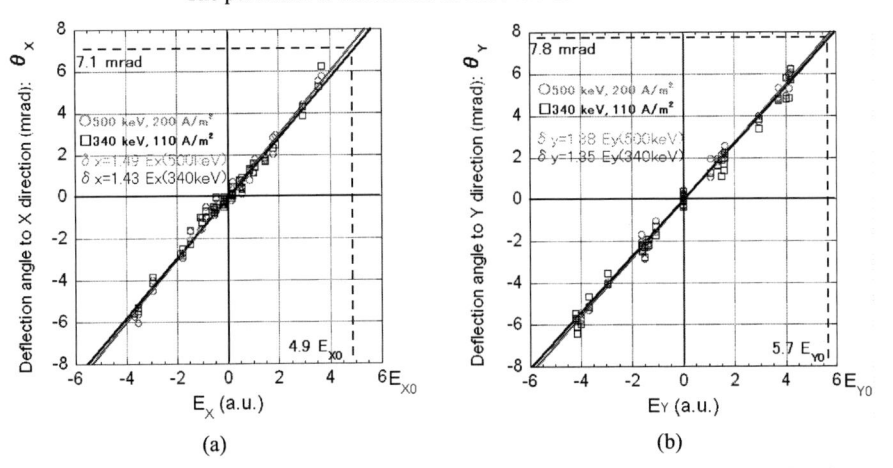

FIGURE 10. Beamlet deflection angle as a function of electric field in (a) x and (b) y directions.

space charge of the neighbor beamlet, E_{X0} and E_{Y0}, as shown in Eq. 4.

$$E_{X0} = \frac{q}{4\pi\varepsilon_0 X_0^2} = C_0 \frac{1}{X_0^2}, E_{Y0} = \frac{q}{4\pi\varepsilon_0 Y_0^2} = C_0 \frac{1}{Y_0^2}, \qquad (4)$$

where X_0 and Y_0 are the pitch of apertures in X and Y. Figure 10 shows the deflection angles was proportional to the electric field. The maximum E_X and E_Y in 10 x 5 apertures were 3.7E_{X0} at C10-L3 and 4.2E_{Y0} at C6-L5. In real aperture arrangement of JT-60U, which is 24 x 9, the maximum E_X and E_Y were 4.9E_{X0} and 5.7E_{Y0}, respectively. It was estimated in Fig.10 that the maximum deflection angles were 7.1 mrad in X and 7.8 mrad in Y. Then, proper aperture offset to compensate these beamlet deflections were estimated to be 0.8 mm in X and 0.9 mm in Y. Note that the stripping loss of negative ions will be taken into account in our future work and required aperture offset will be able to made shorter than these results.

Beamlet Deflection under Magnetic Field

Figure 11 shows the beam footprint obtained by OPERA-3d, at 3.5 m downstream from the GRG with the magnetic field in the extractor. The beam parameter was 340 keV, 110 A/m^2. The beamlets were deflected in ±X directions alternatively at each line due to the dipole magnetic field.

In Fig.12, deflection angles in X direction are shown for the model with the magnetic field. The deflection angles in +X and −X under magnetic field are shown as (2) and (3), respectively. And as a reference, deflection angle without the magnetic field of Fig.9 was also indicated as (1). The differences between before and after including magnetic field are shown as (4) (=(2)-(1)) and (5) (=(3)-(1)). The deflection angle shown as (4) and (5), that is, deflection only by the magnetic field, was about 5.4 mrad and was constant in each aperture column. When the beam energy was 500 keV, the deflection angle by the magnetic field was 4.6 mrad. Thus the deflection angle by the magnetic field was proportional to 1/(beam energy). Proper aperture offsets to compensate these were about 0.65 mm for 340 keV beam and 0.5 mm for

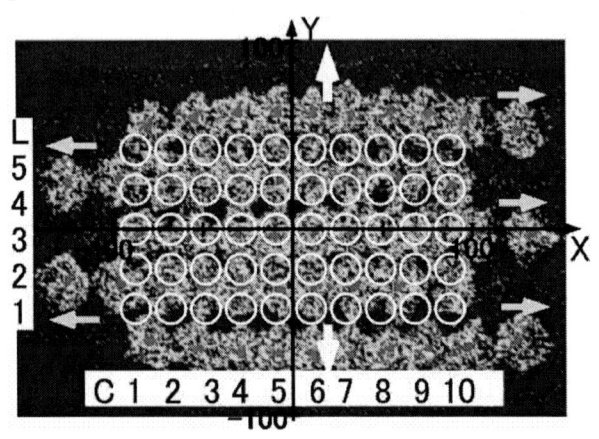

FIGURE 11. Beamlet footprint at 3.5 m downstream from GRG when the magnetic field is applied. Additional deflection from Fig.5 occurred in X. Deflection angle in Y was not changed.

500 keV. This was consistent with the previous experimental results [10], and original design [9].

Figure 13 shows deflection angles with the magnetic field as a function of beam energy. When the aperture offset was set to 0.5 mm, which was proper aperture offset for compensation of 500 keV beamlet deflection by the magnetic field, the deflection angle at 340 keV beam could be suppressed to within 1 mrad. And hence it can be concluded that the 0.5 mm offset is good enough for the present operation of the JT-60U N-NBI to compensate the beamlet deflection by the magnetic field.

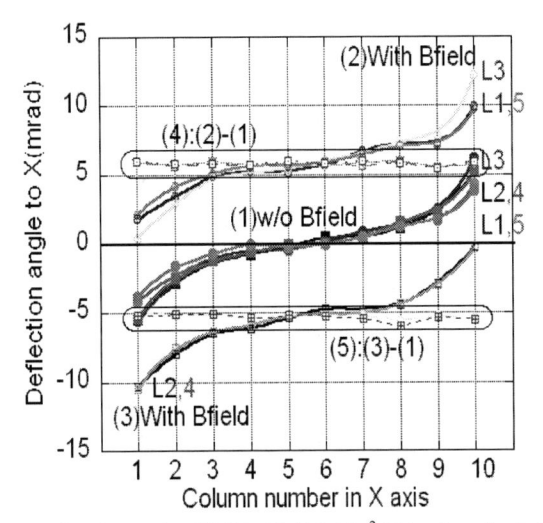

FIGURE 12. Deflection angle of 340 keV, 110 A/m^2 D$^-$ ion beamlet in X direction. Differences of deflection angle between (1) without magnetic field and (2)(3) with magnetic field were about 5.4 mrad ((4)(5)) in all beamlets.

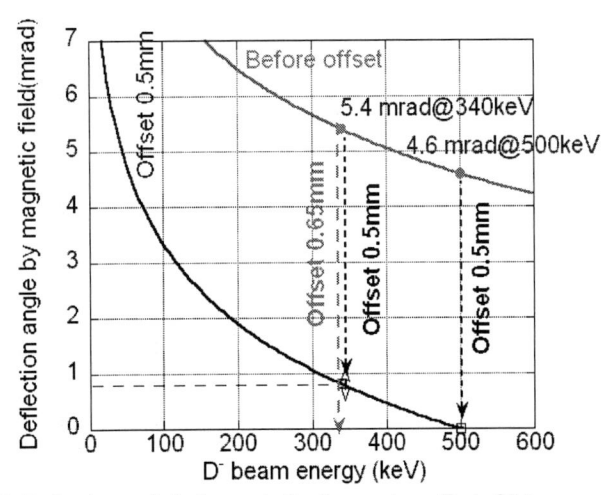

FIGURE 13. Deflection angle before and after the aperture offset of 0.5 mm under magnetic field.
This aperture offset is properly to 500 keV D$^-$ ion beam.

SUMMARY

The beamlet deflections by space charge repulsion and magnetic field were analyzed in the 3D beam analyses, and application of aperture offset steering was discussed for the compensation. The major findings are summarized as follows.

- Aperture offset in the ESG is effective to compensate all beamlet deflections by space charge repulsion.
- Beamlet deflection by space charge and the proper aperture offset steering are independent of the beam energy when the perveance is maintained.
- Aperture offset required in real aperture arrangement of 24 x 9 in the JT-60U negative ion accelerator was estimated from the results in this calculation model with the aperture arrangement of 10 x 5. The maximum aperture offset of 0.9 mm in X and 0.8 mm in Y directions are required at the center apertures in peripheral line and column.
- When the magnetic field in the extraction region was applied, it was confirmed by the present analysis that the aperture offset of 0.5 mm is enough to compensate the beamlet deflection of 500 keV D$^-$ ion beam. This is consistent with the previous experimental results and original design. Even in beam with lower energy like 340 keV, the beamlet deflection angles can be suppressed to < 1 mrad with the aperture offset of 0.5 mm.

REFERENCES

1. N. Umeda et al., Nucl. Fusion **43**, 522-526 (2003).
2. Y. Ikeda et al., Nucl. Fusion **46**, S211-S219 (2006).
3. M. Kamada et al., Rev. Sci. Instrum., **79**(2008) 02C114—02C113.
4. J. H. Whealton, Rev. Sci. Instrum., **48**(11), 1428-1429 (1977).
5. W. L. Gardner et al., Rev. Sci. Instrum., **49-8**, 1214-1215 (1978).
6. Y. Ohara, Japan. J. Appli. Physics., **18-2**, 351-356 (1979).
7. Y. Okumura et al., Rev. Sci. Instrum., **51**(4), 471-473 (1980).
8. NBI facility division and NBI heating group, JAERI-M 94-072 (1994).
9. T. Inoue et al., JAERI-Tech 2000-023 (2000).
10. T. Inoue et al., JAERI-Tech 2000-051 (2000).
11. Vector Fields Co. Ltd., http://www.vectorfields.com/.
12. Y. Ohara, JAERI-M 6757 (1976).
13. K. Watanabe et al., Proc. 13th Symp. On ISIAT '90, Tokyo, 153-156 (1990).
14. T. Inoue et al., JAERI-Tech 94-007.
15. ITER DDD.5.3.
16. Y. Fujiwara et al., Rev. Sci. Instrum., **71** (8), 3059-3064 (2000).

Purification of Radioactive Ion Beams by Photodetachment in a RF Quadrupole Ion Beam Cooler

Y. Liu[a], C. C. Havener[a], T. L. Lewis[b], A. Galindo-Uribarri[a,b], J. R. Beene[a]

[a]*Physics Division, Oak Ridge National Laboratory, Oak Ridge, TN 37831, USA*
[b]*Department of Physics and Astronomy, University of Tennessee, Knoxville, TN 37996, USA*

Abstract. A highly efficient method for suppressing isobar contaminants in negative radioactive ion beams by photodetachment is demonstrated. A laser beam having the appropriate photon energy is used to selectively neutralize the contaminants. The efficiency of photodetachment can be substantially improved when the laser-ion interaction takes place inside a radio frequency quadrupole ion cooler. In off-line experiments with ion beams of stable isotopes, more than 99.9% suppression of Co^-, S^-, and O^- ions has been demonstrated while under the identical conditions only 22% reduction in Ni^- and no reduction in Cl^- and F^- ions were observed. This technique is being developed for on-line purification of a number of interesting radioactive beams, such as ^{56}Ni, $^{17,18}F$, and $^{33,36}Cl$.

Keywords: Photodetachment, isobar suppression, buffer gas cooling; RF quadrupole ion guide.
PACS: 32.80.Gc, 41.75.Cn, 29.27.Eg

INTRODUCTION

The Holifield Radioactive Ion Beam Facility (HRIBF) at the Oak Ridge National Laboratory is an Isotope Separator On-Line (ISOL) facility, providing high-quality radioactive ion beams (RIBs) for studies of exotic nuclei and nuclear astrophysics research. The intensity and purity of the RIBs are of crucial importance. Unfortunately, the RIBs produced using the ISOL technique are often mixtures of the radioactive isotope of interest and isobaric contaminants that cannot be removed effectively by magnetic separation. Many experiments, including some of the most interesting and important, can be compromised by even a modest fraction of impurity ions in the beam. Consequently, development of additional beam purification techniques is necessary. The HRIBF tandem post-accelerator requires negatively charged ions as input. Selective removal of unwanted negative ion species by photodetachment has been suggested for applications in accelerator mass spectrometry [1,2], as well as for purification of RIBs at HRIBF [3]. In this method, a laser beam having the appropriate photon energy is used to selectively neutralize the contaminant if the electron affinity of the contaminant is lower than the electron affinity of the desired radioactive ions. However, in the pioneering work by Berkovits et al. [1,2], the overall degree of isobar suppression reported was too small to be useful.

We have proposed a novel scheme [4,5] that has been demonstrated to be highly efficient and practically suited for use at HRIBF as well as other ISOL and accelerator mass spectrometry (AMS) facilities. In this new scheme, laser-ion interaction is made inside a radio frequency quadrupole (RFQ) ion beam cooler where the ion residence time can be on the order of milliseconds. As a result, the photodetachment efficiency can be dramatically increased. In the initial proof–of-principle experiment [4], 95% suppression of ^{59}Co$^-$ ions was obtained while under identical conditions only 10% of ^{58}Ni$^-$ ions were neutralized. In this paper, we report the latest experimental results on selective suppression of S, O, and Co negative ions.

PHOTODETACHMENT TECHNIQUE

The photodetachment process

$$A^- + h\nu \rightarrow A + e \quad (1)$$

takes place when the photon energy, $h\nu$, is larger than the threshold energy, or the electron affinity (EA) of the atom A, for detaching the electron from the negative ion. Photodetachment, then, can be used to neutralize and hence remove unwanted negative ions from beams. This principle can be applied to isobar suppression if the electron affinity of the isobaric contaminant, EA_1, is lower than the electron affinity of the desired radioactive ions, EA_2, i.e., $EA_1 < EA_2$. In this case, it is possible to selectively neutralize the isobar contaminants by photodetachment with photons of energy $EA_1 < h\nu < EA_2$.

For a given laser power and photon energy, the fraction of negative ions not neutralized by the laser radiation, assuming complete overlap of the ion and laser beams, is given by

$$\frac{n}{n_0} = \exp(-t\sigma\phi) \quad (2)$$

where n_0 is the initial number of negative ions before interaction with the laser radiation, n is number of ions not neutralized by the laser, ϕ is the laser flux (photons cm^{-2}s^{-1}), σ is the photodetachment cross-section (cm^2), and t is the laser-ion interaction time (s). The cross-section for photodetachment is typically on the order of 10^{-17} cm^2. To obtain high photodetachment efficiency and thus a high degree of isobar suppression, very large photon fluxes or long interaction times are required.

Our initial motivation for exploring the photodetachment technique is the desire to purify the ^{56}Ni beam at HRIBF, which is dominated by an order of magnitude more abundant ^{56}Co contaminant. According to Eq. 2, for 99% photodetachment of ^{56}Co$^-$ by 1064 nm (1.1653 eV) laser radiation with a cross-section of 6x10^{-18} cm^2 [2], the product of laser power density and interaction time is ~0.14 (Ws/cm^2), e.g., a 100 W/cm^2 laser power density for a1.4 ms interaction time. However, for 20 keV ^{56}Co$^-$ ions, 1.4 ms implies a spatial laser-ion overlap of over 360 m.

We proposed a new technique of photodetachment in a RFQ ion cooler [4,5], which can substantially improve the efficiency of photodetachment processes, and consequently the degree of isobar suppression. As described in the following Sections, this new technique promises near 100% suppression of isobar contaminants in negative ion beams in a setup compact enough to easily implement at existing accelerator facilities.

DESCRIPTION OF THE RF QUADRUPOLE ION COOLER

A RFQ ion cooler is a gas-filled RFQ ion guide, which has been widely used as an ion beam cooler and buncher for use with radioactive ion beams [6-12]. In such an ion cooler, ions lose energy rapidly in collisions with the buffer gas molecules. The mean kinetic energy of the ions can be reduced to approximately the thermal energy of the buffer gas and the ion trajectories can be confined to a small region near the longitudinal axis of the device. Once thermalized, the ions move at low velocity through the RFQ under the influence of a small longitudinal electrostatic field gradient. Total transit times can be more than milliseconds. This combination of small transverse ion beam dimension and extended ion transit time provides ideal conditions for photodetachment.

A RFQ ion cooler for use with negative ions has been developed at HRIBF [13, 14]. A schematic view of the negative ion cooler is shown in Fig. 1. The quadrupole consists of four parallel cylindrical rods of 8-mm diameter and 40-cm length, equally spaced with an inscribed circle of radius $R_0 = 3.5$ mm. The quadrupole rod structure is mounted inside a Cu cylindrical enclosure (length: 40 cm, inner diameter: 30 mm) with a 3-mm diameter ion beam entrance aperture and a 2-mm diameter exit aperture. Buffer gas is introduced into the quadrupole assembly through an orifice located near the middle of the Cu cylinder. Negative ions are very fragile. Electron detachment (loss of the negative ion) can take place in collisions with the buffer gas if the ion energies are too high relative to the electron affinity of the ion species. In order to reduce losses due to collisional electron detachment, the ions are decelerated to less than 40 eV before being injected into the quadrupole. When the ions reach the exit aperture they will be re-accelerated to high energies.

After being cooled to near thermal energies, the ions move randomly in the axial direction and it is necessary to provide a longitudinal field to guide them out. Four DC electrodes are placed between the quadrupole rods (Fig. 2a). The DC electrodes are tapered at a very small angle with respect to the quadrupole axis, such that the distance between the electrodes changes from 10 mm at the entrance to 11 mm at the exit plane of the quadrupole. When the DC electrodes are negatively biased relative to the quadrupole electrodes, a weak longitudinal field along the axis of the quadrupole is created that gently pushes the negative ions toward the exit aperture. According to simulations conducted using the finite element code ANSYS [15], the axial potential difference across the whole quadrupole length is about 1 V per -20 V bias on the DC electrodes and the average longitudinal field increases linearly with the potential difference between the DC and RF electrodes as shown in Fig. 2b.

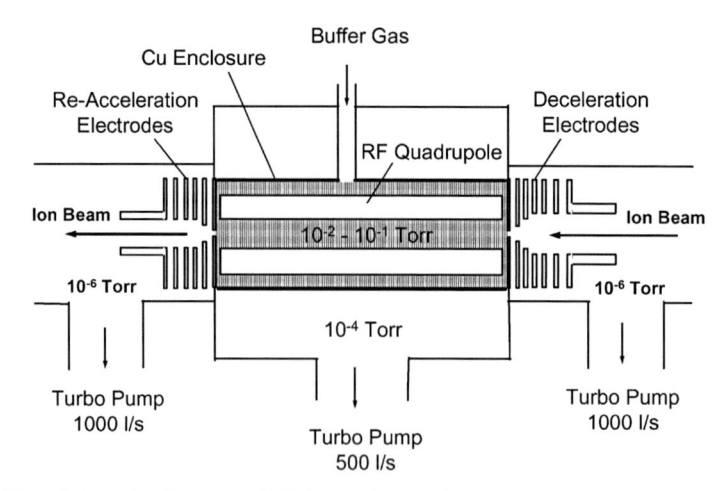

FIGURE 1. Schematic view of the RFQ ion cooler coupled with deceleration and re-acceleration electrodes. Ions enter the RFQ from the right through the 3 mm entrance aperture and exit through the 2 mm exit aperture on the left.

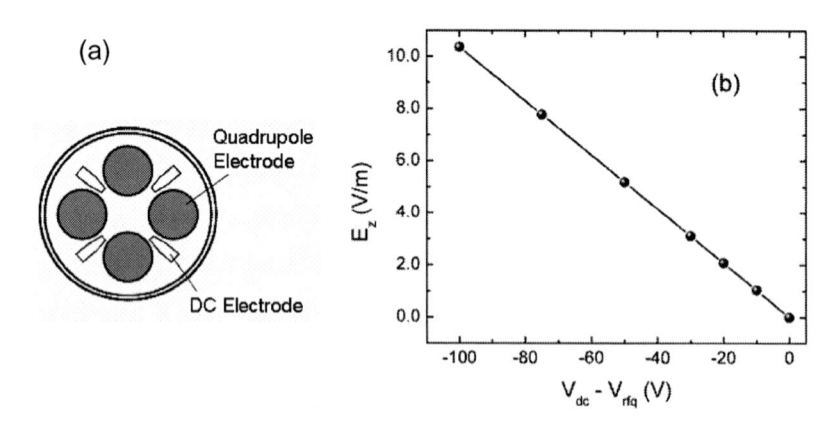

FIGURE 2. (a) Schematic view of the RF and DC electrode assembly. (b) Average axial longitudinal field created by the tapered DC electrodes as a function of the potential difference between the DC and the RF electrodes, as calculated using ANSYS.

The cooler typically operates at RF frequencies of around 2.75 MHz and RF peak voltages up to 500 V, with He as the buffer gas. High transmission for negative radioactive ion beams is required for the HRIBF research program. There are inherent difficulties in cooling and transporting negative ions through a gas-filled RF quadrupole ion guide because negative ions are much more fragile than their positive-ion counterparts. Significant improvements in its performance have been reported [14]. Table 1 presents the cooler transmission efficiencies measured for a number of negative ions.

TABLE 1. Ion Cooler Transmission Efficiency

Negative Ion	Mass (amu)	EA (eV)	Transmission (%)
O	16	1.4611	24 ± 4
F	19	3.3993	38 ± 8
S	32	2.0771	45 ± 5
Cl	35	3.6173	51 ± 8
Ni	58	1.1561	52 ± 12
Co	59	0.6611	43 ± 13
Cu	64	1.2281	52 ± 11

Ion Residence Time in the RFQ Ion Cooler

A Monte Carlo model has been developed to simulate the transport of ions through the RFQ ion cooler [13]. Simulations show that the ion residence time inside the cooler is affected by ion injection energy, buffer gas pressure inside the quadrupole and the longitudinal field strength created by the tapered DC electrodes. Fig. 3 shows the simulated average transit times for ^{56}Co$^-$ ions at different He buffer gas pressure and different longitudinal fields, with typical ion injection energy 17-40 eV. As expected, the time that an ion may spend inside the cooler increases with increasing buffer gas pressure and decreasing longitudinal drift field. The cooler typically operates with He pressures ranging from about 3 to 6 Pa and the tapered DC electrodes biased at 0-50 V more negative than the RF rods. According to Fig. 2 and 3, ion residence times of ≥ 1 ms could be easily obtained.

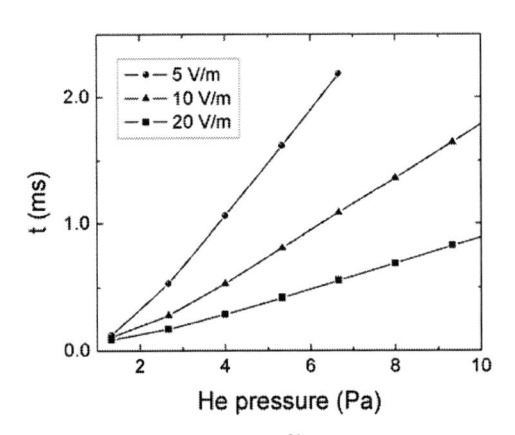

FIGURE 3. Calculated average residence time of ^{56}Co negative ions in the RFQ ion cooler as a function of He buffer gas pressure. The ^{56}Co ions have initial energies of 17-40 eV.

EXPERIMENT

The feasibility of photodetachment in a RFQ ion cooler has been investigated using negative ions of stable isotopes at the off-line Ion Source Test Facility I (ISTF-1) of

HRIBF. Our goal is to study collisional cooling and photodetachment processes in the ion cooler, optimize the cooler design and the optical system for laser beam transportation and coupling, and eventually implement this technique at the HRIBF for purifying isobaric contaminants in RIBs.

There are a number of interesting radioactive beams, such as ^{17,18}F, ^{33}Cl, and ^{56}Ni, needed for studies in nuclear structure and nuclear astrophysics at HRIBF. However, these beams are often dominated by an isobaric contaminant, ^{17,18}O$^-$, ^{33}S$^-$, and ^{56}Co$^-$, respectively, resulting in low-purity radioactive beams not desired for experiments. The electron affinities of these negative ions are F (3.3993 eV), O (1.4611 eV), Cl (3.6173 eV), S (2.0771 eV), Ni (1.1561 eV) and Co (0.6611 eV). In each case, a laser beam having the appropriate photon energy can be used to selectively neutralize the isobar contaminant. Our effort has been focused on developing the technique for these RIBs.

A detailed description of the experimental apparatus has been given previously [4,5]. Briefly, negative ions were produced with a Cs-sputter negative ion source. After acceleration to about 5 keV, the negative ions were separated in mass by a 90° dipole magnetic and the mass-selected ions were sent into the RFQ ion cooler. The ion current before the cooler was measured with a Faraday cup. Ions emerging from the cooler were deflected by an electrostatic deflector to an off-axis Faraday cup detector. For photodetachment, a laser beam was sent into the cooler through the 2 mm exit aperture, traveling collinearly with but in the opposite direction of the ion beam propagation. The laser beam intensity was measured before entering the experimental chamber and after passing through the ion cooler and emerging from a vacuum window at the mass-separation magnet. The interaction of the laser with the negative ions was studied by monitoring the changes in the ion current obtained at the off-axis Faraday cup when the laser beam was on.

RESULTS

Photodetachment in the RFQ Ion Cooler

A pulsed Nd:YLF laser with frequency-doubled output at 527 nm was used for photodetachment of S and O. The photon energy, $h\nu = 2.3526$ eV, is larger then the electron affinity of S$^-$ and O$^-$ but smaller than that of F$^-$ and Cl$^-$. Therefore, S$^-$ and O$^-$ ions were expected to be strongly neutralized by the 2.3526 eV photons with Cl$^-$ and F$^-$ unaffected. The laser provided short photon pulses of 10 – 30 ns at variable rates of 1 Hz – 10 kHz with maximum average power of about 3 W at 3 kHz.

A set of typical data are illustrated in Fig. 4. Fig. 4a shows the fraction of ^{32}S$^-$ ions not neutralized by the laser beam versus laser power obtained at different He gas pressures, plotted in logarithmic scale. The cooler was operated at RF frequency of 2.75 MHz with peak RF amplitude of $V_{rf} \sim 150$ V. The ^{32}S$^-$ ions were decelerated to about 20 eV when entering the quadrupole and the bias voltage on the tapered DC electrodes was optimized for ion transmission. The fraction of the surviving ions was determined as the ratio of the ion current measured at the off-axis Faraday cup when the laser was on to that with the laser off. It can be seen that the ratios, n/n_0, became saturated at gas pressures < 5.7 Pa. One possible explanation is that at relatively low He pressures, the ion trajectories in the radial direction were not sufficiently damped and the spatial distribution of the ions was larger than the laser beam size. When the He pressure was increased to 5.7 Pa, the surviving ion fraction decreased exponentially with increasing laser power, as given by Eq. 2. A factor of 1000 reduction of the $^{32}S^-$ ions was obtained with only about 1 W laser power. This may indicate that the ions were sufficiently cooled and their spatial distribution in the radial direction was reduced to be comparable with or smaller than the laser beam size.

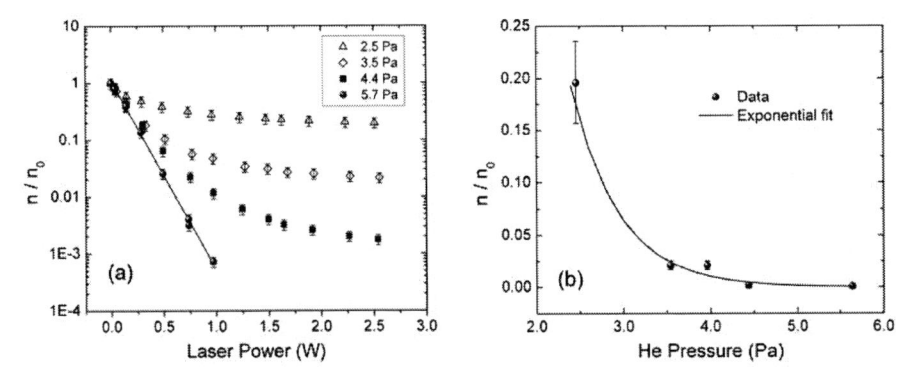

FIGURE 4. (a) The fraction of surviving $^{32}S^-$ ions versus laser power measured at a different He pressures. The efficiency of photodetachment was saturated at lower He pressures. (b) The fraction of surviving ^{32}S negative ions versus He buffer gas pressure at a fixed laser power of 2.5 W. The data fit reasonably well to an exponential decay function.

For a fixed laser power, the photodetachment efficiency also increased exponentially with increasing buffer gas pressure, as shown in Fig. 4b. This implies that the ion residence time in the cooler increased linearly with increasing buffer gas pressure, in agreement with the simulation results (Fig. 3).

The laser pulse rate was variable from 1 Hz to 15 kHz. It was found that more S^- ions were depleted at higher laser repetition rates, even though the average laser power was about the same. Fig. 5 shows the ratio of n/n_0 as a function of the laser repetition rate, obtained at laser power of 2.5 W and two different He pressures. The DC electrodes were biased at about -40 V relative to the RF electrodes, which corresponds to a longitudinal field of $E_z \sim 4$ V/m. In both cases, the fraction of surviving $^{32}S^-$ ions decreased with increasing laser repetition rate but reached a saturation level above about 10 kHz. The results indicate that at low pulse repetition rates, e.g., less than 3 kHz, the laser pulse duty cycle was too low that a substantial fraction of the ions could pass the ion cooler without interaction with the photons. Increasing the pulse repetition rate increased the laser pulse duty cycle and thus the probability for more ions to interact with the photons, resulting in higher photodetachment efficiency. Above 10 kHz, most of the ions encountered at least one laser pulse during their passes through the cooler. Thus, the overall photodetachment efficiency was saturated.

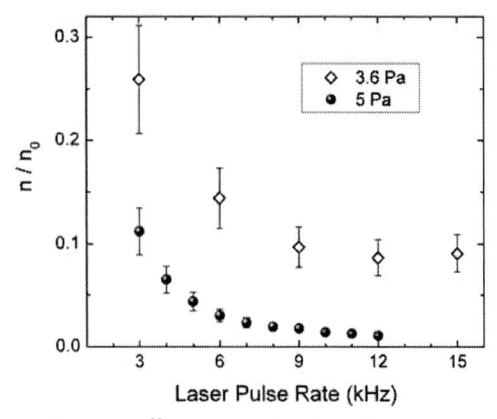

FIGURE 5. The fraction of surviving $^{32}S^-$ ions as a function of the laser pulse rate, obtained at He pressure of 3.6 Pa and 5Pa. The laser power was fixed at 2.5 W.

The efficiency of photodetachment was also affected by the longitudinal field created by the tapered DC electrodes. Fig. 6 shows the fraction of $^{32}S^-$ not neutralized as a function of the laser power, obtained at two different bias potentials on the tapered DC electrodes, $\Delta V = V_{dc} - V_{rfq}$. A larger negative value of ΔV corresponds to a larger longitudinal field that pushes the ions toward exit (Fig. 2b), and in turn a shorter transit time for the ions. Therefore, smaller photodetachment efficiency is expected for a larger negative bias potential on the DC electrodes. This was experimentally observed. The measurements were made at about 6 Pa He pressure in the quadrupole. Again, the data could fit well to an exponential decay and 99.9% of the $^{32}S^-$ ions were depleted with only about 1 W laser when $\Delta V = -10$ V.

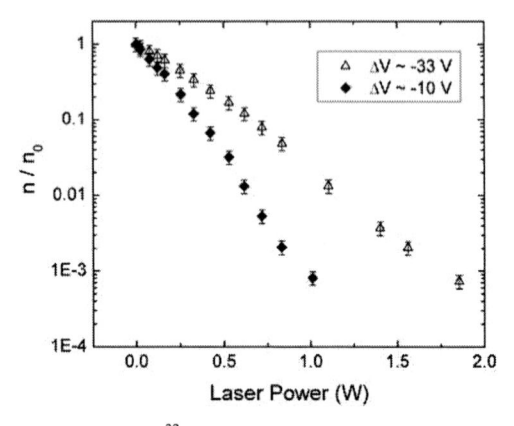

FIGURE 6. The fraction of surviving $^{32}S^-$ ions as a function of laser power, obtained at two DC electrode bias potentials. As expected, larger bias, thus larger longitudinal field, resulted in lower photodetachment efficiency.

The effective interaction time between the laser and $^{32}S^-$ ions under different cooler operation conditions was estimated by fitting the corresponding photodetachment data, n/n_0 versus laser power, to Eq. 2. A photodetachment cross section of 1.0×10^{-17} cm^2 for S^- [1] was used for the calculations. The preliminary results, calculated using a simple assumption of an effective laser beam size of 2 mm in diameter, are plotted in Fig. 7. These data were obtained at DC electrode bias $\Delta V = -3$ V, which corresponded to a very small longitudinal field. The estimated interaction times are between 4 to 8 ms, increasing linearly with increasing He pressure, as expected by simulations (Fig. 3). Further analysis will take into account the photon and ion spatial distributions.

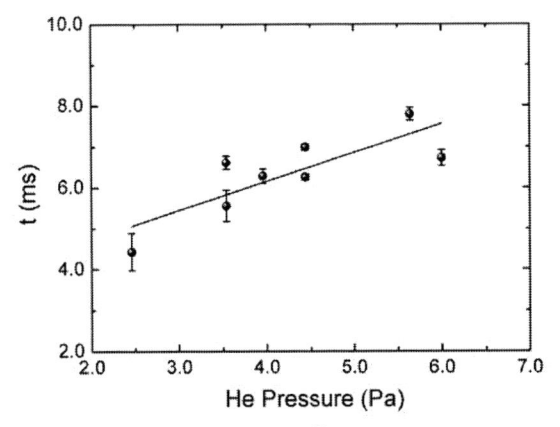

FIGURE 7. The effective laser-ion interaction time for $^{32}S^-$ ions, calculated by fitting the n/n_0 versus laser power data to Eq. 2. The RFQ ion cooler was operated with $\Delta V = -3$ V bias on the DC electrodes.

Selective Suppression of the (S, Cl) and (O, F) Pairs

Near 100% photodetachment efficiency was obtained at ~ 6 Pa He pressure. Fig. 8 shows a temporal recording of the $^{32}S^-$ and $^{35}Cl^-$ current measured at the off-axis Faraday cup. When the 527 nm laser beam of 2.5 W was turned on, the $^{32}S^-$ current dropped immediately from about 6.3 nA to 2 pA, which was the baseline of our data acquisition system. This corresponded to a photodetachment efficiency of about 99.97%, or a factor of 3000 depletion of the $^{32}S^-$ ions. When the laser beam was blocked, the ion current quickly returned to the initial values. The photo-depletion process was fast and reproducible as shown by the data. Under the same conditions, no depletion of $^{35}Cl^-$ beams was observed. The photodetachment efficiency was also measured without the use of the RFQ ion cooler. In this case, the laser beam interacted with 5 keV $^{32}S^-$ beam and only about 1% $^{32}S^-$ ions were depleted by 2.5 W laser power.

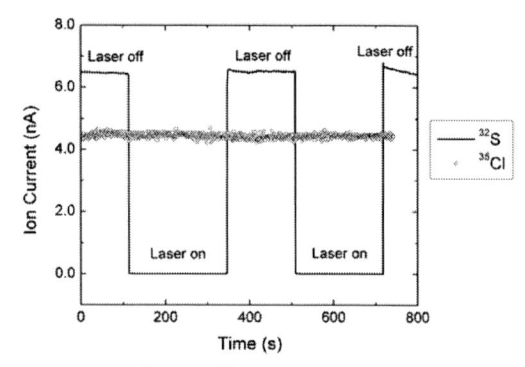

FIGURE 8. Measured intensities of $^{32}S^-$ and $^{35}Cl^-$ ion beams accelerated from the RFQ ion cooler. 99.97% of $^{32}S^-$ ions were neutralized by 2.5 W laser ($\lambda = 527$ nm), while the $^{35}Cl^-$ were not affected. He pressure 6 Pa, $f = 2.76$ MHz, $V_{rf} \sim 200$ V, DC electrode bias $\Delta V = -10$ V.

Similar results were obtained for O^- and F^- ion pairs. As shown in Fig. 9, 99.9% of the $^{16}O^-$ ion beam was depleted by ~2.5 W laser power (Fig. 9a), while no neutralization of the $^{19}F^-$ beams was observed (Fig. 9b).

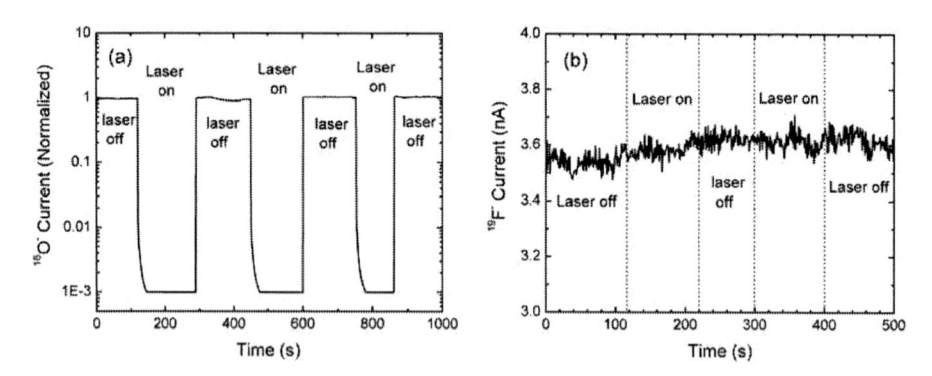

FIGURE 9. (a) $^{16}O^-$ and (b) $^{19}F^-$ currents measured at the off-axis Faraday cup. The $^{16}O^-$ current shown was normalized to the ion current when the laser beam was off. 99.9% of the $^{16}O^-$ ions were neutralized by 2.5 W laser ($\lambda = 527$ nm). No reduction in the $^{19}F^-$ beam was observed. He pressure ~6 Pa, $f = 2.76$ MHz, $V_{rf} \sim 75$ V, DC electrodes bias $\Delta V = -5$ V.

Photodetachment of the (Co, Ni) Pair

A continuous wave (CW) Nd:YAG laser at 1064 nm ($h\nu = 1.1653$ eV) was used to selectively remove the Co^-. The photon energy was well above the electron affinity of Co^- (0.6611 eV), but also slightly larger than the electron affinity of Ni^- (1.1561 eV). Therefore, some neutralization of Ni^- ions was expected. Figure 10 shows the $^{59}Co^-$ and $^{58}Ni^-$ currents measured by the off-axis Faraday cup when the RFQ ion cooler was operated at a He pressure of ~ 10 Pa and the laser beam was turned on and off. About 99.9% photodetachment efficiency for $^{59}Co^-$ was achieved with about 4 W laser power measured at the output of the laser while under the same conditions, about 22%

depletion of Ni⁻ was observed. When the ion cooler was off and the ions were moving at about 5 keV energy, only about 3% depletion in the ⁵⁹Co⁻ current, and no change in the ⁵⁸Ni⁻ current, was observed with 4 W laser radiation. .

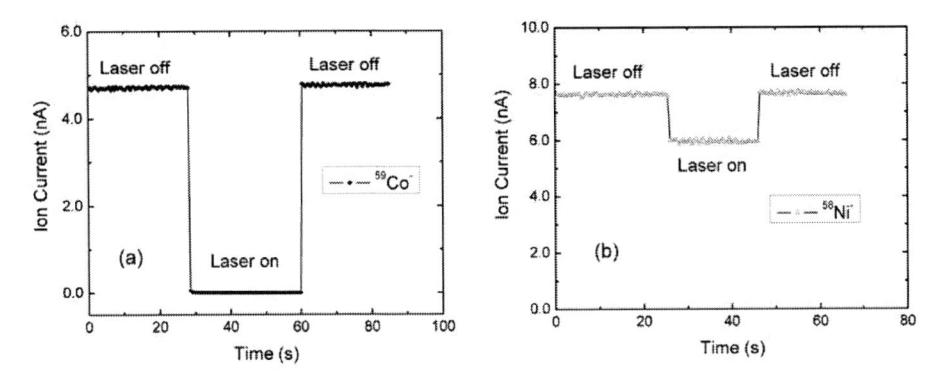

FIGURE 10. Measured intensities of (a) ⁵⁹Co⁻ and (b) ⁵⁸Ni⁻ negative ion beams through the RFQ cooler at a He pressure of ~10 Pa. 99.9% of the ⁵⁹Co⁻ and 22% of the ⁵⁸Ni⁻ ions were neutralized with a CW laser beam of 4 W. RF = 2.75 MHz, V_{rf} ~ 300 V, and DC electrode bias $\Delta V = -40$ V.

CONCLUSION

The feasibility of efficient isobar suppression by photodetachment in a gas-filled RFQ ion guide has been experimentally demonstrated with ion beams of stable isotopes. Laser beams having the appropriate photon energy were used to selectively neutralize the contaminants if the electron affinity of the contaminants was lower than the electron affinity of the desired radioactive ions. Simulation and experimental results showed that the photodetachment efficiency could be dramatically increased when the laser-ion interaction took place inside a RF quadrupole ion beam cooler where the laser-ion interaction time was on the order of milliseconds. For the selected negative ion pairs of (S, Cl) and (O, F), 99.97% of ³²S⁻ and 99.9% of ¹⁶O⁻ ions were depleted with only about 2.5 W average laser radiation of 527 nm photons from a pulsed laser, while the desired Cl⁻ and F⁻ ions were not affected. In the case of the (Co, Ni) pair, 99.9% suppression of ⁵⁹Co⁻ ions was demonstrated with a CW Nd:YAG laser beam of 4W power at 1064 nm, with only about 22% of ⁵⁸Ni⁻ ions suppressed. The depletion of Ni⁻ ions can be avoided by using photons of energy less than the electron affinity of Ni. The laser power used to achieve nearly 100% photodetachment is readily available from existing commercial lasers and the setup is compact enough to easily implement at an existing accelerator facility. Therefore, this technique is very promising for real applications, such as purifying isobaric contaminants in radioactive ion beams for nuclear research or accelerator mass spectrometry. It is being developed for use at the HRIBF for on-line purification of a number of interesting radioactive beams, such as ⁵⁶Ni, ¹⁷,¹⁸F, and ³³Cl. Applications in accelerator mass spectrometry to suppress the ³⁶S contaminants in ³⁶Cl beams will also be explored.

ACKNOWLEDGMENTS

This research has been sponsored by the U.S. Department of Energy, under contract DE-AC05-00OR22725 with UT-Battelle, LLC. CCH was supported by the Office of Fusion Energy Sciences and the Office of Basic Energy Sciences of the U.S. Department of Energy.

REFERENCES

1. D. Berkovits, et al., Nucl. Instr. and Meth. A 281 (1989) 663-666.
2. D. Berkovits, et al., Nucl. Instr. and Meth. B 52 (1990) 378-383.
3. G. D. Alton (private communication, 2003).
4. Y. Liu, J. R. Beene, C. C. Havener and J. F. Liang, APL 87, 113504 (2005).
5. C. C. Havener, Y. Liu, J. F. Liang, H. Wollnik and J. R. Beene, AIP Conference Proceedings, Vol. 925, pp. 346-357 (2007).
6. M.D. Lunney and R.B. Moore, Int. J. Mass. Spectrom. 190/191 (1999) 153-160.
7. F. Herfurth et al., Nucl. Instr. and Meth. A 469, 254-275 (2001).
8. A. Nieminen, J. Huikari, A. Jokinen, J. Äysto, P.Campbell, E.C.A. Cochrane, Nucl. Instr. and Meth. A 469 (2001) 244-253.
9. J. Äysto, A. Jokinen and the EXOTRAPS Collaboration, J. Phy. B: At. Mol. Opt. Phys. 36 (2003) 573-584.
10. A. Kellerbauer, T. Kim. R.B. Moore, P. Varfalvy, Nucl. Instr. and Meth. A 469 (2001) 276-285.
11. S. Schwarz, G. Bollen, D. Lawton, A. Neudert, R. Ringle, P. Schury, T.Sun, Nucl. Instr. and Meth. B 204 (2003) 274-277.
12. G. Darius, et al., Rev. Sci. Instrum. 75 (2004) 4804-4810.
13. Y. Liu, J. Liang, G. D. Alton, J. R. Beene, Z. Zhou and H. Wollnik, Nucl. Instr. and Meth. B 187, 117-131 (2002).
14. Y. Liu, J. F. Liang and J. R. Beene, Nucl. Instr. and Meth. B 255, 416-422 (2007).
15. http://www.ansys.com/.

Characteristics of a He⁻ Beam Produced in Lithium Vapor

N. Tanaka[a], T. Nagamura[a], M. Kikuchi[a], A. Okamoto[a], T. Kobuchi[a], S. Kitajima[a], M. Sasao[a], H. Yamaoka[b], and M. Wada[c]

[a] *Tohoku University, Sendai 980-8579, Japan*
[b] *Harima Institute, RIKEN (The Institute of Physical and Chemical Research), Hyogo 679-5148, Japan*
[c] *Doshisha University, Kyotanabe, Kyoto 610-0321, Japan*

Abstract. A test stand for the proof of principle experiments of He⁰ beam production has been constructed aiming at development of an alpha particle diagnostic system in next generation fusion reactors. The characteristics of a He⁻ beam produced in Lithium vapor has been studied. With an increase in the vapor pressure in the cell, mitigation of space charge in a He⁺ beam and bunching of a He⁻ beam which travels part of the way with the He⁺ beam have been observed.

Keywords: Alpha particle diagnostics, Negative ion beam, Space charge effect
PACS: 52.70.Nc, 52.55.Pi, 29.25.Ni

INTRODUCTION

The confined alpha particle diagnostics in next generation magnetically confined fusion reactors such as ITER is one of the most critical issues in maintaining self-burning plasmas. The use of an energetic neutral helium beam (He⁰) to measure the energy and spatial distributions of alpha particles (He⁺⁺) produced by DT reactions has been proposed [1, 2]. The production of a fast He⁰ beam is therefore of primary importance. The He⁰ beam should be produced from an accelerated He⁻ beam by electron detachment, due to the higher neutralization efficiency compared to that from He⁺. A charge exchange cell (CX cell) filled with an alkali metal vapor is used to generate the He⁻ beam from a He⁺ beam because the He⁻ beam is only produced by the charge exchange process, not by surface or by volume production [3].

There are several critical issues in the production of an intense He⁰ beam. (1) The beam transport: it should be good enough to enable the beams to pass through the small entrance and exit apertures of the charge exchange cell. In addition, the He⁻ beam size should be small enough at the entrance of the accelerator (2) Optimization of the He⁻ production: the liner-integrated density of the alkali metal vapor should be at the optimum. (3) Efficient neutralization to a ground state He⁰: We propose a time of flight (TOF) neutralization using the 10 and 300 microsecond life times of He⁻ [2].

We have been developing a test stand Advanced Beam Source 103 (ABS103) using Lithium as the alkali metal in the CX cell to confirm the feasibility of the He⁰ production method for the real fusion reactor, ITER and studied the beam transportation mechanisms in the system [4-6]. Performance of a compact He⁺ ion source was studied in a separate device before installing it to the test stand, and dependence of the extracted ion current upon the plasma parameters, and that upon the extraction voltage indicated that the source performance was limited by space charge effect [7]. The energy straggling in the CX cell might increase the beam expansion downstream of the bending magnet. It was estimated ~13 eV in case of He$^+$ ions and Lithium atoms, and is within the tolerable level for the beam design [8]. The beam expansion by the multiple scattering was also estimated to be negligible.

The remaining factor that influences the beam transmission is space charge effect. The simulation of the beam transport of ABS103 showed that the He$^+$ beam was transported effectively when the space charge effect was not included in the calculations, but many beam particles were cut off at the exit aperture of the CX cell because of the beam expansion when the space charge effect was included.

Considering these results, we activated the Lithium cell and measured the beam properties of the He$^+$ and He$^-$ beams depending on the Lithium temperature at the entrance of the accelerator to study the space charge effect when the He$^-$ beam is produced.

EXPERIMENTS

The schematic diagram of ABS103 is shown in Figure 1. The ion source is a compact bucket-type ion source and the ion beam is extracted in a DC mode. The bending magnet is a 90-degree double focusing magnet. It enables the beam to have two focal points at the center of the CX cell and the entrance of the accelerator [4]. We employed Lithium as the alkali metal vapor for safety reasons even though the maximum charge exchange efficiency of He$^+$ to He$^-$ is only ~0.5% [9]. The optimum beam energy for He$^-$ production is 10 – 15 keV.

The CX cell consists of a melting pot, orifices, and heaters (Fig. 2). Lithium bulks are contained in the melting pot. The sizes of the entrance and exit orifices are 13 mm and 16 mm respectively. Lithium vapor in the cold area of the baffles condenses and returns to the melting pot. A small fraction of the vapor leaks out of the CX cell, but the alkali metal can be recycled. The heaters on the baffle prevent the solidification of Lithium on the apertures [10]. The vapor pressure of Lithium is;

$$\log_{10} P = 8.012 - \frac{8172}{T} \quad (1)$$

where P is the pressure in millimeter mercury (mmHg) and T is in thermodynamic temperature (K) [11]. It is controlled within 1K by a feedback system combined with thermocouples installed at 2 locations of the cell. The linear integrated density inside the CX cell is estimated to be ~5 x 10^{15} (atoms/cm^2) when we estimate a uniform Lithium vapor of 10 cm at 873 K.

The He$^+$ and He$^-$ beams were scanned vertically by a Faraday cup with a 1 mm slit aperture at ~10 mm before the entrance of the accelerator. The two beams were measured independently by changing polarity of the bending magnet.

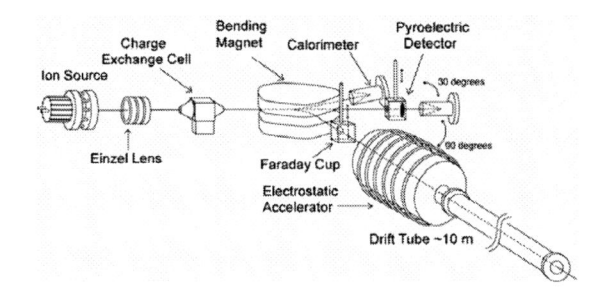

FIGURE 1. Schematic view of ABS103. The ion source is a compact bucket-type ion source. The bending magnet is a 90-degree double focusing magnet.

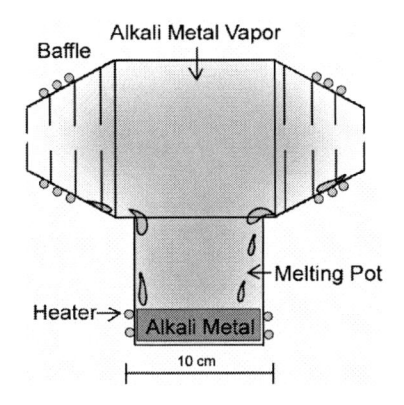

FIGURE 2. Schematic diagram of the Lithium cell. Lithium bulks are contained in the melting pot. The alkali metals can be recycled.

RESULTS AND DISUSSIONS

Figure 3 shows the Lithium temperature dependence of the beam profiles of He$^+$ and He$^-$ beams at I_{arc} =1.0 A, V_{acc} =10 kV, T_{Li}= 773 K - 893 K. The plots in Fig. 3(b), respectively, are each fitted using a Gaussian distribution. The mechanical center was at 42 mm. The He$^+$ beam width was broad and the edge of the beam was cut off in Fig. 3 (a), when the temperature was lower than 793 K. As the cell temperature went up, the width became narrower. On the other hand, the He$^-$ beam spectra showed sharp Gaussian distributions (Fig. 3(b)), and the width increased slightly as the cell temperature went up over the optimum. Figure 4 shows the integrated beam current intensities of Fig. 3 [12]. The current of He$^+$ beam decreases drastically, and that of He$^-$ beam has a maximum value of 0.9 µA at 873 K.

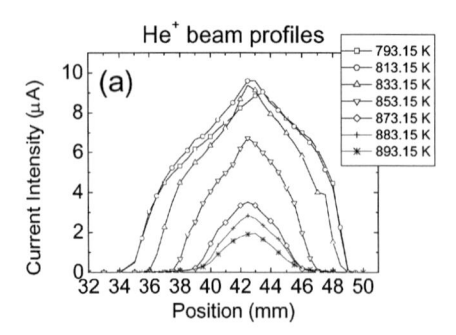

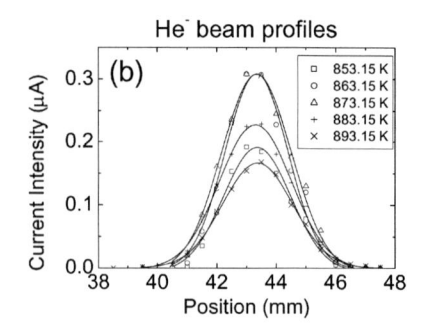

FIGURE 3. Beam profiles of (a) He$^+$ (b) He$^-$ Beam current at each vertical position was detected by the Faraday cup with a 1 mm slit

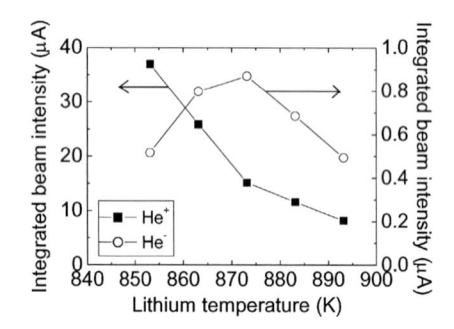

FIGURE 4. Integrated beam current intensity of fig. 3. The optimum Lithium temperature is 873 K.

The half widths at 1/e maximum (w) of the beams are shown in Figure 5 [12]. The w of the He$^+$ beam is ~7 mm at 793 K and it decreases to ~2 mm as the temperature increases. However that of He$^-$ beam does not change much. The half widths of the beams converge to ~2 mm showing the improvement of beam transmission.

The space charge effect on the He$^+$ beam has been studied using a simulation code SIMION [13]. Figure 6(a) and (b) show the calculation results of the He$^+$ beam transportation not considering and considering the space charge effect, respectively [5]. When the space charge effect is not included in the calculations (Fig. 6(a)), the He$^+$ beam is transported effectively. However when the beam is fully affected by the space charge, a substantial fraction of beam particles are cut off at the exit aperture of the CX cell and at the bending magnet chamber that the vertical width is 20 mm as shown in Fig. 6 (b) because of the beam expansion. The real situations are in between the two cases.

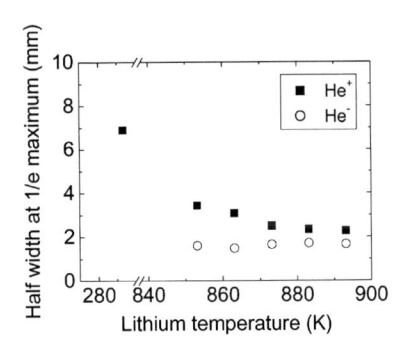

FIGURE 5. The half widths at 1/e maximum. The width of the He⁺ beam decreases as temperature increase while that of He- does not increase so much. The values converge to ~2 mm.

When the cell is activated, the space charge of the beam might be compensated by electrons/ions produced by ionization in the cell (region A in Fig. 6(a)). Downstream of the cell (region B in Fig. 6(a)), both He⁺ / He⁻ beams may undergo the same electric field, E_r. In the ideal case of a laminar flow and a uniform current distribution, the electric field at the beam radius, a, may be expressed as:

$$E_r = \frac{I_t(\eta_+ - \eta_-)}{2\pi\varepsilon_0 \cdot a \cdot v} \quad (2)$$

$$\eta_+ + \eta_0 + \eta_- = 1 \quad (3)$$

where the I_t is the total beam current, v the beam velocity, and $I_t\eta_+$, $I_t\eta_0$, $I_t\eta_-$ are He⁺ / He⁰ / He⁻ beam current, respectively. As the Lithium vapor pressure increases, the E_r, which acts as repulsion force to He⁺ beam and bunching force to He⁻ beam, decreases in the region B, and the hence the He⁺ beam size decreases and He⁻ beam size increases.

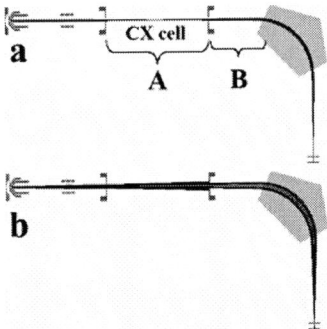

FIGURE 6. Beam trajectory calculated by SIMION [13]. (a) 0% of space charge (b) 30% of space charge.

FUTURE PLANS

Now, we have started the post-acceleration column, on ABS103. Preliminary experimental results indicate the focusing effect of the beam by the post-acceleration. The TOF neutralization experiment will be carried out using a pyroelectric detector that was already developed and calibrated. Then the meta-stable fraction will be measured using a Laser absorption method [14].

ACKNOWLEDGMENTS

This work was supported by a Grant-in-aid from the Ministry of Education, Culture, Sports, Science, and Technology of Japan (priority area 442-16082201). This work was also partially supported by NIFS collaboration programs (NIFS05KCBB006 and NIFS06KCBB007).

REFERENCES

1. D. E. Post, D. R. Mikkelesen, R. A. Hulse, L. D. Stewart, and J. C. Weisheit, J. Fusion Energy **1**, 129 (1981)
2. M. Sasao *et al.*, Fus. Tech. **10**, 236 (1986)
3. Mamiko Sasao Yushirou Okabe, Junji Fujita, Motoi Wada, Hitoshi Yamaoka, and Henry J. Ramos Rev. Sci. Instrum. **61**, 418 (1990)
4. N. Tanaka *et al.* J. Plasma Fusion Res. **2** S1105 (2007)
5. N. Tanaka *et al.* Rev. Sci. Instrum. **79**, 02A512 (2008)
6. N. Tanaka *et al.* AIP conf. Proc **988**, 335 (2008)
7. H. Sugawara *et al.* Rev. Sci. Instrum. **79**, 02B708 (2008)
8. S. Takeuchi *et al.* Rev. Sci. Instrum. **79**, 02A509 (2008)
9. A. S. Schlachter, AIP conf. Proc. **111**, 300 (1984)
10. Instruction book; "Lithium cell for negative ion beam production" (in Japanese), HITACHI Ltd. (2005)
11. M. Maucherat, J. Phys. Radium **10**, 441 (1939)
12. T. Nagamura, N. Tanaka, H. Yamaoka, M. Kikuchi, S. Kitajima, M. Wada, and M. Sasao "Measurement of low energy beam profiles for study of space charge compensation", The 4th Japan-Korea Seminar, Pohang, South Korea (2008)
13. SIMION (www.simion.com)
14. A. Okamoto *et al.* Plasma and Fusion Res. **2** S1044 (2007)

BEAMLINES AND FACILITIES

Physical and Experimental Background of the Design of the ELISE Test Facility

P. Franzen, U. Fantz, W. Kraus, H. Falter, B. Heinemann,
R. Nocentini and the NNBI Team

Max-Planck-Institut für Plasmaphysik, EURATOM Association, PO Box 1533, 85740 Garching, Germany

Abstract. In 2007 the IPP RF driven negative hydrogen ion source was chosen by the ITER board as the new reference source for the ITER neutral beam system. In order to support the design of the Neutral Beam Test Facility in Padua and its commissioning and operating phases, IPP is presently constructing a new test facility ELISE (Extraction from a Large Ion Source Experiment) for a large-scale extraction from a half-size ITER RF source. Plasma operation of up to one hour is foreseen; but due to the limits of the IPP HV system, pulsed extraction only is possible. The extraction system is designed for acceleration of negative ions of up to 60 kV. The start of the ELISE operation is planned for middle of 2010. The aim of the design of the ELISE source and extraction system was to be as close as possible to the ITER design; it has however some modifications allowing a better diagnostic access as well as more flexibility for exploring open questions. The design was also supported by diagnostics and modeling efforts of the processes leading to negative ion production and extraction in a RF source.

Keywords: Neutral Beam Injection, ITER, negative ion source, RF source
PACS: 28.52.Cx, 52.27.Cm, 52.50.Dg, 52.50.Gj

INTRODUCTION

The development of large negative hydrogen ion sources for the ITER NBI system [1] was started in the early 90's in Japan with filamented arc sources as the basis for the design. Filamented sources, however, suffer from regular maintenance periods (twice per year in ITER) due to the limited lifetime of the filaments. Due to the advantages of the RF source — it is in principle maintenance-free — and due to the good experience with the positive ion based RF sources [2] at the NBI system for ASDEX Upgrade and W7-AS, IPP Garching started the development of a RF driven negative ion source end of the 90's, from 2002 on within a framework of an official EFDA contract. The development was very successful [3,4,5]: recently, in July 2007, the RF source was chosen by the ITER board as the reference source [6,7].

The development of the RF driven negative hydrogen ion source is being done at IPP at three test facilities in parallel: Current densities of 330 A/m^2 with H− and 230 A/m^2 with D− have been achieved with the IPP RF source on the small test facility BATMAN (Bavarian Test Machine for Negative Ions) at the required source pressure (0.3 Pa) and electron/ion ratios, but with a small extraction area (70 cm^2) and limited pulse length (<4 s) [3]. The long pulse test facility MANITU (Multi Ampere Negative

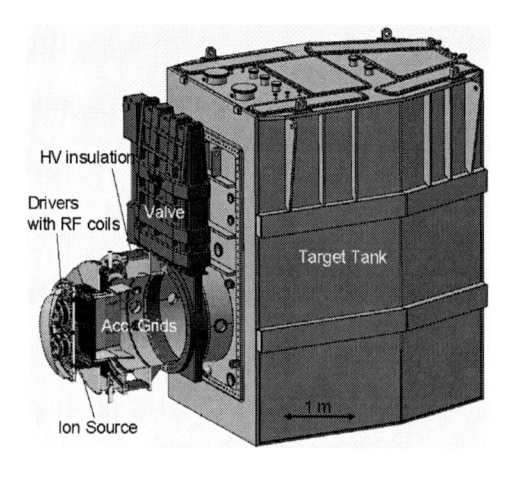

Figure 1. The ELISE test facility

Ion Test Unit) equipped with the same source as it is used at BATMAN but having an extraction area of about 200 cm² demonstrated recently stable one hour pulses; the parameters however, are still below the ITER requirements [8,9]. The ion source test facility RADI, equipped with a source of approximately the width and half the height of the ITER source, aims to demonstrate the required plasma homogeneity of a large RF source [10,11,12]; its modular driver concept will allow a straightforward extrapolation to the full size ITER source.

The results of RADI, however, have limited significance due to the lack of extraction. Hence, IPP is presently designing a new test facility ELISE (Extraction from a Large Ion Source Experiment) for long pulse plasma operation and short pulse, but large-scale extraction from a half-size ITER source. ELISE is an important step between the small scale extraction experiments at BATMAN and MANITU and the full size ITER neutral beam system and supports the design of the Neutral Beam Test Facility in Padua and its commissioning and operating phases. The integrated commissioning of ELISE is planned for 2010, assuming a start of the project end of 2008.

Figure 1 shows an overview of the ELISE test facility. The detailed technical details are given in Refs. [13] and [14]. The aim of the design of the ELISE source and extraction system was to be as close as possible to the ITER design; it has however some modifications allowing a better diagnostic access as well as more experimental flexibility. The paper discusses these design choices and their physical and experimental background that was gained at the present IPP test facilities. Some of these design choices have been also already implemented in the design of the ITER ion source which is presently being finalized by RFX Padua [15].

THE DESIGN OF ELISE

Table 1 shows the main parameters of ELISE. Due to the limits of the IPP HV system, only pulsed extraction during a long plasma pulse is possible. How this will affect the performance of ELISE is not clear; experiments at MANITU showed that at least for a well conditioned source the same performance could be obtained during pulsed extraction during a long pulse [16].

The total maximum available voltage

TABLE 1. Parameters of the ELISE test facility.

Isotope	H, D (limited)
Extraction Area	1000 cm²
Apertures	640, 2x4 groups, 14 mm Ø
Source Size	1.0 x 0.873 m²
Total Voltage	<60 kV
Extraction Voltage	<12 kV
Extracted Ion Current	<25 A for $j_e/j_{H^-}=1$
Acc. Ion Current	<20 A
RF power	<360 kW
Pulse Length:	
Plasma	3600 s
Extraction	10 s every 160 s

corresponds to the pre-acceleration voltage of the SINGAP accelerator design [17] (the MAMUG pre-acceleration voltage is larger, at 200 kV). The total available current is 50 A. Taking the stripping losses of the ELISE grid system into account (~15%, as measured in BATMAN with the SINGAP-like CEA grid system [18]), that current allows an operation at the required parameters with respect to the accelerated current density both in H and D (300 A/m^2 and 200 A/m^2, respectively) at the respective electron/ion ratios (0.5 in H, 1.0 in D) as well as with respect to the required extracted current density in H (330 A/m^2). But in order to achieve the required extracted ion current density in D (290 A/m^2), the amount of co-extracted electrons has to be further reduced to an electron/ion ratio of about 0.8.

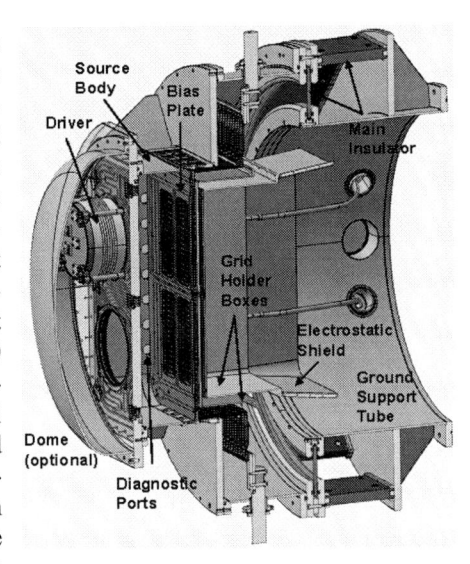

FIGURE 2. Details of the ELISE ion source and the extraction system

Figure 2 shows the details of the ELISE ion source and the extraction system. In contrast to the ITER, the ELISE source will usually be operated in air — with a possibility to operate the drivers in vacuum — to facilitate diagnostic access near the most important region, i.e. the region near the plasma grid [12], and to provide experimental flexibility for e.g. an easy change of the magnetic field configuration. The necessary HV insulation of the source is done with a combination of a glass ring as vacuum boundary and epoxy plates for the mechanical support [13].

The experience at MANITU showed that a source valve is mandatory in order to keep the source performance — when it is cesiated — unaffected during the cryo pump recovery phase. At MANITU, the cryo pump is normally recovered during the night by the control system automatically. The Cs conditioning phase in the next morning was shorter when the pressure, at which the source valve is closed during the recovery phase, was reduced from 10^{-3} mbar to 10^{-5} mbar. This is a strong indication of a Cs poisoning effect of poor vacuum.

Ion Source

At the start of the activities, the design of the ELISE — and the ITER source — was based on the existing large RF source at RADI. However, Langmuir probe measurements showed that the electron density starts to decrease towards the source wall already within the projection of the driver (see Figure 3). Although there is not necessarily a strong correlation of the electron density

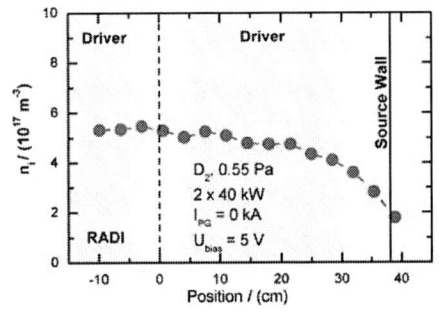

FIGURE 3. Plasma density profile in RADI in 2 cm distance from the (dummy) plasma grid.

TABLE 2. Design power loads for the ELISE source components and the grids. The maximum total power levels are given for full performance operation (P_{RF} = 300 kW, 20 A acc. current).

Component	Amount of RF power	Max. Power Density	Total Power	Comments
Faraday Screens	~40%	300 kW/m^2	120 kW	measured at RADI and BATMAN
RF Coils	~5%	-	15 kW	measured at RADI
Bias Plate	~1%	30 kW/m^2	3 kW	measured at MANITU
Driver Back Plate:				
Eddy Currents	2.5%	-	7.5 kW	calculated
Drivers Plate:			67.5 kW	
Eddy Currents	2.5%	-		calculated
Driver Exit	10%	300 kW/m^2		estimated from FS power density
Surface	10%	100 kW/m^2		estimated from power balance
Losses in RF lines and Transformer	~5%	-	15 kW	estimated
Lateral Walls	20%	100 kW/m^2	60 kW	estimated from power balance
Plasma Grid	~5%	20 kW/m^2	20 kW	measured at RADI (w/o bias plate), including 4 kA current
Extraction Grid	-	32 MW/m^2	400 kW	twice the nominal power density, calculated with transport code
Grounded Grid	-		60 kW	10% of total beam power, BATMAN experience

with the negative ion density for surface production, this decrease of the plasma density is expected to be not of advantage for a homogeneous beam. Hence, the size of the source for ELISE (and accordingly the size of the ITER source) was increased by about 100 mm in both directions having now 873 mm in width and 1000 mm in height.

Furthermore, the diameter of the drivers was increased from 245 mm to 284 mm (inside) and they were shifted radially outwards, so that the inner edges of the drivers are still at the same position. With this change, also the outermost apertures are within the drivers projection leading to a more homogeneous illumination of the grid.

Another still open point is the optimum depth of the ion source. The presently foreseen source depth is 238 mm, as it is the case for the small IPP source; this is about a factor of 2 smaller than the filamented source. In order to have some experimental flexibility, the depth of the ELISE source can be enlarged by another 54 mm by a spacer ring between the source flange and the high voltage ring. An additional row of ports allows diagnostic access at the same position with respect to the plasma grid at the new position.

The expected power levels to the source components are shown in Table 2. As a full source calorimetry at RADI and MANITU is not available up to now (but it is part of the experimental program of the next months), the power loads of the drivers plate and the lateral walls are estimations from the missing power. As can be seen, the power density levels to the source components are rather low (100 - 300 kW/m^2).

For ELISE a cooled bias plate is foreseen with "window frame" like openings around the beamlet groups (see Figure 2). The windows are 12.5 mm larger than the outer edge of the apertures in all directions. The bias plate is insulated against the source body so that the potential of the plate can be changed with respect to the body

— and the plasma grid — as it is the case now at MANITU. First experiments here showed that the electron current could be reduced by changing the bias plate potential [16].

The installation of a bias plate enhanced the extracted negative ion current significantly on the test facility BATMAN. In this test facility the bias plate is electrically connected to the source body potential and extends that potential to the edge of the apertures. The use of this bias plate reduced the necessary filter field strength for suppression of the co-extracted electrons. The latter was not a big problem for hydrogen operation, but was essential for deuterium operation [19].

For the present experimental period, almost all inner copper surfaces of the MANITU source that are heavily loaded have been coated with a thin Mo layer, because there was an indication that sputtered Cu deteriorated the source performance by burying the Cs. First experiments with the Mo coated source surfaces showed indeed almost no copper impurities in the plasma. This is consistent with the sputtering yields: For 100 eV — most probable the upper limit of the energy the ions can gain due to sheath acceleration at the plasma edge in the faraday screen — the self sputtering yield of copper is almost a factor 10 larger than the respective values for tungsten or molybdenum [20]. Furthermore, the sputtering threshold of copper for hydrogen/deuterium is much lower than for tungsten and molybdenum. Hence, an avalanche effect with exponentially increasing impurity content in the source can be expected for copper.

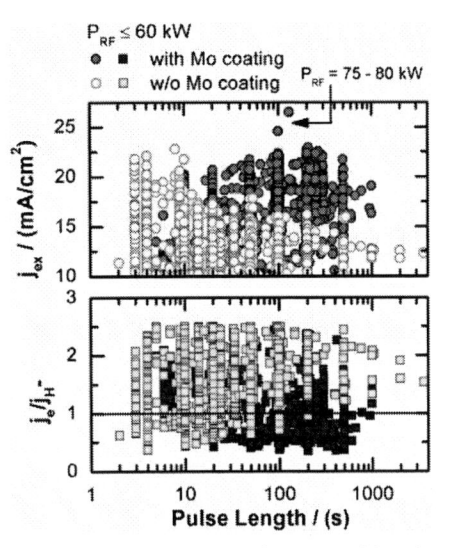

FIGURE 4. MANITU performance with and w/o Mo coating of the source [9].

The performance of MANITU increased drastically with the coated source walls (see Figure 4 and Refs. [9,16]): stable long pulses in hydrogen and deuterium for up to 10 min with respect to ion and electron currents have been achieved up to now with an electron/ion ratio well below 1 and extracted current densities just below the required parameters, the latter limited by the available RF power. Hence, the inner surfaces of the ELISE source — and accordingly the ITER source [15] — will be coated with Mo.

Extraction System

The extraction system for ELISE is a three grid system — plasma grid, extraction grid and grounded grid — which will allow acceleration up to 60 kV. The source is at high potential, while the calorimeter inside the vacuum chamber is at ground potential. The aim of the design was an extraction system as close as possible to the ITER design; it has however some modifications mainly for better diagnostic access for an improved understanding of the physical processes inside the source and of negative ions extraction and for an improved cooling and filter field homogeneity.

The overall size of the grid segments is about the same as the ITER grid segments, but in contrast to ITER, the segments are flat. The position and the size of the beamlet groups are identical to the SINGAP geometry (the MAMUG geometry is slightly different). All grids are mounted on grid holder boxes which are nested inside each other similar to positive ion PINI design. Furthermore, the grids are immersed inside the source body (see Figure 2). This design enables an absolutely flat surface of the plasma grid and allows good diagnostic access from all sides just above the grid surface.

All grids and the bias plate are electrically insulated against each other as well as against the source body to allow biasing of the source against the plasma grid and to measure electrically all grid currents. In order to have some indication of the vertical source homogeneity, the two segments of the extraction grid are also insulated against each other so that the (electron) current on these segments can be measured individually.

Table 2 shows also the expected power loads to the grids. These values are based on measurements at the IPP test facilities. The power load to the plasma grid is rather low (20 kW/m² at maximum) due to the effective screening of the plasma, i.e. reduction of electron density and temperature by the filter field and the plasma grid bias [12]. The total power load of the grounded grid is typically 10% of the total beam power, most probably due to back streaming ions or stripping losses; direct beam interception is not very likely due to the large diameter of the grounded grid apertures [15].

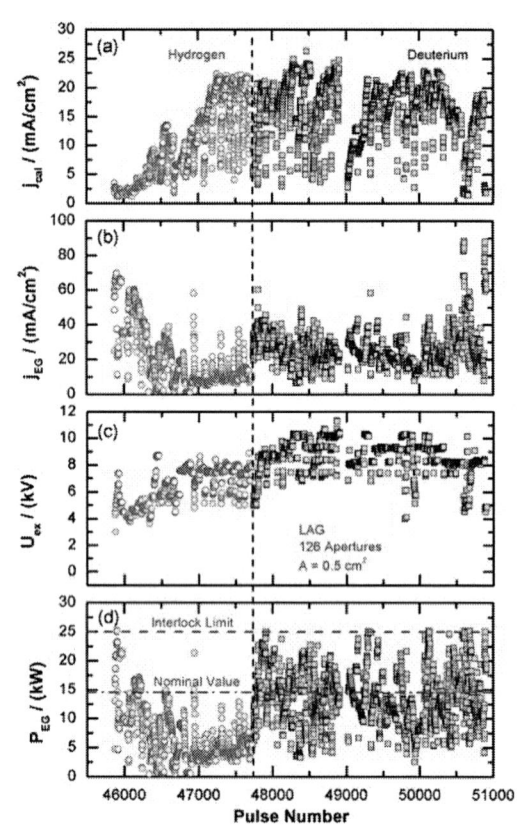

FIGURE 5. Performance evolution of BATMAN for Hydrogen and Deuterium operation, respectively. (a) accelerated current density; (b) current density as measured on the extraction grid; (c) extraction voltage; (d) power load of the extraction grid. Also indicated is the interlock limit and the nominal value (j_e/j_{H^-} = 1, 10 kV).

The most critical component for the source operation and also for the performance is the extraction grid. This is demonstrated in Figure 5 showing experimental data of BATMAN. Especially for deuterium with a much larger electron/ion ratio compared to hydrogen and a higher required extraction voltage, the power load of the extraction grid is much larger compared to hydrogen, both during conditioning phases and high performance operation. As can be seen in Figure 5, the interlock limit of 25 kW (about 200 W/aperture) is frequently reached for deuterium. As this value corresponds to roughly twice the nominal power, the maximum total power and power density for the ELISE extraction grid design was increased accordingly by an improved cooling scheme, in order to have some safety margin and a larger operating window during conditioning phases.

A good source performance and hence high current densities were obtained on BATMAN and MANITU at elevated source wall temperatures of about 40 °C. In order to have some margin, the cooling system of ELISE is considered to operate between room temperature and 60 °C. The water inlet operation temperature of the source and the grid system will be 55 °C according to the actual ITER reference design. The position of the plasma grid apertures is machined with a pre-offset taking into account the different operating temperatures of the extraction/grounded grid (55 °C) and the plasma grid (150 °C). For that case all apertures will be in line, steering by aperture offset is not planned on ELISE. The remaining expansion for the operating range of ±50°C of the plasma grid causes a maximum aperture offset for the outer most apertures of 0.17 mm in vertical direction and 0.56 mm in the horizontal direction. The effect on beamlet deflection has been calculated by KOBRA-3D and was found to be just acceptable [21]. The worst case is for the orthogonal orientation of the electron suppression field (see below) to the filter field and the outermost apertures: here the beam just touches the wall of the extraction grid.

Optimum Magnetic Filter Field Configuration

The filter field of the reference design of ITER consists of a combination of a 4 kA plasma grid current and permanent magnets outside the source. This filter field was optimized for minimum electron and ion deflection in the accelerator and to ensure the necessary $\int B_x dz$ for electron suppression for deuterium operation. For both types of small sources (filamented source and the BATMAN IPP RF source) a typical value of 0.9 to 1.1 mTm was sufficient to achieve the ITER relevant parameters (see Figure 6).

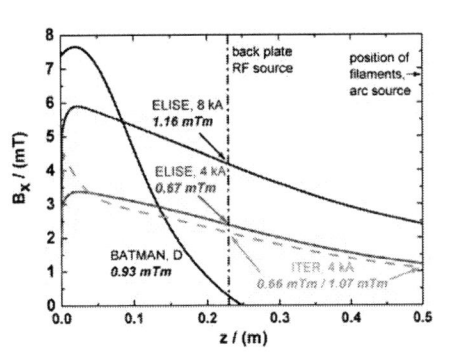

Figure 6: Filter field strength for different combinations of permanent magnets (CoSm, 30x20 mm²) and plasma grid currents and the corresponding values for $\int B_x dz$ for the RF source and the filamented source, respectively. Also shown for comparison is the value for BATMAN for deuterium operation.

The reduced depth of the RF source, however, has consequences for the value of $\int B_x dz$. By using the reference filter field (permanent magnets and 4 kA of plasma grid current) for the ITER RF source, the $\int B_x dz$ is reduced by a factor of about 1.5 compared to the optimum value for BATMAN and the filamented source (see Figure 6). To obtain the optimum value, the plasma grid current has to be increased accordingly with all the consequences for electron and ion deflection in the accelerator. Hence, in order to have some operational margin, the ELISE plasma grid was designed for 8 kA maximum current.

The magnetic filter field in a negative source reduces the co-extracted electron current, but also bents the negative ions that are accelerated by the sheath potential from the plasma grid into the source back to the apertures. However, it is still not clear what the really important parameter is: the $\int B_x dz$, the filter field strength near the plasma grid, or the 3-dimensional structure of the field. Additionally, the electron deflection field from the extraction grid reaches also into the source and leads to further complexity. There is evidence from BATMAN that this field is necessary for a sufficient suppression of the co-extracted electrons. Hence, the design of the extraction grid was made flexible so that the electron suppression magnets can be installed with their respective magnetic field perpendicular (as in BATMAN and MANITU) or parallel to the filter field. The latter case was chosen for ITER in order to minimize horizontal beam deflection; however, zero field regions in front of the plasma grid are created [14] with unknown consequences for the local electron suppression.

Furthermore, due to the increase of the ELISE source size, the horizontal distance of the permanent magnets is larger as it is in the ITER reference field. This results in a reduced field in the centre of the source; the increase of the field towards the outermost apertures is less pronounced, however. This may be an advantage for the homogeneity of the beam. The experience at MANITU with different aperture patterns showed indeed, that the more apertures have been in the region of the increasing filter field, the less beamlets could be accelerated to the calorimeter [22].

Due to the apertures and the beamlet group arrangement of the plasma grid, the current distribution and hence the resulting filter field cannot be uniform. It is not clear up to now whether this is harmless for the plasma homogeneity, but it has certainly consequences for the electron and ion deflection in the accelerator. Hence the design of the plasma and the extraction grid was optimized with an emphasis on a homogeneous temperature of the plasma grid surface — to minimize the risk of inhomogeneous Cs coverage — as well as on a current distribution on the plasma grid as homogeneous as possible. The latter was achieved by optimizing the current density distribution on the plasma grid by introducing pockets into the grid so that the current is forced to flow around the apertures (for details see Ref. [14]). Figure 7 shows the difference of the new ELISE design to the ITER design: the magnetic field ripple is much more reduced; in 20 mm distance from the plasma grid, the magnetic filter field is almost flat.

The issue of the optimum fields in the source — and in the extraction grid — will be certainly a main topic for ELISE as well for the ongoing modeling and experimental efforts at IPP. The flexibility of ELISE — with access to the source near the plasma grid — allows an easy change of permanent magnets outside the source as well of other means of creating a sufficient, simple local filter field without an appreciable field in the accelerator like the plasma grid current solution. Examples may be magnets placed in the plasma grid (pockets) or to introduce

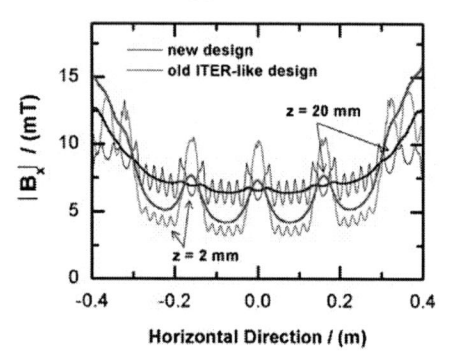

FIGURE 7. Magnetic filter field of ELISE for 8 kA plasma grid current at different distances from the plasma grid for the ELISE design and the ITER design, respectively.

vertical magnet rods between the aperture groups.

Calorimeter

A first conceptual design of a full power calorimeter has been started. Figure 8 shows the calculated power density profile in a distance of 3 m from the grounded grid for a full beam and a beamlet divergence of 2 degree. Even for this rather high divergence the upper and the lower half of the beam are clearly separated due to the large gap between the grid segments (96 mm) and due to the fact that no steering nor focusing is done in the ELISE grids.

The maximum power density is rather low, being about 3 MW/m^2 in the center of the segment. It increases to about 3.6 MW/m^2 for 1 degree divergence. This

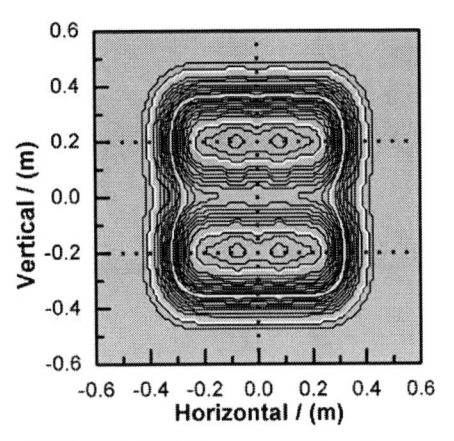

FIGURE 8. ELISE beam profile in a distance of 3 m for a total power of 1.2 MW and a divergence of 2 degree. The maximum power density is 2.9 MW/m^2. The white lines indicate the 10%, 50% and 90% curvatures, respectively. Also indicated is a possible distribution of TC's for beam profile measurements.

together with the pulsed beam gives the opportunity to use an inertially cooled calorimeter, i.e. a copper plate with a cooling circuit at the back of the plate. The temperature increase of a 2 cm thick copper plate for a 10 s pulse with that power density is about 450 K, which is sufficiently low. A cooling loop with about 10 l/min only is sufficient to cool down the calorimeter within the 160 s before the next 10 s pulse.

The separation of the upper and lower half of the segments can be used for beam profile measurements. A possible arrangement of thermocouples is also shown in Figure 8. The thermocouples can be embedded in almost thermally isolated areas of the copper plate with a low thermal contact to the main plate, as it is now the case for the BATMAN calorimeter. These areas can consist of copper or graphite for thermography from the backside.

Diagnostic Setup

The signals/diagnostics at the test facilities in IPP are divided in three groups: (1) *control parameters* for protection and safe operation, as well as for operational parameters; examples are voltages, currents, temperatures, RF power; but also more complicated quantities like the power on the extraction grid, calculated on-line by the control system; (2) *performance parameters* for the source conditioning status and the source and beam 'quality'; examples are the electron/ion ratio, the calorimetric current density, source and beam homogeneity, beam divergence, Cs content; (3) *physics parameters* for better understanding of the processes in the source and as input for modeling; examples are the electron density and temperatures or the negative ion density.

The main diagnostics — the "work horses" — are the measurement of the currents at the various grids and in the HV system, the calorimeter and the optical emission spectroscopy both in the source and the beam. A special feature of the ELISE design is that the two segments of the extraction grid are insulated against each other and the grid holder box (see above) so that the current on each segment can be measured separately. Hence, a rough evidence of the top-bottom source homogeneity can be obtained.

In order to measure the Cs dynamics also in between the discharges (i.e. the gas or vacuum distribution), Cs absorption spectroscopy is foreseen. This was already successfully tested at the University of Augsburg [23]. The use (if or how) of the other physical diagnostics (Langmuir Probes, Laser Detachment and Cavity Ringdown spectroscopy) is still not decided; all three are problematic due to space limitations (for the probes) due to the small HV insulation distances and due to their complexity.

CONCLUSIONS

ELISE is an important intermediate step for the RF source development for the ITER NBI system. With respect to ISTF/NBTF, the ELISE design offers a larger experimental flexibility for the investigation of concept improvements and a better diagnostic access to the extraction region near the plasma grid.

ACKNOWLEDGEMENTS

The work was (partly) supported by a grant (#03FUS0002) from the German Bundesministerium für Bildung und Forschung as well as by a grant (#TW6-THHN-RSFD4) from the European Union within the framework of EFDA (European Fusion Development Agreement). The authors are solely responsible for the content.

REFERENCES

1. ITER EDA Documentation Series No. 24, Plant Description Document, Sec. 2.5.1. IAEA 2002
2. E. Speth et al. Plasma Science and Technology 6 (2004) 2135
3. E. Speth et al., Nuclear Fusion 46(6) (2006) S220
4. P. Franzen et al., Nuclear Fusion 47 (2007) 264
5. A. Stäbler et al., Symposium on Fusion Technology, 2008
6. R. Hemsworth et al., Review of Scientific Instruments, vol.79, no.2, Feb. 2008, pp. 02C109-1-5
7. B. Schunke et al., these proceedings
8. W. Kraus et al., Review of Scientific Instruments, vol.79, no.2, Feb. 2008, pp. 02C108-1-3
9. W. Kraus, Proceedings of the 35nd EPS Conference, 2008
10. P. Franzen et al., Fusion Engineering and Design 82 (2007) 407
11. U. Fantz et al., Review of Scientific Instruments, vol.79, no.2, Feb. 2008, pp. 02A511-1-6
12. U. Fantz et al., these proceedings
13. B. Heinemann et al., Symposium on Fusion Technology, 2008
14. R. Nocentini et al., Symposium on Fusion Technology, 2008
15. D. Marcuzzi et al., Symposium on Fusion Technology, 2008
16. W. Kraus et al., these proceedings
17. R. Hemsworth et al., AIP Conference Proceedings 925 (2007) 290
18. A. Lorenz et al., Technical Report IPP 4/285, Max-Planck-Institut für Plasmaphysik, Garching, 2006
19. P. Franzen et al., AIP Conference Proceedings 993 (2008) 51
20. W. Eckstein et al., Sputtering Data, IPP Report 9/82.
21. R. Gutser et al., these proceedings
22. W. Kraus et al., Fusion Engineering and Design 74 (2006) 337
23. U. Fantz et al., to be published

A Test Stand for Ion Sources of Ultimate Reliability

R. Enparantza[a], L. Uriarte[a], F. J. Bermejo[b], V. Etxebarria[b], J. Lucas[c] J.M. Del Rio[d], A. Letchford[e], D. Faircloth[e], M. Stockli[f], P. Romano[a], J. Alonso[a], I. Ariz[a] and M. Egiraun[a]

[a]*Fundación Tekniker-IK4, Eibar, Spain*
[b]*University of the Basque Country, Dpt. Electricity and Electronics, Leioa, Spain, and Consejo Superior de Investgaciones Científicas, Inst. Estructura de la Materia*
[c]*Elytt Energy, Portugalete, Spain*
[d]*Jema Group, Lasarte, Spain*
[e]*ISIS Accelerator Division, Rutherford Appleton Laboratory, Didcot, UK*
[f]*Spallation Neutron Source, Oak Ridge National Lab, Oak Ridge, USA*

Abstract. The rationale behind the ITUR project is to perform a comparison between different kinds of H⁻ ion sources using the same beam diagnostics setup. In particular, a direct comparison will be made in terms of the emittance characteristics of Penning Type sources such as those currently in use in the injector for the ISIS (UK) Pulsed Neutron Source and those of volumetric type such as that driving the injector for the ORNL Spallation Neutron Source (TN, U.S.A.). The endeavour here pursued is thus to build an Ion Source Test Stand where virtually any type of source can be tested and its features measured and, thus compared to the results of other sources under the same gauge. It would be possible then to establish a common ground for effectively comparing different ion sources. The long term objectives are thus to contribute towards building compact sources of minimum emittance, maximum performance, high reliability-availability, high percentage of desired particle production, stability and high brightness. The project consortium is lead by Tekniker-IK4 research centre and partners are companies Elytt Energy and Jema Group. The technical viability is guaranteed by the collaboration between the project consortium and several scientific institutions, such the CSIC (Spain), the University of the Basque Country (Spain), ISIS (STFC-UK), SNS (ORNL-USA) and CEA in Saclay (France).

Keywords: Test Stand, Reliability, Comparison
PACS: 07.77.-Ka; 07.30.Kf; 41.75.Cn; 41.85.-p

INTRODUCTION

A consortium of research centres and industrial companies, partially supported by both the Basque and Spanish administrations, is carrying out an ambitious research programme that started with R&D work on ion sources. As it is well known, the beam current required from the ion source and LEBT (Low Energy Beam Transport) depend strongly on the beam emittance, because the RFQ (Radiofrequency Quadrupole) transmission decreases rapidly with increasing emittance and increasing beam current. For example, the ESS (European Spallation Source) requirement of a current of 150-

mA at the beginning of the medium-energy beam transport requires an RFQ input current between 85 and 95 mA for a normalized rms emittance between 0.20 and 0.35 π.mm.mrad, which put into different words indicates that developing a low-emittance source is a must. The aim is to develop high-current, low-emittance ion sources and a LEBT that inflicts minimal emittance growth. The first phase of such a research programme which is financed through ministries of Industry and Education & Science [1] is well underway and consists on a test stand able to compare the emittance characteristics of H⁻ arc-discharge sources such as the Penning trap used at ISIS [2] and RF driven sources such as the multicusp H⁻ source being at present in use at the Spallation Neutron Source (SNS) [3]. In the future, other types of sources, even proton sources as the CEA-Saclay [4] will also be able to be mounted and measured in the test bench.

The strategic goal for the coming three years will consist on the construction of a complete accelerator Front-End Test-Stand (FETS) able to diagnose ion beams generated by the set of ion sources referred to above. The effort is conceived as a genuine R&D endeavour which will be financed by both Basque and Spanish Central Governments.

The ion source test stand is being built at the University of the Basque Country. For that, several of its main constituents are being designed, specified and some of them are already being built. It is expected that the test stand will be operative by the end of 2009. The expected beam features on the test stand are summarised in Table 1. It is considered that going beyond 65mA would require a serious effort to compensate for the space charge effects [5].

TABLE 1. Beam features

Parameter	Value
Max Pulse H⁻	65 mA
Max Pulse e-	1A
Pulse Frequency	50 Hz
Duty cycle	6 %

Several parts or main areas can be distinguished in the project: the Faraday cage, the power sources, the ion sources and the diagnostics.

THE FARADAY CAGE

The ion source test stand is situated in a 5 m x 6 m base and 3.5 m high conducting aluminium container, which acts as a Faraday cage to isolate the ion sources from any type of electromagnetic interference and to avoid any accidental access to the high voltage components. The cabin is air conditioned to keep the inside temperature and humidity within acceptable limits. The main reason to choose a closed Faraday cage is to provide a stable environment to the main ion source components, so that the comparisons performed in the different kind of ion sources are reliable and not dependant on ambient conditions.

FIGURE 1. Faraday Cage of the ITUR ion source test stand at the Univ. Basque Country.

The Faraday cage also allows for:
1. Screening the residual electric field of the HV platform (see next chapter) so that it does not affect the rest of the components out of the cage.
2. Preventing human access to platform when in HV operation.
3. Protecting from X-rays or gamma rays, if required.

The roof is removable to allow for easily introducing heavy equipment and it has three 50 cm x 50 cm windows covered by 8 mm diameter hole perforated lids.

The cage is serviced from outside by the following supplies:
1. Hydrogen. Hydrogen gas tanks are situated outside the building and connected to the ion source through a seamless copper pipe that, inside the cage, is rubber made to provide galvanic insulation to the platform and the devices on it. A gas flow meter cuts the supply as soon as it increases beyond a limit.
2. Air. Three pipes of 0,5" diameter at 10 bar pressure have been planned.
3. Power. 400V 200kW
4. Air conditioned. An air condition system will be installed to control temperature and humidity inside the cage.
5. Exhaust tubes for turbopumps. 3 inch pipes outside the building. Air would contain H_2 and Cs, although in very low quantities.

THE PLATFORM

A High Voltage platform of dimensions 2.3 m x 4.3 m is mounted inside the cage and supported on 15 insulators separated some 900 mm from each other. The platform will be set to reach 100kV, thus kept at 800 mm distance from the cage walls and as well as from the ground. Insulators are made of C130 porcelain and are able to withstand 325 kV lightning impulse. The platform is made of an aluminium profile grid covered by aluminium sheet plates and with all edges rounded to prevent zones where the electric field may concentrate. It is calculated to bear more than 4 kN/m² weight.

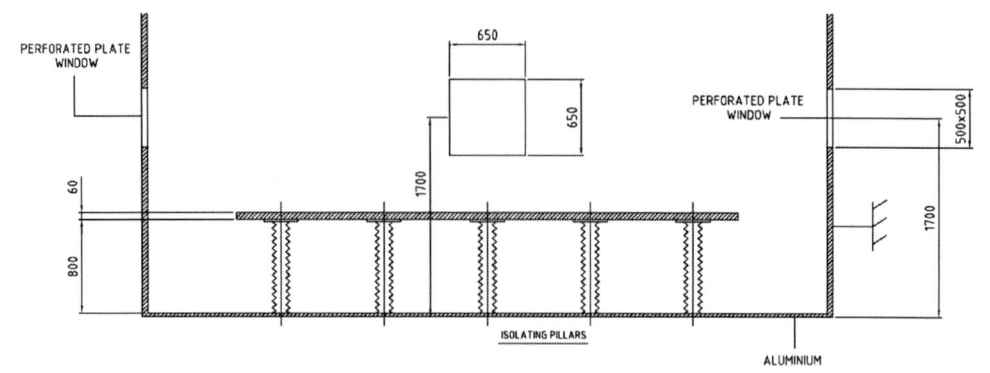

FIGURE 2. Sketch of the HV platform on 800 mm insulators

THE POWER SUPPLIES

Several power supplies are required to run the ion sources test stand. As said, at least in a first stage, the test stand should be able to accommodate two types of ion sources, namely, the H⁻ pulsed source of ISIS (Penning type) and that of SNS (multicusp RF powered). Thus, power sources implemented should be able to generate the energy required for both types of sources.

The DC Platform Power Supply

The DC platform power supply is designed to keep the platform at a maximum voltage of 100 kV in order to generate the potential required to extract and accelerate the ions from the ion source itself. The main features of this power source are gathered in Table 2 below.

TABLE 2. Platform DC power supply.

Parameter	Value
Maximum Voltage	100 kV
Intensity	20 mA
Drop	0,10%
Precision	0,10%
Electron return to extraction voltage	25 kV
Condensators capacitance	1 µF
Polarity change (manual)	Yes
Discharge time	2 s
Charging time	30 s

It should be noted that, as shown in the table, electrons do not return to earth voltage but only to the extraction one, ie, 25kV (post acceleration is as large as 75 kV), so the post acceleration is restricted to the 65mA ion intensity.

As seen, the platform is prepared for changing the polarity in case the source is required to change to a H$^+$ source. This is thought to be a manual polarity change rather than automatic.

In order to keep the drop at 0,1% during the 1,2 ms pulse (50 Hz, 6% duty cycle), a bench of condensators is required to be put in parallel with the DC power supply. Considering that electrons are extracted only up to 25kV, a 1µF capacitance is considered enough. A discharge time of 2 s is estimated necessary from the point of view of safety, so that nobody accesses the platform before discharge is complete. Discharge would be passive with a resistance in parallel with the output capacity that would give time constant (RC) equal to 0.25 s, for example. A circuit breaker will be put in series with the resistance to unload the capacitors. Another circuit breaker will also be installed in case the resistance would not work. This would be an external circuit breaker that would short circuit the condensators and that would be installed in the doors lock, so that the door cannot be opened without the platform been disconnected and discharged.

Along with the capacitors, an isolating transformer is required to supply the systems on the HV platform, which have an earth at platform level. Therefore, it is necessary that the isolation between primary and secondary is at least equal to 100kV. Concerning power requirements, estimations indicate that the maximum required instantaneous power could be around 300 kW, although the average power for the 20 ms cycle would not be higher than 35 kW. Thus, a 100kVA transformer should be enough, assuming that the different elements operating in pulses have enough condensators in their power supplies as to not transferring the consumptions peaks to the mains. Three-phase 400 V seems to be the most appropriate voltage, a secondary in star being required to connect the neutral to the platform earth. Transformation ratio is 1:1. The transformer will be oil filled.

The Plasma Formation Power Supplies

The power supplies required to form the plasma are different for different types of ion sources. The ISIS Penning type ion source has a Pulsed discharge power supply together with a DC one. In the case of ITUR, a single power supply capable of providing pulsed and DC current will be implemented. The pulsed current should have 50 Hz repetition rate and 0.15 duty cycle, whereas the DC current should provide a maximum voltage of 800 V (at 10 mA) and a maximum intensity of 2 A (at 80 V). The characteristics of this power supply are shown in Table 3.

TABLE 3. Discharge power supply (Penning)

Parameter	Value
High Current Discharge Voltage	400 V
High Current Discharge Intensity	80 A
Frequency	50Hz
Duty cycle	0.15
Low Current Discharge Voltage	800 V
Low Current Discharge Intensity	2 A
Power	250 W

The RF ion source requires a couple of discharge power supplies for plasma formation: a pulsed one and a continuous wave (CW) one. These are defined by their parameters in Table 4.

TABLE 4. Discharge power supplies (RF multicusp)

Type	Parameter	Value
Pulsed RF power supply	Power	20-60 kW
	Frequency	2 MHz
CW RF power supply	Power	200 W
	Frequency	13 MHz

The Extraction Power Supply

The Penning type ion source requires, in addition, a pulsed extraction power supply, which is not needed by the RF multicusp ion source. The features of this extraction power supply are gathered in Table 5.

TABLE 5. Extraction power supply (Penning)

Parameter	Value
Voltage	25 kV
Intensity	2 A
Frequency	50 Hz
Duty factor	10 %

This is a critical element and a source of difficult to address problems. In operation, it easily shuts off due to demanding working conditions. So it has to be robust, protected against short circuits and voltages coming in and able to condition the electrodes. In ITUR, an up to date solid state power supply will be built.

THE ION SOURCES

In a first stage, as said, two types of H⁻ ions sources will be tested: the ISIS Penning type and the SNS multicusp RF source. Other sources, even proton ones, are also planned to be tested in the future.

Penning Type H⁻ Ion Sources

A Penning Type ion source will be developed from the sources currently being used in ISIS at the Rutherford Appleton Laboratory. In this type of sources, the beam is generated by discharges on a plasma hold on mutually perpendicular electric and magnetic fields. It is a surface plasma source that produces energetic H⁻ ions on the cathode surface of the discharge chamber. Caesium is used to increase the production of ions. The source works with high energy density emission [6], over 1 Acm^{-2}, far higher than what could be obtained in volumetric sources. Fig. 3 shows the ISIS Penning type source [7].

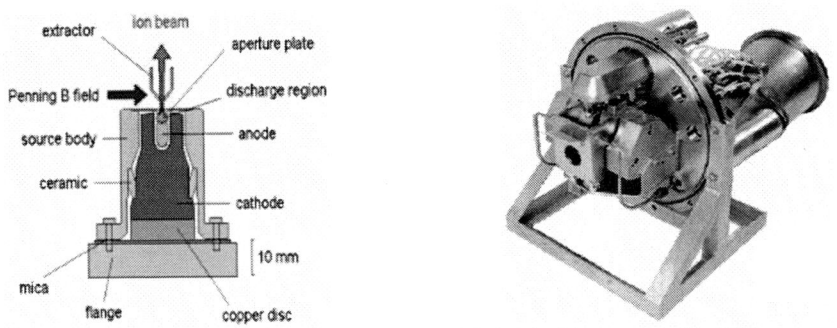

FIGURE 3 – Penning type source: (a) sketch; (b) source used in ISIS

Surface erosion is probably the most important factor limiting its useful life. In the ISIS Penning type source, anode and cathode surfaces are eroded and the erosion rate depends on the electrode material, the uniformity of the Penning discharge and the source operating conditions. All those parameters are being studied in order to maximise its life. Experience shows average duration values of 26 days and a maximum of 49 days [8].

Volumetric H⁻ Ion Sources

This type of sources is being used and tested at the Institute of Applied Physics (IAP) of Frankfurt (Fig. 4b) [9] and the SNS of the Oak Ridge National Laboratory as well as in DESY (Deutsches Elektronen-Synchrotron). The low electron temperature and the high energy densities that can be reached with these sources make them very attractive. The SNS source (see Fig. 4a) is based on a design of the Lawrence Berkeley National Laboratory (LBNL) [10].

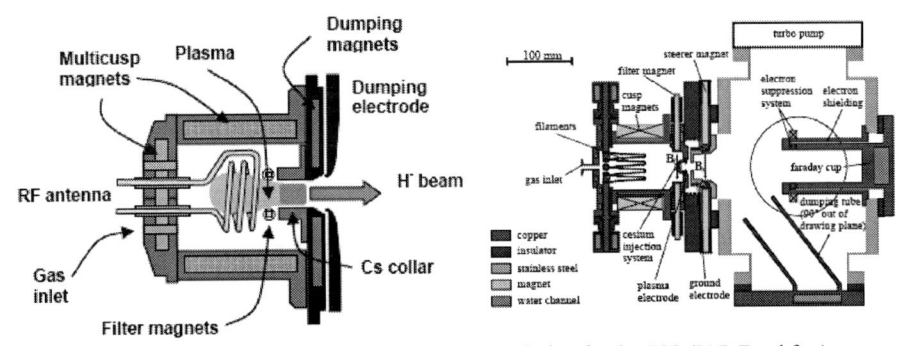

FIGURE 4 – (a) SNS (LBNL) source; (b) Source design for the ESS (IAP-Frankfurt)

The main difference between both types is the way in which energy is transmitted to the plasma. SNS uses a 2 MHz radiofrequency field generated by a RF antenna to heat the plasma. On the other hand, in the IAP design, it is the discharge of a filament arch which keeps the plasma. In both cases, Caesium is evaporated to favour forming negative ions. The DESY source is a multicusp Caesium-free source.

The duration of the source is limited in one case to the life of the antenna and, in the other, to the wear of the filament. At the moment, the life of this type of sources is smaller than the Penning one although developments are carried out to significantly increase it [11].

ECR type H⁻ and proton sources

Although not contemplated in the first version of the ion source test stand, this type of sources should also be able to be tested.

Specifically, Electron Cyclotron Resonance (ECR) type proton sources such as the SILHI in CEA/Saclay have already shown that could meet the requirements of the ESS [12]. It is being reported that durations up to 6 months have already been achieved in CW mode, with good emittance and current values (>100 mA). As mentioned, the ion source test stand will be prepared to easily change the polarity of the platform power supply to allow proton sources to operate.

THE DIAGNOSTICS

A basic set of diagnostic elements have been selected to measure the parameters of the beam extracted from the ion sources. These will be kept in a diagnostic vessel in appropriate vacuum conditions by a couple of magnetic vacuum pumps. Table 6 shows the list of devices that will be used in a first stage.

TABLE 6 Diagnostics

Measurement	Component
Current	Slow Current Transformer
Current	Faraday Cup
Space charge effect	Buffer Gas delivery System, Flow controller
Emittance and profile	Scintillator, Interchangable Pepper Pot, CCD Camera
Degree of stripping	Diagnostic dipole
Energy spread	Retarding Potential Energy Analyser

Current measurement is firstly thought to be carried out by a Slow Current Transformer able to measure long pulses and macropluses up to some milliseconds. A fast current transformer for high frequencies is not considered strictly necessary at the stage of development of the ion source test stand. An extra current measurement will be registered by means of a Faraday Cup that will also act as a beam stop. This second measurement will allow testing the first one.

The space charge effect allows having an idea of the charge distribution in space. For that purpose, a Buffer Gas delivery System will be implemented with its corresponding gas flow controller.

Emittance is a key feature to be measured in a beam in order to assess its quality. There are a number of devices to measure it, such as slit wire scanners, slit grid scanners, Allison scanners and others. Although not as precise as others, an interchangable Pepper-pot type device with its scintillator and CCD camera will be installed in the diagnostic vessel and will allow to measure the emittance as well as the beam profile. The measuring device will be movable so that different measurements can be taken at different z distances from the ion source extraction area, this way overcoming the drawbacks of fixed emittance measuring devices [13].

The degree of stripping, percentage of neutral atoms to ions, will be obtained by a diagnostic dipole. Finally, the energy spread of the beam will be measured by a Retarding Potential Energy Analyser.

These measurements are considered as basic for the ion source test stand, the aim of which is to compare beams produced by different ion sources.

REFERENCES

1. ITUR 2007 - Identification Number PNE-20071027 of the Spanish Industry Ministry.
2. Faircloth, D.C. et al, "The development of the ISIS ⁻H Surface Plasma Ion Source at RAL", *Proceedings of the 18th Meeting of the International Collaboration on Advanced neutron Sources (ICANS-XVIII)*, April 25-29, 2007, Dongguan, Guangdong, China
3. T.E. Mason and L.K. Price, "Spallation Neutron Source Completion Report", June 2006, SNS 100000000-BL0005-R00
4. J. Peters, H.H. Sahling and I. Hansen, "Beam characteristics of the new DESY H⁻ source and investigations of the plasma load", *Review of Scientific Inst.*, 79, 02A523 (2008).
5. A. Ben Ismail, "Etude des effets transitoires de la compensationde charge d'espace sur la dynamique d'un faisceau intense", PhD thesis of the University of Orsay, September 2005..
6. G E Derevyankin, Negative Hydrogen Ions for Accelerators, Proceedings European Particle Accelerator Conference, pp 1450 - 1452, London, 1994.
7. D.Faircloth, "Development of High Performance Ion Sources", presentation by the author at ISIS-CCLRC (www.isis.rl.ac.uk/accelerator/lectures/DLRALJAW/2004/DC_Faircloth.ppt)
8. J W G Thomason and R Sidlow, "ISIS Ion Source Operational Experience", *Proceedings EPAC* 2000, Vienna, Austria, June 2000.
9. K Volk, A Maaser, H Klein, "The Frankfurt H– Source for the ESS", *Proceedings LINAC* 1998, Chicago
10. R.F. Welton, T.A. Justice, S.N. Murray, M.P. Stockli, "Operation of the SNS Ion Source at High Duty-Factor", *Proceedings EPAC* 2004, Lucerne, Switzerland, July 2004.
11. R Thomae et al., "Beam Measurements on the H⁻ Source and LEBT System for the SNS", *Review Scientific Instruments* 73(2) 2002 .
12. R Gobin et al., "High Intensity ECR Ion Source (H⁺, D⁺, H⁻), Developments at CEA-Saclay", *Review Scientific Instruments* 73(2), 2002.
13. S. Jolly, D. Lee, J. Pozimski, P. Savage, D. Faircloth, C. Gabor, "Beam Diagnostics for the Front End Test Stand at RAL", *Proceedings DIPAC* 2007.

Recent Progress in the Negative-Ion-Based Neutral Beam Injectors in Large Helical Device

Y. Takeiri, K. Tsumori, K. Ikeda, M. Osakabe, K. Nagaoka, Y. Oka,
E. Asano, T. Kondo, M. Sato, M. Shibuya, S. Komada, and O. Kaneko

National Institute for Fusion Science, Toki 509-5292, Japan

Abstract. Negative-ion-based neutral beam injection (negative-NBI) system has been operated for 10 years in Large Helical Device (LHD). The injection power has been increased year by year, according to the improvement of the negative ion sources. Up to now, every injector achieves the designed injection energy and power of 180keV-5MW with hydrogen beams, and the total injection power exceeds 16MW with three injectors. In the multi-round aperture grounded grid (GG), the diameter of a round aperture has been enlarged for higher GG transparency. Then, the GG heat load is reduced, as well as in the multi-slotted GG, and the voltage holding ability in the beam acceleration was improved. As a result, the beam energy is raised and the injection power is increased. To improve the anisotropic property of the beamlet convergence condition between the perpendicular and the parallel directions to the slots in the multi-slotted GG, a round-shape aperture of the steering grid (SG) has been changed to a racetrack shape. As a result, the difference of the beamlet conversion condition is much mitigated, and the injection efficiency (port-transmission efficiency) is improved, leading to 188keV-6.4MW injection. The Cs consumption is observed to be proportional to the tungsten evaporation from filaments. The Cs behavior is investigated with optical emission spectroscopy. During the beam extraction, the Cs recycling is dominated by Cs on the backplate, which is evaporated into the plasma by the backstreaming positive ions, and the wall surfaces should be loss regions for the supplied Cs.

Keywords: negative ion source, negative-NBI, Large Helical Device, beamlet steering, cesium seeding, optical emission spectroscopy
PACS: 29.25.Ni, 28.52.Cx, 52.50.Gj

INTRODUCTION

A negative-ion-based neutral beam injection (negative-NBI) system started its operation in 1998 with two injectors in Large Helical Device (LHD), which is the world's largest superconducting fusion machine [1], and the third injector was operational in 2001 [2]. The negative-NB injector is designed as 180keV of the beam energy and 5MW of the injection power, and equipped with two large negative-ion sources. Although at the first three years the actual injection power was as low as about a half of the designed value, the injection power has been increased year by year by continuous R&D in parallel with the operation. In 2007, as a result, 16MW was totally injected into the plasma with three injectors, and one injector achieved 188keV-6.4MW injection. The negative-NBI system in LHD is reliably operated as a main heating device, and indispensable for the LHD plasma experiments [3]. The LHD negative-NBI demonstrates the availability and the superiority of negative-NB injectors toward the next-step fusion machine like ITER [4].

The increase in the injection power in the negative-NBI is ascribed to improvement of the negative ion sources [3,5]. The major development is the discharge optimization for the increase in the negative ion current and the accelerator improvement for the increase in the beam energy. Optimization of the cusp magnetic field in combination with the filter magnetic field led to enhancement of the arc efficiency for the negative ion production [6], and improvement of the arc plasma uniformity by the individual control of the divided twelve outputs of the arc and the filament power supplies contributed to improvement of the injection efficiency [7]. As for the beam acceleration, the multi-slotted grounded grid (GG) was extremely effective to raise the beam energy through reduction of the GG heat load, and resulted in exceeding the designed beam energy [6].

Recently, even in the GG with round apertures, enlargement of the aperture diameter has turned out to be effective in reducing the GG heat load. As a result, the beam energy and the injection power have been increased to 186keV-5.5MW. As for the multi-slotted GG, the anisotropic beam properties between the parallel and the perpendicular directions to the slots have been resolved by using the steering grid (SG) with racetrack apertures. As a result, 6.4MW injection has been achieved due to improvement of the transmission efficiency at the injection port.

The cesium behavior should be investigated for optimization of the ion source operation. Optical emission spectroscopy (OES) is utilized for investigation of the cesium behavior, and a scenario for the cesium recycling is proposed.

In the following, recent development related to the negative-NBI system in LHD is presented from the point of view of the ion source performance.

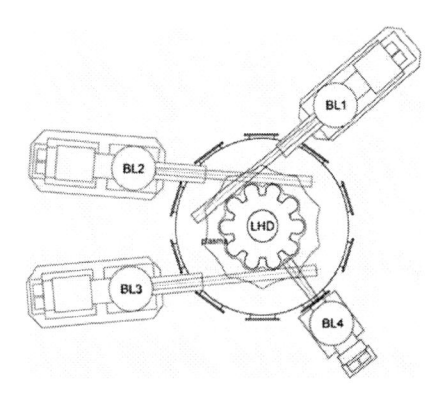

FIGURE 1. Arrangement of the negative-NB injectors in LHD. BL1, BL2, and BL3 are negative-NB injectors for the tangential injection, and BL4 is a positive-NB injector for the perpendicular injection.

NEGATIVE-NBI SYSTEM AND THE ION SOURCES

Three negative-NB injectors are installed on the LHD, which are arranged as tangential injection as shown in Fig. 1. The nominal injection energy and power are 180keV and 5MW, respectively, and the injection species is hydrogen. Figure 2 shows a plan view of the injector. Two large negative ion sources are attached side by side. The accelerated negative ion beams are neutralized, and then, injected into the LHD plasma. The shine-through beam, passing through the plasma, is incident on the beam-facing armor plate installed inside the LHD vacuum vessel, which is located 23m downstream from the ion sources. The injection power is estimated from the shine-through power, which is measured with a calorimeter array on the armor plate [8].

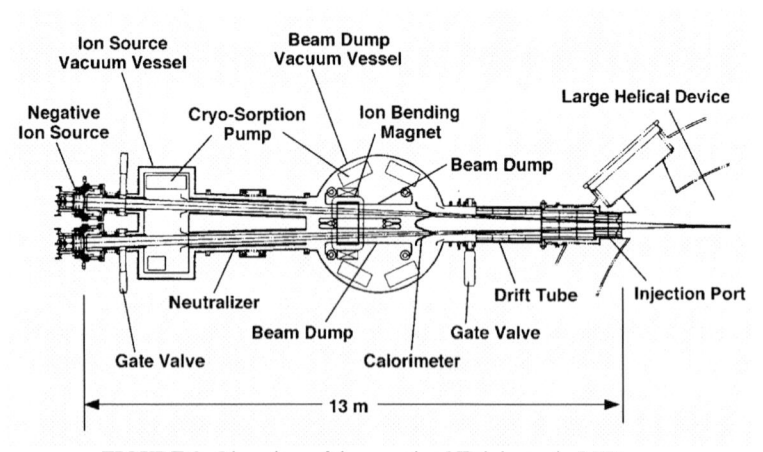

FIGURE 2. Plan view of the negative-NB injector in LHD.

The negative ion source, as shown in Fig. 3, is a filament-driven arc discharge plasma source with cesium seeded, in which the negative ions are produced on the plasma grid (PG) surface with a low work function realized by cesium coverage. The dimensions of the arc chamber are 35cm in width, 145cm in height, and 20cm in depth, and surrounded with line-cusp magnetic field for the plasma confinement. The magnetic filter field is transverse in the width direction in front of the PG. The accelerator consists of 4 grids, PG, EG (extraction grid), SG, and GG. The extraction voltage is applied between the PG and the EG, and the acceleration voltage is between the SG and the GG. The SG is used for the beamlet steering, and the potential of the SG is the same as that of the EG [3,6,9].

The total injection power has been increased year by year. Figure 4 shows the evolution of the total injection power in the negative-NBI system in LHD. Although the total injection power exceeded 9MW in 2001 with adding the third injector, at that time the individual injection power of the injector was limited to 3.5MW due to insufficient voltage holding ability of lower than 160keV. After that, one injector modified the GG aperture shape from multi-round one to multi-slotted one, and the injection energy was much raised to 190keV, leading to great enhancement of the injection power above 6MW. As for the other two injectors, the diameter of the GG round apertures was enlarged, and the injection energy was also raised to 180keV with an injection power of 5MW. As a result, the total injection power was achieved to 16MW in 2007, and every injector exceeds the designed value of 180keV-5MW. In the next section, the effects of the increased grid transparency are discussed with regard to the grid heat load.

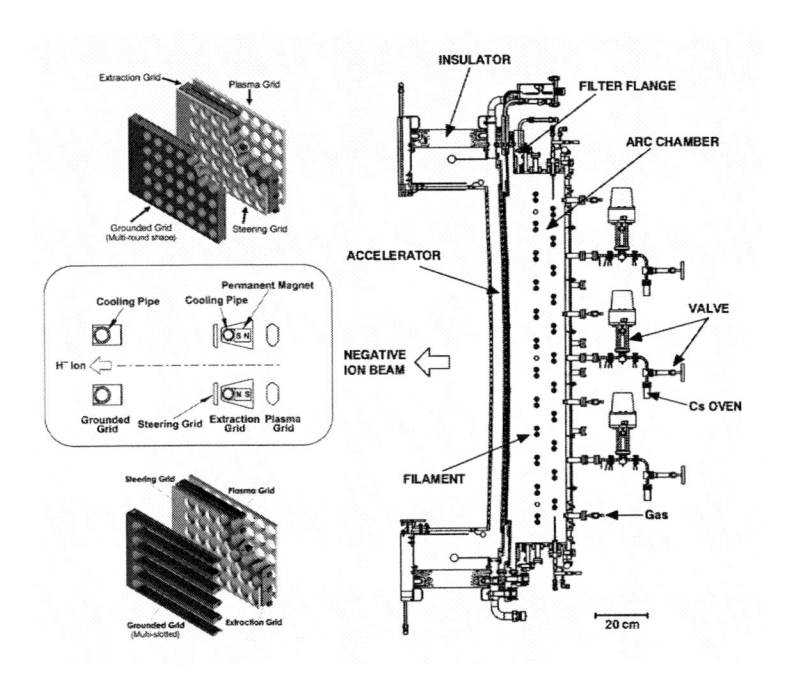

FIGURE 3. Schematic diagram of the negative ion source (right-hand side) and the negative ion accelerator (left-hand side). The multi-round shape GG (left-top) and the multi-slotted GG (left-bottom) are indicated in illustrations of the accelerator.

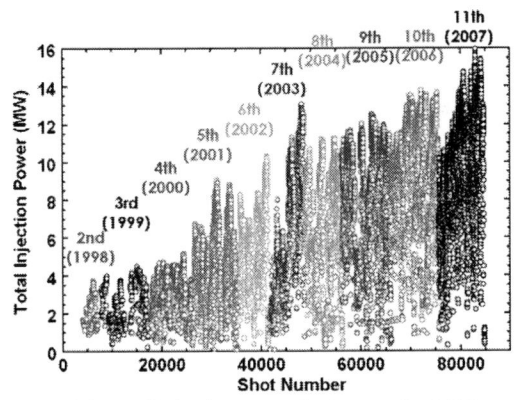

FIGURE 4. History of the total injection power in the negative-NBI system in LHD.

HEAT LOAD REDUCTION BY ENHANCEMENT OF GG TRANSPARENCY

As already reported, the multi-slotted GG is effective in raising the acceleration voltage [6]. A large transparency of 67% lowers the gas pressure in the acceleration gap, leading to reduction of the stripping loss of negative ions. As a result, heat load of the GG is reduced to about a half compared with the multi-round aperture GG with a transparency of 35%. This heat load reduction should contribute to raising the acceleration voltage.

The similar effect is also observed in the multi-round aperture GG. To increase the transparency, the aperture diameter was enlarged to 16mm from 14mm, which corresponds to enlargement of the transparency from 40% to 52% in this case. Figure 5(a) shows the GG heat load currents as a function of the drain current of the acceleration P.S. for GG aperture diameters of 14mm and 16mm. It is found that the GG heat load is much reduced by the enlargement of GG transparency. The heat load reduction led to higher acceleration voltage, and the injection power was increased. A part of the accelerated electrons pass through the GG apertures, and most of these electrons are incident on the electron beam-dump located just downstream from the ion source. The equivalent electron beam-dump currents are shown in Fig. 5(b) for GG aperture diameters of 14mm and 16mm. It is difficult to estimate the absolute value of the electron beam-dump current with water-calorimetry because of structural problems of the electron beam-dump. Thus, the currents in Fig. 5(b) are regarded as relative values. As shown in Fig. 5(b), the passing-through electrons are increased by a factor of about 2 with the enlargement of the GG aperture. The increase in the GG transparency causes a reduction of the stripping loss of negative ions due to the lowering of the gas pressure in the acceleration gap, and also reduces intersection of the accelerated electrons with the GG. The direct intersection of the accelerated negative ions is also reduced by the increase in the GG transparency, and that should contribute to the reduction of the GG heat load as well.

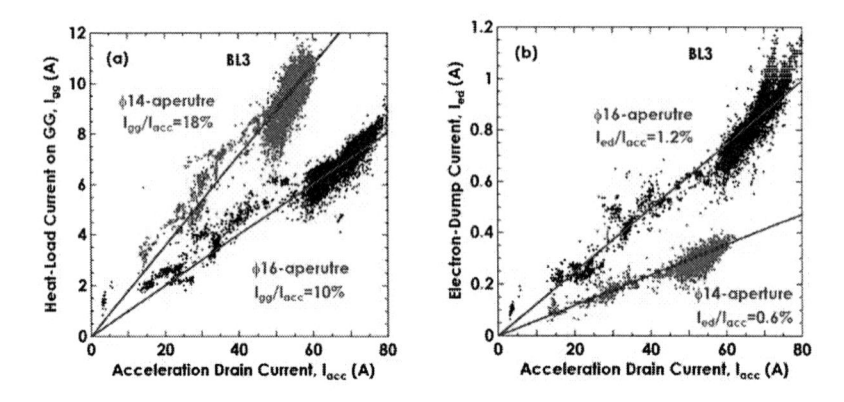

FIGURE 5. (a) Equivalent heat load currents on the GG, and (b) equivalent heat load currents on the electron-beam dump for aperture diameters of 14mm and 16mm in the multi-round aperture GG.

BEAMLET STEERING IN MULTI-SLOTTED GG

Although the multi-slotted GG is effective in raising the acceleration voltage, there is a problem that the optimum condition for the beamlet convergence is different between parallel and perpendicular directions to the slots, due to the anisotropic properties of the electrostatic lens effect at the slots [3,5]. As a result, the whole beam shape is deformed in the perpendicular direction to the slots at the injection port, which is located 13m downstream from the ion sources, and the protection plates for the injection port are occasionally damaged. In the operation, the beamlet convergence condition is made shifted from the optimum one, and thus, the injection efficiency is degraded. To minimize the difference of the optimum condition between the parallel and the perpendicular directions to the slots, the SG apertures have been changed to a racetrack shape from a round one. The racetrack shape works as such an electrostatic lens that the anisotropic properties of the electrostatic lens effect at the GG slots would be compensated. Figure 6 shows the 1/e-half widths of the whole beam profile in the vertical (perpendicular to the slots) and the horizontal (parallel to the slots) directions measured at the calorimeter located 8.6m downstream from the ion source. The beam convergence condition is controlled by the ratio of the acceleration voltage to the extraction voltage, and the optimum voltage ratios were 14.3 and 21.2 in the horizontal and the vertical directions, respectively, for the round-shape aperture SG. As shown in Fig. 6, for the racetrack-shape aperture SG, the optimum voltage ratios are 16.8 and 16.5 in the horizontal and the vertical directions, respectively, and the difference in the optimum beam convergence condition is much reduced. As a result, the deformation of the whole beam shape at the injection port was mitigated, and the port-transmission efficiency was improved. Figure 7 shows the injection efficiency, which is defined as the injection power divided by the acceleration P.S output power, for the racetrack-shape and the round-shape aperture SGs. For the round-shape aperture SG, the voltage ratio in operation deviates from the optimum condition of around 18 to avoid damage of the injection port due to the deformation of the whole beam shape. On the other hand, the optimum voltage ratio is utilized in operation for the racetrack-shape aperture SG, and the injection efficiency in operation is improved more than 10%. As a result, the injection power exceeds 6MW with an energy of 187keV.

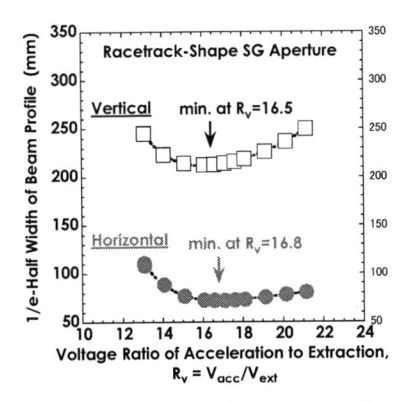

FIGURE 6. 1/e-half widths of the vertical (perpendicular to the slots) and horizontal (parallel to the slots) profiles measured at the calorimeter 8.6m downstream from the ion source, as a function of the voltage ratio of the acceleration to the extraction, in the case of the racetrack-shape aperture SG.

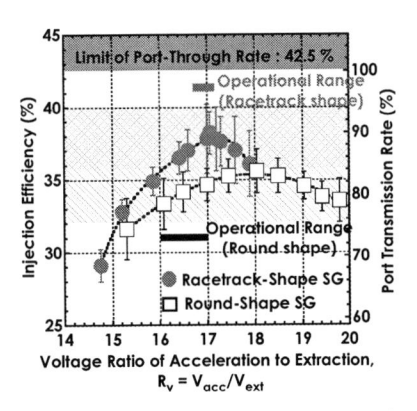

FIGURE 7. Injection efficiency as a function of the voltage ratio of the acceleration to the extraction for the racetrack-shape and the round-shape aperture SGs. The corresponding beam transmission efficiency at the injection port is also indicated on the right-hand vertical axis.

CESIUM BEHAVIOR IN ARC PLASMAS

Cesium is supplied into the arc chamber for the negative ion production, and understanding of the Cs behavior is important to efficient operation of the negative ion source. Tungsten filaments are used as cathodes for the dc-arc discharge, and the tungsten is evaporated during the arc discharge. The amount of the evaporated tungsten is estimated as the weight loss by measuring the weight of W-filaments before and after a series of operation. Figure 8 shows the relationship between the amount of the supplied Cs and the W-filament weight loss. Although the W-evaporation rate is dependent on the operational conditions, such as the arc power and the arc duration, it is found that the supplied Cs amount is nearly proportional to the W-filament weight loss. This result suggests that a dominant contaminant for the cesium should be the tungsten vapor [3,5]. However, a direct evidence to conclude it has not been obtained. To understand the mechanism of cesium consumption, the cesium behavior inside the arc chamber should be investigated.

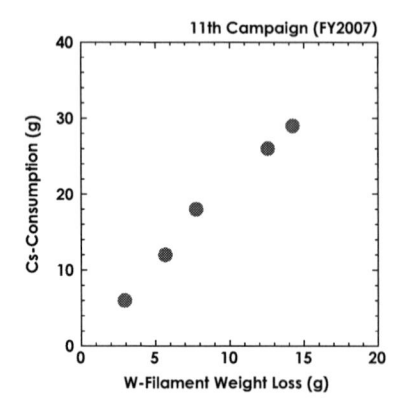

FIGURE 8. Relationship between the amount of the cesium consumption and the weight loss of tungsten-filaments.

Optical emission spectroscopy (OES) is useful to investigate the cesium behavior in the arc plasma as well as the hydrogen plasma properties including the negative ions [11]. Two lines of sight, which observe the discharge region and the negative ion production region (PG region) both in the parallel direction to the PG, are selected to detect the cesium neutral spectrum (CsI:852nm) and the cesium ion spectrum (CsII:460nm). Figures 9(a) and (b) show the time evolution of the spectrum intensities of the CsI neutrals and CsII ions, respectively, in the PG region for long-pulse injection shots with a relatively low arc power of 78kW. The pre-arc duration is 10sec and the beam duration is 20sec. During the injection the acceleration current is nearly constant of 16.5A with an energy of 111keV. Two cases of the initial PG temperature of 180 and 200°C are indicated in the figure, and the PG temperature constantly rose by 100°C for 30sec of the discharge for two cases. Figures 10(a) and (b) also show the time evolution of CsI and CsII, respectively, in the discharge region for the same shots as those indicated in Fig. 10. In the PG region, the Cs neutral intensity is gradually increased, which is not strongly influenced by the beam extraction, and at around t=30sec its increasing rate is much enhanced. On the other hand, in the discharge region, the Cs ion intensity is abruptly increased at the beam extraction, probably because the cesium on the backplate surface is sputtered and evaporated into the plasma by the backstreaming positive ions [12]. After the abrupt increase, the Cs ion intensity continues to increase while it tends to be saturated before the beam extraction.

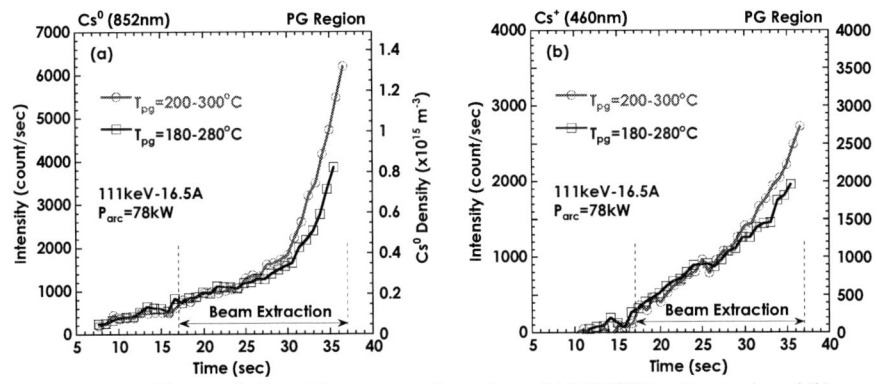

FIGURE 9. Time evolution of the spectrum intensities of (a) CsI(852nm) neutrals and (b) CsII(460nm) ions in the PG region for long-pulse injection shots. In (a), the estimated neutral-Cs density is also indicated on the right-hand vertical axis.

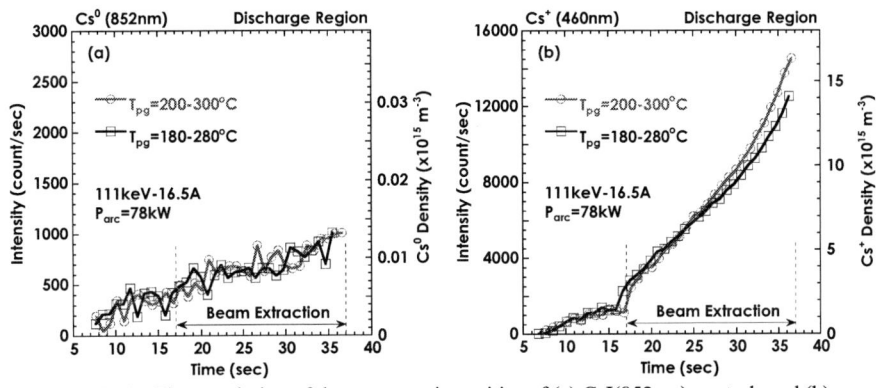

FIGURE 10. Time evolution of the spectrum intensities of (a) CsI(852nm) neutrals and (b) CsII(460nm) ions in the discharge region for the same long-pulse injection shots as those in Fig. 9. The estimated neutral-Cs density and ion-Cs density are indicated on the right-hand vertical axes in (a) and (b), respectively.

The Cs neutral and ion densities can be estimated from the spectrum intensities shown in Figs. 9 and 10, and they are indicated on the right-hand vertical axes. In the PG region, in which the electron temperature is lower than 2eV, the Cs ion density cannot be estimated because the emissivity of the Cs ions is too sensitive to the electron temperature in such a low-temperature range. While the ionization degree is more than 99.9% in the discharge region, the Cs neutral density in the PG region is higher by a factor of 20-100 than that in the discharge region.

In the PG region, the PG support structure is cooled and Cs is much condensed on this cold surface. Thus, the condensed Cs would start evaporation at around t=30sec when the increasing rate is enhanced as shown in Fig. 9(a). Since the electron temperature is not high enough to ionize the neutral Cs, a part of the evaporated Cs diffuses into the discharge region and is ionized. The time evolution of the Cs ion intensity in the PG region seems to follow that of the Cs ion intensity in the discharge region rather than that of the Cs neutral intensity in the PG region. That would indicate that the Cs ions in the discharge region diffuse into the PG region. In the discharge region, the increasing rate of the Cs ion intensity is also enhanced a little at around t=30sec. It is thought that the adsorbed Cs on the chamber wall would start evaporation at that time.

Since the arc power excluding the radiation power is deposited on the cusp lines, only the Cs adsorbed on the cusp lines is evaporated into the arc plasma during the arc discharge. The Cs flowing out from the plasma is then adsorbed on a large area of the chamber wall. During the beam extraction, the backstreaming positive-ions are incident on the backplate, and the deposited power on the backstreaming-beam spots, is roughly estimated at 20% of the arc power [3,5]. Since the total area of the backstreaming-beam spots is larger than that of the cusp lines, it is considered that the Cs recycling is dominated by the backstreaming ions. Thus, the increasing rate of the Cs ion density is much larger during the beam extraction period than that during the only arc discharge period.

According to this Cs-recycling scenario, only Cs on the cusp lines enters the plasma during the arc discharge without the beam extraction, and a larger amount of Cs on the backstreaming-beam spots on the backplate is evaporated into the plasma during the beam extraction. As the pulse duration is extended over several tens of seconds, Cs adsorbed on a large area of the cold surface starts to evaporate into the plasma. Therefore, in the short-pulse injection for a few seconds, it is considered that most of Cs on the chamber wall, except for the cusp lines and the backstreaming-beam spots on the backplate, is not recycled. That means that this area is regarded as a loss region for the supplied Cs, and that the deposited Cs on the loss region would determine the Cs consumption.

SUMMARY

In LHD, the negative-ion-based NBI system is routinely operated for high-power plasma heating experiments. All three negative-NB injectors have achieved the designed performance of 180keV-5MW injection, and one injector has injected 188keV-6.4MW. Total injection power achieved with three injectors is 16MW. Reduction of the GG heat load is essential to improve the voltage holding ability, and enhancement of the GG transparency is effective to reduce the heat load. As well as in the multi-slotted GG, it is observed that the GG heat load is reduced in the multi-round aperture GG with an enlarged aperture diameter and that the voltage holding ability is improved. As a result, the beam energy is raised, leading to an increase in the injection power.

To compensate the anisotropic property of the electrostatic lens effect between the perpendicular and the parallel directions to the slots in the multi-slotted GG, the SG aperture shape has been changed to a race-track shape from a round shape. As a result, the difference of the optimum condition for the beamlet convergence is much reduced between the perpendicular and the parallel directions, and the whole beam injection efficiency is improved, leading to a increase in the injection power.

The cesium behavior is investigated with optical emission spectroscopy (OES). The Cs adsorbed on the cusp lines is evaporated into the plasma during the arc discharge, and during the beam extraction the Cs adsorbed on the backplate is sputtered and thermally evaporated into the plasma by the backstreaming positive ions. The Cs flowing out from the plasma is deposited onto a large area of the chamber surface, and most of Cs on the chamber surface is not evaporated. Since the total area of the backstreaming-beam spots is larger than that of the cusp lines, the Cs recycling is dominated by the evaporated Cs by the backstreaming ions. As the arc duration is extended, a part of Cs adsorbed on cold surfaces starts evaporation. In this Cs recycling model, most of chamber surfaces are loss regions for the supplied Cs. On the other hand, it is observed that the amount of supplied Cs is proportional to the amount of the tungsten evaporated from the filaments. The further investigation is required for the Cs recycling, and it should be clarified whether the tungsten vapor is a dominant contaminant for the cesium or not.

The LHD negative-NBI system now shows high-performance, which is ascribed to the continuous improvement of the negative ion sources. This achievement should contribute to the development of the ITER negative-NBI system.

ACKNOWLEDGMENTS

The authors would like to acknowledge to all members of the LHD experimental group and the technical stuff in the LHD. They are grateful to Professor O. Motojima, Director-General, for his continuous encouragement and support. This work is partially supported by NIFS07ULBB501.

REFERENCES

1. O. Motojima, et al., *phys. Plasmas* **6**, 1843 (1999).
2. O. Kaneko, et al., *Nucl. Fusion* **43**, 692 (2003).
3. Y. Takeiri, et al., *Nucl. Fusion* **46**, S199 (2006).
4. R. S. Hemsworth, et al., *Rev. Sci. Instrum.* **67**, 1120 (1996).
5. Y. Takeiri, et al., "High-Power Negative Ion Sources for Neutral Beam Injectors in Large Helical Device" in *Production and Neutralization of Negative Ions and Beams*, edited by M. P. Stockli, AIP Conference Proceedings 925, American Institute of Physics, Melville, NY, 2007, pp. 211-223.
6. K. Tsumori, et al., *Rev. Sci. Instrum.* **75**, 1847 (2004).
7. K Ikeda, et al., *Rev. Sci. Instrum.* **75**, 1744 (2004).
8. M. Osakabe, et al., *Rev. Sci. Instrum.* **72**, 590 (2001).
9. Y. Takeiri, et al., *Rev. Sci. Instrum.* **73**, 1087 (2002).
10. K. Tsumori, et al., *Rev. Sci. Instrum.* **79**, 02C107 (2008).
11. U. Fantz, et al., *Nucl. Fusion* **46**, S297 (2006).
12. K. Ikeda, et al., *Rev. Sci. Instrum.* **79**, 02A518 (2008).

Status of the Negative Ion Based Heating and Diagnostic Neutral Beams for ITER

B. Schunke, D. Bora, R. Hemsworth, A. Tanga

ITER Organization, Cadarache, 13108 St.-Paul-lez-Durance, France

Abstract. The current baseline of ITER foresees 2 Heating Neutral Beam (HNB's) systems based on negative ion technology, each accelerating to 1 MeV 40 A of D⁻ and capable of delivering 16.5 MW of D^0 to the ITER plasma, with a 3rd HNB injector foreseen as an upgrade option [1]. In addition a dedicated Diagnostic Neutral Beam (DNB) accelerating 60 A of H⁻ to 100 keV will inject ≈15 A equivalent of H^0 for charge exchange recombination spectroscopy and other diagnostics. Recently the RF driven negative ion source developed by IPP Garching has replaced the filamented ion source as the reference ITER design. The RF source developed at IPP, which is approximately a quarter scale of the source needed for ITER, is expected to have reduced caesium consumption compared to the filamented arc driven ion source. The RF driven source has demonstrated adequate accelerated D⁻ and H⁻ current densities as well as long-pulse operation [2, 3]. It is foreseen that the HNB's and the DNB will use the same negative ion source. Experiments with a half ITER-size ion source are on-going at IPP and the operation of a full-scale ion source will be demonstrated, at full power and pulse length, in the dedicated Ion Source Test Bed (ISTF), which will be part of the Neutral Beam Test Facility (NBTF), in Padua, Italy. This facility will carry out the necessary R&D for the HNB's for ITER and demonstrate operation of the full-scale HNB beamline. An overview of the current status of the neutral beam (NB) systems and the chosen configuration will be given and the ongoing integration effort into the ITER plant will be highlighted. It will be demonstrated how installation and maintenance logistics have influenced the design, notably the top access scheme facilitating access for maintenance and installation. The impact of the ITER Design Review and recent design change requests (DCRs) will be briefly discussed, including start-up and commissioning issues. The low current hydrogen phase now envisaged for start-up imposed specific requirements for operating the HNB's at full beam power. It has been decided to address the shinethrough issue by installing wall armour protection, which increases the operational space in all scenarios. Other NB related issues identified by the Design Review process will be discussed and the possible changes to the ITER baseline indicated.

Keywords: ITER, Neutral Beams, Ion Sources
PACS: 28.52, 52.55.Fa, 29.25

Introduction

In the current ITER baseline, four operating scenarios are foreseen with up to 110MW of auxiliary heating power using a variable heating mix (TABLE 1). In all scenarios a substantial contribution (up to 45%) will come from neutral beam

	Start up	Scenario 1: Elmy Hmode I	Scenario 2: Elmy HMode II	Scenario 3: Hybrid	Scenario 4: Steady state
	Power [MW]	Power [MW]	Power [MW]	Power [MW]	Power [MW]
NB	33	33	50	50	50
IC	20	40	20	40	20
EC	20	40	40	40	20
LH	0	20	20	0	40
Total	73	133	130	130	130

TABLE 1. DESIGN SCENARIOS FOR ITER OPERATION [1]

injection. The scenarios will certainly be re-assessed in the future as research progresses and new results from scaling and modelling introduce corrections to the scenarios. Flexibility and upgradeability therefore has to be an important factor in the design of any heating system. In the past modelling effort has understandably focussed on the burn phases of ITER, but now start-up and commissioning scenarios are being assessed in detail to prepare for the first ITER plasmas. It is planned to end the ITER construction phase with the demonstration of the first plasma at 2 MA. This will be followed by 5 years of start-up consisting of hydrogen and helium plasmas, which will allow plasma commissioning and testing of all major components at full power without unnecessarily activating the machine. Access to the H-Mode should be assured to allow testing of the divertor components. The expert panels taking part in the ITER Design Review 2007/2008 have stressed the key role to be played by the heating and current drive systems, including the neutral beams, in achieving and controlling plasma performance in ITER. As the routine availability of 73 MW of auxiliary power is paramount, it is suggested that a power margin of 10 MW is already made available in the start-up phase to compensate for eventual down time of un-optimised systems. It is likely that most of the initial operations will be devoted to mapping of the operational space, and this should not be limited by the heating and current drive systems.

The Heating Neutral Beams for ITER

The current baseline of ITER includes 2 Heating Neutral Beam (HNB) systems based on negative ion technology, each operating at 1 MeV 40 A D$^-$ ions and delivering up to 16.7 MW to the ITER plasma. Both injectors will be operational for the first ITER plasma scheduled in mid 2018. A 3rd HNB injector is foreseen as an upgrade option to meet the requirements of operating scenarios 2 to 4. Each

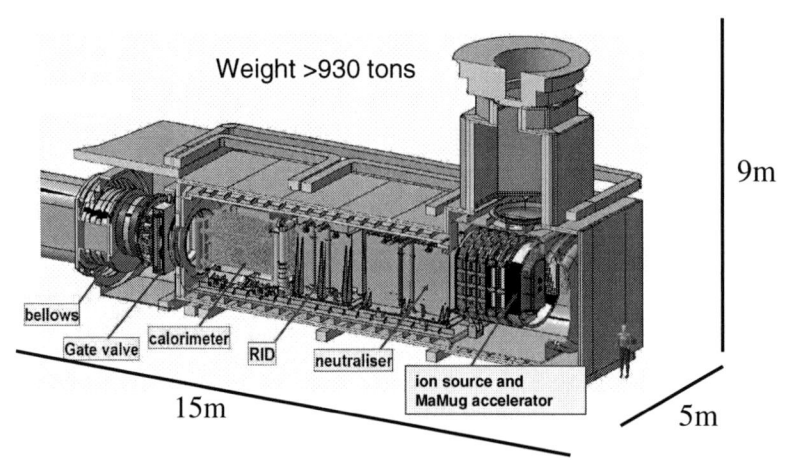

FIGURE 1. Schematic of the HNB injector

HNB injector (Fig. 1) consists of an RF ion source and a 5 stage accelerator, a neutraliser, a residual ion dump (RID) and a calorimeter. The latter is a moveable beam dump which, in its normal position allows the beam to pass directly to ITER, but which can be moved to intercept the beam downstream of the RID, allowing the injectors to be commissioned and tested independently of ITER operation. Cryopumps maintain an appropriately low pressure inside the vacuum vessels. The primary vacuum boundary consists of the beam source vessel (BSV), the beam-line vessel (BLV) and the 1 MV bushing (see figure 1). The ion source and accelerator are located in the BSV and the beamline components in the BLV. The high voltage (HV) bushing is the interface between the SF_6 insulated transmission line from the power supplies and the ion source and accelerator in vacuum in the BSV. It also provides the passage of coolant and gas lines from the transmission line to the ion source and the accelerator. To shield the beam line from the magnetic field from ITER, ferromagnetic shielding and active correction and compensating coils (ACCCs) are foreseen.

The ITER baseline has been modified in 2007 to accept the RF ion source, based on the Garching design (Fig. 2) [2], as the reference ion source for ITER. The main advantages are the expected significant reduction in caesium consumption and a reduced maintenance frequency because the regular replacement of the filaments required for the arc driven source is no

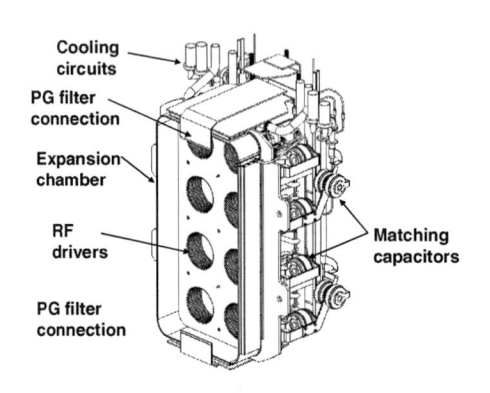

FIGURE 2. The RF Source for ITER.

longer required. The latter is very important in the ITER context as the ion source will become highly activated during DT operation of ITER and such maintenance has to be done by remote handling, which is difficult and long since letting the source up to atmospheric pressure will require cleaning the then contaminated Cs from the ion source and plasma grid. Maintenance of the RF driven source becomes necessary when the quantity of Cs in the source becomes excessive, either because of safety considerations or because the ion source performance deteriorates. The maintenance frequency will be assessed once the Cs consumption in normal operation is established.

It has been previously shown that in short pulse operation (≈4 s) ITER relevant H⁻ and D⁻ current densities can be extracted from the RF driven source [2]. Recently stable operation for 1 h pulse operation [3] with an accelerated H⁻ current density of ≈120A/m^2 and <0.5 electrons extracted per accelerated H⁻ has been achieved.

In May 2008 the decision has been taken to select the Multi Aperture Multi Grid accelerator (MAMuG) [4, 5] as the ITER reference accelerator. MaMug has demonstrated superior voltage holding capability and, more importantly, significantly lower electron power exiting the accelerator than the alternative SINGAP [4] accelerator. The HNB power supplies have been designed to be compatible with both accelerator options should the decision have to be corrected in the future.

The baseline operating parameters of the ITER NB systems are summarized in Table 2. The required parameters (e.g. 1 MeV 200 A/m^2 for the HNB) have not been met simultaneously in today's NB testbeds and it is acknowledged that substantial R&D is necessary to meet the ITER specifications.

	HNB	**DNB**
Beam Power	16.7 MW	3.6 MW excluding duct losses
Beam Energy	1 MeV (D⁻) / 870 keV (H⁻)	100 keV (H⁻)
Extracted Current	40 A (D) / 46 A (H⁻)	60 A (H⁻)
Current density	200 A/m^2 (D⁻) / 300 A/m^2 (H⁻)	300 A/m^2
Current density uniformity	± 10%	± 10%
Divergence	7 mrad	7 mrad
Pulse Length	≤ 3600s	5Hz. 1/6 of ITER pulse

TABLE 2. NEUTRAL BEAM SPECIFICATIONS

The Diagnostic Neutral Beam for ITER

The DNB provides the dedicated probe beam for the CXRS diagnostic, which is the only diagnostic system capable of providing absolute measurements of He density profiles in the ITER plasmas. To concentrate R&D effort and necessary tools it had been decided that the DNB should follow the HNB concept wherever possible (see figure 3). Some DNB performance parameters are even more stringent than those for the HNB, as can be deduced from table 2. The beam quality in particular has an important impact on the fraction of the beam that will reach the ITER plasma, and thus on the signal to noise ratio of the diagnostics.

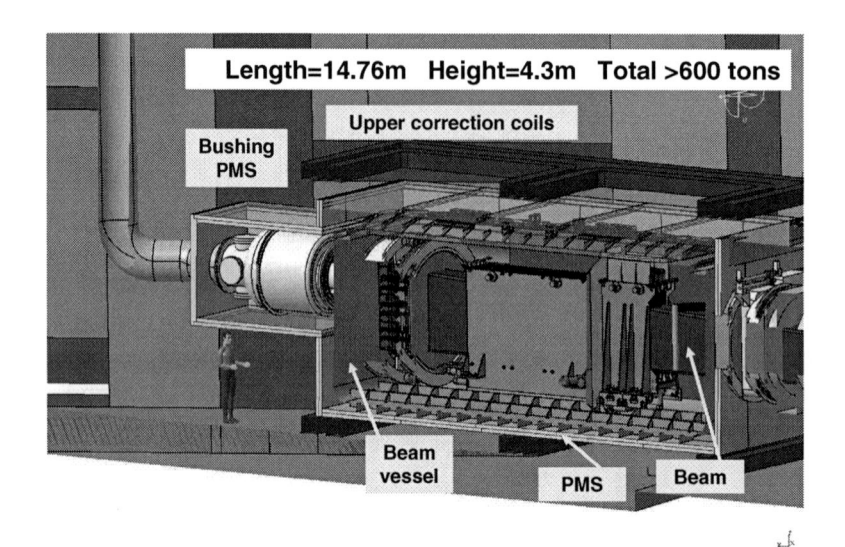

FIGURE 3. View of the DNB

The DNB will operate in hydrogen for all ITER plasmas, the target parameter for the divergence is <7 mrad. A low divergence is necessary to minimise both the aperture size in the vessel wall and the heat loads in the beam duct. The DNB will be modulated at 5 Hz with a duty cycle of ~1/6 mainly limited by technical constraints, namely the fatigue life of the components. The DNB accelerator is much simpler than that of the HNB, having only one acceleration stage and an acceleration voltage of 100 kV. The DNB is presently designed to use the same RF ion source as the HNB, but note that the accelerated current is 60 A (H$^-$). The increase in the accelerated current compared to that from the HNB will be achieved partly by the reduced stripping losses in the shorter accelerator ($\approx$14% compared to $\approx$30%) and partly by operation of the ion source at higher power. The latter is not possible with the HNB as the power to the extraction grid from the co-extracted electrons becomes excessive. The DNB will operate exclusively in hydrogen for all ITER scenarios. As the co-extracted electron current is found to be significantly lower with H$^-$ than with D$^-$, higher power operation, giving higher H$^-$ currents, is possible without overloading the extraction grid.

As with all high power negative ion sources, a magnetic filter is needed in the source to restrict the co-extracted electrons to acceptable levels. In the HNB ion source this is produced by the combination of a 4 kA current passing vertically through the plasma grid and columns of permanent magnets either side of the plasma grid on the wall of the ion source, which are magnetised to create a field traversing the source that adds to the field from the current in the plasma grid. Recent analyses by M. Singh [6] have demonstrated that this magnetic filter is not viable for the DNB as the long range field from the filter leads to an adverse deflection of significant parts of the accelerated beam. Investigations have recently started of several alternative magnetic filters, and it has already been shown that one of the options appears viable. The final choice will be made after all the options have been studied in detail.

The magnetic shielding system, consisting of a passive magnetic shield and ACCCs, will have to be optimised to minimise the effect of the residual field on the H⁻ beam before it is neutralised (increased divergence and beam deflection).

Originally it was intended to inject the diagnostic beam exactly perpendicular to the ITER plasma at near the equatorial plane of the machine. However the injection angle of the DNB has been modified following concerns expressed at the ITER 2007 Design Review that ripple trapped ions could both pose a hazard to the inner wall below the injection point and perturb the diagnostic measurements. Minor adjustments to the NB cell and repositioning of the DNB allowed a ~6° injection angle to be obtained. Additionally the design of one of the ITER blanket modules has been modified in order to have a much larger aperture in the blanket through which the diagnostic beam will be injected. The port in the vacuum vessel has been modified and the opening enlarged to accommodate this change and the detailed design of the re-arranged and enlarged DNB duct between the injector and the torus is now underway. In parallel the beam optics has been optimised to maximise the throughput with the new arrangement.

The low beam energy makes operation of the DNB possible even at the low plasma densities achievable with 2MA current. The diagnostic tool should therefore be available for all ITER operating scenarios from the hydrogen phase onwards.

NB R&D Issues

The ITER NB systems will be provided by in kind procurement: the EU and Japan are jointly responsible for the first two HNB injectors, and India for the DNB. The ITER International Organization will provide the necessary integration effort. The requirements for the NB injectors for ITER represent a large step forward from present day injectors. A detailed risk assessment carried out in the framework of the 2007 Design Review had highlighted the main technical issues to be solved. The importance of a testbed capable of demonstrating the ITER size injector had been emphasized, and confirmed as the main risk mitigation measure for the neutral beams for ITER. The Neutral Beam Test Facility (NBTF) has the mission to complete the development of the full size HNB and DNB ion sources and to develop and test all critical elements of the HNB system. To ensure that this development can be carried out as rapidly as possible the NBTF will actually consist of two facilities: the Ion Source Test Facility (ISTF) and the MegaVolt Test Facility (MVTF) which will operate independently. The ISTF will finalise the development of the HNB and DNB ion sources. It will consist essentially of a beamline made up of a vacuum vessel, cryopumps, the ion source, a 100 kV accelerator, a calorimeter and various diagnostics. The extraction and acceleration of the 100 keV beam will enable the beam uniformity, current density and the 100 keV beamlet divergence to be measured correctly. It will be capable of operation in either hydrogen or deuterium. The MVTF will be essentially a complete HNB injector. It will use a power system, ion source accelerator and beamline components essentially identical to those of the HNB. It will be capable of full power, full pulse length (3600 s) operation in either hydrogen (46 A H⁻ beams accelerated to 870 keV) or deuterium (40 A D⁻ beams accelerated to 1 MeV).

Among the known issues that have to be addressed are: beam uniformity, high current and high current density extraction for long pulses, control of the co-extracted electron current, voltage holding, beam optics and the performance of the RID and calorimeter. Progress on most of these issues is pending as the number of 1 MV test beds in the world is small (2), and those available (the Japanese test bed at JAEA and the European test bed at Tore Supra) are limited to low currents (<0.1 and 0.5 A respectively) and short pulses ($\approx$1 s) [7, 8].

The NBTF is to be constructed in Padua, Italy and operated by the EU with participation from Japan, India and other ITER parties. The MVTB will have real 1 MeV handling capability and demonstrate the full scale ITER neutral beam. The ISTF will concentrate on the HNB ion source development first in order to have

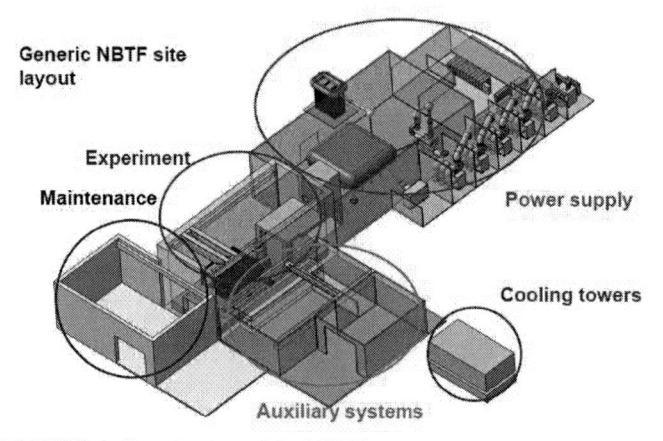

FIGURE 4. Generic view of the NBTF

that completed as far as possible before the MVTF starts operation. The ISTF should start operation in 2011 and the NBTF in early 2014. The design of the NBTF facility has progressed (see figure 4) and the site preparation is due to start soon.

Work in the NBTF will address the specific risks that have been identified, such as the homogeneity of the RF ion source; avoidance of uncontrolled breakdowns in the accelerator, and control of back-streaming ions. Material fatigue of high heat flux components will be studied and reliability analysis undertaken. Viable 1 MV technology has to be developed including adequate isolation transformers and transmission lines with 1 MV holding capability. On the ITER beamline diagnostics will be reduced to the necessary minimum for control and feedback purposes, whereas the NBTF will incorporate extensive diagnostic tools, most of which still have to be designed.

In preparation for the NBTF, research will continue in Garching, where the next step – a test bed with a half size ion source with ion extraction - is being designed and will start operation in 2009. India has also prepared an accompanying negative ion program to support the DNB development. For this a quarter size negative ion source will be procured from Garching. Experiments in support of the DNB and the ISTF can start in the Indian testbed end of 2009.

ITER Plant Integration

The NB systems are installed in the neutral beam cell, which occupies most of the northern part of the equatorial level of the tokamak hall. They will be installed through a door on the north wall of the ITER building, to be closed once nuclear

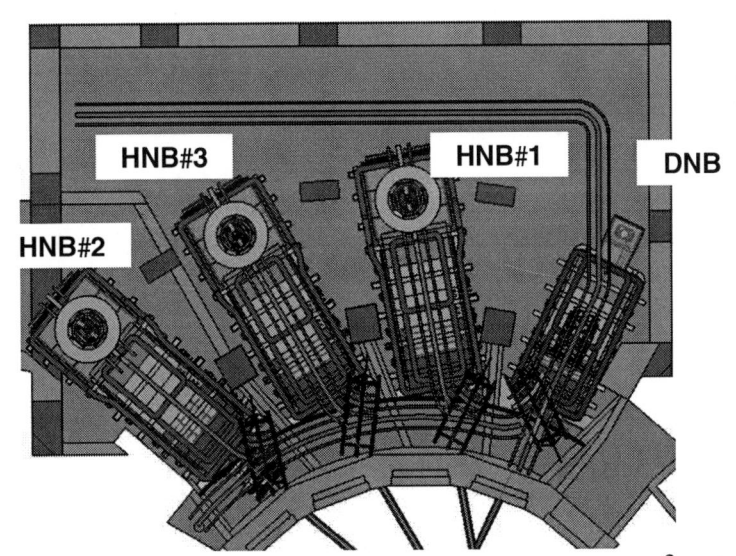

FIGURE 5. Lay-out of the neutral beam cell showing (left to right) the 3 HNB injectors and the DNB.

operation starts. The injectors are shown in Fig. 5, including the 3rd injector, foreseen as an upgrade option. Installation and access logistics require to install the first two HNB injectors on equatorial ports 4 and 6, and the optional future 3rd HNB injector on port 5. The DNB duct crosses that of HNB1, with consequent complexity in the design. A remote handling (RH) monorail crane will be used for installation and maintenance operations. The injectors are connected directly to the torus vacuum via the neutral beam ducts, which will be equipped with a suitable liner to withstand the power load due to direct interception of divergent parts of the beam or re-ionised particles deflected onto the liner by the magnetic field form ITER. Passive magnetic shields (PMS's - made of 150 mm thick iron) and pairs of active correction and compensation coils above and below the beamline are foreseen to reduce the field from the tokamak inside the injectors. The PMS's also act as radiation shields for the NB cell, minimising the neutron and gamma radiation entering the NB cell. A support structure is installed around the PMS's to prevent movement of the PMS's and the beamline vessels in the case of an earthquake.

Installation and maintenance logistics have necessitated several changes from the original 2001 design, as has site adaptation requirements. When the PMS's are closed man access to the neutral beam cell is possible, but restricted because of activation of the injectors and the shielding. If the PMS of an injector is opened, no man access will be possible as the injector components will be highly activated. Thus maintenance and interventions on the beamline components, the ion source and the accelerator will have to be carried out by remote handling (RH). A maintenance scheme has been developed whereby the beamline components are removed or installed via an overhead crane after opening the top lid of the beamline vessel. The beamline vessels had to be re-designed to have a rectangular shape in order to accommodate the removable lids. Consequently the beam source vessel, the PMS, and the cryopumps had to be re-designed to have a corresponding rectangular shape. Because the HV bushing is installed at the top of the beam source vessel, the ion source and the accelerators are maintained (horizontally) from the rear of the injector. To facilitate this, the back of the beam source vessel incorporates a large rectangular flange that is almost as large as the beam source vessel itself. Two smaller flanges are incorporated in the rear flange that allow the replacement of empty with full Cs ovens.

The NB injectors are directly coupled to the tokamak and represent the extension of the first confinement barrier for radioactive materials. Therefore an absolute valve had to be added between the injector and the tokamak. This valve will be closed only during injector maintenance, when the injectors will be open, to avoid contamination of the NB cell. The valve has to be all-metal as organic materials cannot withstand the radiation environment. During cryopump regeneration, or when the injectors are not operating, the fast shutter will isolate the injectors from the tokamak. The detailed design of the valve and the beam ducts is ongoing, with the HNB-DNB cross-over and the valve representing technically challenging engineering tasks.

A major change to the NB design concerned the modification of the NB power supplies. Re-locating the high voltage deck outside the tokamak building and using air-insulation allows easier access to electronic control equipment [10], thus minimising eventual system downtime for repairs

Start-up issues and ITER Baseline changes

The start-up phase in hydrogen will allow commissioning of all sub-systems without unnecessarily activating the machine. During this phase achievable plasma densities will be low, which would have prevented the neutral beam injection from efficiently delivering power to the plasma at full beam energy as the shine through would exceed the power density limit on the inner wall of the tokamak (P<0.5 MW/m^2). It has been decided to install additional wall protection (wall armour), rather than the alternative of modifying the NB system to operate at 500 keV at about half power. The HNB's will now be operated in hydrogen at up to 870 kV. The HNB's will use the same accelerator during H operation as intended for D operation. At 870 keV the accelerated H- current at perveance match will then be 46 A, and the beam power will be 40 MW. Under these conditions the injectors will able to deliver the full 33 MW beam power from two injectors during H or He operation of ITER. This change to the ITER baseline also increases the range of plasma scenarios accessible in DD and DT. A technically feasible wall armour solution has been identified, based on hypervapotron technology. The wall armour will be capable of accepting ≤4 MW/m^2, which will allow operation over the foreseen range of ITER operating scenarios.

The existing duct liner design has been reviewed and judged not maintainable. A new, maintainable, conceptual design has been proposed, based on a lighter modular construction.

ITER requirements state that the HNB's should be capable of injecting at a vertical angle of ±9mrad from the nominal beam axis. It is calculated that tilting the beam at the extreme position up or down can lead to ≈1.5 MW loss in the duct of each injector, and that any beam misalignment vertically combined with the full tilt could damage the liner. To fully comply with the requested beam tilting of ±9 mrad, the height of the NB port and the corresponding blanket openings would have to be increased by >200 mm. To avoid overloading the liner, thermocouples have to be incorporated in the critically loaded sections.

Installation of the HNB and the DNB injector will start as soon as the tokamak building is available, allowing sufficient time for commissioning and debugging. The installation work on all injectors will progress in parallel as the same site infrastructure and tools will be used. The complete installation and off-plasma commissioning will take around 4 years. During this time the work in the NBTF will continue. It is expected that a number of changes and modifications will originate from the NBTF, and that these changes will then be replicated in the HNB. Provisions have therefore been taken to ease access to the beamline components during the commissioning phase.

Most of the NB commissioning can be carried out without plasma as the HNB and DNB have a calorimeter to perform the off line commissioning. Plasma is only needed for commissioning and conditioning of the duct. The power can be injected as soon as the required plasma densities can be obtained. With first plasma mid 2018 the NB commissioning will be completed toward the middle of the H phase for the 2 HNB injectors and the DNB. With long flat top plasma capability the NB pulse can be extended up to the design value of 3600 s.

Summary and Outlook

The design activities aimed at providing NBs for ITER are progressing and several changes have been accepted into the ITER baseline recently. Both the RF ion source and the air insulated power supply will guarantee better maintainability of the NB systems. The R&D necessary for the design of the 1 MeV neutral beams for ITER and the ion source development will be provided by the EU in the NBTF in Padua, with participation of Japan and India. The development and testing at the NBTF strongly mitigates the risk and it will clarify the performance of all the injector components. The detailed design phase for the ITER injectors started at the beginning of 2008.

References

1. ITER PID, https://users.iter.org/users/idm?document_id=ITER_D_2234RH
2. E. Speth, et. al., *Nucl. Fusion* **46** 220 (2006)
3. W Kraus, et. al., *Rev. Sci. Inst.* **73** 02C108 (2008)
4. D. Boilson, et. al., *Rev. Sci. Inst.* **73** 1093 (2002)
5. T. Inoue, et. al., *Fusion Eng. Des.* **A 56–57** 517 (2001)
6. M. Singh, ITER Task Report C3PP41FI
7. T. Inoue, et. al., *Fusion Eng. Des.* **82** 813 (2007)
8. L. Svensson, et. al., *Fusion Eng. Des.* **66-68** 627 (2003)
9. A. Staebler, et. al,, "Development of an RF-driven ion source for the ITER NBI system", to be published in Fusion Eng. Des. [2009]
10. E. Gaio., et. al., *Fusion Eng. Des.* **83** 21–29 (2008)

List of Participants

Agostinetti	Piero	Consorzio RFX	piero.agostinetti@igi.cnr.it
Alonso	Jesus	Fundación Tekniker-IK4	jalonso@tekniker.es
An	Younghwa	Seoul National University	ayh1800@snu.ac.kr
Ando	Akira	Tohoku University	akira@ecei.tohoku.ac.jp
Annaratone	Beatrice Maria	PIIM CNRS/Universite de Provence	bma@mpe.mpg.de
Bacal	Marthe	Ecole Polytechnique	bacal@lptp.polytechnique.fr marthebacal@free.fr
Baturin	Volodymyr	National Academy of Sciences of Ukraine	baturin@ipflab.sumy.ua, belska@ipfcentr.sumy.ua
Bechu	Stephane	CNRS / LPSC	stephane.bechu@ujf-grenoble.fr
Becker	Reinard	J.W.Goethe Universitat Frankfurt	rbecker@physik.uni-frankfurt.de
Boeuf	Jean Pierre	LAPLACE CNRS	jpb@laplace.univ-tlse.fr
Boilson	Deirdre	DCU	deirdre.boilson@cea.fr
Boudreault	Ghislain	Copenhagen University Hospital	ghilt@pet.rh.dk
Carrere	Marcel	Universite de Provence	carrere@up.univ-mrs.fr
Cartry	Gilles	Universite de Provence / CNRS	gilles.cartry@univ-provence.fr
Cavenago	Marco	INFN-LNL	cavenago@lnl.infn.it marco.cavenago@lnl.infn.it

Chaibi	Oualid	CEA Cadarache	oualid.chaibi@cea.fr
Cojocaru	Gabriel	TRIUMF	cojocaru@triumf.ca
Crowley	Brendan	UKAEA	brendan.Crowley@jet.uk
De Esch	Hubert	CEA Cadarache	hubert.de-esch@cea.fr
Delvaux	Jean-luc	Ion Beam Applications s.a.	delvaux@iba.be
Devynck	Pascal	CEA Cadarache	pascal.devynck@cea.fr
Dure	Franck	Universite Paris Sud XI	franck.dure@u-psud.fr
Engeln	Richard	Eindhoven University of Technology	r.engeln@tue.nl
Enparantza	Rafael	Fundación Tekniker-IK4	renparantza@tekniker.es
Faircloth	Dan	STFC	Dan.Faircloth@rl.ac.uk D.C.Faircloth@rl.ac.uk
Fantz	Ursel	Max-Planck-Institut fur Plasma Physik	ursel.fantz@ipp.mpg.de, fantz@ipp.mpg.de
Franzen	Peter	Max-Planck-Institut fur Plasma Physik	peter.franzen@ipp.mpg.de
Fubiani	Gwenael	CNRS	gwenael.fubiani@laplace.univ-tlse.fr
Fukumasa	Osamu	Yamaguchi University	fukumasa@plasma.eee.yamaguchi-u.ac.jp
Fumelli	Michele	CEA Cadarache	michele.fumelli@wanadoo.fr
Gabriel	Onno	Eindhoven University of Technology	o.g.gabriel@tue.nl

Grisham	Larry	Princeton University	lgrisham@pppl.gov travel@pppl.gov
Gutser	Raphael	Max-Planck-Institut fur Plasma Physik	raphael.gutser@ipp.mpg.de
Hanstorp	Dag	University of Gothenburg	dag.hanstorp@physics.gu.se
Hatayama	Akiyoshi	Keio University	akh@ppl.appi.keio.ac.jp
Hemsworth	Ronald	ITER Organization	ronald.hemsworth@iter.org
Hutter	Thierry	CEA Cadarache	thierry.hutter@cea.fr
Inoue	Takashi	JAEA	inoue.takashi52@jaea.go.jp
Jacquot	Claude	CEA Cadarache	jacquotclaude6231@neuf.fr
Kamada	Masaki	JAEA	kamada.masaki@jaea.go.jp
Keller	Roderich	LANL	roderich@lanl.gov
Khemliche	Hocine	LCAM - CNRS/UPS	hocine.khemliche@u-psud.fr
Kisaki	Masashi	Tohoku University	kisaki.masashi@jaea.go.jp, masashi.kisaki@ppl2.qse.tohoku.ac.jp
Kohen	Nicolas	CNRS	nicolas.kohen@laposte.net gwenael.fubiani@laplace.univ-tlse.fr
Kolev	Stanimir	CNRS	Stanimir.Kolev@laplace.univ-tlse.fr
Kovari	Michael	UKAEA	michael.kovari@jet.uk,
Kraus	Werner	Max-Planck-Institut fur Plasma Physik	Kraus@ipp.mpg.de wrk@ipp.mpg.de

Krylov	Alexander	Kurchatov Institute	krylov@nfi.kiae.ru
Kumar	Ajeet	Raja Ramanna Centre for Advanced Technology	ajeetk52@yahoo.co.in ajeet@cat.ernet.in
Kuppel	Sylvain	Keio University	shiruvan@ppl.appi.keio.ac.jp
Lawrie	Scott	STFC	scott.lawrie@rl.ac.uk S.R.Lawrie@rl.ac.uk
Lee	Seok-geun	Seuol National University	ssugny1@snu.ac.kr
Lemoine	Didier	LCAR, CNRS-Universite Paul Sabatier	didier.lemoine@irsamc.ups-tlse.fr
Lepetit	Bruno	CNRS	bruno.lepetit@irsamc.ups-tlse.fr
Lettry	Jacques	CERN	jacques.lettry@cern.ch Lauriane.Bueno@cern.ch
Liu	Yuan	ORNL	liuy@ornl.gov
Longo	Savino	BARI University & IMIP/CNR	savino.longo@ba.imip.cnr.it
Lotte	Philippe	CEA Cadarache	Philippe.Lotte@cea.fr
Matsushita	Daisuke	Keio University	matsushita@ppl.appi.keio.ac.jp
McAdams	Robert	UKAEA	Roy.McAdams@jet.uk
McNeely	Paul	Max-Planck-Institut fur Plasma Physik	p.mcneely@ipp.mpg.de
Minea	Tiberiu	LPGP / CNRS	tiberiu.minea@u-psud.fr, tiberiu.minea@pgp.u-psud.fr
Mochalskyy	Serhiy	Laboratorie de Physique des Gas des Plasmas	Mochalskyy@gmail.com

Oka	Yoshihide	NIFS	oka@lhd.nifs.ac.jp, oka@sc.starcat.ne.jp
Peters	Jens	DESY	jens.peters@desy.de
Sanin	Andrey	Budker Institute of Nuclear Physics	sanin@inp.nsk.su
Schiesko	Loic	Universite de Provence	loic.schiesko@univ-provence.fr
Schunke	Beatrix	ITER Organization	beatrix.schunke@iter.org
Scrivens	Richard	CERN	Richard.Scrivens@cern.ch, Lauriane.Bueno@cern.ch
Shivarova	Antoniya	Sofia University	ashiva@phys.uni-sofia.bg
Simonin	Alain	CEA Cadarache	alain.simonin@cea.fr
Singh	Mahendrajit	ITER India	mahendrajit@gmail.com
Stockli	Martin	ORNL	stockli@ornl.gov
Surrey	Elizabeth	UKAEA	elizabeth.surrey@jet.uk
Svensson	Lennart	CEA Cadarache	lennart.svensson@cea.fr
Taccogna	Francesco	IMIP-CNR	francesco.taccogna@ba.imip.cnr.it
Takeiri	Yasuhiko	NIFS	takeiri@nifs.ac.jp
Tanaka	Masanobu	ITER Organization	masanobu.tanaka@iter.org
Tanaka	Nozomi	Tohoku University	nozomi.tanaka@ppl2.qse.tohoku.ac.jp

Taniguchi	Masaki	JAEA	taniguchi.masaki@jaea.go.jp
Tarnev	Khristo	Technical University Sofia, DPF	tarnev@tu-sofia.bg
Tarvainen	Olli	University of Jyvaskyla	olli.tarvainen@jyu.fi
Thomas	Dan	ITER Organization	dan.thomas@iter.org
Tsankov	Tsanko	Sofia University	tsankov@phys.uni-sofia.bg
Tsumori	Katsuyoshi	NIFS	tsumori@nifs.ac.jp
Tuske	Olivier	CEA Saclay	otuske@cea.fr
Welton	Robert	ORNL	welton@ornl.gov

Author Index

A

Agostinetti, P., 325
Allen, J. E., 31
Alonso, J., 461
Ando, A., 291
Annaratone, B. M., 31
Antoni, V., 149, 325
Ariz, I., 461
Asano, E., 282, 470

B

Bacal, M., 38, 47, 55, 74
Bates, M., 243
Baturin, V. A., 208
Béchu, S., 47, 74
Becker, R., 319
Beene, J. R., 431
Belchenko, Yu., 214
Berger, M., 265, 275
Bermejo, F. J., 461
Bès, A., 74
Blondel, C., 385
Bora, D., 480
Bretagne, J., 374

C

Cabaret, L., 385
Capitelli, M., 3, 65
Carmichael, J., 181
Carrere, M., 84
Cartry, G., 84
Cavenago, M., 149, 325
Chacon-Golcher, E., 161
Chaibi, W., 385
Crisp, D., 181

D

Dairaku, M., 335
de Esch, H. P. L., 309, 325, 335, 353
Del Rio, J. M., 461
Delsart, C., 385
Diomede, P., 3
Drag, C., 385
Dure, F., 374

E

Egiraun, M., 461
Engeln, R., 22
Enparantza, R., 461
Etxebarria, V., 461

F

Faircloth, D. C., 243, 253, 461
Falter, H., 451
Fantz, U., 265, 275, 297, 451
Franzen, P., 265, 275, 297, 451
Fröschle, M., 275
Fubiani, G., 325, 353
Fukano, A., 38
Fukumasa, O., 109, 118

G

Gabor, C., 243
Gabriel, O., 22
Galindo-Uribarri, A., 431
Geros, E. G., 161, 199
Goulding, R. H., 181
Grisham, L. R., 364, 412, 421
Gutser, R., 265, 297

H

Han, B. X., 181, 223, 395
Hanada, M., 344, 412, 421
Hatayama, A., 38, 55
Havener, C. C., 431
Heinemann, B., 275, 297, 451
Hemsworth, R., 480
Holmes, A. J. T., 402

I

Ikeda, K., 282, 470
Ikeda, Y., 412
Inoue, T., 335, 353, 421
Ivanov, A., 214

J

Johnson, K. F., 161
Jones, T. T. C., 402

K

Kamada, M., 344, 412, 421
Kaneko, O., 282, 470
Kashiwagi, M., 335, 353, 421
Katsonis, K., 374
Keller, R., 161, 199
Kikuchi, M., 443
Kisaki, M., 344
Kitajima, S., 443
Kobayashi, K., 344
Kobuchi, T., 443
Komada, S., 282, 470
Komuro, J., 291
Kondo, T., 282, 470
Kraus, W., 265, 275, 451
Kulevoy, T., 149
Kumar, A., 137
Kuppel, S., 38, 55
Kurennoy, S., 191

L

Lawrie, S. R., 243, 253
Layet, J.-M., 84
Lee, D. A., 243
Lemoine, D., 74
Letchford, A. P., 243, 253, 461
Lewis, T. L., 431
Lifschitz, A., 374
Lishev, S. St., 127
Litvinov, P. A., 208
Liu, Y., 431
Longo, S., 3, 65
Lucas, J., 461

M

Marcuzzi, D., 325
Matsushita, D., 38, 55
Maynard, G., 374
McAdams, R., 89
Minea, T., 374
Moon, C. H., 291
Mori, S., 118
Murray, S. N., 181, 223

N

Nagamura, T., 443
Nagaoka, K., 282, 470
Nakano, T., 118
Newland, D., 223

Nocentini, R., 297, 451

O

Oka, Y., 282, 470
Okada, J., 109
Okamoto, A., 443
Oohara, W., 109, 118
Osakabe, M., 282, 470

P

Paunska, T. V., 12, 99
Pelletier, J., 47, 74
Pennisi, T. R., 181, 223
Perkins, M., 243
Peters, J., 171, 236
Petrenko, S., 149, 325
Pilan, N., 325
Pozimski, J. K., 243, 253
Pustovoitov, S. A., 208

R

Riedl, R., 275
Rigato, W., 325
Riz, D., 309
Romano, P., 461
Rouleau, G., 161, 199

S

Sakamoto, K., 335
Sanin, A., 214
Santana, M., 181, 223
Sasao, M., 344, 443
Sato, M., 282, 470
Savage, P. J., 243
Schiesko, L., 84
Schneider, R., 65
Schram, D. C., 22
Schunke, B., 480
Senecha, V. K., 137
Serianni, G., 149, 325
Shibuya, M., 282, 470
Shivarova, A. P., 12, 99, 127
Shoji, T., 282
Simonin, A., 374, 385
Singh, M., 325
Sonato, P., 325
Speth, E., 275
Stäbler, A., 275

Stelzer, J. E., 161
Stockli, M. P., 181, 223, 395, 461
Surrey, E., 89, 402
Svarnas, P., 47
Svensson, L., 309, 335, 353

T

Taccogna, F., 65
Takeiri, Y., 109, 118, 282, 291, 470
Tanaka, N., 443
Tanaka, Y., 344
Tanga, A., 480
Taniguchi, M., 335, 353, 421
Tarnev, K. Ts., 12, 99
Tarvainen, O., 161, 181, 191, 199
Tauchi, Y., 109
Tobari, H., 335
Tsankov, T. V., 99, 127
Tsankova, T. V., 12
Tsumori, K., 109, 118, 282, 291, 470

U

Umeda, N., 335, 353, 421
Uriarte, L., 461

V

Vadjikar, R. M., 137
van de Sanden, M. C. M., 22
van Harskamp, W. E. N., 22
Veltri, P., 149, 325

W

Wada, M., 443
Watanabe, K., 335, 421
Welton, R. F., 181, 223
Westall, M., 253
Whitehead, M. O., 243, 253
Wise, P., 243
Wood, T., 243, 253
Wünderlich, D., 265, 275, 297

Y

Yamaoka, H., 443

Z

Zaccaria, P., 325
Zaugg, T. J., 161, 199